SAFETY AND HEALTH
Management Planning

SAFETY AND HEALTH
Management Planning

TED FERRY, CSP, PE, EdD

 VAN NOSTRAND REINHOLD
New York

Copyright © 1990 by Van Nostrand Reinhold

Library of Congress Catalog Card Number 90-45218
ISBN 0-442-00470-2

All rights reserved. No part of this work covered by the copyright hereon may be reproduced or used in any form or by any means—graphic, electronic, or mechanical, including photocopying, recording, taping, or information storage and retrieval systems—without written permission of the publisher.

Manufactured in the United States of America

Published by Van Nostrand Reinhold
115 Fifth Avenue
New York, New York 10003

Chapman and Hall
2-6 Boundary Row
London, SE1 8HN

Thomas Nelson Australia
102 Dodds Street
South Melbourne 3205
Victoria, Australia

Nelson Canada
1120 Birchmount Road
Scarborough, Ontario M1K 5G4, Canada

16　15　14　13　12　11　10　9　8　7　6　5　4　3　2　1

Library of Congress Cataloging-in-Publication Data

Ferry, Ted S.
　　Safety and health management planning / Ted Ferry.
　　　　p.　　cm.
　　Includes index.
　　ISBN 0-442-00470-2
　　1. Industrial hygiene—Management.　2. Industrial safety—
Management.　I. Title.
RC967.F43　1990
658.3'82—dc20　　　　　　　　　　　　　　　　　　　　　90-45218
　　　　　　　　　　　　　　　　　　　　　　　　　　　　　CIP

Contents

Preface / xiii

1   Safety and Health: A Management Function / 1
    The Safety and Health Movement / 1
    Occupational Disease: A Larger Problem / 9
    The Senior Manager as an Expert / 11
    References and Bibliography / 12

2-1   The Safety Function / 13
    Is Safety Really First? / 13
    Management Roles in Attaining Safety
       Objectives / 17
    Is Compliance Enough? / 24
    Labor Forces / 27
    Summary / 27
    References and Bibliography / 28

2-2   Budgeting for Safety / 29
    Time Factors in Budgeting / 29
    Practical Reasons for Budgeting / 30
    Developing the Optimum Budget / 30
    Methods of Determining the Budget / 30
    Developing the Budget / 32
    Budgeting the Total Effort / 36
    Workplace Safety / 37
    Environmental Safety / 42
    Product Safety / 43
    Administration / 44
    Summary and Conclusions / 46
    References and Bibliography / 48

2-3   Cost/Benefit Analysis for Health and Safety in Business / 49
    Health and Safety Motivators / 149
    Origin and Definitions / 49
    Making a Cost/Benefit Analysis / 51

Summary / 56
References and Bibliography / 57

3 Planning and Organizing the Safety and Health Program / 58
The Planning Function / 58
Policy / 59
Organizing and Operating Functions / 62
Organizing the Safety and Health Function / 67
Summary / 74
References and Bibliography / 74

4-1 Controlling the Safety Program / 75
Communications and Information Systems / 75
Safety and Health Audits / 79
Safety and Health Committees / 82
References and Bibliography / 86

4-2 Microcomputers in the Safety and Health Department / 87
Introduction / 87
Summary / 110
References and Bibliography / 110

5-1 Safety and Health Training and Education / 111
The Need for Training / 111
Getting Started / 113
Internal Safety and Health Training / 118
External Safety and Health Training / 121
Training and Education Requirements for Safety
   Professionals / 125
A Matter of Education / 127
References and Bibliography / 129

5-2 Employer and Employee Training / 130
Introduction / 130
What OSHA Requires in Training and Education / 132
The Training Process / 139
Appendix: Resources to Aid the Trainor / 153
References and Bibliography / 157

6 New Installations / 158
Basic Readiness / 158
Safety and Health Responsibilities / 162
Safety and Health Specifics / 168
The Role of Change / 169

Emergency Preparedness / 171
The Manager's Safety and Health Role / 173
The Manager's Safety and Health Checklist / 173
References and Bibliography / 174

7   **Maintenance Programs and Safety** / 175
Program Coverage / 175
Maintenance and Accidents / 185
Building Maintenance / 186
Conclusion / 188
References / 189

8   **Job Process Control and Safety** / 190
Introduction / 190
The Main Elements of Job Process Control / 190
People Requirements / 191
Hardware Requirements / 200
Managerial Requirements / 201
Procedural Requirements / 204
Interfacing the Elements / 206
References and Bibliography / 208

9-1 **Engineering Controls: A Comprehensive Overview** / 209
Introduction / 209
Engineering Controls vs. Personal Protective Equipment / 210
Cost/Benefit Analysis / 224
Government Regulatory Demands / 229
Areas of Application / 231
Conclusion / 234
Bibliography / 234

9-2 **Practical Applications of Engineering Controls** / 237
Introduction / 237
Energy: A Cause of Accidents / 237
Special Problems / 239
Policy and Plan / 243
Energy Barriers / 244
Dealing with Risk / 245
Classes of Hazards / 246
Machine Guarding / 247
Purchasing / 251
The Cost of Safe Processes / 252
Bibliography / 254

10 Management's Stake in Accident Investigation / 255
The Impact of Accidents / 255
Who Are the Stakeholders? / 259
Priorities for Investigation / 265
How to Ensure a Good Investigation / 267
Bibliography / 267

11 Workers' Compensation and the Cost of Safety / 269
The Development of Workers' Compensation / 269
The Benefits of Workers' Compensation Insurance / 272
Workers' Compensation Administration / 277
Future Trends in Workers' Compensation / 280
Controlling Workers' Compensation Costs / 281
Bibliography / 282

12-1 Design and Purchasing / 283
Design / 283
Phases / 289
Costs / 299
Guidelines / 300
Product Liability and Documentation / 301
Purchasing / 304
Conclusion / 306
Bibliography / 306

12-2 Purchasing Controls / 308
Purchasing Controls: What's It All About? / 308
Fundamental Purchasing Concepts / 308
Selecting Sources of Supply / 311
Transportation of Goods / 320
Disposal Exposures and Problems / 321
Hazardous Materials Management / 322
General Materials / 325
Tools and Equipment / 326
Summary / 327
Bibliography / 328

13 Materials Handling / 329
Introduction / 329
Materials Handling, Safety, and Efficiency / 329
The Cost of Materials Handling Accidents / 330
Ergonomic Optimization / 331
Materials Handling Function / 333
Some Principles of Materials Handling / 334

CONTENTS ix

       Equipment / 335
       Materials Storage / 345
       Materials Flow / 345
       Hazardous Materials / 347
       Conclusion / 350
       References and Bibliography / 350

14 Record-Keeping Function / 351
       Why Record Keeping? / 351
       Record Keeping Under OSHA / 352
       Consumer Product Safety / 362
       Medical Records / 364
       Environmental Protection Agency (EPA) / 365
       Chemical Hazard Communication / 365
       Department of Transportation (DOT) / 366
       Fire Reports / 367
       Nuclear Regulatory Commission (NRC) / 367
       Relief Coming? / 367
       Conclusion / 368
       Bibliography / 368

15-1 Medical Services / 370
       Elements of an Industrial Medical Service / 370
       Medical Staff and Facilities / 370
       Basic Functions / 373
       Ancillary Functions / 376
       Special Programs / 380
       Physical Facilities / 388
       Company-Community Interaction / 389
       Medical Case Management / 391
       Wellness / 392
       Management's Role / 397
       Bibliography / 398

15-2 Stress in the Workplace / 399
       Introduction / 399
       Body Response / 399
       Predisposing Factors / 400
       Work Environment / 401
       Contributing Factors / 404
       Preventive Measures / 407
       Management Stress / 409
       Conclusion / 412
       Bibliography / 412

15-3 Off-the-Job Safety Program Management / 413
  Justifying Off-the-Job Safety Programs / 413
  A Matter of Environment / 413
  Off-the-Job Comparison / 414
  Loss Exposure Identification / 414
  Bibliography / 433

16-1 Industrial Hygiene / 435
  Introduction / 435
  Industrial Hygiene and Corporate Management / 435
  Responsibilities of Health and Safety Professionals / 437
  Stress in the Workplace / 439
  Environmental Factors and Work Processes / 440
  The Hygienist as Evaluator and Protector / 440
  Industrial Hygiene: Some Historical Notes / 441
  Conclusion / 449
  Bibliography / 450

16-2 Radiation Safety / 451
  Introduction / 451
  Administrative Program / 452
  Operational Program / 453
  Record-Keeping Program / 457
  Education and Training Programs / 458
  Conclusion / 460
  Glossary of Terms / 460

17-1 Emergency and Disaster Planning / 463
  A Matter of Survival / 463
  Types of Emergencies and Disasters / 465
  Civil Preparedness: A New Perspective / 477
  Planning Steps / 479
  Dealing with the Media in Emergencies / 483
  Conclusion / 484
  Bibliography / 484

17-2 Emergency and Disaster Planning / 485
  Plan Elements / 485
  Plan Functionaries / 499
  Annex and Appendices / 505
  Conclusion / 506
  Bibliography / 506

17-3 Safeguarding Computer Systems / 508
  Introduction / 508
  Fire / 508

Water / 511
Environment / 512
Electrical Power Protection / 514
Contingency Planning / 516
Computer Virus / 517
Computer Crime / 519
Conclusion / 519
Computer Protection Checklist / 519
Bibliography / 521

18-1 **Hazardous Waste Management: A Problem of Growing Concern** / 523
Introduction / 523
Legal Aspects of Compliance / 524
Social and Legal Influences / 530
Identification of Stakeholders and Interfacing Considerations / 531
Conclusion / 533
Bibliography / 533

18-2 **Hazardous Waste: Responding to Chronic and Acute Episodes — A Systems Safety Approach** / 534
Introduction / 534
General Considerations for a System Safety Program / 535
Responsibilities and Authority / 536
Creating an Effective Program / 542
Conclusion / 548
Bibliography / 548

18-3 **SARA — Superfund Amendments and Reauthorization Act** / 550
Title III — Emergency Planning and Community Right-to-Know Law / 550
Special Terms Associated with SARA / 551
Title III — Community Right-to-Know (CRTK) Act: Sections 301-304 and 311-313 of SARA / 551
Right-to-Know Overview / 563
Summary and Observation / 567
Bibliography / 568

19 **Women in the Workplace** / 569
Introduction / 569
History of Women in the Work Force / 570
Demographics — Who Is the Working Woman? / 571
Employment — Where Do Women Work? How Much Do They Earn? / 574

Occupational Safety and Health Considerations of the Working Woman / 576
Summary / 583
Bibliography / 584

20 Legal Aspects / 586
Safety and Risk / 586
Setting Acceptable Safety Levels / 586
Federal Occupational Safety and Health Act (OSHA) / 588
Workers' Compensation / 590
Product Liability / 591
How to Minimize Liability / 597
Conclusion / 601
Bibliography / 601

21 Regulatory Considerations / 602
Introduction / 602
Administrative and Civil Law / 602
Resource Conservation and Recovery Act (RCRA) / 607
Toxic Substances Control Act (TSCA) / 611
Federal Occupational Safety and Health Act (OSHA) / 613
Consumer Product Safety Act (CPSA) / 613
Conclusion / 617

Author Index / 619
Subject Index / 623

Preface

Many safety and health books are written for the supervisor and line manager or for the student making an in-depth study of safety and health. While both of these are necessary and useful, they leave a large need unfilled—the need of the senior line and staff managers. These persons require a comprehensive overview of why safety and health have a natural place in corporate objectives—*written from management's standpoint*. This book targets the safety and health management functions of middle and top management and the safety and health professional.

Central to this standpoint is the belief that management's first concern must always be to do business profitably. This book shows the senior manager how safety and health programming embraces profitable ways of operating effectiveness to achieve corporate goals.

Some of the areas the book explores include:

Safety and health management analysis techniques
Cost reduction through safety and health management
Developing effective operating policies
Allocating safety and health responsibilities to different organizational levels
Program planning, development, control, and evaluation
Legal and regulatory aspects of safety and health management

In each case, the book points out how safety and health can be integrated into total management programs and objectives. At times the information is presented from several perspectives, drawing on the special knowledge of persons considered experts in their field. The senior manager thus receives the essential and varied views needed to make intelligent, farsighted decisions that can have important consequences for his or her organization. The material is easily read and consulted, without professional safety and health jargon or lengthy discussions.

While this book includes helpful references and some aid for line managers and staff, it is not a supervisor's handbook. Rather, it is a practical guide for the senior line or staff manager.

The coming decade will be one of great change, technical challenge, and sophisticated problems. This book is a step toward the mastery of workplace and environmental hazards.

Ted Ferry
Edmonds, WA

Acknowledgments

The following nationally and internationally recognized experts contributed material to this book. They include:

Dr. Paul S. Adams, Ann Arbor, Michigan
Dr. Charles I. Barron, Northridge, CA
Lewis Bass, J.D., Mountain View, CA
Michael F. Biancardi, Wausau, WI
Judith Erickson, A.B.D. Irvine, CA
Dr. Leo Greenberg, Israel
Dr. Earl D. Heath, Towson, MD
Raymond A. Kuhlman, deceased
Charles B. Mitchell, Pleasanton, CA
T. C. Noble, San Leandro, CA
Dr. James O. Pierce, Rancho Palos Verdes, CA
Dr. Jack B. ReVelle, Orange, CA
Dr. Walter F. Wegst, Los Angeles

This book stems from a limited-edition, two-volume manual, supplemented with regular changes and additions over 7 years, and published by the Merritt Company of Santa Monica, California. Special recognition goes to Harry Borowka, Macgregor Eadie, and Mary Ann Georgio of the Merritt Company for their encouragement and help in making this book possible.

SAFETY AND HEALTH
Management Planning

Chapter 1

Safety and Health: A Management Function

Let us begin by stating that safety and health programs are valuable and efficient business tools. However, the senior manager must understand what makes a safety and health program profitable, and how it should be adjusted to fit his or her own organization and case. The background for that understanding is not only a good place to start but a vital link in creating a sound, intelligent executive approach to safety and health.

THE SAFETY AND HEALTH MOVEMENT

How concerned a country is about safety and health is closely linked to its level of industrialization, its type of government and its social outlook. In countries that place great value on individual life, there is a greater degree of concern. The safety movement in the United States illustrates this point, as this chapter will show. The American reader may be surprised to find that in many aspects of safety and health, there are countries that have long been the leaders and in some cases are now well ahead of the United States. Conversely, in some countries we think of as industrially advanced, little has been done from the safety and health viewpoint. Among the emerging nations, we find everything from complete commitment to disdain for the very idea.

In several respects, Canada and many European countries have advanced approaches to safety and health and have used them more rapidly than the United States. A computer network ties together the nations of the European Community (EC), allowing each member access to the safety and health expertise of all. A person in France can use a personal computer terminal to query a data center on a hazardous material for which England has the expertise. An answer comes from the terminal in short order. Countries as far away as Algeria, Austria, Australia, New Zealand, South Africa, and Singapore are so innovative and forward-looking that their concepts deserve universal acceptance. Finland and Sweden lead the way in construction safety. Nor should we neglect advances in the Eastern Bloc, particularly in the area of worker health care. Under the auspices of the International Labor Organization, there is continuous emphasis on the health aspects of the workplace, especially in emerging nations.

2 SAFETY AND HEALTH MANAGEMENT

These differing levels of safety and health awareness do not just happen. They evolve gradually as the country itself evolves. The growth of the safety and health movement in the United States illustrates the process.

The Beginning in the United States

In 1776, the year of its independence, the United States was a nation of planters and farmers. There were a few skilled workers. Industry as such was primarily "cottage industry," wth occasional roving jacks-of-all-trades to repair wagons, pots, guns and the like. At this time there was little machinery in use in the thirteen original states, but this situation was soon to change. Since world markets were unable to provide goods during the American Revolution, it became important to develop local industries to meet these needs. Textile mills were the first to develop. Working conditions were generally poor. The 7-day work week with 14-hour days was standard. With a plentiful labor supply and tough, profit-minded management, improvement in the environment of the workplace was slow in coming. Labor unions in those years strove to obtain benefits for widows and orphans rather than worker safety. For the most part, safety and health were largely ignored during this period and into the mid-1800s. The country itself grew from small farming communities into large agricultural areas. A period of great social change began, along with rapid industrial development, bringing with it many safety and health problems.

The Victorian Period

The period from 1850 to 1900 was a traumatic one for the United States, but it laid much groundwork for the safety and health movement.

The Civil War (1861-1865) produced an unexpected surge of industrial development as the country sought to meet the demands of a war economy. Railroads developed and, together with the waterways, became great highways for moving materials, encouraging even more rapid growth. With these rapid developments, political tension and social upsets were commonplace. The working man bore the brunt of these industrial problems, laboring in hazardous working environments. We have no idea of the total number of laborers killed and maimed in work-related accidents, and the concept of occupational diseases was not even recognized. If a factory worker was injured or killed, no one paid the survivors or family anything. Usually they became burdens on society. These intolerable conditions brought forth reformers who lamented the human toll of the workplace and began to agitate for improvements. Workers' compensation had become established throughout Germany under Bismark, and American reformers urged that it be adopted in this country.

During this period, coal mining in the United States flourished, but it took a terrible toll of accidents, injuries, and deaths. Anthracite coal mining in Pennsylvania was particularly dangerous. Among railroad workers, there were few who had not lost a finger or a limb.

During this period, labor unions were becoming increasingly active. However, their militant opposition to management ended in setbacks, which also delayed safety and health progress. A few states did experiment with safety laws for the workplace, and there were occasional signs of progress. Surveys made at the time, however, showed a terrible toll of accidents and, increasingly, health problem losses in the workplace. The country was moving rapidly toward full industrialization, and by the 1880s was truly an industrial nation. At about this time there was agitation for an 8-hour day and other improvements in working conditions. But again there was great strife based on opposition to union activity, and hopes for reforms in the workplace were dashed once more.

In spite of all these troubles, the safety and health movement did manage to gain the passage of laws to improve workplace conditions. At the turn of the century, the United States was ready to engage in genuine safety and health activities.

The Early Twentieth Century

The next 50 years saw some growth in the safety and health movement. Under President Theodore Roosevelt, many workplace studies were made to determine working conditions. There was new—and generally enforced—legislation for the workplace, including mines, boats, and railroads. A few managers began to recognize that workplace conditions affecting safety and health were tied directly to profits. Reducing accidents and injuries was seen as lowering expenses. In general, however, U.S. industry had a long way to go. In Europe many good safety and health concepts existed, but the United States was slow to adopt them, calling them "foreign" and "unsuitable."

It took some dramatic—and sometimes tragic—events to wake up the U.S. public to the need for more protection in the workplace and more stringent safety regulations. A terrible fire at the Triangle Shirtwaist Factory in New York City in 1911 produced just such a response. A shocked public learned that the deaths of over 100 workers were directly caused by unsafe conditions. More alarming was the discovery that locked exit doors had trapped workers in the burning building. Public and legislative pressure caused management to take a closer look at the need for worker protection and safety.

World War I gave yet another boost to the expanding industrial structure and, in some unexpected ways, aided the safety and health movement. Oddly enough, the war initiated the use of personal safety equipment. A safety helmet (based on the soldier's trench helmet) was developed, and a respirator (based on the poison gas mask) came into industrial use. Even the first-aid kit was a wartime development. By the 1920s companies had started competing for safety awards, and many new safety laws were put into effect. By the late 1930s, the American National Standards Institute had developed nearly 400 safety standards for industrial use. To illustrate how effective this activity was, in 1930 the accident frequency rate was 17.5. With the increased emphasis on safety, by 1940 this rate had dropped to 11.04. Many of the advances made during this time could be traced to the activities of America's first female Secretary of Labor, Frances Perkins. However, not all the gains were due to government action, since the private sector was now far ahead of the government sector in safety and health matters.

Although the first U.S. safety books in common use appeared in the 1920s, it was not until 1931 that H. W. Heinrich's *Industrial Accident Prevention* came to be a standard reference.

Now in its fifth edition, the book's original ideas have withstood the test of time and, for the most part, are still valid. However, this fact also raises the point that we may now be managing safety and health using the ideas developed 60 years ago and little changed since!

World War II provided another impetus for increased safety and health activity. Without such an effort, the United States could easily have lost the war due to accidental losses at home and abroad during this period of scarce resources and greatly increased industrial activity. The accident frequency rate climbed back to a wartime high of 15.39 at one point. Safety and health action was needed. In response, educational programs were developed and safety became a required way of doing business. Unfortunately, after the war, most of this activity died off; schools dropped their safety courses, and the emphasis on safety gradually diminished. The country settled back into the doldrums.

A guiding light behind U.S. safety activity has been the National Safety Council. It began with a group of men committed to improving working conditions who met in 1912 to discuss industrial safety. Known as the First Cooperative Safety Congress, some 200 attendees focused on an unacceptable accident rate of 83.1 per 100,000 population. Their resolution led to the 1913 formation of the National Safety Council for Industrial Safety, later called the National Safety Council (NSC). The 62 founding member companies are, for the most part, still prominent in the business and industrial world. By 1922 their annual convention was international in scope, drawing 3500 attendees and featuring 230 speakers. The new national interest and voluntary participation in safety achieved great results. The NSC continued to lead the voluntary U.S. safety movement and in 1953 received a federal charter by an act of Congress. This charged the Council with building public interest in preventing accidental death and injury. The NSC has now marked its 75th birthday.

Uncertain Times for Safety

Despite the decreased emphasis on safety and health activity during the 1950s, the accident rate dropped to 6.96 in 1955. Obviously, the safety programs developed during the war years had had an effect. This reduced accident rate enabled safety personnel to show management how safety paid off. Reducing lost work time, they pointed out, meant higher productivity and hence greater profits. But many experienced safety workers were now close to retirement age; others were not particularly effective in motivating management. They applied only basic methods, making little attempt to meet changing workplace conditions. New, progressive and innovative ideas were slow in coming. As a result, management's attitude was one of reserved acceptance, with a failure to see safety and health personnel as a valuable resource. This made it harder to find talented recruits for the "dead-end" career field of safety.

Gradually, however, new safety and health ideas began to take hold. Health concerns became increasingly important to the public at large. Forward-looking, capable people began to enter the field. By the mid-1960s, companies were carefully reviewing their safety concepts in a period of soul searching and some experimentation. Safety and health emphases started coming from different directions. A man named Ralph Nader burst on the scene to put industry and management on trial for their allegedly poor safety practices. The public itself became increasingly aware of workplace safety and health hazards. In reponse to this growing social concern, Congress passed many laws. Then, in the early 1970s, the Occupational Safety and Health Act (OSHA) came into being. Safety thus became largely mandated and suddenly respectable, as any concept can be when it is forced on business and industry. While acceptance was at first reluctant, it became more widespread as the Occupational Safety and Health Administration (OSHA) became highly visible, with corps of experts and the emergence of state OSHA programs. In the meantime, new standards and laws were either recognized or passed. Today at least 60 federal laws relate to safety and health.

To the everlasting credit of senior management and a breed of new safety professionals, safety and health were seen as valuable resources if properly managed. Safety and health professionals had finally attained positions of importance in industry. Thus, during a mild recession in the mid-1970s, while the rest of the workforce was reduced, safety and health professionals were retained or even added to corporate management. They were finally being recognized as contributors to efficient and profitable operations.

Initially, the new safety programs consisted mostly of inspections. That is, the answer to all safety problems was to inspect the facilities and operations, look for equipment faults and hazards, and try to correct them. This approach was far from adequate, as it soon became obvious that people, as well as equipment, were involved in most accidents. A more comprehensive approach involved emphasizing the correction of both unsafe conditions and unsafe behavior. During the mid-1980s, OSHA developed programs promoting good safety performance and gave national recognition to companies for their efforts. By 1987 the emphasis had shifted to wide publicity of large penalties for highly visible major companies. OSHA expected this policy to serve as a deterrent to safety violations for hundreds of thousands of smaller companies. Fines that once consisted of mere "wrist slapping" now total millions of dollars. Even large fines that were once routinely settled for a few cents on the dollar are now often crippling. In the meantime, accident and injury rates have remained virtually unchanged over the most recent 5- to 7-year period.

During this time, it became apparent that occupational disease was also an important aspect of health and safety and had to be controlled. Problems such as noise and toxic substances in the workplace became prime targets. A broader viewpoint began to emerge. Safety and health personnel sought and found tools for their profession in other disciplines, improving safety practices to some degree. It became more acceptable to talk about defining safety and health in terms of management's roles and responsibilities, as well as in terms of hardware and facilities.

One result of this increased emphasis on health in the workplace has been a shift in the training of OSHA personnel primarily for health and a workplace inspection system concentrating on health aspects. The research mandated by the OSHA act and done mostly by the National Institute for Occupational Safety and Health, under the Center for Disease Control, has concentrated on the epidemiological aspects of mishaps. Thus research is extensive on health in the workplace but is nearly nonexistent in the closely related area of injury prevention.

Another view of the OSHA era is summarized by the U.S. Chamber of Commerce in commenting on the OSHA Act's reform in 1987:

> During OSHA's 16 year history, despite its massive regulatory efforts, which required vast industry expenditures, its impact on workplace fatality, injury, and lost worktime rates has been disappointing. Although OSHA has been extremist in rulemaking and overzealous in establishing the legitimacy of its purpose, current OSHA leadership is bringing positive administrative reforms. However, it needs more authority, which can be expanded or clarified only by legislation, to reform administratively and to codify administrative reforms to make them more consistent and permanent.

The Chamber of Commerce does not fully support this view, but it does point out that more legislation has its backers. Since this statement was made, OSHA's leadership has changed; the most apparent result is the heavy increase for fines for violations.

The era after OSHA does not mean that OSHA will be gone. It refers instead to industry's placing its emphasis on other areas, such as safety and profit, mission accomplishment through safety, safety as a management function, and so on.

As we pass through the "OSHA era," better management approaches and a new breed professionals are helping to make safety and health programming a more effective, practical, and profitable way to do business. Once adept at one or two management styles, these professionals are now familiar with several types or the all-inclusive contingency style. Even this does not keep up with today's workplace as evidenced by the popular (for now) concept of "chaos management," which thrives on rapid solutions of workplace problems. The complex problem of dealing with chaos is getting attention from a special breed of researchers armed with equations and computer graphics. They have found patterns where others see only disorder. Centers for studying and ordering chaos are now established in such diverse places as the Los Alamos National Laboratory, the Institute for Advanced Study in Princeton, New Jersey, and the University of California, Berkeley. The application of their methods to orderly chaos management is not yet widespread, but this is only a matter of time.

The modern safety and health professional is nearly 10 years younger than his predecessor, often comes from another field, and has a sophisticated educational background. The number of colleges and universities that offer baccalaureate and graduate safety degrees has nearly tripled in the last 10 years, and these degrees often appear in job descriptions and position announcements. Recruiting firms now have specialized departments seeking safety and health candidates, just as for other professional positions. Professional certification and licensing are widespread, often as a condition of appointment, and result in higher salaries.

Today's safety and health professionals are more qualified to act as a valuable staff resource. Their increased value is seen in their starting and average salary levels which are competitive with those of most professions.

Where Do We Stand?

The problems facing business and industry today are largely shaped by the dramatic increase in government regulations on workplace safety and health. Especially in the last decade, required reports, notifications, and statistics alone have placed large administrative burdens on organizations. These regulations reflect today's greater social pressure for change. To some extent, they are also a response to earlier management's failure to provide a safer and healthier working environment. However, big business and industry are a part of society. Their viewpoints must also be adequately represented as government wrangling and public soul searching result in new legislation. Safety and health legislation should not be just the result of social pressures and the efforts of do-gooders. It should also be an honest part of sound business management, benefiting both the worker and the company.

The government's attitude toward management responsibilities as seen in recent legislation and court decisions. Several corporate managers have received large fines, and some have gone to jail in cases where safety and health violations were deemed criminal, particularly where fatalities occurred. In Los Angeles, for example, a full-time staff of 10–12 have been working in this area.

The most visible new official attitudes relate to hazardous materials spills and contaminated former storage areas. Obligations totaling millions of dollars are now being passed to both former and new owners or operators.

There is no question that any responsible corporate officer can be held criminally liable. The issue is merely the extent of the criminal liability and the type of violations involved. This is not a matter of legislative theory. Although courts do acquit corporate managers in some cases, there is a clear precedent for liability, and nearly all normal and often logical defenses have been overridden.

With such powerful forces for compliance and action in the background, finding effective ways to institute safety and health practices in our organizations becomes much more urgent.

What other trends in workplace safety and health will soon influence us? In most countries with strong safety and health legislation, the emphasis is now on the health aspects of the workplace. New technologies are constantly producing new sets of hazards which threaten the safety and health of workers. The number and complexity of these hazards force management to rely increasingly upon external sources of information, such as government agencies, suppliers, research organizations, and professional and scientific institutions. Seldom will someone bring us the information, however. This means that we must seek out and develop sources from which to obtain it. We must recognize that safety is a dynamic, ever-changing profession. This point was made by Goldsmith (1987), who emphasized seven issues, developed over the last decade, that were of little concern 10 years earlier. They had become national issues and would be of major importance for the following decade:

1. Information systems
2. Tort reform
3. Right-to-know legislation
4. Worker notification

5. Substance abuse
6. Acquired immunodeficiency syndrome (AIDS)
7. Employee wellness programs

These are only samples of the new issues, and a decade from now they will be different. This is partly because of the information explosion and the new information available to us.

The lack of accurate information on which to base national action is seen in the U.S. Bureau of Labor Statistics (BLS), which in 1987 started a year-long detailed research study of workplace reporting to determine the accuracy and completeness of workplace accident and injury statistics. The direction came from a National Science Research Council report.

The Council had made an 18-month study of the BLS Annual Survey of Occupational Injuries and Illnesses and other reporting systems. Of 24 recommendations to the BLS, 14 dealt specifically with safety and health and were accepted by the BLS. Part of the BLS survey in the coming year will research and follow up on these recommendations, most of which have to do with better reporting and changes to reporting forms. Major changes will not likely appear soon, but they should significantly improve the quality of data available to the BLS.

While the strength of the labor movement has dropped in recent years, labor unions are taking more prominent roles in safety and health matters. They have developed extensive in-house safety and health expertise, often filling state and local safety and health authority positions based on their recognized expertise.

New contractual negotiations center on safety- and health-related issues as never before and give workers a voice in their union and company. The intent is to woo workers back to the labor fold. There is a noticeable trend not only to seek more worker participation in these areas, but also to train and encourage workers to take more responsibility for their actions. Astute management may anticipate, recognize and respond to these changes in a manner which will increase resources and improve production, rather than just responding to enforced government regulations.

OCCUPATIONAL DISEASE: A LARGER PROBLEM

The difficulty in identifying new health hazards in the workplace is only one problem in this area. We cannot yet, for example, link many diseases with the workplace exposures that may cause them. What about long-standing hazards and occupational diseases that already exist? In the United States, it appears that the amount of workers' compensation benefits paid out for occupational disease-related cases is approaching 10 percent of the total. Exact and comparable figures are not available. Of the 400,000 workers with occupational illness who receive workers' compensation benefits each year, over 14 percent have stress-related disorders. With an average direct cost of $20,000 each, these conditions are a significant and rapidly growing expense.

The toughest problems are yet to come. For example, under right-to-know legislation, we must advise workers about workplace hazards we cannot even identify. "Occupational stress" has changed from an insignificant term to a condition that often represents much of the money paid out for workers' compensation. AIDS appears to be one of the largest health threats in coming years, and evidently workers will continue to work through the last stages of the illness. Treatment is both lengthy and costly. Asbestos victims have already won court fights, and companies will be forced to pick up the workers' share of that costly disease, which has already brought several firms to their financial knees.

The right to know concept involves making public information about hazards and environmental monitoring tests results, record keeping, and allowing workers access to their own medical reports and similar information. It is thus essential that all management personnel down to the supervisory level understand the new regulations and required procedures to carry them out. This legal duty cannot be avoided, and ignorance of it could be very costly.

The size of the occupational health problem is best judged by knowing that, as of this writing, to the American Conference of Governmental Industrial Hygienists has developed the threshold limit values for only 750 chemical substances, while thousands are in use in the workplace. In California alone, for example, several new chemical substances come into use every day. Only a handful of them have strict regulation in the United States. While there is circumstantial evidence that 80 to 90 percent of cancer-producing agents are chemical in nature, they are seldom identified further and are nearly impossible to monitor. This challenge is more than the senior manager or any one industry can handle alone. We need a partnership between business, the scientific communities, the medical communities, and the government. Meanwhile, the problem must be approached by the individual manager who starts with the correct procedures, including workplace monitoring, which is increasingly required.

How Exposure Limits Are Set

A good deal of workplace monitoring centers on the concept of "threshold limit value (TLV)." Standards for most of the physical, biological, ergonomic and chemical hazards that require monitoring use TLVs in some way. The TLVs recommend limits of exposure, depending on the concentration and length of exposure. For example, the TLV may be expressed as "so many parts of the substance per million parts of air" or as "so many particles absorbed over an 8-hour work day." In the United States, these TLVs are generally developed by the American Conference of Governmental Industrial Hygienists. Each year the conference updates and publishes a booklet giving the TLVs for substances. Several other countries base their programs on TLVs as well. Unfortunately, there is no guarantee attached to a TLV and no definite line between "safe" and "dangerous." The "acceptable safe" amount this year may be judged grossly "unsafe" next year.

In the meantime, tens of thousands of chemical substances arrive at plants each day in unlabeled containers, without any TLV indicated. More governmental labeling regulations will partly solve this problem, at least for the container's contents. The responsibility to warn workers about the hazardous potential of those substances lies with the user/employer, whether the substance's manufacturer gave that information or not. Moreover, where legislative controls exist, the user/employer must monitor to ensure that exposures are not exceeded.

Safety On and Off the Job

Not all the executive's safety and health problems are found in the workplace. Roughly 75 percent of health and safety injuries and illnesses happen outside the workplace. These outside problems may not appear in the direct costs of workers' compensation or damaged equipment. However, the failure of trained employees to appear for critical jobs has many "hidden" and not-so-hidden costs, such as delayed production, uncompleted deliveries, training costs for new employees, and so on. These costs emphasize the need to consider and include off-the-job safety as a target as well.

THE SENIOR MANAGER AS AN EXPERT

Senior managers cannot be experts on all workplace safety and health hazards. However, they should know which ones can adversely affect their company in terms of market share, profits, standing in the field, operation, and so on. Except for smaller companies, this does not mean that the senior manager (or staff) must be an expert in safety or health; that is the function of the firm's safety and health advisors. Senior management may not even need in-house safety and health expertise if the nature of the operation can be adequately served by government agencies, insurers, or consultants, or if there are no unusual hazards.

The economic concerns of the company override most other considerations. Thus, to merit the senior manager's attention, safety and health concerns must either impact or have a potential impact on the company's economic health or survival.

If the manager or staff advisors understand the potential of safety- and health-related problems, that may be enough. There lies the catch. How many senior staff members or managers know the hidden cost of minor accidents or what they cost in terms of the manufactured product or sales?

12 SAFETY AND HEALTH MANAGEMENT

How many of them can relate health costs to unsafe working conditions? How many relate extended workers' compensation costs to poor handling of claims in the organization? How many see the relationship between poor product performance and inadequate safety considerations in the initial design? This is where the executive must be an expert, understanding the potential impact of safety and health problems on the company.

REFERENCES AND BIBLIOGRAPHY

Cheney, Carol, "NSC Celebrates 75th Anniversary," *Occupational Health and Safety,* October 1987, pp. 26-23, 33, 42.
"Combating Dermatitis—The $40 Million Itch," *Occupational Hazards,* March 1980, pp. 52-55.
Findlay, James V., *Safety and the Executive,* Loganville, GA, Institute Press, 1979.
Garner, Charlotte A., "Let's Put Safety on the Appraisal Form," *Occupational Hazards,* March 1980, pp. 75-78.
Gleick, James, "New Images of Chaos That Are Stirring a Science Revolution," *Smithsonian,* December 1987, pp. 122-134.
Goldsmith, Phillip E., "Risk Management: Emerging Topics, *Proceedings of the 1987 Professional Development Conference, American Society of Safety Engineers,* Boston, MA, June 1987.
Heinrich, H. W., Dan Petersen, and Nestor Roos, *Industrial Accident Prevention,* 5th edition, New York, McGraw-Hill, 1980.
King, Ralph W., *Industrial Hazards and Safety Handbook,* London, Newens-Butterworth, 1979.
Knox, N. W., and R. W. Eicher, *MORT User's Manual SSDC-4,* Washington, D.C., Government Printing Office, 1976.
Peters, Tom, *Thriving on Chaos,* New York, Alfred A. Knopf, 1987.
Petersen, Dan, *Techniques of Safety Management,* 2nd edition, New York, McGraw-Hill, 1978.
Pope, William C., *Systems Safety—Management Series,* Alexandria, VA, Safety Management Information Systems, 1976-1980.
Schulman, Brian, "Lingual Techniques Avoid Mislabeling Stressful Workplace Environments," *Occupational Health and Safety News Digest,* December 1987, pp. 1-4.
Threshold Limit Values and Biological Exposure Indices for 1986-1987, Cincinnati, OH, American Conference of Governmental Industrial Hygienists, 1986.
U.S. Chamber of Commerce, *Congressional Issues,* Washington, D.C., 1987.
U.S. Department of Labor, "Norwood Plans BLS Action to Improve Occupational Safety and Health Data," *News—Bureau of Labor Statistics,* press release, October 16, 1987.
U.S. Department of Labor, BLS Reports on Survey of Occupational Injuries and Illnesses in 1986," *News—Bureau of Labor Statistics, USDL—87-492,* press release, November 12, 1987.
Weaver, D. A., "TOR Analysis," *Directions in Safety,* Springfield, Il., Charles C. Thomas, 1976.
Webber, Ross A., *Management: Basic Elements of Managing Organizations,* revised, Homewood, IL, Richard D., Irwin, 1979.

Chapter 2 / Part 1

The Safety Function

IS SAFETY REALLY FIRST?

Many years ago, it was fashionable to quote the slogan "Safety First." Let us put that belief to rest by emphasizing what top management has always known: Safety is not and cannot be first. It is profits, mission, product, service and the strength and growth of the business that are first. Safety is important because it can either support or damage these things; it is not something separate, a function by itself. Safety and health considerations touch all functions of the organization and should be fully integrated into those functions, not merely imposed as an afterthought.

We want efficient production to maximize profits. Efficient production requires facilities, equipment, materials, and personnel, and management is responsible for supplying them. In the process of supplying them, certain support services are also needed, such as engineering, purchasing, product inspection, maintenance, research, and many others. All these support services have safety and health aspects. Incorporating safety and health into support services allows us to involve safety and health at a function level, as an integral part of the operation. Consider personnel functions, for example. Personnel selection, placement, training, employee relations and motivation all involve safety and health considerations. Our purpose could be stated as building safety and health controls into each of these personnel functions and continually auditing them to ensure that the controls are adequate. The same concept applies to any organizational function.

Safety Problems: Symptoms of Management Failure

The business organization is a system composed of many subsystems. There are financial systems, personnel systems, production systems, maintenance systems, etc. Any accident, injury, unsafe act, workplace health problems or unsafe conditions show a failure in the system. Safety and health professionals act as systems evaluators, advising management on how the functional parts of the system should operate in regard to safety and health. Safety thus used becomes a resource to get the job done, not just a regulatory burden that hampers the system.

Modern accident/injury and health theory subscribes to the idea of multiple causation. That is, there are always several causes that allow accidents to

14   SAFETY AND HEALTH MANAGEMENT

happen. In multiple causation, we trace all such factors to their underlying causes. When doing this in depth, we find that management actions or inactions are always contributing aspects.

Take the simple example of a ladder in a stockroom that breaks when a clerk climbs to reach a high storage bin. We might typically find that the ladder had not been properly and regularly inspected (a lower management error). Asking why this happened, we might find that there was no procedure for regular ladder inspection (a management failure), that an inspection system for other equipment was not in effect (a management failure), that such ladders were inferior in quality and should not have been purchased in the first place (a staff failure), and that funds to purchase proper equipment were not available (staff and management failures). A closer review of the entire operation might show that these management and staff failures were also causing similar problems throughout the organization, including costly delays and production problems, worker unrest and labor group concern over the use of unsafe equipment. From a simple broken rung we thus detect manys causes (multiple causation) and many cases of poor management and staff work. Once we recognize and correct such conditions, the entire operation will improve. Thus the safety function can help improve the entire system.

Safety as a Management Function

There is a little difference between the safety function and any other organizational function. It starts with management setting achievable goals and objectives, which are achieved the same way that other functions are. It is important to understand that safety and health functions compare to other management functions such as quality, cost, quantity, and production controls. It is not enough for management merely to support these functions; they must lead the way. The safety functions, too, should have management's active leadership, not just passive support. Safety expertise should be available as a resource for management. If this expertise is not available, management should either assume the burden or bring in professionals from the outside. In reality, a mix of the two is probably best.

We have sought to show that safety and health matters are management concerns and functions. As with any other function, management directs the safety and health effort by setting realistic goals and by planning, organizing and establishing controls for the program. While management may assign responsibility and authority for safety and health to line managers, and hold them accountable for the results, it must assume final accountability and responsibility. This includes giving credit for a job well done.

Line and staff managers, of course, can and should be held accountable for those safety and health functions which they control. To use our previous case, we cannot hold the purchasing department responsible for the stockroom clerk's action of climbing the faulty ladder. However, the department is responsible for the error of buying an inferior and possibly unapproved product if they have the money and directions to purchase proper equipment. Perhaps we cannot hold the safety office responsible for a storeroom accident over which they have no direct control. But we can hold them responsible and accountable if inadequate purchasing guidance exists. Likewise, it is their responsibility to conduct equipment inspections and advise management on using certain guidelines and procedures. The safety office cannot be held accountable for line functions, but it is responsible for its failure to advise and warn line and staff.

Sizing up the Safety Function

To start, a company's involvement in safety can be evaluated by asking these questions:

1. Is there a company safety policy? What is it? Who knows it?
2. How is safety built into the organization?
3. Where is the safety function found in the firm? To whom does the safety manager report?
4. What are the line and staff relationships on safety and health?
5. Where is the health function located in the organization? To whom is the health manager responsible?
6. How are the safety and health responsibilities defined?
7. How does management hold the line and staff accountable for safety and health? Specifically, how are they graded?
8. How is senior management held accountable and rated in terms of safety and health performance?
9. How is the staff measured in terms of safety and health performance?
10. How are supervisors measured in terms of safety and health performance?
11. Does senior management relate safety and health functions to profit and production in a positive manner?

These questions require more than simple answers, so be ready for an in-depth look at your company.

Safety and Avoidance of Loss

Management consultant Peter Drucker (1982) gives us a new perspective on safety when he says, "It is the first duty of business to survive. The guiding principle is not the maximization of profits but the avoidance of loss."

Avoidance of Loss. "Avoidance of loss" describes the safety function very well. Operating errors produce accidents, injuries and economic losses. These operating errors and unplanned events occur mostly when human and physical resources are badly handled.

Note that the causes of operating errors and unplanned events may have little relationship to the results. That is to say, the results (accidents and injuries) are largely fortuitous. The results themselves should not dictate the amount of effort we expend to bring about changes in operating procedures. Chance, a reflex movement, or a broken machine part all contribute to the results. We can have hundreds of resource-consuming foulups, goof-ups and mistakes without having a recordable accident or injury. Chance alone may make a difference. The results of unplanned events are unpredictable. However, recording the number of undesired, unplanned events is an excellent way to measure managerial performance—much better than tallying the injury rate. Merely keeping track of injuries or mishaps may not give us the full picture on how many unplanned events ineffective management allows. Finding operational errors is a good management technique for loss avoidance and a sound safety approach.

A manager should see unwanted or unplanned events in terms of

Skill deficiencies
Poor supervision
Fiscal mismanagement
Unwise decision making

Unwanted events, in turn, translate indirectly into

Excessive maintenance
Cost overruns
Slippage
Wastage
Downtime

These easily recognized management terms fully apply to safety and health functions.

Loss Control. "Loss control," as some practitioners refer to safety and health management, also reflects this view. It holds that safety and health management should "locate and define operational errors involving incomplete decision-making, faulty judgment, administrative miscalculations, and plain honest-to-goodness bad management and individual practices leading to incidents which downgrade the system; and provide sound advice to management on how such mistakes can be avoided" (Pope, 1976-1980).

The circle is complete. Safety and health management does exactly what management wants to do in any of its other functions. Its aims and goals are the same as management's general objective.

Management vs. Supervision

The important role of the supervisor in any operation is not debatable. However, this is not a book for supervisors; it is for senior management.

The supervisor seldom has time to be a creative manager. This individual's duties are most often prescribed, and are so demanding that he or she must concentrate on those responsibilities that senior management emphasizes the most. The supervisor is the last in a line of managers and is often the person who, while not given the resources to do his job properly, is still held responsible for worker safety and health. Finally, while closest to the actual work tasks, he or she may have the least opportunity to control the safety and health aspects of those tasks. Later in this manual, we will discuss what to expect of supervisors in terms of safety and health performance. At this point, let us focus instead on what management should expect of itself.

MANAGEMENT ROLES IN ATTAINING SAFETY OBJECTIVES

The ideal safety and health program should have the same goals as any other corporate function. Let us look at what foremost management experts Peter Drucker and Warren Bennis say about the multiple objectives of business and discuss some of the safety implications:

1. *Profitability.* This objective refers to the manager's desire to maximize profits (or avoid losses). Reducing production, personnel and equipment costs makes operations more profitable and takes up organizational slack. Accidents and injuries cost money, thus reducing profits. Safety, therefore, has an impact on profits that must be understood.
2. *Market Standing.* This refers to a company's position compared with that of competition. Safety and health issues can adversely affect product or service marketing and lower a company's market standing. Liability suits, rumors of an unsafe product, and newsworthy accidents that occur within the firm could endanger this standing. The greatly reduced market standing suffered by the Chevrolet Corvair is a good example.
3. *Productivity.* This is directly influenced by time-consuming accidents which shut down a production line, delay deliveries, incapacitate skilled operators, destroy essential equipment and the like. Less easily measured but just as harmful to production is the loss of efficiency and morale that develops and remains after a serious accident or incident.
4. *State of Resources.* Accidental resource losses reduce the overall stock of resources. Money may have to be diverted from more profitable pursuits to eliminate needless safety and health losses.

5. *Service.* A company's market position and profitability may rest heavily on the quality and speed of the service it supplies. When safety and health problems occur and must be explained or defined, the ability to give quality service to customers is reduced. An unsafe product, moreover, can result in major service costs if repair or replacement is necessary.
6. *Innovation.* The development of new products and services is basic to company growth. The increase in new standards and regulations and the need to include costly safety and health considerations radically affect innovation from concept through design and production. The cost of meeting these standards must be taken into account if the company is to innovate profitably.
7. *Social Contribution.* This is synonymous with public responsibility, and nowhere do safety and health considerations have more impact or more far-reaching effects. Consider manufacturing processes that, knowingly or unknowingly, pollute the environment or result in the distribution of products which later prove to be unsafe. Consider, too, the cost to society (and, in turn, to business) of rehabilitating an injured worker and supporting his or her family, as well as the social cost of losing a skilled, productive member of society.
8. *Identification.* This involves a management and staff commitment to the same organizational objectives. Today the safety and health issues surrounding a process or product can seriously divide management and staff. When they both understand that safety is tied directly to the prime objectives of service, profit or mission, real identification is achieved and organizational goals can be realized.
9. *Social Influence.* This refers to distributing power and influence within an organization to achieve expectations. Effective safety and health performance should become an expectation. To achieve this expectation, responsibility must be distributed throughout the organization, along with accountability.
10. *Collaboration.* This refers to the need to resolve internal conflicts and problems. Safety and health issues may surface at all levels and may involve conflicts or problems between staff and management, departmental interests, and so on. Sound management focuses on safety and health as a common goal influencing mutual interests and achieves needed cooperation at all levels.
11. *Adaptation.* Every organization continually needs to meet changing environmental and technological requirements and changes in market taste and needs. Management must recognize that it faces a new climate of public awareness about safety and health, as well as a new regulatory climate. Having a reputation for credibility and integrity requires greater safety and health awareness by management.

12. *Revitalization.* This is an ambiguous objective that refers to the need to change the organization periodically, even though things are running smoothly, since a firm cannot remain a leader by merely maintaining the status quo. A new product, a new production method, or even a change in organizational structure always involves new safety and health problems, real or potential. These must be fully integrated into management's total safety and health programs.
13. *Integration.* This is the need to overlap objectives and responsibilities for smooth operation. It may be crucial to achieve safety and health functions, which cannot be confined to a single department. These functions may require multilevel and interdepartmental action to keep them from "falling through the cracks."

Some governmental organizations, public institutions and nonprofit organizations consider themselves mission- rather than profit oriented. However, they still must be run on a businesslike basis to achieve that mission. The 13 objectives listed above also apply to them.

While management is tasked with safety and health roles, we must recognize that management roles themselves are not always clearly defined. Today's managers have many hard-to-define roles, often without clear boundaries or lines of authority. They often operate in isolation from subordinates and cannot manage effectively merely by looking at results. Their authority seldom equals their responsibilities. Safety and health issues add to their problems, since most of these issues require time and expertise to recognize and solve. Managers need all the help available.

Management and Safety and Health Problems

At one time, most safety programs consisted of such resources as committee meetings, posters, slogans and slide shows. This suggested that safety was not a very demanding or sophisticated concept. Some managers today still think that warnings to be careful, wear hard hats and read labels are the way to keep safety and health problems under control. Nothing is further from the truth. Effective safety management is becoming increasingly complex and sophisticated, and concerns management on every level.

One fairly sophisticated technique for solving safety problems, for instance, identifies over 1,500 causal factors in accidents, mostly related to management! With hundreds of thousands of chemicals in common industrial use, and regulations in effect on only a few of them, the potential for health problems is beyond easy grasp. Management may not have the answers to these problems, but it can take progressive steps to help keep them under control. Almost without exception, these approaches require management involvement, if not management initiation.

This suggests at least two things:

1. The safety and health manager must be highly knowledgeable about general management ideas and practices.
2. Today's manager must have a broad appreciation of how various safety and health roles fit into his or her normal management functions.

What Is Adequate?

While top management may be equal to the task of taking clear-cut, decisive steps to solve problems, neither the problems nor the solutions are always clear. Most are not. This is particularly true of safety and health problems. Few of these problems can be solved with a simple "yes or no," "true or false," "right or wrong" answer, since few of them are clear-cut.

For example, take the question, "Do safety controls in the warehouse work?" A simple "yes" or "no" answer might tell the manager nothing at all. "Yes" might imply that there are no unrecognized or uncontrolled warehouse hazards, that all safety systems and procedures are carefully monitored and followed, and that no possible injury risk exists. A "no," on the other hand, could be just as meaningless in management's terms. It might mean that equipment is not properly inspected or that employees are not regularly instructed in safe lifting and handling procedures, or even that management has never even determined the safety and health status of the warehouse. Thus, "yes" or "no" answers neither help management pinpoint specific problems nor provide a tool to measure realistically how successful any part of the system is.

A useful approach employs the concept of "less than adequate (LTA)." It measures safety and health performance against an acceptable standard. LTA is a suitable measure where a definite "yes" or "no" requires some reservations or where neither is entirely true or false. It allows management to focus on areas classed as LTA in order to bring them up to an acceptable level. It implies that management will review the standards of what is "acceptable" or "adequate" to the organization.

How Is Management Involved?

The reader may not yet be entirely clear on how completely the solution of safety and health problems hinges on management. To state the issue in its simplest terms, once the manager identifies and sizes up a safety and health problem, he or she:

1. Either takes proper corrective actions or becomes guilty of oversights and omissions.
2. Either takes the best-known action to eliminate known hazards or decides that nothing can be done.

THE SAFETY FUNCTION   21

How these two factors relate to management functions and objectives is shown in Figure 2-1, which broadly outlines an adequate management system and specific control factors.

The figure only briefly describes the senior management, middle management and staff roles, but it does show where assuming risk, oversights and omissions enter the system. Specific corrective actions in management systems and specific controls relate to management's initiative and directives.

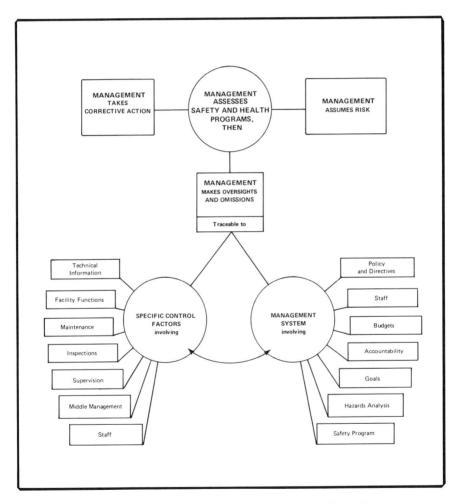

Figure 2-1. Management either takes corrective action or assumes risks and makes oversights and omissions related to control factors and the management system.

Accidents and Hazards — Management Failure

When forced to economize, management reduces costs by streamlining operations, cutting back on equipment purchases, reducing stockpiles, tightening up on maintenance, increasing efficiency of production and cutting down on overhead personnel. Safety and health activities are often among those considered a luxury. The senior manager who sees it as a marginal activity overlooks its potential for great monetary savings and improved productivity, without costly expenditures and without cutting down on operations, personnel or services.

This systematic way to achieve savings stems from the simple assumption that all accidents and hazards reflect management failures. Radical as it may seem, this approach is widely used in industry. It can be used in all types of organizations. These same signs of management failure can cause many production losses, accidents and hazards. Investigation shows that these failures often go on for years, continually causing large, cumulative losses without anyone being aware of them. They may be simple inefficiencies, errors or oversights. A system developed by William Fine routinely analyzes accidents, hazards, near misses and poor operation as a means of finding these inefficiencies, oversights and errors by management.

Hypothesis: It is arbitrarily assumed that management will be responsible for the causes of every accident and the existence of every hazard.

The rationale for this hypothesis is that, in investigating every accident, one always finds some management involvement or activity that might have prevented the event. Thus the investigation of each hazard, accident and injury aims to identify the underlying management involvement. By management, we mean all levels of management above that of first-line supervisor. It includes all staff and support functions such as personnel, supply, maintenance, security and safety. The first-line supervisor is excluded because we believe that the key to prevention and correction lies above that level. This does not lessen the supervisor's role, but instead places the emphasis on power of correction where it realistically belongs.

Applying the System. The method consists of investigating each event with management introspect. We seek to answer a single question at every turn, and in connection with every fact and circumstance uncovered. The question is: "Where did management fail?" We seek to find what management did or did not do that permitted the event. Sometimes the management failures will be obvious. In others, they will be remote and indistinct, but they will always be present. This technique is not confined to accidents or injuries. It is just as appropriate in any unusual event which suggests that the organization does not perform as it should. Thus, this process, formalized for safety and health, has wide use in all areas.

There is no reason to feel defensive about a system that deliberately seeks management failures or, more precisely, failures in the firm's structure and systems. It is sound business, however hard on the ego, and does more good than harm. In the end, it benefits management's real interests. Prompt attention to inefficiencies found by the technique shows management's concern and interest in personnel. It wins appreciation and boosts employee morale. The benefits of seeking the management role are seen in two examples from actual industrial cases that illustrate the technique in operation. Note that the system is useful whether used to discover the cause of an accident (Case 2) or in a case where waste went on for several years without an injury to bring it to management's attention (Case 1).

Case 1—The Improvised Scaffold. This case involved an improvised scaffold constructed of stepladders, planks and beams, fastened with rope. There was no question that it was unsafe and could cause a serious injury or even a fatality. Simply correcting the unsafe condition by buying a new scaffold would have been simple. But investigating the condition to find the underlying management errors yielded some surprising results.

It was found that a properly manufactured scaffold was not handy because someone in a higher management position wanted to operate at the lowest possible cost, avoiding new equipment purchases. It was also found that the improvised scaffold had been assembled and disassembled for use at different job sites at least weekly for the past year, and perhaps for a much longer time.

Each time the improvised scaffold went up or down, it took three to four men about 3 hours. (This adds up to 1,872 worker-hours per year!) In use it required frequent inspection and tightening, which took more worker-hours. Moreover, the men working on it were never at ease.

Compare this to the situation where a new scaffold was bought. Two men could easily erect a scaffold this size in one-half hour (only 52 worker-hours per year). Simple math shows that an investment of $1,000 for a new scaffold could save $10,000 to $15,000 in annual labor costs. Also, it would greatly improve the safety status, efficiency and morale of the work crew. (Note: Inflation has shrunk that figure. Both the costs and the savings have now doubled.)

This case dramatically brought to top management's notice how much its "penny wise, pound foolish" policy was costing. They were saving a few dollars by using improvised equipment but incurring large hidden costs. Reviewing and changing this penny-pinching policy throughout the organization brought about large savings while eliminating many potentially costly hazards.

Case 2— The Balky Generator. The second case shows what this approach found after an injury involving repair work on a large generator in a machine shop. Because of limited space, the mechanic doing the repair had to work in an awkward position. As he exerted pressure on a wrench, he bumped his head on a projection on the side of the generator and received a laceration.

The routine approach and corrective action would consist of the supervisor warning the worker to be more careful, to move materials and to give himself more working space. The management involvement investigation found that the generator was old and decrepit, and should have been overhauled or replaced years ago. Department management had avoided or postponed this action for reasons unknown. This generator had had to be repaired at least once a week for the past year or more. Repair time averaged 3 hours each time.

The generator supplied power for operating five other machines. When it was shut down, two or three machinists or operators had to wait or find other odd jobs while it was repaired. Thus, there was a cost or loss of at least three employees (the repairman and two or three others) for 3 hours once a week. The loss was 450 worker-hours, or about $13,500 per year, because of management's failure to replace the generator when it became uneconomical to continue using it.

Further investigation found that the generator could be eliminated by using an alternative power source from two other nearby generators. The cost would be about $300 for electrical wiring and materials. Exploring management's role in this simple first aid case eliminated more than a source of potential injuries. For only $300, management realized savings of about $13,500 per year by eliminating the repetitive repair work.

In both cases, a brief review or the usual brushoff would not have yielded the benefits achieved. Safety investigation should not simply be profit motivated, but the moral is clear. In general, management's role in safety and health hazards is merely symptomatic of general inefficient operations. Investigating with the question "Where did management fail?" in mind can save the company money in the long run. It can pinpoint problems that are not safety issues but that do affect purchasing, quality control, personnel training and production policies. Of course, the investigation system is only half of the story; action is not complete until the proper level of management takes corrective action.

More detailed information on this approach, with additional examples and methods for implementation, is available from the author.

IS COMPLIANCE ENOUGH?

Most companies tend to think of safety and health strictly in compliance terms—doing the minimum required by law. The theory is that compliance alone is quite costly, and exceeding these requirements is even more costly.

In fact, the reverse may be true. Over time it is more profitable to operate safely. Minimum compliance yields minimum benefits.

A publication of the American Society of Safety Engineers discussed the actual payoff for exceeding government standards. Using the Alliance of American Insurers as its source, *Professional Safety* stated:

> If your business meets that standards set by the Federal Occupational Safety and Health Administration (OSHA), you have taken a step toward practicing loss control. But, you haven't completed the journey.
>
> Meeting the standards set by OSHA helps to create safe working conditions, but federal rules do not stipulate the type of management, employee training and organizational and planning methods needed for a comprehensive loss control program.
>
> When a small-business owner practices loss control, it means he has not only met legal requirements for workplace safety, but has also examined this whole operation: management decisions, employee training, supervision, controls, auditing, and plant design. The goal is to cut losses from work accidents and illnesses. A hypothetical example explains how effective loss control works.
>
> A carpenter is cut when a band-saw blade breaks. At the time of the accident, he is cutting wet scrap material to make tote boxes. To prevent the same type of accident from happening again, his employer installs a guard on the machine. The guard will meet OSHA requirements, but that may only be part of the solution. The employer should also ask himself: "Could the injury have been prevented if I were aware of all facets of operations that might have contributed to the accident?"
>
> An investigation of the accident by the employer should look beyond what government standards demand. The employer should ask himself what other, perhaps less obvious causes could have contributed to the mishap. On answering those questions, he may find that this is not the only accident that could have been avoided if more attention had been given to certain managerial facets of his operation.
>
> In fact, the employer may find that many worker errors which cause a productivity loss or property damage could also have been avoided if supervisory or training duties had been properly followed. Out of every 300 to 400 errors made, one will result in any injury that will keep someone off work. About one out of 10 errors will cause some property damage, according to the Alliance.
>
> For example, investigating the blade-saw accident should mean that the employer asks himself the following questions:
>
> Should tote boxes be made on site in the first place?
> Was the carpenter properly instructed in the use of the saw?

Was the storage of the scrap material in a well-designed place?
Was the saw purchased and installed without a guard?
Did the supervisor fail to observe the hazard?
Did management require a documented monthly inspection?
If inspected, why was the machine permitted to operate?
Was the employee told he was violating the safety rules by operating the saw without a guard?

If the answer to one or more of these questions is "no," chances are the employer should examine the management leadership, employee training programs, supervisorial and equipment purchasing methods practiced in his plant.

Perhaps that scrap material was being stored in a place where it could get wet. Perhaps the carpenter was not trained in the proper handling of the equipment. Perhaps no guard was on the machine because the person who bought it for the company was not aware of guard requirements. If any or several of these factors had been corrected, the accident might never have happened. That is what loss control (safety) is all about.

This example shows that any loss control program begins with a serious investigation of the entire working environment. The employer must look at the working environment, the machines or equipment used and the workers with any eye for possible hazards.

If an employer sees materials, tools, or similar objects that could fall on people, he can either prevent the object from falling or buy hard hats for those who must work in the hazardous areas.

If the employer cannot prevent injuries by changing the environment of the machines first, then he should make sure everyone exposed to the hazards has the right protective equipment.

The loss control specialist or Safety Professional can be the most help in controlling overall losses. He is interested in more than federal and state safety requirements. He has a vested interest in reducing conditions or situations with inherent risk. He can tell the employer:

How to correct the problems, investigate accidents and errors and keep up daily records.
How to maintain machines to avoid breakdowns that may cause injuries.
How to install guards on machinery that protect the operation without slowing down production.
How to set up a first aid program.
How to improve his auto-accident record.
How to find out whether chemicals or other substances used at this establishment may have health implications for employees.

This advice is available through the small-business owner's insurer. But, the motivation to practice loss control must come from the employer.

LABOR FORCES

In ending this chapter on management's relationship to safety and health problems, we would be remiss if we failed to mention the role of organized labor. There is a growing trend for government, industry and organized labor to work together on safety and health problems. Several countries mandate such joint efforts. There is a tendency to strengthen labor's role and activities through:

1. The right of a worker to refuse to work in an unsafe environment,
2. The use of joint safety and health committees, and
3. Giving new rights to union safety and health representatives.

These rights will strengthen labor's role in safety and health matters and in the organization.

Many unions now have highly qualified safety and health specialists on their staffs. One of the largest labor organizations has its own safety and health institute for the benefit of its membership. With this labor role, it is possible that management and supervisors will not know as much about safety and health matters as the union membership. Thus, management can lose the initiative on safety and health matters. The answer is to have properly informed and trained management and supervisory personnel.

The climate of union involvement requires management-labor cooperation for workplace safety and health. Since over 80 percent of union contracts contain safety and health provisions, the role of organized labor must be part of today's business strategy.

SUMMARY

Safety and health movements are well established, and have an economic and social impact that involves management from the chief executive down. Today management feels safety and health pressures from government, society at large, the work force, organized labor and the marketplace, as well as from their own managers and staff. The new technologies are creating new safety challenges and health concerns, and a better-educated and trained work force has a greater concern for safe working conditions. Finally, management must realize that safety and health programming is not only inevitable, but can also be a profitable management tool. What must be done and how managers should do it are the subjects of the remaining chapters.

REFERENCES AND BIBLIOGRAPHY

Drucker, Peter F., *Managing in Turbulent Times,* New York: Harper & Row, 1982.

Fine, William T., *A Management Appraisal in Accident Investigation,* Silver Springs, MD: U.S. Naval Surface Weapons Laboratory, 1976.

Findlay, James V., and Raymond L. Kuhlman, *Leadership in Safety,* Loganville, GA: Institute Press, 1980.

Know, N. W., and R. W. Eicher *MORT User's Manual SSDC-4,* Washington, DC: Government Printing Office, 1976.

Petersen, Dan, *Techniques of Safety Management,* 2nd edition, New York: McGraw-Hill, 1978.

Pope, William C., *Systems Safety Management Series,* Alexandria, VA: Safety Management Information Systems, 1976-1980.

Professional Safety, Park Ridge, IL: American Society of Safety Engineers, July 1980.

Chapter 2 / Part 2

Budgeting For Safety

Safety, like any other important corporate function, must have and be accountable for a budget. Budgeting is a part of the planning function and must be done as early as possible to be effective. The budget must be developed and agreed to by all the concerned parties before implementation.

Although most safety professionals may have little or no formal training in budgeting, and although their experience is likely to be limited to a few instances of previous similar work, it is important that they participate in this effort. They know what must be done for an effective safety effort and what is desirable but optional. Budgeting without the safety professional's input will result in a budget that does not adequately reflect the needs or best interests of the safety organization or the company.

The level of experience is very important when developing a budget. Experienced persons may be able to work more quickly and correctly than inexperienced ones. However, the higher salary of the experienced person may offset the difference in the budget. In addition, there is always some degree of inaccuracy in budgeting when one considers only the hours required to do a given task. If the task is done by a more highly paid professional than was first budgeted for, the hours can be correct but the dollar cost higher than expected.

TIME FACTORS IN BUDGETING

The budgeting task should be assigned to the level most qualified person, and the minimum time needed to complete it determined in order to decide when it must be finished. The budget must be tied to certain milestones to be trackable. If there are no milestones, the budget becomes difficult to spread correctly and even harder to control once it is in place. This is where budgeting for the level of effort seems to be appropriate. Otherwise, the temptation to use a level-of-effort budget should be avoided, if possible. Setting the minimum completion time should be tempered by the knowledge that there are some things beyond the control of the safety program that can impact the schedule and cause a task to take longer than first planned. For this reason, it is usually wiser to budget for average time to completion rather than minimum time.

Materials and equipment needed to complete the tasks must also be in the budget if the budget will reflect the true cost of doing business. Material costs can be very significant when considered on an annual or semiannual basis. Equipment costs should also include replacement and repair costs. The cost of in-house repair, as opposed to the use of an outside vendor, must be evaluated to determine which is more cost effective.

PRACTICAL REASONS FOR BUDGETING

Budgeting is one of the most important tasks done by safety professionals. Without enough funding, any job is almost impossible to do. With over-budgeting, unnecessary tasks are done and the costs of the safety program are higher than necessary. When the costs of a program are higher than budgeted for, the cost overrun can be traumatic for everyone. Tasks must be taken out or cut back, personnel may be laid off, managers must spend time and resources to defend their efforts, and the program suffers. Cost overruns can also adversely affect the company's image and that of the people involved.

DEVELOPING THE OPTIMUM BUDGET

The best budget is one that allows all necessary tasks to be done with the right people at the correct time. It matches the skills of the individual who doing the task with the time and materials needed to complete it. The optimum budget must allow for revision and necessary research to develop documentation.

It must also be foremost in everyone's mind that a budget is a planning document for the proper disbursement of resources. It deals not only with personnel and dollars but also with the outlook of the company. The approved budget reflects the perspective of the company's management.

METHODS OF DETERMINING THE BUDGET

There are two major types of budgets: fixed and variable. The more common one is the fixed budget. It plans for a specific period of time and is updated and compared with actual results at the end of the period. The time period may be lengthened or shortened, depending upon the needs of the company.

The variable budget provides for the possibility that the actual costs may not be the same as the planned costs. This allows for restructuring of the budget to align it with actual costs. This allows for restructuring of the budget to align it with actual costs. This budget obviates the need for contingency budgets but requires a very good accounting system.

For the purposes of this discussion, we will consider the fixed budget, but the concepts hold true for the variable budget as well. There are several methods to determine the budget that should be allocated for safety. Three methods will be briefly discussed.

One method is to define program elements or tasks. Task definition is a good way to determine the budget. If a task can be defined, it can be properly budgeted for; if it cannot be defined, then it cannot be adequately covered in the budget. This method requires much up-front detail in describing the total program in manageable steps.

A second method provides for a specific percentage of the total company budget. A variation of that method is to allocate a percentage of the engineering, operations or administration budget to the safety program budget. This is an easy way to budget but it is not as good, because it does not determine how to do any one task or define the cost of doing it. This method requires little effort in developing the budget but is useless in identifying where resources will be used.

Level of effort on an as-needed basis is a third way safety may be budgeted. This is the "I'll call you when I need you" method, but it rarely allows the company to do the tasks correctly and on time. This means that no matter what the tasks are, the allocated budget does not change to accommodate them. There is no measurable effort in terms of significant accomplishment.

Of these three methods, the task definition one is the best way to identify and allocate where to spend and control company resources. It also provides an easy way to identify the problem area when an overrun or underbudget situation occurs. For this reason, the task definition method of budgeting will be discussed at length.

First, it must be decided which tasks are to be budgeted. In any safety effort, there are certain tasks that must be completed to ensure that the product and the workplace have the best level of safety. To ensure that the programs which the safety organization must implement are fully covered, it is probably best to divide the tasks.

An effective, comprehensive safety program can be broken into four major categories:

1. Workplace safety
2. Environmental safety
3. Product safety
4. Administration

These major categories can be further broken down into subcategories which we will call "safety programs." The various safety programs necessary for any company could include:

1. Safety management
2. Industrial hygiene
3. Hazardous materials/waste
4. Radiation safety

5. Hearing conservation
6. Safety training
7. System safety
8. Electrical safety
9. Construction safety
10. Personal protective equipment

This is not a comprehensive list. It does not include the subcategories of each of these programs, nor does it indicate the order of priority or importance. That is dependent upon the product or services of the company and should be considered when developing a task definition budget. All these factors may not be pertinent for any one company, but many of them will be appropriate for most organizations. There may also be other tasks, not identified above, that must be budgeted for and controlled. The basic requirements for any safety program depend on what the company needs for a safe and productive workplace and to produce a safe and effective product. As a first step in the budget process, the safety professional must determine what resources will be required for each task in the overall safety program.

Each of these programs requires that certain tasks be performed to reach its goals. The tasks serve as the basis for the budget. Identifying the scope of these tasks is the hardest part of budgeting.

DEVELOPING THE BUDGET

There are several considerations that must be kept in mind when budgeting for safety. Among the most important of these are the personnel assigned and the equipment, materials and time required to complete a task. These are closely interrelated and have a direct effect on the budget.

Personnel budgeting commonly takes two forms. One portion of the budget is the time, in worker-hours, needed to complete a task; the other is the dollars required to pay for the level of skill required to do the task. Therefore, identification of the skill level required to do each task is very important to the budgeting process.

When the budget is developed, considerations include the type of personnel needed and the difficulty of the task. Two items to consider when discussing personnel are availability and level of experience. Commonly, safety organizations worry only about the worker-hours needed to complete a task, but this leads to a false sense of security when assigning personnel. If a lower-level person is designated in the budget but a higher-level person does the tasks, the budget may go into overrun. Conversely, if a lower-level person does tasks that require a higher level of skill, the time needed to complete the task may exceed the allocated budget and impact the program schedule.

The availability of highly skilled personnel to do the safety tasks can be a major concern. If the tasks are complex, highly skilled personnel will be required to accomplish them in a timely fashion. However, the skilled personnel may not be available due to priorities and requirements of other programs. The personnel necessary to service a program must be identified early and an effort made to secure their services for the task.

Outside services such as periodic physical examinations, laboratory services and hazardous waste disposal costs must be included in the budget. The calibration of instruments must also be specified.

Some type of contingency budget will ensure that unexpected or unplanned tasks are completed. Contingency funds should be used only in an emergency.

It is very important to budget the total task. A common mistake is to budget only the personnel necessary to do the task and to budget the equipment and materials for the total program or department separately. Figure 2-2 is a form which outlines the items for each task.

Budget reduction in a company is always a possibility. The contracts expected may not materialize, the product may be late, or the lawsuit may require more resources than expected. The task definition method allows the budget to be reduced by specific tasks. Budgets are often cut by a certain percentage, with no reduction in the number or scope of the tasks. With a well-defined task definition budget, the effort can be reduced as appropriate. This allows the safety professional to get concurrence from company management to delete tasks and reduce budget requirements.

During development, the budget should be spread over the length of each specific safety program. Most companies budget annually; however, many times various safety programs are for shorter durations. This budget spread should cover the entire period that safety will be involved and should be tied to specific milestones to ensure that the coverage is complete and correct. The budget should be spread as needed to accomplish specific tasks. The budget should put the extra personnel where and when it will be needed. Figure 2-3 shows a typical system safety program budget spread tied to product design reviews, manufacturing and testing. It should be noted that most of the effort is expended in the early stages of the program. This is known as "front loading" of the system's safety effort and considered to be the most effective method.

Other types of ongoing safety programs will have a different budget spread. The peaks relate to activities relevant to doing various tasks within the program. Figure 2-4 shows an industrial hygiene program whose milestones refer to specific air and noise monitoring activities. Note that it is also slightly front loaded. This is common with increased costs to acquire equipment and the training necessary to use the equipment properly.

Although the budget is normally developed on the basis of worker-hours needed to complete a task, worker-days should be used, where possible. An equivalent measure should be based upon a 40-hour-week basis or at least on an 8-hour day. It is difficult to use only part of a person's time and still keep

TASK BUDGET

TASK: _____

EXPECTED DATE OF COMPLETION: _____

DATE OF FIRST BUDGET REVIEW: _____

DATE OF SECOND BUDGET REVIEW: _____

ITEM	MIN.	AVE.	MAX.	QUOTE	ESTIMATE AT 1ST REVIEW	2ND REVIEW
Manpower (numbers)						
Skill Level (yrs. of exper.)						
EQUIPMENT (List) ($)						
MATERIALS (List) ($)						
OTHER (List) ($)						
Time to Complete (man-days)						
Contingency ($)						

Figure 2-2. Budget planning form.

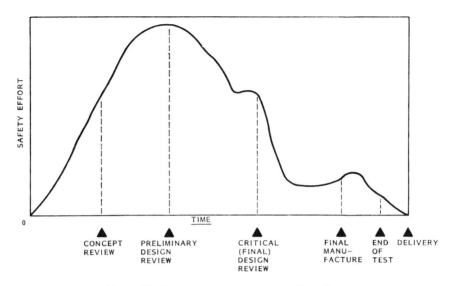

Figure 2-3. Typical budget spread for a safety effort.

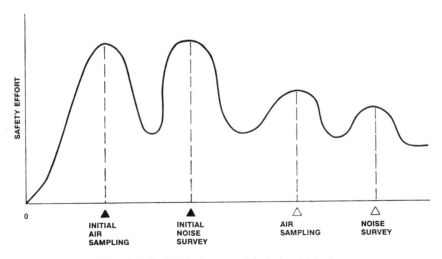

Figure 2-4. Typical budget spread for industrial hygiene.

35

that person active and interested in the program. If a person is working on several programs, his or her effort is diluted, making it less effective.

Each task to be accomplished should have a basis for the budget estimate. Developing a basis for the estimated budget can be a difficult and frustrating task. There are several methods for doing this. This chapter deals only with the methods used with task definition budgets.

The ideal basis for a budget estimate is history. If another program with similar tasks has actual data on the hours expended to complete the tasks, and this data is available, that is the best situation. More commonly, the estimate will be based upon the judgment of the individual who is preparing the budget, because there is little or no available documented historical data. Even when historical data is available, it is generally not well documented or is not in a format that is easy to use.

The engineering judgment method of justifying the budget is always subject to interpretation. However, if a task is broken down into the smallest possible parts, budgeting becomes much more straightforward. The judgment of the individual may be questioned, but if the tasks are well defined and defendable, there is little to argue over.

There must be a task description and a basis of estimate for each task. The task should be described in detail so that a nonsafety person can understand it. It should be as specific as possible, and should list the requirements and state where they may be filled. It should show what will be done, why it is needed and the scope of the task. The description should also state the skill level of the person required to do the task.

It has been suggested that a "fault tree" approach could aid in this effort, that is, a "top-down" approach which breaks down the task into its necessary parts and then subdivides those parts. Figure 2-5 is an example of this approach.

When discussing the tasks to budgeted for in the safety area, keep in mind that each program is different and that there are various ways to reach a budget total.

BUDGETING THE TOTAL EFFORT

Although task determination can be done in a top-down manner, all budgeting must be done on a "bottom-up: basis. In order to determine where a company will spend its resources, it must be clear what it will receive for that investment. Many tasks are required in order to have an effective and complete safety department. Each of these tasks needs a budget to ensure that it can be accomplished effectively. Earlier, the safety tasks were broken down into four separate categories. These categories are very broad and should be broken down into subcategories to be effective in the budgeting process.

BUDGETING FOR SAFETY 37

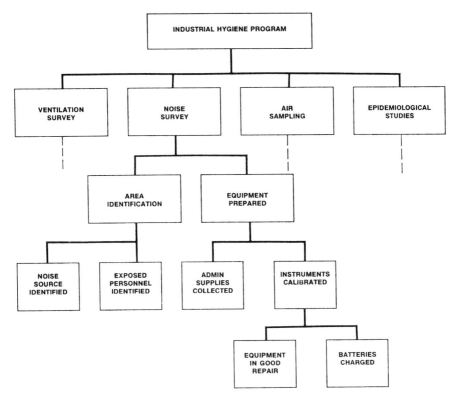

Figure 2-5. Top-down approach to tasking.

WORKPLACE SAFETY

An effective workplace safety program needs several subprograms to be effective. These subprograms should include those required by law and those that make good sense in doing business on a day-to-day basis. They will now be discussed briefly.

Eye Protection

In order for an eye protection program to be effective, a survey of eye hazards must be made. Once the eye hazards are found, controls must be established. When the controls are in place, the personnel must be trained to use them. This procedure is typical of most safety-related programs. The cost of replaceable materials, such as safety glasses, goggles, and face shields, must be considered. The time required to monitor, train the personnel and document the program must be established. Outside services needed to provide prescription safety glasses, fit nonprescription glasses, etc., should be specified in the budget. One important effort to include is the time needed to identify and control those eye hazards that can be eliminated by a good engineering effort.

Industrial Hygiene

Industrial hygiene programs provide the quantitative evidence needed in court to prove that hazardous conditions in the plant are properly controlled. They provide the best way to identify a potential problem early enough to correct it before someone is injured or an OSHA citation is issued. Accurate, independent evaluation of the work site can be performed effectively and efficiently by an in-house industrial hygienist or by an outside consultant. Either is appropriate, depending upon the desire of company management; a highly paid, well-qualified, full-time industrial hygienist may not be justifiable. However, the convenience of being able to pick up the phone and get an industrial hygienist on the scene immediately may be well worth the investment. The outside consultant cannot provide the instant service that a company may require or desire, but it can provide the service in a planned mode very effectively.

The budget for an industrial hygiene program must include the equipment necessary to do the required tasks. This typically includes air sampling pumps, noise survey equipment (such as sound level meters and noise dosimeter), gas detectors, radiation survey meters, ventilation survey equipment, heat stress equipment and various other support items.

The cost of sample analysis must be included. Very few companies can afford the cost of setting up a comprehensive, accredited industrial hygiene laboratory; therefore most send their samples to a certified laboratory. Costs depend upon the complexity and the time necessary to run the sample. This must also be provided for in the budget.

Radiation Safety

The basis for all radiation safety programs is the monitoring of personnel exposure. Personnel dosimetry is usually handled by an outside firm which provides radiation monitoring badges and processes them for a set fee. The cost of this service varies with the number of badges used or how often they are processed. Some companies want a radiation safety effort large enough to justify a full-time health physicist on staff, but most rely on the safety department to provide this support.

To ensure that no contamination by radioactive materials exists, area surveys are done on a regular basis. This usually requires portable monitoring equipment and wipes.

The equipment requires regular calibration and the wipes must be analyzed, usually by an outside service on a fee basis. The monitoring equipment may be leased from the firm doing the calibration or purchased and sent out for calibration a scheduled basis. The time and materials necessary for these area surveys can be considerable.

Much documentation is needed to manage a radiation safety program. Responses to requests from other companies on their employees who have worked at the organization's facility must be handled quickly. Exposure records must be requested from other companies on new employees if they have had occupational exposure to radiation.

Regulatory audits require much time to provide the agency with the needed data. Since these audits occur on a regular basis, they should be planned and their cost incorporated in the budget.

Hearing Conservation

With the knowledge that noise-induced hearing loss can result in a disability claim, it becomes imperative to reduce noise to acceptable levels and provide protection to employees where noise reduction is not possible. A hearing conservation program identifies sources, gives a preplacement audiometric examination for all employees, provides periodic audiometric testing for employees working in high-noise areas, fits and issues hearing protection devices, and provides ongoing noise surveys and evaluation of all work sites. Audiometric examinations are usually done by an outside firm but can be done inside with the proper equipment. Wherever they are done, they require properly calibrated equipment used by a trained and certified audiometric technician.

Hearing protection comes in various sizes, shapes and materials, and the cost of this equipment can vary dramatically. A program which includes custom-fit ear protection devices is more expensive than the use of off-the-shelf materials. However, the former may be preferable because of greater acceptance by employees.

Sound level monitoring equipment can be expensive, and the complexity of the equipment depends upon many workplace factors. The commonly used sound level meter combined with personal dosimetry provides excellent results in most environments. But if there is a need for octave band analysis or for measuring impact noise, the cost rises quickly. Calibration by in-house laboratories is possible but usually not feasible.

Safety Training

An important element in any comprehensive safety program is training. Safety training should be designed to prevent accidents and teach the employee the hazards associated with his or her job.

When budgeting and training, there are several factors that must be considered. The instructor's time in teaching the class must be taken into account. If it is an outside instructor, the procedure is fairly simple; a fee is paid and a service is rendered. For inside service, one must account not only for the time needed to present the material but also for the instructor's preparation time.

The Hazard Communication requirement brings into sharper focus the need for a safety training program. The law now requires the training of employees in the hazardous nature of their jobs. This training is ongoing and must be done on a periodic basis.

Refresher training is necessary for employees to remain up-to-date on new developments and for recertification. If the company has a requirement that all electricians must be (CPR) cardiopulmonary resuscitation certified, then there is an annual requirement.

The materials necessary for a training class must also be considered. These include handouts, slides, quizzes, samples and other such materials. The strong use of audiovisual aids, such as films and video cassettes has opened a new perspective on training. These films can be purchased and shown over and over, with little instruction required. The technique does require specialized equipment, and this may be a necessary added expense to include in the budget.

Some organizations have developed safety training correspondence courses to train employees in specific hazards and operations. These can be very effective and can be tailored to a specific location. They can also be expensive and require much updating and research time.

Construction Safety

New construction projects for facility renovation or for new buildings and facilities are usually well planned. The safety effort required for this activity must also be planned. The problems encountered on construction sites are sometimes unique and different from those in a manufacturing effort. A consultant or a safety engineer specifically trained in construction safety may be necessary for a large, complex activity. In either case, the cost must be budgeted.

If the company is the prime contractor and provides all the personnel and materials, the safety department must be alert to potential problems involving personnel who are not familiar with construction sites.

Regular safety inspections of the site to ensure maintenance of work site safety are needed. They should also ensure that personnel use proper protective equipment, such as head, eye and hearing protection devices. This equipment may have to be acquired separately, and its cost should be included in the budget.

Construction safety orders have different requirements than general industry safety orders and may require some research by the safety department to understand them fully. The time necessary to do this research should be included in the budget.

Subcontractors may cause specific safety problems due to the increased use of poorly trained personnel or inadequate equipment. The safety department must budget extra time for each subcontractor to ensure that the safety requirements on this particular job are understood.

A review of drawings is very important to the safety of a construction project. The safety engineer can identify particularly hazardous operations and can ensure that he or she is present to make certain that they are conducted in the safest manner possible.

Forklifts and other vehicles used on construction projects have specific requirements which must be met. These include rollover protection, seat belts and backup alarms. The safety professional should ensure that these vehicles have this equipment before use.

A common safety problem on construction sites is inadequate or inappropriate scaffolding. Safety department personnel will be required to spend much time making sure that the scaffolding meets the regulations. Cranes and other lifting devices also have special inspection and maintenance requirements that the safety professional must be aware of. The time needed to do the inspections and to solve problems must be considered in the budget.

Crane Safety

Cranes and hoists need special handling. They require routine inspections by operators and also by another person qualified to inspect them. Nondestructive testing may be needed periodically to ensure that they are safe to operate. This may require the use of a consultant or an outside service.

Respiratory Protection

Most manufacturing companies have some personnel who must wear respirators on a periodic basis. The costs of this program must be listed in the budget. All personnel who wear a respirator must pass a physical qualification examination. This can be done by a company physician or by a contract physician.

Training in the limitations of the respirator and proper respirator selection must be provided. These employees must be provided with a respirator that fits their faces. Qualitative or quantitative fit testing must be provided whereby the individual wears the respirator in a test atmosphere. Depending upon the type of testing done, the equipment can be expensive.

Different sizes and types of respirators must be available for fitting and testing of personnel. Consumable materials for the testing and the cleaning of equipment after testing must be included in the budget.

ENVIRONMENTAL SAFETY

The impact of industry on the environment comes in many forms: noise, hazardous materials, oil spills, ionizing and nonionizing radiation or thermal energy. Their sources must be identified and controlled, and the various programs needed to to this must be established and used. This requires the time and effort of the safety department.

Hazardous Materials

Materials used by the company must be identified, their hazards noted, and controls placed to ensure that they do not harm personnel or the environment while they are in the plant. The safety department must be a part of the purchasing activity to ensure that the hazardous materials coming into the plant are identified before arrival. This ensures that proper storage facilities exist and that there are procedures for safe use of the materials. For many manufacturing companies (and, in some states, all companies), a Material Safety Data Sheet (MSDS) must be on file for all hazardous materials received in a plant, or one must be ordered. If the company manufactures a chemical product, an MSDS must be developed and provided to the users. The maintenance of a Hazardous Materials Inventory is one very effective way of knowing what materials are in the plant and what hazards they pose. This inventory must be updated periodically. The labeling of all chemicals is an essential part of any hazardous materials program. It must tell users what hazards exist and how they can avoid them.

A hazardous materials spill plan must be developed and used. This can require a significant amount of equipment, materials and personnel. Personnel must be trained to respond correctly to different types of spills. Another approach is to hire an outside contractor to respond to spills. This can be very costly, but it is sometimes well worth the expense because it does not require the company to train its own personnel and maintain the equipment.

Hazardous Waste

Every company regardless of its size, must be concerned with the proper disposal of hazardous wastes. Large companies typically have full-time staff to do nothing but manage hazardous waste programs. Smaller firms, usually classified as "small generators" because of the small quantity of waste they create, use the safety department to manage the program. The mandatory use of the "manifest system" requires that all waste be identified and a chain of custody established for it. This manifest system and the cradle-to-grave responsibility for waste require a significant amount of time and effort, which must be budgeted.

Annual reporting of waste usage to federal or state regulatory agencies can take a significant amount of time and documentation. If hazardous waste is stored at a company site for any time period, it may require a significant amount of contingency planning. Planning is needed for the proper response to a spill of hazardous waste in order to prevent it from becoming a problem for personnel or the environment. These plans may require a consultant who is an expert in their preparation or much work by in-house personnel. Such plans should be updated annually or as significant changes occur. Their development, implementation and documentation must be budgeted.

Oil Spill Control

If a company stores a significant amount of oil at its facility, it may be required to have a Spill Prevention, Control and Countermeasures (SPCC) plan. This plan, mandated by federal regulations, requires much research and documentation. It involves inspections of tanks and other storage vessels, and documentation of these inspections. If a spill occurs, The SPCC directs how the company will respond to the clean-up and how it will provide funds for clean-up costs. Documentation and implementation of SPCC plans can be extensive and expensive.

PRODUCT SAFETY

An effective product safety program must identify all the potential hazards in the product or service, warn the user that they exist, and follow the product from cradle to grave.

System Safety

An effective system safety program covers the product from concept through disposal. The safety engineer should review the drawings, prepare a preliminary design and operations hazard analyses, ensure that the labeling and operations manuals describe the hazards, and provide documentation that is appropriate and detailed. This can be a minimal effort for many products which have a good history, but can be extensive and time-consuming for new product lines. Analytical testing may be required to verify information in the hazard analysis. Stress analyses may have to be run, and various types and levels of inspection performed, to ensure that the product is made as designed. The effective system safety program requires that the safety engineer review this data and ensure that it meets the safety requirements. Time for consultation with legal personnel, design engineers, operations personnel, and repair and maintenance personnel must be budgeted to ensure that all aspects of the product are evaluated.

Marketing

Marketing activities can cause a significant problem for effective produce safety programs due to misleading or inaccurate advertising. The product safety engineer should review the marketing literature to ensure that it does not mislead the user or show the product used in a manner in which it was not intended. He or she should also review the disclaimers for faulty or inaccurate statements. This requires time and documentation, as well as coordination with marketing, advertising and legal personnel. This time and documentation must be budgeted.

Recall

If a product is recalled, the safety engineer must have some method to identify all the users of the product. This requires the maintenance of a database of users or distributors. Before the recall, the safety professional should be consulted, or should consult the legal department or corporate counsel, on the proper wording of the recall.

Litigation

All current and pending litigation should be included in the budget. Time for personnel to give depositions, testify in court, or perform research activities should be budgeted. If there is a settlement out of court, a budget for the costs of this settlement should be included.

ADMINISTRATION

One common mistake made in the budgeting process is a failure to budget for the administrative part of the program. The following subsections discuss several items that should be considered when budgeting for the administrative tasks in a safety department.

Safety Management Program

The safety management program must include the tasks necessary to supervise, evaluate and counsel personnel assigned to the safety department. These can include nontechnical people, such as secretaries and clerks, and technical personnel, such as industrial hygienists, health physicians and safety engineers. It should include the costs of acquiring a new person, either for replacement or for growth of the department. The tasks associated with acquiring a new person should include advertising, recruiting, interviewing, hiring, orientation, and training. Each of these tasks should be broken down into its subtasks. For advertising, that means the work of writing the job description, reducing it to an advertisement, and getting it approved. These tasks could also be broken into smaller increments, depending upon the difficulty of getting an advertisement approved by the company.

The costs associated with planning, organizing and using new products must also be included. A hazardous process has a direct and significant effect upon the personnel of the safety department. The time it takes the safety professional to review the process, evaluate personnel hazards, evaluate the environmental impact, evaluate and recommend controls and plan the monitoring strategy can be very significant.

Materials required for the successful implementation of a safety program can be a serious cost factor if they are not planned. When considering materials, the budget for safety management should include the administrative supplies and services necessary for the smooth operation of the safety department. Materials should be budgeted for on the basis of annual use. This might seem to be a minor cost, unworthy of much effort, but in the area of documentation the safety department must be exceedingly thorough. The cost of the reproduction paper, for example, can be staggering when the company is large and the distribution list contains 40 to 50 people. A 20-page document becomes an 800-page distribution when 40 copies are needed. Sometimes the cost of reproduction may exceed the effort needed to generate the original work.

Equipment acquisition or repair must be included in the budget. The repair of a copying machine no longer covered by a maintenance agreement or repair contract may represent a greater cost than its replacement with newer and better equipment. The need for a computer in the safety department to allow for efficient database management is becoming well recognized. However, the costs of maintenance and supplies necessary for the effective use of a computer may not be so readily apparent. Many computer systems use a floppy disc drive system and have a printer attached. The cost of software, on-line services, floppy discs, computer paper, ribbons, floppy disc storage containers and annual maintenance costs must be considered in the budget.

Accident Investigation

To have an effective safety program, accidents and incidents must be investigated. For the purposes of this discussion, an incident is considered a "near miss," an unplanned event that did not hurt anyone or damage any equipment but could have under different circumstances.

In any effective accident investigation, there is the need for equipment that must be included in the budget. This equipment usually consists of a camera, a couple of lenses, a tripod, a camera bag, and the normal accessories for photography. The cost of film and processing must also be included.

Other supplies can include tape measures, flashlights, drawing sets, equipment cases, and so on.

Emergency Evacuation

Every effective program provides for the safe evacuation of personnel from the facility. This requires the generation of a detailed emergency plan. Such a plan should include medical, fire and other emergency assistance. The research, development and implementation of such a plan are commonly assigned to the safety department. This requires detailed planning, and when the plan is finished, it should be tested.

The evacuation plan may require the purchase of communication equipment, emergency medical equipment, special aids for the handicapped, and special tools for rescue. It also requires periodic drills to test the plan. These activities must be stated in the budget.

Coordination with local agencies that provide support can be time-consuming, requiring many meetings to ensure that they understand the facility, what they can support, and the hazards they may have to deal with in an emergency.

Some companies recognize that there may be no effective way to evacuate personnel in certain situations. For example, a massive earthquake, such as the 1985 quake in Mexico City, can paralyze all emergency services for a period of time. Many companies are considering how to provide food, a good source of water, shelter, and emergency medical support for their personnel in this type of emergency. If a company decides that it needs to provide for this situation, the planning and implementation costs can be very high.

It does not matter whether the emergency plan is detailed, general, or comprehensive, or deals only with specific problems. It requires time and materials to develop and place in use. The above tasks do not represent the entire safety effort, nor was that the intent of this discussion. These tasks are common to almost any safety department and should be considered when developing the company's safety budget.

Figure 2-6 is a table abstracted from Olson (1982) which shows the number of hours typically assigned to specific system safety tasks. This table may not be applicable to every company's system safety program, but it does delineate the type of detail that is needed in task definition budgeting.

SUMMARY AND CONCLUSION

The budgeting process in vital to every program. The individual responsible for the safety department's budget must learn to describe the tasks in such a way that they are defensible and realistic. The task definition method of budgeting provides this detail. The best data available for budgeting is historical, but in lieu of that, good judgment and use of experience must suffice. The key to successful task definition is to break down each task into

REFERENCES AND BIBLIOGRAPHY

Anton, Thomas J., *Occupational Safety and Health Management,* New York: McGraw-Hill, 1979, pp. 60-75.

Donnelly, James H., Jr., James L. Gibson, and John M. Invancevich, *Fundamentals of Management,* Dallas, TX: Business Publications, Inc., 1987, pp. 80-81.

Olson, R. E., *System Safety Handbook for the Acquisition Manager,* prepared by the Aerospace Corporation for the Space Division, Air Force Systems Command, Document Number SD-GB-10 December, 1982, p. 17.

BUDGETING FOR SAFETY

TYPICAL WORK BREAKDOWN*

Task No.	Task Description	Estimated Man-hours	Task Assignment
4.1.1	System Safety Program Management	4160**	System Safety Manager
4.1.2	• Customer single point of contact, 2 TI meetings per month @ 1 day each.	684	"
	• Approval of safety critical documents, 200 documents @ 1 hour each.	200	"
	• MGMT/T & program element coordination, 6 meetings/week @ 1 hour each.	624	"
	• Internal audits, 3 @ 24 hours each.	72	"
	Subcontractor audits, 6 @ 24 hours each.	144	"
	• Industrial safety interface	400	"
	• Preparation for safety reviews, 4 reviews @ 2 days each.	64	"
	• Participation in safety reviews, 4 reviews @ 16 hours each.	64	"
	• Preparation for readiness reviews, 3 reviews @ 16 hours each.	48	"
	• Trade study reviews, 16 meetings @ 2 hours each.	32	"
	• Configuration control board activities, 4 meetings/month @ 2 hours each.	192	"
	• Review & approval of accident risk assessments, 200 documents @ 2 hours each.	400	"
	• Safety certification of safety critical documentation, 75 documents @ ½ hour each.	37.5	"
4.3	• Integration function		
	• Prepare and coordinate integrated system safety plan, etc.	200	"

* Breakdown shown is for a 2-year program based on Para. 4.1.1 of MIL-STD-1574A. Other tasks specified in the standard are not included.
** For major system development efforts assignment of less than one fulltime safety engineer is unacceptable.

Figure 2-6. Typical work breakdown for system safety tasks.

the smaller portions that can be budgeted for, and budget against a specific measurable effort.

The budget, if properly used, can be the basis for effective planning year after year. Once a total budget exists, it may change very little on a year-to-year basis. If there is a major change, the task definition method of budgeting allows the changes to be incorporated in the overall company or organization budget in a responsible manner. It also makes it much easier to prepare subsequent budgets because most of the work is done and there is a detailed history to use in correcting the estimates.

Chapter 2 / Part 3

Cost/Benefit Analysis For Health and Safety in Business

This chapter discusses cost/benefit analysis and cost effectiveness analysis as they relate to safety and health matters in business. Technical details are avoided; however, those seeking more detail can read the literature listed in the references at the end of this chapter.

This discussion of cost/benefit analysis relates to the multiple ways that health and safety may impact a business by occupational illness and injury to people and damage to or loss of property. Therefore, program areas include occupational health and safety, off-the-job safety, product safety, fire safety, and control of the environmental, highway safety, and public incidents that may have liability exposure for an organization.

HEALTH AND SAFETY MOTIVATORS

One or more of the following considerations provide the impetus for health and safety improvements; moral/social, legal/regulatory, and economic. The most clear-cut, nonarguable fact is legal/regulatory. A manager may disagree with the need for, reasonableness of, and practicability of a legal safety and health requirement. However, the decision is out of the manager's hands unless a variation or waiver of compliance is sought. In complying with legal requirements, analyses of costs and benefits may help the manager to choose the proper course of correction/improvement, but not to determine the need for compliance. Fortunately, federal regulators must conduct cost/benefit analyses of proposed regulations or standards before imposing them on the public. This somewhat eases the problem of determining the cost burdens of requirements over which we have little control.

There are many claims (some well documented) for the strong correlation between low incidence of occupational diseases and accidents and high operating efficiency. However, the fundamental reason and motivation for good health and safety practices should be social/moral. Here, other things besides the moral values of the manager and his or her organization must be considered. It is also necessary to think of those affected by health and safety decisions: employees, labor unions, purchasers of products/services, the community, the public at large, and the government. Here, cost considera-

Table 2-1. Gas tank evaluation.

Benefits
Savings: 180 burn deaths, 180 serious injuries, 2,100 burned vehicles
Unit Cost: $200,000 per death, $67,000 per injury, $700 per vehicle
Total Benefit: 180 × ($200,000) + 180 × ($67,000) + 2,100 × ($700) = $49.5 million

Costs
Sales: 11 million cars, 1.5 million light trucks
Unit Cost: $11 per car, $11 per truck
Total Cost: 11,000,000 × ($11) + 1,500,000 × ($11) = $137 million

Source: Dowie (1977); and *Mother Jones Magazine*

tions are useful in selecting the improvement measure to be applied, not in making the decision to improve or not improve. Table 2-1 shows a simple cost/benefit comparison from an in-house memo of an automobile company. One might infer that these figures influenced a decision not to change the gas tank design.

Obviously, the cost/benefit factors considered in this example were identified after several incidents occurred and legal settlements were made. What is not shown are other benefit losses, such as goodwill, and questions about quality in consumers' minds about other products bearing the company's brand name. Also, the monetary award for loss of life or serious injury can be set higher or lower at the whim of a sympathetic jury or because of the lack of skill of legal counsel.

Because of the growing recognition that health and safety measures contribute to operating and product efficiency, they are becoming increasingly a part of plant process, product design/redesign, and employee-management relations. In many instances, minimum health and safety standards, required by law or demanded by societal values, are met. But health and safety improvements may also lead to general improvement of operations or products. Cost evaluation can then be the major reason for selecting the best way to achieve further improvements in health and safety, plant efficiency, and product acceptance.

In summary, where conditions fall short of legal and moral standards, cost/benefit or cost/effectiveness evaluations should not decide the need for improvement; they should merely help to show the mode of improvement.

ORIGIN AND DEFINITIONS

James L. Riggs, in *Essentials of Engineering Economics,* traces the origin of cost/benefit analysis back to 1844. He relates its required uses in this country to the Rivers and Harbors Act of 1902 and the Flood Control Act of 1936. Use of such analyses in health and safety decision making appears to have started about 1965. At that time, government procedures required the Department of Health, Education and Welfare to make monetary estimates of the effects of competing programs. Research shows that, since then, government departments, and other analysts and researchers, have used cost/benefit analysis to help make health and safety decisions on matters of widespread scope. However, its current use for business safety and health appears to be nonexistent or, at best, highly limited.

Although the terms "costs/benefit analysis" and "cost/effectiveness analysis" are often used interchangeably, there is a distinct difference between them.

Cost/Benefit Analysis

In cost/benefit analysis, the value of all cost items for a given project is compared with the value of all benefit items for the same project. Thus, a real or "fabricated" (depending on how hard the value data is) comparison can be made to see if a project can pay for itself. Since cost/benefit analysis can be made for many projects, it can also be used to determine which project/objective has the highest benefit/cost relationship.

Cost Effectiveness Analysis

In cost/effectiveness analysis, monetary values for cost items alone are gathered for projects with the same objectives/benefits. Thus, the process helps to select one of several methods which will provide the objective(s) at the best cost. The monetary values of the benefits (with all the vagaries that can accompany such valuations) need not be found. They are constant for all projects. However, cost/effectiveness analysis does not determine if any project will pay for itself.

Although the processes differ from one another, both help the manager:

Choose from various courses of action and
Establish a framework to measure the economic performance of a program.

MAKING A COST/BENEFIT ANALYSIS

R. A. Donvito, in *The Essentials of a Planning-Programming-Budgeting System,* these items as part of a cost/benefit analysis:

Objectives of the program
Financial requirement
Expected benefits
Uncertainty about cost/benefit estimates
Uncertainty about the time when costs are expected to be incurred and benefits are expected to result

He also suggests two approaches to making selections:

Fixed budget
Assumes availability of a fixed level of funds and selects a course of action to achieve maximum benefits.
Fixed effectiveness
Assumes a specific objective to be accomplished and selects a course of action to minimize the cost of achievement.

Table 2-2. Steps performed in cost/benefit and cost/effectiveness analyses.

Step I	Identify the stimulus or problem for conducting economic analysis
Step II	Specify the objectives or goals to be achieved by the program under study
Step III	Determine the alternative means of attaining those objectives or goals
Step IV	Enumerate the costs and benefits or effects for each alternative
Step V	Assign monetary values to the costs and benefits or assign units of effectiveness
Step VI	Perform discounting
Step VII	Calculate the cost/benefit or cost/effectiveness ratio
Step VIII	Compare these ratios by use of a decision matrix and feed this information into the decision-making process

Source: Ossler (1984).

Table 2-3. Cost centers used in economic analysis of occupational health programs.

Personnel	Professional, clerical, administrative salaries, and fringe benefits
Supplies	Records, printed materials, desk supplies, clinical supplies
Capital expenditures	Sound booth, spirometer
Facility	Space rental, utilities
Lost production time	

Source: Ossler (1984).

Table 2-4. Units of effectiveness used in economic analysis of occupational health programs.

Absenteeism rate
Health care services utilization
Life style habits; smoking, obesity, seat belt use, alcohol ingestion
Incidence, prevalence of work- and non-work-related injuries and illnesses
Workers compensation claims: number, severity
Number of accidents
Productivity

Source: Ossler (1984).

Table 2-5. Cost centers and benefits used in the 1980 A. D. Little study.

Costs
  Physical facility, utility services, capital equipment and fixtures, depreciation
  Supplies, journals, printed materials
  Training costs
  Employee time lost by utilizing services

Benefits
  Direct
    Decreased cost of physical examinations
    Salary savings from decreased lost worktime for onsite (versus offsite) health care
    Savings on workers compensation claims
    Savings based on OSHA recordable events
    Savings based on first aid log data
    Reduced nonoccupational health care costs
  Indirect
    Reduced absenteeism
    Reduced labor turnover
    Increased worker productivity

Source: A. D. Little, Co., *The Economic Cost to Society of Motor Vehicle Accidents, Reference Number DOT HS806346,* January 1983.

C. C. Ossler, in "Cost-Benefit and Cost-Effectiveness Analysis in Occupational Health," provides (Tables 2-2 to 2-5) step-by-step methods and examples of factors used in cost/benefit and cost/effectiveness analysis.

Problems/Controversy

In health and safety matters, a field with a scant database, and in which some cost and benefit valuations are elusive, problems in making and using cost/benefit analysis abound. Much controversy centers on two issues:

1. The lack of reliable data needed to make cost and benefit studies
2. What monetary valuations, if any, should be placed on human life, pain and suffering, and well-being (Tables 2-6 and 2-7 give figures for traffic situations.)

Specific areas with valuation problems include:

Improved well-being
Better personal relationships
Improved personal capabilities
Improved productivity
Improved product performance
Discovery of product problems unrelated to the health and safety problems

54 SAFETY AND HEALTH MANAGEMENT

Table 2-6. Average cost of traffic fatalities.

FACTOR	AVERAGE COST
Medical	$ 1,370
Productivity	236,865
Property damage	3,406
Legal and court	13,394
Coroner and medical examiner	168
Emergency	290
Insurance	12,523
Public assistance administration	576
Government programs	135
	$268,727

Source: A. D. Little Co., *The Economic Cost to Society of Motor Vehicle Accidents*, Reference Number DOT HS806346, January 1983.

Table 2-7. Average economic costs for motor vehicle injuries.

ABBREVIATED INJURY SCALE	AVERAGE COST	
5	$190,010	Most serious injuries, e.g., spinal cord damage
4	43,729	
3	10,260	
2	4,592	
1	2,276	Least serious injuries, e.g., bumps, bruises, cuts

Source: A. D. Little Co., *The Economic Cost to Society of Motor Vehicle Accidents*, Reference Number DOT HS806346, January 1983.

Lower absenteeism
Fewer production interruptions

Contributing to the problems are variables that affect health and safety valuations. Risk acceptance differs with people and situations. For example, a volunteer smoker may consider smoke from a machining operation hazardous. Additionally, health and safety matters may have social and/or political overtones that cause valuation differences.

Basically, the problem is that data that appears to be solid may be significantly soft and arguable.

In reviewing the recommendations drawn from the cost/benefit analysis, it is well to consider costs which would be incurred now or at some future time. The analysis might consider the availability of resources, inflation, and the potential for increased risk of harm or damage resulting from failure to take needed corrective action. While some projects can be delayed without appreciable loss of benefits or cost escalations, benefits which could come from better maintenance programs may far outweigh any increased costs.

Misuses

Like any other helpful system or procedure, cost/benefit analysis for health and safety decision making may be misused or misapplied.

First, there are relatively straightforward situations with ample precedent and fundamentally agreed upon legal/regulatory or moral/social requirements. To "manufacture" more justification via a cost/benefit or cost/effectiveness analysis, contributes nothing but extra expense and wasted time.

Second, it is possible to be lured into using so-called hard numbers to avoid unpalatable (and maybe expensive) health and safety actions with ample moral/social justification.

Uses/Adaptions

Occupational illness and injury programs pose many questions of control that require answers from management. These include:

Will the program pay for itself? In what period of time?
Is control of a hazard affordable? That is, what will be the cost of correction, and what funds are available?
Which hazard(s) produce the most serious exposure?
Which corrective measure(s) produce the most exposure control, or the most exposure control with available funds?
Which corrective measures under consideration cost least? Most?
Which exposure problems are most urgent?
Which corrective measures are most acceptable to all audiences concerned?

These and other questions, may not be answerable, even in part, by economic analysis. But where uncertainties exist about priorities, costs, or efficiencies, cost evaluations should be done. Too often, there are none. Whether there are more complex economic evaluations such as cost/benefit or cost/evaluation analysis depends on the degree of uncertainty, the level of expenditures, and the time and expense needed to produce data and make the evaluation. With the humanitarian goals of programs/actions and the abstract nature of such activities, health and safety should be more subject to cost evaluations both before and after programs begin. Allocating and efficiently using money to underwrite safety programs (as contrasted to expenditures for competing nonhealth and safety activities) and the selection of efficient ways to use allocated funds demand more economic analysis.

For possible use in business and industry, the following health- and safety-related economic evaluations using cost/benefit or cost/effectiveness analysis (or components of them) are visualized:

1. Determination of the total health and safety budget. Here sizable funds are needed in many firms and may require a full cost/benefit analysis, particularly when costs and benefits relate to on-the-job incidents; off-the-job safety; workers' compensation insurance; health and accident insurance; public safety and public liability matters, fire prevention and protection, and disaster control; the health and safety aspects of employee relations, union relations, public relations, community relations, and government relations (including OSHA); production interruption; and discovery of production inefficiencies.
2. Allocation of funds to program components within the overall health and safety program, such as engineering controls, employee information/education, management information, protective equipment, supervisory training, sales training, product user information, and product redesign.
3. Simple, direct cost evaluation (simplified cost/effectiveness evaluation) for the selection of hazard control alternatives for a given exposure.
4. Tracking of illness, injury, workers' compensation, and other costs relating to a specific exposure before and after controls are implemented.

These are just a few of the ways that cost/benefit and cost/effectiveness analyses might be used or adapted. Considering the process will undoubtedly stimulate the development of others that are useful and reasonably practicable. But, since overuse can divert time and money from needed exposure controls, a cost/time evaluation is needed to determine the extent to which this tool may be most beneficial to a business.

SUMMARY

Cost/benefit and cost/effectiveness analyses are related but different processes.

Health and safety use of these analyses has been limited to evaluation of widespread national problems and research.

Don't use cost/benefit and cost/effectiveness analyses or any cost evaluation to determine the need for controls where legal/regulatory or social/moral standards exist.

Don't use these analyses to replace simple, straightforward, economically acceptable (based on experience) solutions.

Use these analyses or adaptions of them when there are sizable expenditures and questions about priorities, costs, and efficiencies.

REFERENCES AND BIBLIOGRAPHY

Biancardi, M. F., "The Cost/Benefit Factor in Safety Decisions," *Professional Safety,* November 1978, pp.

Biancardi, M. F., *Handbook of Occupational Safety and Health,* Lawrence Slote, ed. New York: Wiley, 1987.

Dardis, R. A., "A Critical Evaluation of Current Approaches to Lite Valuation in Cost-Benefit Analysis," *The Journal of Consumer Affairs,* Summer 1981, pp.

Dardis, R. A., and C. Zent, "The Economics of the Pinto Recall," *The Journal of Consumer Affairs,* Winter 1982, pp.

Donvito, P. A., *The Essentials of a Planning-Programming-Budgeting System,* Washington, DC: U.S. Department of Commerce/National Bureau of Standards, July 1978.

Dowie, M., "How Ford Put 2,000,000 Fire Traps on Wheels," *Business and Society Review.*

Fine, W. T., *Mathematical Evaluations for Controlling Hazards, Proceedings of the National Safety Congress,* 1973.

Gagne, F. Jr., "Risk Assessment Valuable Approach to Cost Assessment," *National Safety Congress Aerospace Newsletter,* October 1987, pp.

Grimaldi, J. V., and R. H. Simonds, *Safety Management,* Homewood, IL: Richard D. Irwin, 1975.

Harvey, C. M., "Cost-Benefits Study of a Proposed Daylight Running Lights Safety Programme," *Journal of Operational Research Society,* 1983, pp.

Kelman, S., "Cost-Benefit Analysis, An Ethical Critique," *Across the Board,* July-August 1981, pp.

Klarman, H. E., "The Road to Cost-Effectiveness Analysis," *Milbank Memorial Fund Quarterly,* Fall 1982,

Litecky, C. R., "Intangible in Cost-Benefit Analysis, *Journal of Systems Management,* February 1981, pp.

Lowrance, W. H., *Of Acceptable Risk,* Wm. Kaufmann, 1976.

Moll, K. D., and D. P. Tihansky, "Risk-Benefit Analysis for Industrial and Social Needs," *American Industrial Hygiene Association Journal,* April 1987.

O'Neill, B., and A. B. Kelly, "Costs, Benefits, Effectiveness, and Safety: Setting the Record Straight," *Professional Safety,* August 1975, pp.

Ossler, C. C., "Cost-Benefit and Cost-Effectiveness Analysis in Occupational Health," *Occupational Nursing,* January 1984, pp.

Recht, J. L., *How to Do a Cost Benefit Analysis of Motor Vehicle Accident Countermeasures,* National Safety Council, September 1966.

Riggs, J. L., *Essentials of Engineering Economics,* New York: McGraw-Hill, 1982.

"Why the Pinto Jury Felt Ford Deserved $125,000,000 Penalty, *Wall Street Journal,* February 14, 1978, p.

Whipple, C., *Risk Analysis, Proceedings of the National Safety Congress,* 1977.

Chapter 3

Planning and Organizing the Safety and Health Program

A house, to use a cliché, is only as strong as its foundation. This is equally true of the safety and health program: It will never be more effective than the foundation it is built on. Establishing this foundation is senior management's proper task. Long before the program goes into effect, senior management can do much to determine its ultimate success—or failure. This chapter will concentrate on the necessary advance steps to be taken in developing a safety and health program. These are:

The planning function
The organizing and operating functions

THE PLANNING FUNCTION

The planning function is commonly broken down into four major areas: (1) forecasting, (2) goal and objective setting, (3) policy making, and (4) budgeting. Forecasting the market and the future direction of the organization is done at the highest levels of management and must precede the other steps. Setting goals and objectives then defines the organization's future direction more specifically. Policies are next established to carry out these goals and objectives. Providing for them through budgeting is the necessary last step. In planning for safety and health programs, these four areas should operate as interrelated activities, culminating in a successful program.

Specifically, planners must consider the following points in planning safety and health programs:

1. *Appropriateness.* Executive management must consider the future goals and objectives of a company in order to develop a program that will be appropriate to them.
2. *Accountability.* Senior management must assume general responsibility and accountability for the program.
3. *Delegation.* Safety and health duties must be logically delegated throughout the management structure, though senior management retains final responsibility and accountability.

4. *Key Personnel.* The key persons responsible for carrying out the plan should be identified. This is usually a function of position.
5. *Definition.* The safety and health relationships between management, staff and line personnel should be clearly defined. Usually this is done through job descriptions.
6. *Measurement.* There should be a method of ensuring that each member knows his or her exact safety and health responsibilities and an effective means for measuring what he or she has done.
7. *Support.* All management and staff must provide adequate support to the safety and health function.
8. *Identification.* There should be detailed plans for identifying, controlling, and auditing workplace safety and health hazards.
9. *Monitoring.* There should be a system to monitor individual, divisional, and corporate performance in safety and health matters.
10. *Information.* Managers and staff should provide information and results from this monitoring to senior management for decision making.
11. *Training.* Safety and health training needs should be clearly identified and periodically reviewed.
12. *Resources.* Executive management should provide enough resources to carry out the safety and health program through budgeting, allocation, and personnel. These resources constrain program implementation.
13. *Implementation.* Middle management should assign and direct personnel to carry out the safety and health program.
14. *Objectives.* Middle managers and staff must specify future conditions for which safety and health policy will be needed.

After preliminary planning is done and the decision has been made to go ahead with a safety and health program, senior management can develop a safety and health policy to confirm its commitment and interest.

POLICY

Policy and objectives are not the same. Policy is general in nature, while program objectives and goals are specific and spell out the expected results. Policy gives notice that the program will be carried out; objectives state the results expected as policy is carried out. The policy may be to prevent injury to employees; the specific objective is to reduce the number of lost-time accidents to 25 percent of their present level. We have not explained exactly how this is to be done. It is up to individual managers to see that it is done and to explain the process and goals to their subordinates. Their interpretation may not apply to the entire company, but only to their operation or department. The lower the level of implementation, the more specific it becomes. In the end, it all depends on policy.

Policy Problems

We have all heard the reasons that policies are weak or ineffective. Before trying to develop a new safety and health policy, we should review these reasons, if only to help us avoid obvious pitfalls.

1. Policies are an excuse for not taking action or not making decisions ("That's not company policy").
2. Policy is sometimes so weakly worded or generally written that it has no solid meaning and calls for no real commitment to anything.
3. Policies and rules are sometimes confused. Policies express a general thought that guides management and leaves room for discretion. Rules support policy by giving specific directions.
4. Much corporate safety and health policy defers to public opinion but is neither realistic nor convincing. For example, "Safety first" is a noble slogan, but it overlooks the obvious fact that the company must actually say, "Profits and mission first!" Employees and the public can see through this contradiction.
5. A safety and health policy is only one of many policies. Too often these policies clash with each other or are not coordinated. A policy of "Safety first," for instance, is obviously out of step with another policy calling for "Efficient production at the lowest cost." Safety and health policies should be designed to complement and support other company policies, and should never do otherwise.

What Is a Good Policy?

A timely, well-prepared policy will avoid most of the problems listed above. A good policy Will establish the character of a company, instead of letting it be fashioned by external and fortuitous events. The reasons listed below illustrate why a good safety and health policy is important. An effective safety and health policy is needed:

1. To allow managers to delegate safety and health activities without having to consider problems previously solved.
2. To support the need for safety and health actions throughout the company, rather than confining them to the safety and health manager's activities.
3. To promote unified thinking among managers so that they can see safety and health problems as part of the total corporate structure. They become members of the team, instead of isolated individuals with problems.
4. To strengthen management's position and ease its tasks. A policy agreed to by the corporate head takes pressure off the operating managers, since it represents a total corporate viewpoint.

PLANNING AND ORGANIZING THE PROGRAM   61

5. To allow improved decision making. Policy crosses functional lines and provides a common base for solving problems that do not fall within a particular manager's area of expertise.
6. To encourage managerial teamwork in solving common problems, since all managers are equally affected by the policy and equally responsible for following it.

A safety and health policy is a communication tool and a corporate statement. The more clearly it communicates, the more effective it will be. The following are points to remember in creating a policy that communicates effectively:

1. Be brief. Explain as succinctly as possible what the company will do or try to do in regard to safety and health. Think in terms of a telegram rather than a dissertation.
2. Be inclusive. State the policy in the broadest terms possible, so that it includes all that management wants to do.
3. Think ahead. Take a long-range view so that the policy won't be outdated soon. It should not be necessary to change the policy every time management determines a new objective or direction.
4. Involve others. Involve other managers in framing policy in order to represent consolidated opinions and reduce resistance to the policy.
5. Put it in writing. To be effective and stable, a policy must be committed to writing. Oral policies quickly become vague or distorted, or cause disagreement. One can readily refer to a written policy for clarification.
6. Communicate it. Copies of the written policy should be distributed to every manager in the organization. This will ensure a greater degree of consistency and application.
7. Go to the top. Get the policy approved by the highest corporate authority to give it the power and acceptance it needs.
8. Manualize it. Include it in the corporate manual so that it becomes an integrated part of the organization's direction.

All the characteristics of an effective safety and health policy are probably best illustrated by an actual example of a good policy. Considering that we want the policy to be briefly stated, general in nature, and applicable to all, there is no better example than the 55-word safety policy of the United States Department of Defense. Considering that it is one of the world's largest corporations, this brief statement is a masterpiece:

> It is the policy of the Department of Defense that active, continuing, aggressive, and effective accident prevention and safety programs designed to protect DOD personnel from injury or occupational illness, and government equipment and property from loss or damage, will be conducted

within all DOD components and will support the guidelines cited in [manual cited here].

These few words lay the groundwork for implementing throughout all the armed forces what is considered to be one of the finest, most comprehensive safety and health programs in existence. Your company may not be as large or complex as the DOD, but the example is a good one.

Policy states the will of management, but it does not give details. A fuller explanation of the policy is usually needed, but it should not be part of the policy itself. Other documents, such as implementing directions, reports, or manuals, may be used. They should explain:

1. Why there is a policy.
2. How the policy will be carried out.
3. Who the policy covers, who has responsibility and authority for it, and who will give it timely support and direction.
4. When the policy will be implemented.
5. Where the policy will be in effect in the organization.

In summary, then, a policy statement is a broad plan of action. Good policy formulation is not easy. It must be understood by those who are responsible for formulating it and by those who will follow it.

ORGANIZING AND OPERATING FUNCTIONS

Organizing and operating functions involve an enormous number of activities and responsibilities in the corporate structure. Three particular areas have an important bearing on safety and health programming:

Responsibility and authority
Establishing the safety function
Safety committees

The misunderstandings surrounding these three areas are legion.

Responsibility and Authority

Much confusion exists in current discussions of "authority" and "responsibility," as these terms are abstract and hard to define. Some authors say that they are the same, while others make clear distinctions between them. We will use them interchangeably.

The statement of policy has no value unless the policy objectives relate to practical, achievable results. Actual achievement directly relates to a clear definition of the responsibilities of every person and function involved in the safety and health program. A clear allocation of responsibilities will ensure that:

1. Policy and objectives are integrated into the overall organizational objectives.
2. The legal basis for programs is clearly defined and someone will take care of them.
3. Information is collected and analyzed for management decision-making and control purposes.
4. Roles and responsibilities are translated into training needs.
5. Accountability for safety and health, and awareness of the program, are promoted.
6. Recruitment and selection of new personnel will conform to appropriate criteria for safety and health purposes.
7. A clear channel or line of communication for all safety and health problems is defined and known to all.

A management hierarchy concept and varied levels of management relationship to safety performance are presented in Figure 3-1.

Authority and responsibility at any level are related directly to those at adjoining levels of operation. More precisely, when a mishap or injury occurs, it is always traced to a higher level of responsibility. Conversely, if appropriate procedures and plans are not in place, the result is a mishap or injury at a lower level. This relationship is shown in Figure 3-2.

In developing specific safety and health responsibilities, we might do well to consider Table 3-1, taken largely from the work of James V. Findlay (1988). The list presents some typical items of responsibility for various functions and positions, along with an example of failure to accept those responsibilities.

The Supervisor

The supervisor in charge of a job is not included in Table 3-1, unless he or she is included under "Production." The reason for this omission is that the supervisor's role is too complex to be shown by a single item on a list. The

These are all roadblocks to effective safety and health programming, yet most of the burden for workplace safety and health is on the supervisor.

	Mgt. Level	Output	S&H Output
Top mgt.	Strategic	Policy	S&H policy
Middle mgt.	Management	Procedures	S&H resources & procedures
First level mgt. & Operating employees	Operating Supervisory	Performance	S&H conditions & performance

Figure 3-1. Management level related to safety and health output.

64 SAFETY AND HEALTH MANAGEMENT

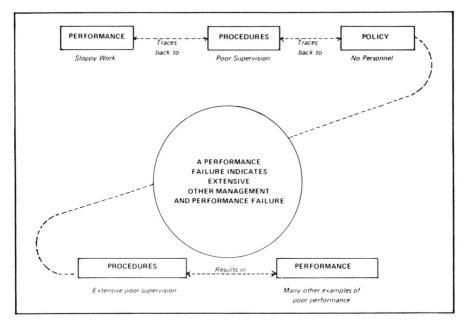

Figure 3-2. Relationship of policy, procedures, and performance.

supervisor is an important part of the safety/health program, in charge at the site where mishaps and injuries occur. However, several factors reduce the supervisor's role and its success. Nowhere is the difference greater between what the responsibilities and authority of a position should be and what they actually are. Consider that the supervisor;

1. Generally lacks the status of a manager.
2. Usually has too many detailed responsibilities.
3. Follows priorities set by management.
4. Rarely has proper staff support.
5. Seldom has enough authority to support his or her accountability and responsibility. For instance, the supervisor;
   a. May not be able to hire, fire, or discipline workers.
   b. Cannot promote or demote workers.
   c. Seldom helps make rules that his or her workers must follow.
   d. Seldom helps to develop work standards for the job supervised.
6. Must fully support company policies and goals, right or wrong.
7. Is constrained by cost/benefit studies on safety equipment.
8. Has limited control over working conditions.
9. Often lacks adequate parallel or upward communication channels.

Table 3-1. Safety and health responsibilities by position.

POSITION	TYPICAL RESPONSIBILITIES	TYPICAL IMPACT OF FAILURE
Senior executive	• Exerts policy direction • Reviews control information • Delegates safety and health responsibility/authority • Makes budgetary allocations	Safety and health actions "fall through the crack."
Line management	• Conducts operator and other training • Supervises workers to ensure correct working procedures • Communication awareness of hazards • Provides employee consultation	New employees have extraordinarily high injury rates.
Design/engineering	• Assesses and designs out safety hazards • Considers safety and health regulations • Performs environmental monitoring of the workplace • Designs or recommends engineering controls	Inadequate engineering control of dust results in prosecution.
Purchasing	• Purchases required protective equipment • Ensures that supplier provides servicing instructions • Acquires maintenance manuals • Ensures that item complies with safety and health specifications • Secures hazard information from suppliers	A machine is purchased that fails to comply with safety and health regulations and must be shut down.
Finance	• Costs safety and health-related expenditures • Projects costs to comply with safety and health legislation • Budgets for safety and health training	Accident costs significantly escalate in one department.
Production	• Ensures use of protective equipment and safe work practices • Oversees proper housekeeping procedures • Ensures compliance with safety and health regulations • Effectively communicates hazards to workers	There are government prosecutions for failure to comply.

(continued)

Table 3-1. (continued)

POSITION	TYPICAL RESPONSIBILITIES	TYPICAL IMPACT OF FAILURE
Quality control	• Maintains products to set standards, obviating hazards • Assesses safety and health hazards related to substandard products	Excessive scrap and rework seriously affect production quotas.
Administration	• Keeps abreast of current safety and health legislation and information • Arranges for preemployment medical examinations • Establishes induction skill, safety and health training programs • Maintains legal, medical, and other records	A fellow worker is injured due to a known but unreported hazard.
Maintenance	• Identifies critical preventive maintenance requirements • Determines critical parts lists and replacement availability • Schedules routine preventive maintenance • Maintains maintenance records	A serious accident results because a worker is not trained to handle new machinery.
Sales/marketing	• Ensures that literature contains necessary safety information • Properly represents limitations on use of equipment	A user is electrocuted, resulting in legal action against the firm.
Safety and health staff	• Informs management of safety and health problems • Advises purchasing department of new safety standards • Advises personnel department of available safety and health training	Management is unaware of an unsafe condition until after an accident occurs.
Employee	• Conducts work activities in compliance with rules • Reports unsafe conditions to supervisor • Reports any injury or accident to supervisor	A replacement part is unavailable, resulting in a serious accident when the machine malfunctions.

Organized Labor

Another important role not listed in Table 3-1 is that of organized labor, which has a vested interest in worker safety and health. While "labor" here refers to formal, organized labor, all workers have become more sensitive to and aggressive about promoting safety and health concerns. Thus, this section applies to all workers, organized or not. It is good practice to work closely with the unions, and this closeness depends largely on the will and attitude of management. Because of the increasing union role in safety and health matters, the program needs labor support. While safety and health are seldom primary goals at the bargaining table, they are often important secondary goals. Once workers achieve financial and security goals, safety and health may be their most important concern. Attitude changes in past decade have brought safety and health from a secondary to a primary concern for most workers. Current items of contract negotiation include the following:

1. Safety/health committees will be joint labor/management ventures.
2. Union members will have a say about the frequency and number of safety and health committee meetings.
3. The union must be notified of all lost-time mishaps.
4. Management will supply protective clothing and equipment.
5. The union will be notified whenever grievances result from work-related injuries.
6. The union will be notified of contemplated personnel policy changes that may affect work conditions so that it may comment.
7. Safety and health reports and problems will be channeled to union representatives.

The law now requires some of the above items, especially those related to workplace health. While many of them call for management to release charges, they also call for union and worker commitments. The astute manager now recognizes that labor participation can advance safety and health objectives. If management and labor bargain in good faith on these issues, the result is a win-win situation. When organized labor takes a hand in making workers more safety and health conscious, the final results benefit the company as a whole and management's objectives in particular.

ORGANIZING THE SAFETY AND HEALTH FUNCTION

How do we know when the company needs a safety and health position or department? There is no clear answer that will hold true for every organization, but the following questions may help the manager in making the decision:

1. Have you had several close shaves, coupled with a high accident rate?
2. Has there been an uncomfortably high number of operational problems,

vehicle crashes, tort claims, workers' compensation claims, reported injuries, and lost-time accidents?
3. Is there unusual external pressure in the form of safety inspections, new legislation, or letters or visits from your insurance carrier?
4. Are your insurance rates rising rapidly?
5. Have you had bad publicity in the news media involving safety and health matters?
6. Are workers pressing for safer working conditions or for a greater role in safety and health matters?
7. Have there been enough production errors to call for evaluation of several staff functions?
8. Is your company planning to open a new facility, install new work processes, add new product lines, or radically alter existing ones?
9. Are you aware of pending legislation that may change the safety and health standards that apply to your operations?
10. Is your company planning to reorganize, reassigning management roles and responsibilities or making major changes in department functions?
11. Have enough safety, industrial hygiene, wellness, health, environmental, workers' compensation, liability, regulatory, and similar problems been parceled out to different offices to call for grouping functions for an organized approach?

A "yes" answer to any of these questions suggests that top management should be concerned and should undertake a thorough examination of the area involved. Some of these items show that the company already has a safety and health problem that is not being handled through ordinary channels. Others point to potential or future problems to solve before they develop into full-blown crises. The examination of these questions may convince management that they need a regular safety/health department or position, or that safety and health practices in existing departments must be reviewed and sharpened.

Locating the Safety and Health Function

Where should the company locate the safety and health function for maximum benefit? Its location has an enormous impact on results, since it determines, in part, whether peers and subordinates will accept the rules and recommendations of safety and health personnel. Regardless of where the function is placed, it must make it easy to work with everyone in the company. This need should be obvious when you consider the things that the safety function involves:

Dangerous products
Safe packaging

Labeling and directions
Public and worker health
Environmental controls
Control of dust, gases, and fumes
Hazardous substances or operations
Employee hiring and training

Of course, the safety and health person would like to report directly to the chief executive, as all staff members do. However, not everyone can work with the chief executive on a daily basis, nor is this necessary. Instead, the function should be located where safety and health personnel can move freely throughout the organization, cross channels to solve problems, and carry the credibility of a highly placed superior. The safety and health person's real success will be linked directly to the office to which he or she reports.

Safety Staffing

The question of how may persons to have in safety and health positions is closely related to the management style practiced. One multiplant organization, with one of the best safety and health records in the world for dangerous operations, has only a small group of safety and health professionals. These professionals travel and give advice to the designated safety/health persons at the plant level. Each plant-level safety/health person is a high-caliber, promising young manager designated for big things in the organization. The manager may spend from 1 to 5 years as a full-time safety/health person. Senior management feels that there is more to be learned about the company in that position than in any other.

Another multiplant operation in Southern California, with over 5,000 worker, employs a very respected safety/health person. However, he is the only one on staff. In both of these cases, management is fully committed to safety and health programs, and carries its fair share of the safety and health load.

Not far away are three separate research laboratories, each employing between 110 and 150 people. Two of them have full-time safety and health persons, and the third has a part-time safety professional. Two of the firms also have a full-time environmental engineer, a function sometimes handled by safety/health persons. We also know of many plants with 1,000 to 5,000 employees that have full-time safety and health departments with anywhere from 3 to 10 full-time employees.

There is no absolute rule on how many safety and health persons a company needs. However, if you feel comfortable with numbers, here is a general guide.

70 SAFETY AND HEALTH MANAGEMENT

NO. OF EMPLOYEES	SAFETY AND HEALTH STAFFING
50-100	One collateral duty person
100-500	Two collateral duty persons and a safety/health committee
500-1,000	One full-time professional and two safety committees
1,000-1,500	One full-time professional and two collateral committees
1,500-2,000	One full-time professional, one full-time assistant, and at least four committees
2,000-5,000	One full-time professional, three full-time assistants, and about 10 safety committees.

Collateral duty safety and health persons have safety and health responsibilities as only part of their duties. They are often full-time employees with almost any other function, such as personnel, production, engineering, or quality control. Both collateral duty and assistant safety and health personnel require training. The collateral person requires at least a week of instruction but cannot provide professional assistance. The assistants should be trained, over time, to be professionals. Chapter 5 gives more detail on the types of training and education involved.

Some companies have special problems, such as dust, chemicals, radiation, truck fleets, hazardous chemicals, and so forth, requiring specialists in those areas. Remember, no matter how expert or professional the safety person is, management cannot forget its own responsibilities. With the right professional expertise, however, management can get sound guidance and advice.

What kind of person should the company look for when hiring a safety and health professional? Since he or she will be a key member of staff and an advisor to senior management, it is wise to secure the services of a good safety and health consultant. Have the consultant work with both top management and personnel to clarify the company's needs before seeking to fill the position. Don't hire a safety and health professional until you know exactly what your company wants and needs. Even the most qualified person in your local area may not be right for your specific needs.

Elements of a Safety and Health Program

From *management's viewpoint*, a good policy for a safety and health function might be "To provide for a system of continual measurement and appraisal of administrative oversights, leading to discovery of imperfect performance that results in accidental loss and has an adverse effect on the best utilization of personnel property." A safety and health program to back up that policy might include:

1. Safety and health administration. Management must develop and use strategies to involve others in the program.

2. Safety standards. Appropriate rules, regulations, standards, and procedures, to supplement those imposed by law, may need to be developed and disseminated.
3. Plant inspection. Audits and inspections should be regularly scheduled.
4. Mishap investigation. Proper investigation includes analyzing, reporting to management, record keeping, and correction of deficiencies.
5. Safety/health and training. Deciding who needs training and why is a vital part of the safety/health manager's job. He or she should also be able to advise management on how to proceed.
6. Safety and health research and engineering. This function recognizes safety/health hazards and proposes solutions, including where and how to place them in action.

This list, reflecting management's interests, is only a beginning, but it is a systematic way to start building a safety and health program. Beyond this, help from a professional may be needed. The chapters that follow expand on various parts of safety and health programs.

The Safety and Health Manual

The safety and health program should be incorporated into existing instructions. This is generally better than writing a separate safety manual, which automatically sets it apart from everyday operations. Do this with the support of managers and their staffs, since it involves and supports their functions. In practice, this integration becomes somewhat complicated, since managers must also be able to isolate safety and health functions for some purposes. Integrating the safety program into the existing operating instructions, of course, does not mean that management should not prepare separate safety and health booklets, instructions, or materials for operations that require extra attention by the worker. These materials may be a part of the complete safety and health program, but they should not be confused with the corporate safety and health program itself.

The following are guides to manualization:

1. *Establish the authority* for the safety and health program, whether it is company policy or governmental regulation.
2. Publish the safety and health policy over the *chief executive's signature*.
3. Establish *safety and health responsibilities.* Although the responsibilities of all parties should be included, normally the chief executive assumes general responsibility. He or she should direct implementation and participation of the program. The executive officer should personally participate to reduce operating errors and managerial oversights that could lead to accidents and injuries.

4. Spell out the *responsibilities of the legal firm* representing the company and that of the staff.
5. Specify *line responsibilities*, being careful to point out where the buck stops, so that responsibilities are not continually passed on until the supervisor is left to deal with them alone. While operating responsibilities are passed down the line, remember that higher management remains accountable and responsible for safety and health matters.
6. Spell out *supervisory responsibilities,* such as seeking out and correcting work errors, hazards, mechanical defects, malfunctioning equipment and property, substandard work conditions, and environmental problems. *State what channels are open* to the supervisor if the problems are beyond his or her resources and control.
7. Discuss and define the *employee's role* in observing and following safety and health rules for the protection of all.
8. Discuss the *safety and health manager's function and role.* Make it plain that appointing a safety and health manager does not relieve line managers, staff persons, or supervisors of any safety and health responsibilities, and is a resource to help solve problems.

Manualization of the safety and health program can cover much more, of course, depending on the company's needs and approach to management. However, it should contain statements about the following areas:

1. Organization. The program should be presented as an integral part of the management system on an organizationwide basis.
2. Scope. Spell out the program's full scope. Stress that it should continually identify the causes and costs of operating errors.
3. Safety and health program areas. Identify all safety and health program areas, including employee safety and health, vehicle safety, off-duty safety and health, public safety and health, fire protection, wellness programs, hazardous waste, and so on.
4. Staffing. Provide for staffing of the safety and health functions to ensure the appointment of adequate collateral or full-time safety and health personnel.
5. Advisors. The manual may provide for safety advisors to give needed input in restructuring the program.
6. Safety and health council. A company safety and health council (committee) should be picked as a focal point for prevention activities and a management advisory group. Provide for:
   a. Representation of all operating units.
   b. Advance announcement of meetings.
   c. Special meetings when needed.
   d. Appropriate agenda items.
   e. Consideration of items from subordinate committees.

7. Committees. Safety and health committees should be established by each major organizational unit. These committees should use the safety and health council (item 6 above) to send agenda items that cannot be resolved at their own level. These items should include:
   a. Composition of each committee.
   b. Appropriate activities to undertake.
   c. Procedures for reporting and record keeping.
8. Labor. Provide for the participation of labor, whether organized or not.

Safety and Health Roles

The safety and health program should also define some safety/health roles and functions in the organization. At the least, it should contain the following:

1. Overview. Explain the relationship between the safety and health functions and the management structure and company objectives.
2. Personnel. Provide for the selection, recruitment, and training of safety and health personnel. Include career development opportunities and proper personnel evaluations. Also include provisions for the use of contractual arrangements and in-house services.
3. Rationale. Present a rationale relating the safety and health programs to improved operations, lower costs, and attainment of company objectives. Include the necessity of compliance with applicable government standards (such as those of OSHA and the EPA) or public health regulations.
4. Records and research. Stress that effective safety and health management requires accurate accident/incident/injury/illness data.
   a. Define the required procedures for investigating, analyzing, and reporting accidents, injuries, mishaps, near misses, and similar problems.
   b. Spell out who has the responsibility for undertaking these investigations and reports.
   c. Identify means of carrying out special research projects to identify costly safety and health problems.
5. Review. Provide for an organizational review of certain safety and health problems. Show how to identify them.
6. Finances. Ensure that there are enough funds for safety and health functions. Indicate how provisions will be made for planned and unplanned events. Show the basis for budgeting.
7. Public. Provide for public information and protection, along with in-house safety and health.
8. Recognition. Show the types of incentive, reward, or recognition programs needed to encourage safety and health activities and to recognize substantial contributions.

74 SAFETY AND HEALTH MANAGEMENT

While this is not a complete list covering every organization's needs, it is a useful checklist for all.

SUMMARY

Proper planning and organization are essential to any effective safety and health program. While the material in this chapter may seem complex, it is basic to the foundation of a sound program. The executive can thoroughly review and be aware of the steps listed here *before* developing a new safety and health program or radically changing an existing one. Time-consuming though the process may be, it has the potential to save millions of dollars in lost resources, increased mission success, upgraded quality, and operating efficiency.

REFERENCES AND BIBLIOGRAPHY

Donnelly, James H., et al., *Fundamentals of Management,* 6th edition, Plano, TX: Business Publications, Inc., 1987.

Ferry, Ted S., "3 P's in Safety Policies, Procedures and Performance," *Professional Safety,* Park Ridge, IL: American Society of Safety Engineers, November 1976, pp 26-29.

Findlay, James V., and Raymond L. Kuhlman, *Leadership in Safety,* Loganville, GA: Institute Press, 1980.

Friedman, Milton, and Rose Friedman, *Free to Choose,* New York: Harcourt Brace Jovanovich, 1980, pp. 229-234.

Heath, Earl D., and Ted Ferry, *Training for Workplace Safety: Improved Performance Through Training,* Goshen, NY: Aloray, Inc., 1990.

Pope, William C., Selected monographs on Safety Management Information Systems, Alexandria, VA.

Chapter 4 / Part 1

Controlling the Safety Program

COMMUNICATIONS AND INFORMATION SYSTEMS

Planning and organizing a safety and health program may be management's first safety and health task. However, most of management's safety and health responsibilities become apparent only after the planning and organizing stages are complete and the program is in use. Controlling and directing the program then becomes management's main task. Achieving coordinated action, continuous monitoring of performance, evaluating feedback and results, and revising the program to meet developing needs—these are only a few of the manager's tasks.

The control process rests solidly on a base of effective communications: relaying information down the line, gathering needed feedback, and receiving the information from lower levels of the company. There must always be a two-way system to communication reaching all the way from the president down to the supervisor and from the supervisor back to the top.

Communicating Roles and Responsibilities

All persons above the worker level must know and understand their role and responsibility for safety and health. They must also be aware of the roles and responsibilities of other functions and persons, since these may overlap or depend on each other. Rarely are managers effective in isolation. Moreover, they can fulfill their responsibilities only if they know and understand them. How can these roles be communicated? Among other ways, they can be:

1. Included in job and position descriptions.
2. Made a part of entry training and briefings.
3. Integrated into appraisal forms.
4. Manualized in company handbooks.
5. Referred to in company statements.
6. Specifically mentioned in job assignments.
7. Incorporated in job and position definition reviews.
8. Used as a rationale for coordinating problem-solving efforts.

As new problems develop or new safety and health measures appear, this communication process must be repeated to make sure that all levels of management are aware of their safety and health responsibilities. In briefings, meetings, and reviews, higher management should seek to find out if lower management levels grasp and understand these communications, and clarify them where necessary.

Securing Proper Input and Information

Communicating down the line is only half of the process. Perhaps even more important is securing proper input and information from lower levels. In Chapter 2, we discussed an analysis technique for using safety and health problems to pinpoint management problems in general. Successful use of this system requires a good deal of information and input, including:

1. Death and injury rates.
2. Environmental risks involving the work force and property.
3. Public or political pressures regarding safety and health matters.
4. The potential loss from the worst event that could happen.
5. Average annual losses by company, division, and department.
6. Frequency and severity rates for accidents and injuries.
7. Identification of departments or functions with special risks.
8. Impending legislation or regulations.
9. Comparison of the company's safety and health performance with industrywide performance.
10. Items having a significant or unusual cost impact.
11. Notable changes in safety and health performance (either better or worse).

Intelligent and effective management decisions require this type of input. It allows the manager to compare actual performance with goals and objectives, to review resource allocations and priorities for dealing with the situation, and to consider alternative actions. Management can then accept the situation as satisfactory or take steps to remedy it.

As the above list shows, much information and feedback is involved. It is not possible or good practice for the manager to attempt to become an expert on and deal personally with each item. Intelligent screening and allocation are necessary. Top management should separate the items and send them to the proper managers for action. It can eliminate those with little significance and pass the rest through the organization for routine action. Sometimes a systems specialist must furnish details before such a decision is possible. Perhaps a risk/loss analysis is needed or the manager needs more information. Constant attention will alert the manager to any trends in hazards or risks, or to any new situations that have occurred since the last analysis. Only in this way can early preventive action be taken.

Watching for trends is particularly critical, because it points to areas of change. The change may be for the better (as when a safety system is being properly managed), or it may be for the worse (showing problem areas that may need review). Changes may be detected through:

System performance statistics.
Field monitoring activities.
Appraisals, audits, and inspections.
Specific factors such as:
  Personnel changes.
  Hardware changes.
  Management changes.
  Environmental changes.

The wise manager knows that he or she needs relevant facts in all these areas and will see to it that such information is furnished on a regular and timely basis—not just when a problem develops. It is not hard to merge this information system into the reporting and feedback systems that already exist on every management level. Such channels already provide information on sales, costs, quality, production, and so forth. Often the safety and health factors listed above can simply become part of the regular department or division reports. When special problems do occur or trends appear, the manager can secure more information as needed.

The information and feedback system is, of course, a means of furnishing the data a manager will need. However, raw data alone is rarely useful. The manager should also be familiar with techniques for analyzing the data to discover what it means and what actions are needed.

Many safety and health analysis techniques are available for this purpose. However, these sometimes sophisticated techniques require the services of an expert. Moreover, they may be limited because:

1. Some risks cannot be detected by the worker or first-line supervisor on the site.
2. Some hazards and risks can only be found by the worker or first-line supervisor. How does the manager get information from them?
3. Hazards not directly associated with the worker may be detectable only by an expert or specialist (as when the hazard cannot be detected by the physical senses).

Realizing the limitations of the feedback system is necessary in providing for an adequate information base. Given the above information, adequate feedback requires:

1. Regular feedback from inspections or audits made by someone other than the first-line supervisor.
2. Regular feedback from the worker and supervisor.

3. Regular feedback from experts or specialists (such as those concerned with air quality sampling or noise monitoring).

Establishing Standards for Measurement

One of management's most vital roles in the safety and health mission is to establish standards against which performance can be measured. Only then can one meaningfully assess safety performance and take corrective action. The standards themselves are influenced by many things, some of them outside the control of the manager. Among these are:

Legislative actions.
Technological changes.
Product liability trends.
Sociopolitical pressures.
Increased insurance and compensation costs.

The law mandates some standards, industrial associations or labor unions set others, and still others are dictated by the state of the art itself. The manager should be fully aware of all these standards and integrate them into his or her measurement and control systems.

Certain other standards of measurement can be directly controlled and determined by management. These include:

Acceptable frequency of lost-time and other injuries.
Acceptable costs of insurance and compensation claims.
Damage to property and equipment.
Results of inspections, environmental monitoring, and mishap investigations.
Lost production data.
Departmental budget averages which reflect production and maintenance costs.

Other pertinent inputs capable of measurement and integration into an total control system are:

Labor turnover.
Tardiness and absenteeism.
Productivity level.
Grievances and morale.
Budget overruns.
Waste and rework levels.

Extensive research and examination may be needed before management can set generally accepted standards. There is no single rule that can be applied here, as every organization differs in its definition of what is "standard" or "acceptable." The above lists simply serve as a guideline to managers in reviewing and setting standards.

An adequate standard will carefully consider all the items mentioned above as a bare minimum, plus whatever other items the company considers pertinent, such as customer complaints, returns or rejections of finished goods, quality control reports, and so on.

SAFETY AND HEALTH AUDITS

One of the keys to information feedback, and one of the major control devices, is the safety and health audit or inspection. Properly conducted, it can provide much of the needed input on the state of the safety and health program.

We will use the term "audit" throughout this chapter, though the reader may easily substitute the word "inspection." An audit, however, implies a more thorough examination of conditions than the ordinary "walkaround" inspection.

The Audit Process

The audit is a managerial responsibility, requiring management follow-up to ensure correction of deficiencies. Management and staff at every level take part, both on the giving and receiving ends of audits. A safety and health audit is no different from any other type of audit. Managers should not wait for safety expertise before conducting an audit, nor should they wait until an outside consultant (such as an insurance inspector or OSHA team) requests one. Outside auditors of this sort are generally looking for different things (and will advise management accordingly), such as fire safety or regulatory compliance. This may be quite different from what the manager wishes to achieve. What the top manager really wants is the most efficient operation. His or her objectives and those of the outside auditor may be related, but they are seldom identical.

Relying on safety and health personnel shifts the burden from line management, but the line manager should still be audited by top management. There should be an established schedule for auditing every line manager. By law, employees are not usually held responsible for safety and health matters. The first-line supervisor can be held responsible only for what is specified in writing. However, line managers can, with instruction, make audits a part of their regular control system for monitoring supervisors' performance and assessing their own. A problem with this system is that line managers cannot objectively grade their own performances. This problem can be resolved by making the manager part of the audit team or having other managers make the audits.

Of these two possibilities, the idea of making the line manager part of an audit team is preferable. There are several reasons for this:

1. The line manager is aware of day-to-day problems.
2. The line manager is in constant touch with employees, the main source of information about unsafe conditions.
3. The line manager knows the processes, tools, materials, and equipment that would be unfamiliar to an outsider.

In addition, placing line managers on the audit team gives them the opportunity to appraise the performance of supervisors who report to them.

Audit Management

Audit management involves five key functions:

1. Description. The audit process should be thoroughly described, preferably by manualization. The description should include a list of responsible parties, their duties, procedures, and expected performance.
2. Definition. Audit management should define what needs to be audited (i.e., what will be especially looked at and examined).
3. Designation. Audit management should clearly specify who will make the audit. It should consider and define the respective roles of line, staff, and middle management in the audit process.
4. Communication. Audit management must work out reporting procedures to ensure two-way communication. All inspection forms should go to the appropriate manager, with copies to safety and health personnel.
5. Record keeping. It is necessary to make audits a matter of permanent record. The specter of future lawsuits involving cumulative trauma, delayed injuries, product liability, and so on places new record-keeping burdens on management. Also, current legislation requires certain safety and health records to be kept; more may be required in the future.

Authority to Shut Down

Should the audit team or individual inspectors have the authority to stop work when they find an unsafe condition? If the group or person represents appropriate line management, certainly. If possible, such shutdowns should be a function of line management. Shutdown actions by staff or safety personnel should be taken only in case of imminent danger, where death or serious injury could result if operations continued. This action might then be justified, but such authority must be used with great care. The safety manager sometimes has this authority, but should use it with caution. Including the line manager on the audit team eases this problem.

Type of Audits

Audits can be divided roughly into four types: (1) periodic, (2) intermittent, (3) continuous, and (4) special. Each type has particular uses and applications.

1. Periodic audits. These happen on a recurring basis at regular intervals. Auditors may inspect the same or different items each time. These audits are particularly valuable for inspecting processes or functions where changes occur often.
2. Intermittent audits. These occur at irregular intervals, without forewarning. An unannounced OSHA visit, a city sanitation inspection, or a surprise visit from top management are examples.
3. Continuous audits. Some operations must undergo continuous auditing, particularly where legal standards exist. Routine quality control sampling, monitoring of air quality, radiation or toxic exposures, temperature, or humidity levels are examples.
4. Special audits. These are done for a single purpose. An electrical or construction inspection, an insurance rating visit, or a municipal fire safety check are examples. Management may arrange for a special audit after a new system goes on line or to gather specific information on a problem.

Basic Steps in Making an Audit

Proper planning and procedures, as the following guidelines show, are essential to a successful audit.

1. Plan ahead. Prepare all details well in advance before starting. Prepare for known problems. Have all the materials needed for the audit, including records, writing materials, checklists, safety equipment, meters and other measuring devices, reference regulations and company procedures, and records from the last audit (particularly recommended corrective actions).
2. Communicate. Contact the proper line management upon arrival. Ask for someone to accompany the audit team. Go over the findings with the supervisor and/or manager before leaving. Answer any questions or clarify any ambiguities that may arise at that time.
3. Follow up. After the audit, written reports of the findings go quickly to management and the safety/health department. Other required reports and forms should be completed promptly and filed or forwarded as

directed. Follow-up includes the paperwork and office work, and any later activity to confirm corrective actions. This may include a follow-up visit or inspection.

Help in conducting audits may come from outside the organization: from the insurance carrier, local fire department, federal or state consultants, the nearest OSHA office, or professional consultants. Where the necessary expertise for a thorough audit is not available in-house, it may be proper to seek outside help. A perfunctory or haphazard audit is little better than none at all, and does not give top management the control it seeks or the help it needs.

SAFETY AND HEALTH COMMITTEES

Few organizational devices generate as much debate as the committee. Great controversy exists over how successful committees are in dealing with any corporate function. Despite the doubt about any committee's real value, the idea of a safety and health committee is ingrained in management's usual approach to safety and health on the job. These committees are mandated by government regulation and often required by the union contract. Thus, if for no other reason, they deserve discussion. If skillfully constitute and properly administered, these committees become one of the manager's most effective operational and control devices. How valuable a safety committee is depends largely on management itself. Depending on management's approach, the safety committee can be a valuable two-way communication device with broad powers for action or a forum for small and insignificant talk.

Advantages and Disadvantages

Most managers know the various advantages and disadvantages of committees as operating tools. However, there are peculiar features of safety and health committees that make a discussion of their specific advantages and disadvantages appropriate. We will concentrate on them from a safety and health standpoint.

Advantages. A properly constituted and administered safety and health committee offers the manager several advantages:

1. It can serve as a communications channel for safety and health policy and practices throughout every level of an organization.
2. It can give management the benefit of multiple judgments and group opinions on proposed and actual safety and health practices.
3. It can foster understanding and cooperation where matters cross departmental lines.

4. It can accomplish more than one expert working alone.
5. It increases active participation in programs and problem solving.
6. It can lead to consensus, which may be necessary to start new practices.
7. It multiplies safety and health efforts and allows members from various departments to share their knowledge and expertise with each other.
8. It can bring the skills of several different people to bear on a single department's problem.
9. In small operations, it is sometimes the only device needed for handling all safety and health needs.

Disadvantages. There are also many disadvantages in relying on safety and health committees. The manager should be fully aware of these, since most of them can be easily remedied once recognized.

1. Management may try to use the committee to shift safety and health responsibility, avoid making decisions, and avoid taking action.
2. It may be difficult to provide adequate leadership with the requisite safety and health expertise and to provide enough management support.
3. Committee actions may not be recognized, appreciated, or properly valued in terms of their effect on safety and health.
4. Committee recommendations may not, or cannot be, acted upon or responded to by management.
5. Members may not be represented. For instance, workers or union members may be excluded, or a department with an unusual number of problems may not be represented.
6. Scheduling problems may keep the committee from meeting on a timely basis. This may postpone action on crucial safety and health matters.
7. Committees may not have adequate guidelines on what to do or how to proceed. Someone from a higher management level must take the time to prepare and relay such guidelines.
8. Proper records of meetings may not be kept or passed on to the proper management level for action or recognition.
9. The safety and health committee, if not properly organized and chaired, may waste the participant's time on trivial issues that one person could solve more effectively.

Most of these problems can be solved or avoided by proper management action in the first place. If the committee is properly mandated, organized, recognized, and directed, and if two-way communications are open between the committee and top management, none of these problems need develop. In short, these disadvantages are all factors that management can control.

Organizing the Safety and Health Committee

There are many ways of defining and organizing safety and health committees. They may be authoritative, advisory, or informative. They may be ad hoc committees informally convened to solve one problem, or they may be permanent and formally constituted as part of the organizational structure. We will concentrate on the permanent committee.

Membership. On an advisory safety and health committee, properly selected members can be effective even without special knowledge or qualifications. Expertise in managerial decision making may be helpful but is rarely available in operating committees. It is probably better that the committee members simply be thoroughly familiar with the sectors of the company they represent, knowing their operations, problems, and particular needs.

How many people should be on the committee? Most companies tend to select representatives of all major departments and staffs so as not to slight anyone. However, this often entails too many people for effective action or good communication. The larger the group, the more difficult and ineffective it tends to become. The ideal size for a committee for information exchange and decision making is 5 to 7 persons. This allows maximum two-way participation and input of valuable views. The question may then become, for some managers, how to represent, say, 15 departments on a committee of 7. Here are some suggestions for dealing with this problem:

1. Select someone from the department with the most potential for mishaps and injuries (they are going to have a problem).
2. Select someone from the department with the poorest mishap/injury record (they already have a problem).
3. Select someone with a demonstrated interest in and enthusiasm for safety and health matters (they will work on the problem).
4. Select someone from the department with an outstanding safety and health record (they will have experience that others can benefit from).
5. Select someone who represents the interests of the workers, such as a labor representative (they have a vested interest in preventing safety and health problems).
6. Supplement these members, as needed, with experts or temporary members to solve specific problems. Such persons need not be regular members to add to the committee's success.

Chairperson. The chairperson should be a respected leader who will keep committee meetings on track, properly direct and follow up on problem-solving actions, and handle people skillfully. The chair should be able to draw out quieter people, keep more vocal ones from dominating the meeting, and encourage all to contribute.

The regular safety and health manager *should not be the chairperson.* The chair may, at some levels, be the secretary/recorder or merely an advisor or observer. From that position, the safety/health manager can ensure that the proper items are considered and acted upon and that recommendations get action. If the safety/health person is the chair, he or she tends to dominate the meeting. The idea is for others to take roles as members of the safety/health team.

Operating Levels. Ideally, there should be a safety/health committee at every operating level. These levels interrelate on a regular basis, help solve each other's problems, and bring lower-level problems to the attention of the next higher level. Finally, the committee brings senior management's attention to those things that cannot be solved at a lower level.

Where this ideal arrangement is not possible, a plan must be made to disseminate the committee's findings or recommendations to all levels. A method is also needed to allow all levels to give the committee input, bring problems to it's attention, or pass on information from above.

In a larger organization, the ideal is a top management committee, which might be called the "policy" safety and health committee. At the next lower level, there is a management "advisory" safety and health committee. Finally, at the operating level, committees will represent the major functions, departments, or divisions. The minutes of the lower-level committees move up through the other two committees until the chief executive receives them, probably in condensed form. This way, safety and health problems not solvable at a lower level move up until they reach a level where they can be solved. The higher levels can then pass recommendations down the line. Moreover, the chief executive can see if his or her mandate is clear, if committees know their responsibilities and tasks, and if the safety/health program works as expected.

Some Practical Considerations. All meetings should be scheduled in advance for a convenient time. Having them at regular times (every fourth Tuesday at 9 A.M., for instance) will help avoid conflicts. Do not overschedule meetings. Allow enough time between meetings to follow up on actions. Once a week is probably too often for most groups, while every 2 months is not often enough. Monthly meetings are most often used. This ensures that members do not spend too much time attending meetings and that actions recommended at the last meeting are well underway.

Minutes are kept and include the names of attendees, those who make motions, and those who are assigned to take action. Making this information a matter of record is essential for follow-up and helps management assess the performance of individual committee members.

If there are not enough agenda items to make a meeting worthwhile, *cancel it.* Nothing destroys a good committee faster than the thought that members are wasting their time. If most committee members are absent, the meeting should be canceled. In either case, the meeting should be rescheduled if possible.

When to Use a Committee

Here are some general guidelines in determining when to use a safety/health committee. Use the committee when:

1. Information or input comes from several different sources.
2. The full support and understanding of different departments are needed to carry out policies.
3. The action needed requires three or more management functions to be coordinated.

On the other hand, there are times when a safety/health committee is not practical. This is true when:

1. Speed and time are important in making a decision.
2. The problem can be handled by one or two directly concerned people.
3. The problem is plainly the responsibility of a single person or function.
4. The decisions required are routine or trivial.

This section has not covered all the types of safety and health committees, or how any particular organization may choose to incorporate them in its organizational structure. It has, however, given valuable guidelines for the senior manager. A little attention to properly setting up and using a safety and health committee can pay large dividends. By contrast, a carelessly organized or badly handled committee is not only a waste of time but a liability in successful safety and health action.

REFERENCES AND BIBLIOGRAPHY

Briscoe, G. J., J. H. Lofthouse, and R. J. Nertney, *Applications of MORT to Review of Safety Analysis,* Idaho Falls, ID: System Safety Development Center, 1979.

Findlay, James V., and Raymond L. Kuhlman, *Leadership in Safety,* Loganville, GA: Institute Press, 1980.

Nertney, R. J., *The Safety Performance Measurement System,* Idaho Falls, ID: System Safety Development Center, revised 198.

Pope, William C., Selected Monographs, Alexandria VA: Safety Management Information Systems, 1973-1988.

Chapter 4 / Part 2

Microcomputers in the Safety and Health Department

INTRODUCTION

Since this chapter on microcomputers and safety was first written, there have been many new developments. Developments in the microcomputer field occur more rapidly than they can be recorded and printed. Thus, this chapter represents only a slice of time in the fast-moving world of computers and safety. However, for those who are not knowledgeable about computers, it is necessary to cover a few basics.

An important milestone in the computer industry was the entry of IBM into microcomputers with its PC, PC/XT, and PC/AT micros. This produced both a standard of microcomputer construction and a standard for software. It also established a large marketing base. It was then worthwhile for software houses to invest in software writing projects because they knew that the potential market was large.

One might wonder what the differences are between a microcomputer, a minicomputer, and a mainframe. The definitions are sometimes unclear, but try this: one person can move a microcomputer, two or three persons can move a minicomputer, and four or more are needed to move a mainframe. However, even this definition is now out of date, since improvements in computers now allow microcomputers to do what mainframes formerly did.

It is hard to believe that microcomputers did not exist 15 years ago and were popularized barely 10 years ago. Today the question many managers ask themselves is not whether a microcomputer can help them in their activities, but how it can help them. In the occupational safety and health field, there is increasing interest in this tool and a desire to put it to use profitably.

Microcomputer Applications

There are many readily identifiable microcomputer applications for the occupational safety and health department. The first is obvious: office management programs such as those for word processing, database management (filing), and spreadsheets. These will be described later.

It is also possible to automate part of the workplace monitoring functions. Rather than have a safety/health department periodically conduct a survey in a given work area, the instrument can be stationed permanently in the area, connected to the computer, and take readings whenever desired. These readings are then processed automatically and evaluated. This provides full-time surveillance, data collection and recording, recording of exposures, and sounding of alarms as required. This arrangement has been in use for some years, but for critical areas only, because of its high cost. Now, however, the cost of monitoring personnel far exceeds the cost of monitoring equipment, and the microcomputer approach is more viable.

Computer-assisted learning (CAL) teaches users about a subject, and there are several programs available on safety and health. England's Royal Society for the Prevention of Accidents (RoSPA), in collaboration with Datasolve Education, has developed a self-study computer-based program dealing with health and safety in the office. RoSPA markets the package and keeps the course content up-to-date. The course is designed for any commercial environment where there is a need for a convenient and effective way of providing safety and health training. It is a two-hour package that deals with anticipating accidents, preventing them, defining hazards, and so on. The cost is about £80 from the Occupational Safety Division, RoSPA, Cannon House, The Priory, Queensway, Birmingham B4, England. Many other safety/health training programs are now entering the market, many of them interactive with the user, not merely tutorials. These developments highlight the great promise that microcomputers hold for the pursuit of occupational safety and health and the way in which their use may change its practice radically.

Substituting for Specialists

Occupational safety and health practitioners have specialized training, which makes them rare and expensive. The OSHA Act substantially increased employers' responsibilities in these areas. This created a strong demand for these specialists, and with it the safety and health department's personnel costs. On the other hand, recent economic recessions and strong competition in the international markets has forced employers to keep a tight lid on expenses. This has affected safety and health activities through the layoff of staff and the curtailing of overall activities. The person in charge of the company's safety and health functions must therefore set priorities for maintaining activities at an extensive level, even if less thoroughly than before. It is an almost impossible decision to make, but proper microcomputer software may alleviate the situation.

Some persons perceive the microcomputer as simply a more capable electronic calculator. According to this perception, one can write programs to do various calculations, but the person using the microcomputer must still have a full understanding of how and what data to enter and the significance of the results obtained. Using the program saves some computation time.

Many conventionally written computer programs are of this type; they are meant to ease the work of the trained specialist.

It is also possible to design software for persons with limited professional competence. This is done with user instructions within programs, error trapping (checking errors in data input), easy wording of instructions and prompts, a clear and logical building of programs, and the checking of intermediate and final results. Every effort is made to ease the use of the programs.

Under this scheme, the user is guided to the right program for each instance, and is told what data is needed and when and how to enter it. All results are accompanied by explanations. At the same time, reference data needed for the programs is included, so that its use is transparent (not noticeable) to the user.

Using this approach, a person with little safety and health training, perhaps a clerk-typist, may use a certain program, enter data and produce results. Sometimes these will be final results. In other cases, the results are intermediate ones for later thorough examination by a specialist, whose time need not be wasted on these simpler activities. In a similar manner, the less trained person can extract reference data whose location and significance are unknown. The result is that a lower-level person, using the microcomputer, can produce output which normally only a highly trained, upper-level person can derive. Information of this type may go to persons who are not in the safety and health department but who periodically need its help for data research, computations, and the like. Once this safety and health knowledge is sent to corporate headquarters, it can be readily available at almost no increase in cost, even to plants at remote locations.

The safety and health advisor generally has two questions on the use of computers: How will they affect my job? and What adverse effect will they have on the safety and health of users in the organization? The formal information that a safety and health advisor might be concerned with includes:

Internal and local information
    Inspection reports
    Accident reports
    Audit reports
    Committee minutes
    Safety/health training records
    Safety/health policies
    Memoranda, etc.
External information
    Regulations and standards
    Insurance reports
    OSHA inspection reports
    Legal precedents
    Journal articles

90 SAFETY AND HEALTH MANAGEMENT

Computers augment the human brain. They can store, retrieve, sort, reconstitute, assess, and transmit large amounts of data very quickly. A microcomputer is a tool for use by noncomputer professionals (but computer professionals also use them). It is generally like any computer except for its low cost, smaller size, simple installation, lack of need (in general) for special skills, and no linkage to other computers (but it might be linked).

Microcomputer Safety and Health Software

The capacity of a safety and health system should be enough to contain at least 3 years' worth of records in a single file. It should be able to produce summary tables consisting of at least 10 columns and 20 rows. A word processing capability should be available, either integrated or internal. A good program would contain the training records and would automatically select candidates based on the time since their last training. One source lists an arrangement for storing safety and health information, as seen in Figure 4-1.

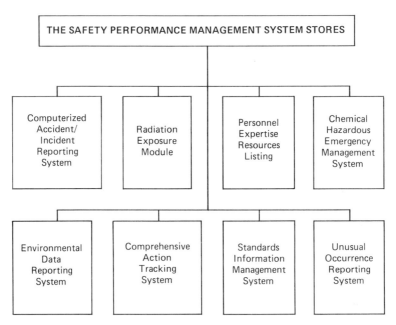

Source: Mort-Based Safety Professional/Program Development and Improvement, May 1986, prepared by R.J. Nertney for the U.S. Department of Energy, Idaho Operations Office. Published by System Safety Development Center, EG&G Idaho, Inc., Idaho Falls, Idaho 83415.

Figure 4-1. A system for storing safety and health information.

The Safety and Health Performance Management System Store

Another system divides company losses into four categories based on the Computerized Accident Incident Reporting System (CAIRS):

1. Property damage, including vehicles
2. Injury/illness
3. Program impact
4. Other intangibles (image, reputation, etc.)

This system can provide a basis for actuarial analysis in three areas: injury/illness, property damage (fire and non-fire), and vehicle damage.

Office Management Software

Many daily tasks of the safety and health department are of a general type, done in millions of other workplaces. A safety/health department must deal with various mundane bookkeeping matters, such as materials inventory (whether office or safety and health equipment), bills to pay, and the like. There has been much software activity in this area. Many vendors advertise attractive packages, and their prices vary widely. Anyone wishing to buy one of these packages should get full details of the input they require and the output they provide. If possible, one should obtain a trial demonstration copy. More specific details are provided in this chapter.

Word Processing

A major activity of any department is generating text, such as reports, letters, and publications. All of them need a good word processing program. The manager should select a package that best serves his or her needs. Some of these packages may be highly touted but are not easy to use. Others are of high quality for special users and combine features which may or not be right for the manager's purpose. Check whether superscripts and subscripts are easy to use and whether mathematical formulas can be written. A key point is ease of learning how to use the software. If special training and continued assistance are required, seek something better.

At this writing, the dominant word processing software packages for IBM-compatible machines are Microsoft Word, Word Perfect and Wordstar 2000. The widely used Macintosh products use other software, but later models can work with IBM-compatible (MS-DOS) software. However, an estimated 90 percent of personal computers still use the MS-DOS operating system. New word processing software programs are constantly being introduced and existing ones upgraded. Because best-selling systems succeed based on good performance, and their costs are similar, the guiding principle in purchasing should be the program's compatibility with other programs used or expected to be used in the company. This is not always easy to determine, since several systems may be in use or major changes planned without an actual implementation date.

Many word processing packages come with spelling checkers; they are indispensable. The package includes a dictionary against which words are compared and misspellings noted. While not perfect, they are very useful, and one should obtain this capability; the larger the dictionary, the better. It should also allow the addition of special terms to the glossary. Some software houses also offer a thesaurus, and this too, should be obtained.

It is often necessary to send the same report or letter to a group of people. If the mailing list is in a file, it is most convenient to be able to issue instructions to have the document automatically addressed to readily identifiable persons on the list. A merging program is often available with word processing packages, but word processing can easily go beyond the simple treatment of text.

Organizations which originate much literature, in the form of training brochures and the like, may want the material to appear more professional than simple typewritten text. Typeset quality is now available at a reasonable price. Two things are needed: (1) several typefaces (fonts) and (2) the ability to print this material in letters with sharp edges (not blurred, as with dot-matrix printers). Software packages are available for most microcomputers that allow one to create any font one wishes. One can then preview on the screen the exact makeup of a page. Next are desktop laser printers, which are now available for less than $2,000. Recent improvements and wide adoption of 24-pin dot-matrix printers give near-letter-quality printing and provide many typefaces in various sizes. These 24-pin printers are often discount priced at less than $400.

Database Management Systems (DBMSs)

The DBMS is a filing system, though a good software package goes beyond that capability. An important first requirement of a DBMS is that its manual be user friendly. The point is that for such a system to be useful, it must be flexible. But flexibility makes it complicated, and the inexperienced user may well lose his or her way.

A DBMS allows you to file items in a manner that lets you sort or search easily for any one item or group of similar items. A list of safety equipment suppliers is a case in point. Within your total filing system, which includes labs, consultants, expert witnesses, and the like, you may want to locate, for example, any manufacturer of rubber boots in a given city. A good DBMS will have listings classified by several characteristics, so a search for any one or a combination of any of them will be straightforward. In choosing a DBMS package, be very selective to be certain that it meets your needs.

A proper database is a complex mix of data and programs to manipulate information. The software packages themselves seldom produce a valuable database. They produce good indexing and filing systems, but the user must insert data into them to make them more valuable.

On-Line DBMSs. Industrial catastrophes, like those that occurred in Bhopal, Flixborough, and Mexico City years ago, can be prevented with the help of an international data bank on major accidents and dangerous chemicals. The Major Hazard Incident Data Service (MHIDAS) is a joint project of the British Health and Safety Executive (HSE) and the Safety and Reliability Directorate (SRD) of the United Kingdom Atomic Energy Authority (UKAEA). Some MHIDAS data goes back 25 years. It already has 2,600 items of professionally assessed information on major hazards, and details of about 200 serious accidents in the chemical industry will be added each year. Close ties exist with the International Labour Office in Geneva. MHIDAS is based at the UKAEA Directorate at Warrington, Cheshire.

The U.S. government's Registry of Toxic Materials (RETECS), formerly known as the Toxic Substances List, produced for (NIOSH) and cooperating federal agencies, is updated every 3 months. The file contains nearly 290,000 entries, of which 80,000 are the names of unique chemicals, with their associated toxicity data, and 269,161 are synonymous names. Toxic data from the scientific literature is included.

NIOSH makes available to the public its computerized occupational safety and health data base, NIOSHTIC, which has over 125,000 citations on workplace safety and health literature. The database is updated quarterly. It is available commercially in the United States from two vendors, DIALOG Information Services and Pergamon InfoLine. The cost for searching the databases is reasonable, and printouts are available. NIOSHTIC is available in other countries through a variety of services.

Bell Canada's Corporate Safety and Health Group has made extensive use of the Canadian Centre for Occupational Health and Safety (CCINFO). Having access to CCINFO's electronic database provides them with rapid consultation and a quick reference tool. They search the material safety data sheets (MSDSs) for a product in the TRADE NAMES database and then use the CHEMINFO database to obtain more information. "I used to have to leave the office to get information and maintain an extensive library. Now I've been able to set up my own documentation center," say users.

TRADE NAMES is the largest publicly available database of its kind in the world, containing over 20,000 MSDSs. More than 350 Canadian companies now have access to it, free of charge through CCINFO. CCINFO now has 10 databases designed to meet various informational needs in occupational safety and health. For more information on TRADE NAMES, call On-Line Services at 1-800-263-8267.

There are also many private safety- and health-related databases. One of the best known is Van Nostrand Rheinhold Company's Chemtox. Using a desktop computer, Chemtox's information on over 3,200 regulated chemicals can be accessed. In minutes one can get all the information needed to comply with several acts and to make personalized MSDSs. The service is not cheap; it costs about $2,500-3,000, and annual updates are extra.

Spreadsheets

"Spreadsheet" is the term used to describe software packages that can help with financial and mathematical modeling. Health and safety applications include:

Budgeting and estimating.
Accident costing and analysis.
Costing of first aid and medical services.
Faster routine calculations.

While spreadsheets are the most difficult packages to use, they can be invaluable when the material has much numerical content. Some spreadsheet programs can convert the data to graphs.

Several computers, when linked together, form a network. The network allows exchange of information between and within organizations, to be used as electronic mail, for example. A parallel development is centrally located safety and health databases, from which one can extract much information. Some of the computerized safety and health sources already mentioned fall into this category. Be aware that not all of these packages are easy to use; hence a trial run is recommended. Moreover, access to safety and health databases is not cheap. One pays for every second of hookup. While reading and trying to understand the material (and this may take time), one's phone bill can quickly soar.

Worldwide, there are hundreds of information sources available to the microcomputer user with the right hardware, software, and telecommunications services. PRESTEL is a well-known European example; others are HES-Line and CHEMABS. In theory it is possible to get information from or exchange information with anyone's computer.

The Visicalc program hit the microcomputer field like thunder. It was the first program of its sort, and as soon as users tried it, they realized how much they needed it. Since then, there have been many copies and improvements. Visicalc is a large program that gives users the equivalent of a large sheet of paper with many rows and columns in which values can be inserted. The value of any cell is dependent on the values in other cells. Thus, using one cell, one can create many "what if" scenarios, insert new values, and see how they will affect the rest. One can, for example, increase one item in one's budget and see its effect on others. The possible applications grow with one's familiarity with the program. Lotus 1-2-3 displaced Visicalc as the spreadsheet leader shortly after its introduction in 1983. Lotus 1-2-3, today's standard, is already being challenged by products such as Professional Plan, which both augments and replaces it.

Integrated Packages

A problem with the above programs is that if one is using one of them and wants to "peek" at the results of another, this cannot always be done. For example, in writing a report, one may want to quote certain data from the spreadsheet program. New, integrated software allows the user to shift from one program to another without disturbing the work that is set aside, and in some cases, to peek into the other report. This feature, called "windows," is a near-standard for the modern user.

The computer acts as a file clerk. It can answer questions only if it is asked the right questions and has the information. For some things the computer is inappropriate. It may take hours for a good computer to project the relative strength of a particular metal, but only a few moments to stretch a bar of ductile steel in a testing machine to the breaking point. This computer approach would be impractical for every conceivable kind of metal.

The computer is both a blessing and a curse. While making computations beyond human endurance, it can also make them beyond verification. A misplaced plus or minus sign can cost millions of dollars, as it did in requiring the rebuilding of several nuclear power plants in an earthquake precautionary measure in the U.S. a few years ago.

Two Examples of Software Packages

No computer can operate without a software program. For safety and health persons, a computer without appropriate software is not a computer. There are now several software packages designed for the safety/health manager.

Microsafe. Microsafe is a microcomputer package designed by Bowring Protection Consultants for IBM-compatible hardware. It allows the user to store a wide range of data on incidents, analyze the information, and display the results in tabular or graphic form.

The package combines powerful analytical techniques with a "user-friendly" mode of operation. Inexperienced operators go through the programs with a series of "help" messages. There are different modules for Microsafe that allow a variety of activities:

1. Injury Recording and Analysis
2. Damage Recording and Analysis
3. Hazard Recording and Analysis
4. Near Miss Recording and Analysis
5. Lost Time Costing

6. Repair and Replacement Costing
7. Hazard Monitoring
8. Communication and Terminal Emulation
9. Hazard Data Bank

To further illustrate the use of this package, here is some information that can be developed out of the first module listed above, "Injury Recording and Analysis." This module allows one to:

1. Record details of injury accidents and save them for later use.
2. Amend records at any time and print copies as needed.
3. Search for a single value or range in any number of fields on record.
4. Summarize injury records in a wide variety of ways and display the values in tabular form.
5. Convert any row or column into a bar graph to aid the presentation of numerical information.
6. Produce monthly, annual, or cumulative incidence rates.
7. Create continuation records so that more information can be added to injury records.
8. Browse through saved records, displaying them on screen and printing them as required.
9. Link any number of records in a chain, including records saved in other modules.

Besides being a comprehensive means of recording accident data, this first module is useful to:

1. Summarize injury records to identify patterns and compare safe performance.
2. Search for groups of records, injuries, employees, etc., to test the theory that a problem area may exist.
3. Maintain the necessary records of reportable injuries and certain dangerous occurrences required to be reported by law.

The preceding discussion shows the capabilities of only one module. It is not a product endorsement but a sample of today's safety and health software. This respected company has extensive consultative experience and is linked with one of the better-known and respected companies in safety and health. To help potential customers fully appreciate the ways in which Microsafe can benefit them, a demonstration diskette is available. Contact Bowring Protection Consultants, The Bowring Building, Tower Place, London, EC3P 3BE, England, or phone 01-283-3100, ext. 3899.

HAZ-MAT. Another representative type of software is HAZ-MAT. To learn what to expect from a hazardous materials control system, we'll briefly review it. For a start, HAZ-MAT is easy to place in a computer. Five minutes after receiving the diskettes, one can be working. Introductions give an excellent overview of the system and answer many questions before starting. HAZ-MAT:

1. Provides extensive data and reports for (SARA) Tier 1 and 2 reporting.
2. Provides MSDS lists by EPA hazard class.
3. Maintains hazardous materials inventories.
4. Maintains alphabetical lists of hazardous materials.
5. Documents all training requirements by individual or job description.
6. Documents all employee training.
7. Documents all employee right-to-know requests.
8. Prints customized hazardous materials container labels.
9. Generates request letters to manufacturers for missing MSDSs.
10. Monitors manufacturers' compliance with requests for MSDSs.
11. Documents disposal of hazardous materials.
12. Provides emergency response procedures and information.

The system quickly gets you to where you want to be. HAZ-MAT is menu driven. This means that the screen lists a few numbered items, and you pick a number. Here is an *extract* of the starting menu:

```
HAZARDOUS MATERIALS      1. Master Information Form
                         2. Inventory Form
                         3. Facility I.D. Information for 2.

REQUESTS FOR MSDS's     10. MSDS Request Letters to Manufac-
                             turers.
                        11. Setup Information for 10.

EMPLOYEE RECORDS        20. Employee Master Form
                        23. Employee Master — Terminated
                             Employees.

BLANK, PRINTED FORMS    30. Menu of Blank Forms.
```

That gives you the idea. Just enter your selection, and you will get a more detailed menu. Suppose that we are interested in "reports" (not shown above). There is a reports menu. Hit one key and you get this:

```
1. MENU OF HAZARDOUS MATERIALS REPORTS
2. MENU OF SARA REPORTS AND INGREDIENT REPORTS
```

3. MENU OF EMPLOYEE REPORTS AND BLANK FORMS
4. MENU OF MSDS REQUESTS TO MANUFACTURERS

The first item is a menu of Hazardous Materials Reports. Touch "1" on the keyboard and you get a menu—Hazardous Material Reports. A sample of that menu shows the following:

1. Hazardous Materials Lists—Abbreviated—Menu
2. Hazardous Materials Lists—by Manufacturer—Menu
3. Hazardous Materials Lists—Detailed
4. Hazardous Materials Synonym List
5. Hazardous Material Container Labels—Menu
6. Hazardous Material Inventory Worksheet (with Chemicals)
7. Hazardous Material Inventory Worksheet (Blank)
8. Hazardous Material Inventory by Site, Bldg, Rm.
9. Hazardous Material Inventory by Chemical
10. Hazardous Material Inventory by Chemical with Totals.
11. Hazardous Material Expiration Date Report
20. Hazardous Material Removal/Disposal Reports—Menu
30. Facility Address and Contact List

Perhaps the most helpful menu for the neophyte is the Help and Troubleshooting Guide:

1. HOW TO USE HAZ-MAT—OVERVIEW
2. ENTER DATA HELP
3. PRODUCE REPORTS HELP
4. FILE SERVICES HELP
5. ERROR MESSAGES & REMEDIES
6. GENERATING MSDS REQUESTS TO MANUFACTURERS
7. GENERATING CONTAINER LABELS
8. MAINTAINING EMPLOYEE RECORDS
9. ADDING EMPLOYEE TRANSACTION CODES
10. GENERATING INVENTORY WORKSHEETS
11. MAKING BACKUPS OF DATA

Help on nearly any aspect of EPA/SARA and OSHA record keeping and reporting is available at the touch of a key. Even if you're computer shy, it is easy to work seriously with this system. If the person doing the record keeping knows nothing about safety and health, he or she can't go wrong with a little professional guidance. The price is good: $899. Other systems may cost slightly less, but any system that does this much may cost much more. The idea here is to show how easy it is to use a good system, how it reduces the workload, and how it reduces the uneasiness we feel when receiving computerized answers to our problems. The company offers a low-cost demonstration diskette for trial in your workplace. Contact MC2—Management and Communications Consultants, P. O. Box 14372, Oklahoma City, OK 73113. Enclose $6, and the diskette is yours. You can also call 1-800-526-5222.

Resources

The rapidity of change in computer software technology and terminology precludes our giving a list of resources that will be valid by the time this book is in print or for any reasonable period afterward. For current safety- and health-related software, we suggest that you review safety and health journals and periodicals. The American Society for Safety Engineers' *Dictionary of Safety Related Computer Resources,* cited at the end of this chapter, presents its latest resources.

Readers unfamiliar with computers, microprocessors, and electronic data processing will find the following discussion of interest, whether or not hands-on involvement with these devices is expected. Those who already deal with the electronic marvels—or devils—may find certain ideas of use and interest.

Computer Terminology Basics

A microcomputer contains several major components. The heart of the microcomputer is the processor or microprocessor, which is a very small, complicated integral circuit that determines the computer's fundamental characteristics. The "instruction sets," or the functions in this tiny chip, and any microcomputer based on this processor have certain capabilities and limits. Microprocessors commonly used are the 6502, the 8088, the 68000, the 80286, the 80386, etc. At this writing, all but the last, the Intel 80386, are considered obsolescent by many. The most popular trade magazine in the world, *PC Magazine,* shares this opinion. The OS2 operating system is taking advantage of the superior qualities of the 80286 and 80386 but has not yet had great success. (These numbers are given so that they will become familiar to the reader.) For one thing, OS2 uses much more costly random access memory (see the next paragraph), often megabytes, to work with it.

No computer, of whatever size, can function without "memory," that is, a means for storing data and operator-issued instructions. There are two kinds of memory. In "random-access read/write memory (RAM)," both the user and the microcomputer have free access to "deposit" (write into), as it were, and retrieve instructions, data, and results. In contrast, there is "read-only memory (ROM)," where data or instructions are *written in* with a special device; henceforth, the material can only be *read out*. The contents can be changed only by special means, even if power is disconnected from ROM. In contrast, RAM loses its contents once power is disconnected, unless it is of a special type. ROMs are generally written by the computer manufacturer and contain some essential instructions. For microcomputers intended mainly for home use, ROMs often contain the BASIC language, a word processing program, and games.

The more memory reserve a computer has, the larger the assignments it can undertake and perform quickly; but if there is "memory overflow," the microprocessor is forced to stop. However, RAM can't be unplugged and sent to another user for transfer of data. hence, there has always been a need for a large memory device.

The simplest and cheapest one is the tape cassette. It has two major faults. First, it is slow. More important, it searches for information sequentially, not randomly. If the information sought is at the end of the tape, one must run the entire tape cassette past the "read head" to locate the desired data. Thus "access time" is high. High-speed "tape drives" go with large mainframe computers, but these are very expensive.

Recording Media

A simple, fast, and cheap solution to this problem is the "floppy diskette," a flexible plastic disk coated with magnetic recording material. A "read/write head" is part of the "disk drive" (one can draw a strong parallel with a record, changer, and tone arm). To use it, insert the diskette into the drive and close its door. To protect the magnetic surface, the diskette is covered by a cardboard jacket. Like a record, the diskette is rotated, at a rate of 300 RPM or higher. That way, a given item can be located quickly (randomly, that is) without reading through all the preceding items. Thus, access time is quite fast. Although 8-inch diskettes were once used widely, 5¼-inch diskettes are replacing them. These, in turn, are being supplemented by even smaller 3½-inch diskettes. Diskette capacity is constantly growing. However, for more "storage" capacity and faster access time, scaled, rigid "hard" disks or "Winchester technology drives" are common.

RAM, ROM, and mass memory devices are rated by their capacity in terms of thousands, millions, or even billions of bytes. The number of times a computer goes on and off is a bit, and eight bits is a byte (1,000 bytes = 1 kilobyte [kb]; 1 million bytes = 1 megabyte [mb]; 1 billion bytes = 1 gigabyte [gb]). Commonly today, business and personal microcomputers are equipped to handle up to 640 kb and floppy diskettes of 360- or 1.2-mb high density. For many purposes, this diskette capacity is inadequate; one can purchase a hard disk with a capacity of 30 mb or more. Forty megabytes, the equivalent of

110 floppy diskettes is now standard, with 65 mb gaining ground. Unfortunately, while the hardware can handle large RAM, the software is not as capable. Nevertheless, newer personal computers, including laptops, feature far higher megabytes of RAM.

To use diskette drives and/or Winchesters requires a "disk controller." In some microcomputers, this is a separate "printed circuit board" with the necessary components. In other cases, it is part of the microprocessor board. In either case, there must be a controller.

The nuts, bolts, sheet metal, electronic components, drives, and so on are the "system hardware." The instructions used, whether located in RAM, in ROM or on diskette, are the "software."

Basis of Communication

A computer is worthless unless you can communicate with it. That is, the user must be able to enter commands and data, and the computer must display results and messages. This is the "input/output (I/O)" set of operations. The user inputs, or enters (instructions and/or data), through a keyboard, which is part of a "terminal." The output appears on a screen or "video display," or prints out on a "printer." It is common for the keyboard to connect to the microcomputer, while a separate "monitor," which looks like a TV set, attaches to the computer. It may be single color (monochrome) or multicolor as in a TV set. All those pieces of hardware connect to the computer by "cables" and are called "peripherals."

Input and output can occur through external storage devices (e.g., disk drives) or another computer. Moreover, there is increasing use of "light pens" and "mice." The first, which looks like a pen, connects to the microcomputer by a cable. It can be used to "draw" pictures on the screen or to select one of several choices presented. The "mouse" works the same way, but its appearance and operation are different. It looks somewhat like a mouse and is moved about on a desktop. An arrow or triangle on the screen corresponds to the mouse's location. By moving the mouse, one moves the cursor. One can then "draw" on the screen and make a selection from among several choices presented.

Software

Because the computer recognizes only certain patterns, it is very tedious to prepare instructions in "machine code." For this reason, "higher-level languages" are needed, including FORTRAN, BASIC, PASCAL, and COBOL. Each has a certain use, and none are good for all possible uses. There are two basic steps in writing a set of computer instructions, or "program." One first decides on the plan of attack (e.g., a formula), or "algorithm," and then writes the "code," or the instructions in a specific computer language. The computer is not a self-starter; it must be provided with instructions on how to do everything. The applications program may tell it to request data from the user (issue "prompts"), then extract data from a disk, make some computations, and finally, print out the results. But who will instruct the computer how to draw information from the disk and get the printer to run properly?

Every desired action requires certain instructions. Those instructions which tell the microcomputer how to manage its affairs are called "system software," the "operating system," or the "disk operating system (DOS)." A variety of these systems are available, the most commonly known being CP/M, PC-DOS, MS-DOS, and UNIX. Speaking of disks, know that when you record something on one, you are "writing a file," which resembles, in concept, a file folder in a filing cabinet.

Many persons think a microcomputer is like a household appliance which you plug into a wall socket, turn it on, and watch it run perfectly. Nothing could be further from the truth. Almost every microcomputer manufacturer uses a unique approach to writing on diskettes and, very likely, its own version or dialect of BASIC or another language. Unless the diskette placed in a drive is for that specific computer, it will probably be illegible. Diskettes need to be written under the same operating system. CP/M will not ordinarily read an Apple DOS diskette, and vice versa, although programs to make this possible are coming into use. Even if this problem is overcome, the language used must conform to that in which files are written or must be converted to that language. This highlights the very frustrating problem of "compatibility" of software between computers. It is safe to assume that if a program or diskette is not written on an identical, compatible computer with the same language and drives, it will not run on another one unless special conversion software is used.

Microcomputer Environments

Given these basic concepts, we can focus on putting microcomputers to practical use, that is, the mode of usage. The simplest "stand-alone" or "single-user." Here the computer exists by itself, unconnected to any others. It can be made to communicate with another computer, located elsewhere, through a "modem" and a telephone line or direct connection through a cable. By using appropriate communications software, the two computers can establish a "link" and carry on a dialogue, passing files to one another.

However, in many workplaces, there already exists a minicomputer or mainframe, besides one or more microcomputers, each with certain strengths and weaknesses. The mainframe's major disadvantage is its very high hourly cost. The microcomputer, in contrast, is inexpensive but limited by its slowness in system access time and memory capacity, while the mainframe is fast, with a large memory. A common approach uses "distributed processing" whereby the big computer stores and retrieves from its large "data bank" and crunches numbers (large-scale computation), at which it is especially good. The microcomputer inputs information, a slow operation, and does intermediate data processing. Thus, each type of computer is used to best advantage. With all this, the microcomputer can still operate in a completely independent mode.

Color is not just a fancy option. Apart from making the screen more attractive the color display helps in at least two ways:

1. It makes it easier to find what you want on the screen. If the contents of a record are in one color and the content labels are in another, the record contents are much easier to find.
2. Specific types of information can be color coded. For example, all warning can be displayed in red and standard text in green.

Added Versatility

All too often, a microcomputer can only do one task at a time. Thus, if material is printing (a slow operation) or a long computation is underway, nothing else can be done on the computer. This wastes the user's time. More and more microcomputers are now capable of "multitasking," or doing several tasks at once. This requires a certain minimal memory size, plus the proper operating system with a faster CPU speed.

Where more than one microcomputer is needed in an area, substantial savings can be achieved with a "multiuser" system. In such a case, there is a "master" microcomputer and "slave" terminals connected to it, with each user allocated part of the system's facilities. The master unit contains the guts of the microcomputer, whereas the slave units are "dumb terminals," totally dependent on the master. For this purpose, the master requires a large RAM, a hard disk drive, and appropriate software, based on a "timesharing" arrangement.

We mentioned earlier that one can communicate with another microcomputer with a modem. This is a box, sometimes built into the computer. With the right software, the modem allows the user to telephone another person, who also has a modem. Then the computers talk to each other. In this manner, one can transmit programs and data, even over long distances, at rates of up to 2,400 bits per second. Direct wiring or other expensive modems, under certain conditions, allow even faster transmittal speeds of up to 96,000 bits per second. This allows field personnel, such as salespeople or compliance officers, to contact the headquarters computer, extract information (e.g., a legal interpretation, prices), and input data, such as violations or orders, while in the field. Whether the headquarters computer is a microcomputer or a mainframe is not important as far as the basic arrangement is concerned.

In this arrangement, there is a one-to-one relationship between the computers. It is possible to create a "network," like a main feeder pipeline, to which one can connect, for tapping into "data banks" or a "bulletin board." The way this works is that an organization arranges access to its database by those who have made the proper financial arrangements. A telephone connection is not the only way or the best way to make the connection. After paying an initial fee, if applicable, one receives an account number, a password, and a telephone number to dial. The user then makes the telephone connection, "signs on" by entering the account number and password, and then requests the information of interest. One normally pays by the second of telephone connection. There are many "public networks," including TOXLINE of NIOSH for toxicity data, and HAZARDLINE and MEDLINE for obtaining medically related information. The bulletin boards permit messages and electronic mail to be sent between subscribers.

Suppose that there are several microcomputers in a department or plant, many using the same programs and/or files. It is wasteful to provide each computer with duplicate files. If they could share these resources, significant savings could be achieved. This is the impetus behind "local area networks (LANS)." While each microcomputer is fully independent, it can connect to any device in the network, including microcomputers, printers, fax machines, telexes, and so on, and make use of it. Thus, if there is a given file in the hard disk on a certain person, the information can be withdrawn, processed, and even updated. Newer modes of applications appear daily.

Which Microcomputer to Buy?

This is one of the most common and most difficult questions raised by many persons. There is no easy answer: one must analyze every case and need, even consult an expert, and then decide. Perhaps some of the points mentioned here will be helpful in reaching the correct decision.

The first basic choice is to select between an IBM (or IBM-compatible) and a non-IBM computer. IBM is the largest computer company in the world, with an enviable reputation for product quality and service. In the microcomputer field, however, IBM is a newcomer. Until its appearance, there were several microcomputer manufacturers, each with its own operating system, diskette-writing format, and so on. Some of these microcomputers were of excellent quality and advanced design, but some of them were weak financially and some have already disappeared from the scene. The IBM PC, when it appeared, was a great commercial success, although technically it was far from the latest word. Many potential corporate users were ready to gamble on IBM because of its excellent reputation and because they were using other IBM equipment. They were not afraid that IBM would disappear, leaving them with an orphan computer, as had sometimes occurred. Hence, the PC enjoyed overwhelming success and continues to do so. As a result, a de facto standard is in place, and outside vendors can aim their product at a large market. There is a large selection of IBM software and hardware enhancements made by non-IBM manufacturers. This has encouraged many firms to produce their own copies of the PC or PC-compatible microcomputers at a lower price than the original PC.

Among the non-IBM manufacturers, there are several microcomputer pioneers that produce highly advanced products. The largest of these is Apple, with its LISA, Macintosh, and IIc models. Other firms have more advanced products than the IBM series of personal computers. Thus, in deciding which microcomputer to buy, several important features must be considered.

1. First and foremost, for which microcomputer is the software you need written? There is no sense in buying microcomputer A when the software runs on microcomputer B. This is of particular importance in the occupational safety and health field, for which only a small selection of programs is now available. If a major motive for having a microcomputer is to use the safety software, then this must be given top priority.
2. Do you intend to use the microcomputer as a stand-alone machine or as part of a larger system? Where no special connections are planned, any microcomputer which satisfies your needs at a reasonable price may be bought. For more complex use, it is important to buy a microcomputer which can be used in the manner planned.
3. How important is it to have the latest technology, as opposed to the security of the manufacturer's staying power? If possible, an expert on the subject should be consulted for guidance.

Laptop and Pocket Computers

Laptop computers, sometimes called "portable computers," come in a wide variety. Their number has increased dramatically in the last few years as professional work tools for them have been developed. A few years ago, the most popular model sold for over $1,000. Now a better machine sells for far less. Hundreds of support programs and peripheral devices are available for those early successful models. They have spawned a new generation of laptops that compete successfully with desktop microcomputers. Their prices range from a few hundred dollars to $3,000. Add-on devices enable them to handle nearly any problem that requires easy portability, including computer-assisted design, inventory control, and interaction with mainframes. So successful are the newer models that they are often used instead of a desktop microcomputer in the office environment.

Contributing to the laptop's increased usefulness are the 3½-inch diskettes. Contributing to its versatility are the later-model computer chips, also available in the laptops. They can handle up to 5 mb of RAM, and 20 mb of hard disk capability is common in several models. If one can pay up to $4,000, even more capability can be found. These laptops are convenient, but their screen legibility is sometimes wanting. Pay close attention to this feature. The same features in a desktop generally cost less.

Some machines even smaller than laptop microcomputers can fit in a coat pocket and have ports for a printer and a modem. Some are specially designed for harsh working environments. This exciting equipment can perform computations and is programmable. What sets these micros apart is large memory in a small physical volume, and they can connect directly with a desktop microcomputer. You can literally carry in your pocket a wealth of information, such as that in many MSDSs or data needed for industrial hygiene functions on site.

Bundled Configurations

There is another point to keep in mind. Unless you have a very good reason to buy a microcomputer because of its unique features, or even if you are buying an IBM, use caution. Some manufacturers, such as IBM, Apple, and Compaq, sell the hardware at the base price, but all software, including the operating system, must then be bought separately. Other manufacturers include "bundled software" in the price of the microcomputer. This can be worth $2,000 or more if you buy it yourself.

It is common practice in the microcomputer field to issue a 3-month guarantee with the equipment, although some manufacturers do better. Many firms offer a service contract, which should be evaluated. Two points should be borne in mind when planning to buy. First, it is common practice for manufacturers to announce the release of their latest new product with a mockup, although a working model may not be available for 6 months or more. You should not order equipment until you have seen it on several dealers' shelves. Second, do not be the first user of a new product.

While the latter is tempting, it is also dangerous. Because of the intense competition among microcomputer manufacturers, they tend to rush products to the market with bugs, or flaws, whether in design, construction, software, documentation, or all of these. Sometimes, after being introduced to the market, products fail so badly that they are withdrawn. In any event, it is quite likely that during the "debugging" process you will have some frustrating experiences in trying to get faulty equipment repaired. It is much better to let others serve as the manufacturer's test bed and buy a proven product yourself.

In evaluating bundled packages for safety and health, consider both "core" and "tailored" items. Core items refer to the date, name of the injured person, location, etc. Tailored items refer to mishap details and may be tailored to suit a particular organization. Consider also:

1. The information is provided for. Skipping information leaves gaps in the records.
2. Invalid items should be rejected by the system. For example, a time of 1,185 hours would not be valid.
3. To enable rapid retrieval, one should be able to find any item quickly by accessing any other item of information on the record.
4. One should be able to summarize accident, injury, and illness causes quickly and in several ways.
5. One should be able to summarize data in graphic form. Many people find it hard to interpret numerical tables but can handle bar charts and histograms.

There are "user groups" for almost every microcomputer made. They will normally publish a newsletter or magazine with information aimed at a given microcomputer family. It is often worthwhile to become a member of such a group, for you will find in it much information, including specialized accessories and software. They may have a local group whose meetings you will want to attend. More experienced members are often helpful in sharing their knowledge and experience with you.

Other Software Users. As the safety and health function slowly becomes integrated with other organizational functions, the availability of related software increases. Environmental, health, and risk management functions have their own software, which often complements the safety function. For example, the Risk Management Information Systems (RMIS) has more than 20 programs, most with a safety and health tie-in. They deal with mishap costs, workers' compensation, cost/benefit analysis to justify expenditures, etc. Data used by risk managers that help the safety professional:

1. Double-check plant OSHA record-keeping procedures against workers' compensation records.
2. Defend safety budgets to their bottom-line-oriented managers to show how highly targeted programs can be cost-effective.
3. Spot repeaters (employees with more than two workers' compensation claims per year) for a special emphasis program.
4. Collect data on the cost of workers' compensation claims or on old machines to convince management to replace obsolete machines with newer and safer ones.

Not only do the various RMIS systems allow for input of most standard accident data, they also allow data to be collected from other computerized databases (e.g., insurers' or workers' compensation claims administrators), which can be downloaded into the RMIS. This is only one coordinated action needed to computer-merge safety requirements and functions with other operating departments.

How to Buy

One can enter a computer store and buy a microcomputer, normally at list price, or from a mail order house at a large discount. What to do? First, know what you need and how to operate a microcomputer system. If you do know this, or if there are people in your firm who can help you, then the dealer will not do much for you except charge a higher price. If however, you need guidance before buying or after the computer arrives, then it is better to use a consultant or buy through a dealer, even if this costs more. Note that the dealer must be professionally competent and have a good reputation.

Anyone with the funds can open a computer shop, whatever his or her background and capabilities. As a result, buyers may find salespersons who know little more than they do and dealers who are unwilling or unable to give good after-sales service. In contrast, there are dealerships with very knowledgeable staffs, both in software and hardware, who have excellent reputations. They generally do not offer discounts, but one good bit of advice or a fast repair job can quickly pay for itself. If your planned application is not conventional, and if you do not have in-house technical capability, obtain recommendations on a good dealer and obtain his or her experience. If, on the other hand, people within the firm can and will help you, then savings of 20 percent or more can be found when buying through a discount house. PC clones are now selling for $400–$800. Color capabilities, modems, and the like cost somewhat more. Leading Edge, Blue Chip, and Amstrand are three Korean firms that dominate the clone market, but other offshore firms are also well known. While some of the clones do not have the support system one desires, they force down the prices of IBM microcomputers that have been models. Some of these clones have improved considerably on their IBM model. Similarly, software has nosedived in price. It is not unusual to find software programs that are roughly one-tenth of their original price only a year or two earlier.

Friends may refer you to a reputable mail order or discount dealer. If you open one of the several hundred microcomputer magazines, such as *BYTE, Microcomputing,* or *PC World,* you will find many firms offering their wares at discount prices. Some of these may be fly-by-nights, but many are legitimate. They can save you substantial sums, including perhaps the applicable sales tax, which by itself can total several hundred dollars.

While you may have good reason to hesitate buying hardware from a discounter, there should be nothing to keep you from buying software from such sources. There is now cutthroat competition among software houses. The difference between the list and discount prices can be 30 percent or more. For a software package with a list price of $500, a $200 or more discount is significant.

Future Applications

One might assume, from the low cost and widespread use of computers in schools and offices, that we are all computer literate. This may not be true of safety and health professionals. The percentage who use computers directly may be only 5 to 10 percent. This may be due to lack of safety- and health-specific software, adverse attitudes of safety and health professionals, or simply the feeling that the computer is not needed for the job.

The lack of software will likely persist. Commercial software houses are unlikely to risk entering the safety/health market when there are other, more lucrative markets. One encouraging note is that the writing of software itself is made easier by new programs designed for that purpose. It may be that computer-oriented safety and health professionals will write sophisticated programs that go beyond simple mishap recording and tabulating of simple data like that from a government report.

There is a general need for systems analysis in producing software products. The many reason include:

1. More widespread use of computers in human/machine systems for controlling task performance,
2. Potential failures resulting from incomplete instructions, and
3. Insufficient communication between safety engineers and computer personnel on software uses and potential system hazards.

In designing software programs, we must always consider the question "How can the hardware or machine fail?" While software maintenance takes time and is expensive, it is the retrofitting and redesign of mechanical devices, which could have been more economically designed into the process and equipment at first, that costs more.

It is feasible to run system safety processes on software to ascertain their correctness. Computerized preliminary hazard analysis, failure mode and effect analysis, and fault tree analysis procedures exist to do this. For example, the fault tree can be used to consider if:

1. The software/hardware does not execute its control function.
2. The software/hardware generates erroneous outputs.
3. The software/hardware executes correctly but is timed incorrectly.

A software package must recognize error possibilities in the inputs, processes, and outputs. Input errors include (1) no input and (2) erroneous input. Output errors can arise from (1) computer failures, (2) computer equipment malfunctions, and (3) transmission line failures. Processing failures include (1) an interpretation or logic error in the code, (2) incorrect signal generation, and (3) unexpected code. Fault tree processes could be used to isolate these problems.

Personal computers also offer computer-aided design tools to the 80 percent of the designers who do not have access to mainframes. Design drafting in two dimensions is 80 percent of the design work and can easily be done on microcomputers. One of the more interesting developments is the growing use of "compact" disks that store great portions or data, and sometimes a complete book. These compact disk systems usually offer an updating service that periodically provides a revised disk at an annual cost. Another piece of hardware is needed to "read" and interact with the disk on one's computer. Interactive compact disks that allow one to "talk" with the computer are available for safety and health training. This permits a "student" at a computer to take instruction and give answers via the compact disk. Mandated training in hazard materials handling is one such use of the compact disk.

SUMMARY

This chapter has presented an overview of microcomputers in the safety and health professions. Starting with computer terminology and computers themselves, the chapter looked at the computer environment, as well as the versatility and economics of computers. Many uses for computers were discussed, as were hardware and software and their costs.

Underlying nearly all aspects of this discussion was software, the critical ingredient for accomplishing anything in computerized safety and health systems. Software will continue to maintain it's dominant role as we build computer literacy and ever better microcomputers.

REFERENCES AND BIBLIOGRAPHY

"Dataguide Resource Finder," *Occupational Health and Safety,* January 1981, p D-1.

Directory of Safety Related Computer Resources, Volume I, *Software and References,* Des Plaines, IL: American Society of Safety Engineers, 1987, $15 ($10 members).

Directory of Safety Related Computer Resources, Volume II, *Databases,* Des Plaines, IL: American Society of Safety Engineers, 1987, $15 ($10 members).

Directory of Safety Related Computer Resources, Volume III, *Systems and Hardware,* Des Plaines, IL: American Society of Safety Engineers, 1987, $15 ($10 members).

"Guide to Comprehensive OH&S Information Management Systems," *Occupational Health and Safety, April 1984,* pp.

Spence, Lowell C., "What Is RMIS, You Might Ask?," *Professional Safety,* July 1989, pp. 21-23.

Resources for material in this chapter include Dr. Leo Greenberg of Israel and Tug Gokaydin of the University of Denver.

Chapter 5 / Part 1

Safety and Health Training and Education

THE NEED FOR TRAINING

In today's era of rapid change, safety and health functions change as quickly as any other function. To name a few examples, safety and health training must keep up with new processes, changing economic and environmental constraints, emerging social and governmental pressures, and higher worker expectations. There is a growing need for every person in the organization to have some sort of safety and health training in order to respond intelligently to these changes. There is also a long-overdue emphasis on discovering inter- and multidisciplinary solutions to our problems of identifying and controlling workplace hazards. This complex environment makes training and education indispensable.

What is the difference between training and education in safety and health? "Education" usually means learning to think and to apply one's mind to new situations. It involves every phase of mental development. We associate education with a broad, total program of formal and informal preparation resulting in professional growth and a higher level of competence.

"Training," on the other hand, is used in a much narrower sense (though it can include some education). It aims at specific problem areas rather than an overview or broad background. Training centers on applying what is learned in a practical and immediate manner. Thus, training is for those who must personally *do* something, as opposed to those who plan, develop, and provide procedures for doing something. Education is generally more applicable to management, while training is for those who must carry out specific tasks. In other words, we educate the manager via a degree in business but train a worker to operate a forklift.

There are two further considerations in today's work situations which we must keep in mind when discussing training and education:

1. Today's work force has a higher level of education than ever before, and can capably handle a higher level of training.
2. Training of some sort is useful for all *levels* of management in order to develop specific solutions for specific problems.

112 SAFETY AND HEALTH MANAGEMENT

Because of its more immediate and practical application, most of this chapter deals with safety and health training rather than education.

Governmental Requirements

Sometimes the manager will have no choice about whether to provide safety and health training. It will be mandated or strongly suggested by regulations or standards. The OSHA standards have set the tone for today, particularly in Part 1910 of the Code of Federal Regulations, which requires worker training in the following areas:

Blasting and explosives
Chain-saw operations
Cranes and derricks
Employee responsibilities
Equipment operations
Forging
Gases, fuels, toxic operations
Hazardous materials
Medical and first aid
Power presses
Powered trucks or motor vehicles
Respirator use
Signs, danger and warning instructions
Welding

In practice, this list is supplemented by sound management practices, self-imposed standards, insurance carrier requirements, and other factors. Similar or even more stringent requirements exist in most American or European countries, and more various other requirements are in force in other areas of the world.

Training and Human Behavior Modification

Most of the training available in safety and health tries to modify human behavior. This approach centers on conclusions about the cause of unsafe actions. It assumes that people do unsafe things because:

They don't know that they're doing a task incorrectly.
They misunderstand the instructions.
They don't consider the instructions important.
They have not received specific instructions for the task.
They may have forgotten the right way to do a task.
They may find it awkward to follow instructions.

SAFETY AND HEALTH TRAINING AND EDUCATION

They deliberately disregard instructions.
They may have never learned the proper way to do the task.
They may be unable to overcome bad work habits, despite instructions.

Training is a partial answer to any of these problems. The type of training needed to overcome them is usually given in-house and falls into three categories:

1. Job safety analysis
2. Job instruction training
3. Over-the-shoulder coaching

All three approaches are widely used. They are generally effective if properly done. The choice and applications of these methods are normally not the tasks of senior management, but of those at the operating level. The NSC has many publications that can guide the reader in choosing and applying any of these methods.

While training is a partial solution to the problem of unsafe acts, management error and oversight also play a part. Top management has a vested interest in safety and health training.

GETTING STARTED

The many levels and types of activities involved in safety and health require a rather complicated program. This applies equally to the training program. Making a complex program of any kind manageable depends largely on a clear statement of the program's objectives. This is certainly true of managing a complex safety and health training program. The objectives must include the entire organization and all functions. While approval and sometimes actual development of the training program's objectives comes from the top, it is the staff and management levels that bear most of the training burden.

To be effective, certain basic requirements apply to any type of training:

1. It must relate to identified needs and persons within the organization.
2. It must relate to organizational policies and objectives as a whole.
3. It must be both practical and acceptable to those involved.
4. Its various levels and aspects must be interrelated. (For example, new-employee training meets an identified need, but would take place only after the training needs of supervisors are met.)
5. It must be assessed periodically to determine how successfully it meets its objectives.

Program Development

The NSC has identified six steps in the development of any safety and health training program. These steps can serve as an excellent guideline.

1. *Identifying training needs.* The biggest problem in training is determining when there is a need. Signs of a need include high accident and injury rates, many near misses, worker grievances or wildcat strikes, frequent equipment breakdowns, wastage and excessive scrap, customer complaints, and changes in equipment and processes. The exact nature of the training problem must be identified; otherwise, training may be provided in the wrong area or on the wrong operational level.
2. *Formulating training objectives.* Specific training objectives must be developed to solve specific problems. Beyond that, the development of training objectives, course objectives, and testing and evaluation methods is a job for the training staff.
3. *Gathering materials and developing course outlines.* This step will involve a commitment of resources from top management and staff. Is in-house training or an outside source the most economical method? Can we develop course materials in-house, or are suitable materials available from outside sources?
4. *Selecting training methods and techniques.* This is another step best left to the training staff or to consultants with the proper expertise. Experience may have shown that certain methods or techniques are particularly good for teaching certain people certain kinds of tasks or behaviors. The specialist is in the best position to know this.
5. *Conducting the training program.* This involves the actual use of a curriculum, timetables, teaching environment (the proper place, lighting, and seating if the training is not done on the job), supplies, instructors, and incentives. These are all matters for the training staff. If there is no training staff, management may have to develop one or use a competent outside training consultant.
6. *Evaluating the program.* It is essential to provide a method to evaluate the success of any training program. Even if training is mandated, this is desirable. Such evaluation ensures that the training program serves the company's objectives and meets the letter of the law. Tests, on-the-job evaluations, evidence of decreasing accident or wastage rates, and many other methods of evaluation may be applied. Management must make sure that such evaluations take place periodically and that higher management receives the results.

Training Program Content

What should a training program for safety and health contain? This will vary according to the person, group, or level addressed. The organization's complete safety and health training program should contain all of the following elements. Specific training programs within the overall program should include as many as are pertinent to that level of operation.

1. Concepts and philosophies of environmental health
2. Concepts and philosophies of accident prevention
3. Company safety and health policies
4. Plant processes
5. Oral and written communication skills
6. Job safety analysis and job instruction training
7. Accident investigation procedures
8. Relationship between safety and health and company objectives
9. Accident, incident, injury, and illness reporting
10. Legislative and reporting requirements
11. Personnel selection and appraisal techniques
12. Job placement and training activities
13. Hazard detection techniques
14. Operating procedures and rules
15. Gathering and interpreting safety and health information
16. Personal protective equipment
17. Human aspects of safety and health
18. Emergency medical treatment
19. Emergency planning and procedures

These subjects have wide application, as safety and health functions are involved in any company activity. Different operating functions may need different kinds of knowledge about a given area. For example, an assembly line worker does not need detailed information about personnel selection or information-gathering techniques, but does need to know how standards of performance apply or whom to inform of an unsafe act or condition. The senior executive needs to know what governmental regulations require of the company, and workers need to know their rights under the same legislation. The manager needs to know how to develop an effective plan to deal with emergencies; the worker must know what specific actions to take in an emergency. As you can see, most of the subjects listed do have some application to every operating level.

Training and Educating the Safety and Health Staff

We assume that a professional safety and health person has had more education than training, since his or her object is professional development. Such education is generally available and qualified persons are in the marketplace, though in short supply. A serious restriction to safety and health staffing, however, is the lack of safety and health management instruction in educational institutions. The education is still largely inspection/engineering oriented, and does not emphasize the administration of safety and health programs or general corporate management aspects. In effect, education has not adopted the systems concept of looking at the total education package required. Solving this particular problem is complicated by the scope and the interdisciplinary nature of safety and health, which, in its entirety, is beyond the capability of any one school, or even a group of colleges or universities. Two remedies are available: educating safety and health personnel to be managers or educating managers to be safety and health personnel. Either is a formal, time-consuming process. Top management should encourage both safety and management personnel to learn as much as possible about the skills and disciplines of the other.

Training Responsibilities by Operating Function

The exact training needs of an individual or group in an organization depend on its function. An understanding of these functions is essential in developing the safety and health program. The following list attempts to summarize the responsibilities of each operating function.

Executive and Senior Management Safety and Health Functions:
1. Accept responsibility for safety and health programming.
2. Develop, agree on, promulgate, review, and evaluate safety and health policies.
3. Determine safety and health program objectives.
4. Allocate resources to attain those objectives.
5. Develop, disseminate, review, and evaluate safety and health programs.

Staff Safety and Health Functions:
1. Develop the following:
    a. Operating rules and standards
    b. Accident investigation procedures
    c. Personnel selection and training procedures
    d. Useful information and reporting formats
    e. Emergency planning procedures
    f. Operating objectives
    g. Training requirements and resource allocations
2. Ensure that safety and health considerations are included in concept and design phases.
3. Interpret policy for line management.
4. Provide information for management decisions.
5. Review and evaluate performance.

Line Management Safety and Health Functions:
1. Communicate objectives and goals.
2. Work with the staff in developing objectives, rules, emergency plans, processes, and equipment design.
3. Develop strategies to comply with rules and regulations.
4. Develop job procedures with staff supervisors.
5. Interpret information for preventive application.
6. Implement procedures for the use of personal protective equipment.
7. Provide feedback:
    a. To senior management and staff on objective performance.
    b. As a basis for reviews and reports on accidents, injuries, and illnesses.
8. Investigate, evaluate, and report on accidents, injuries, and illnesses.
9. Take corrective action on conditions uncovered through accident investigations.
10. Evaluate emergency plans.
11. Coordinate training needs and make persons available for training.

Supervisor Safety and Health Functions:
1. Understand the reasons for accident prevention and health measures.
2. Understand, adopt, and ensure compliance with safe practices.
3. Implement government requirements, company policies, and procedures.
4. Develop safety awareness in workers.
5. Supervise workers to ensure the use of safe, healthy practices.
6. Brief workers on and discuss safe, healthy practices.
7. Arrange for own and workers' training.
8. Inspect to identify unsafe practices and conditions.
9. Inform management of unsafe practices and conditions beyond own control.
10. Ensure use and maintenance of personal protective equipment.
11. Arrange drills or tests of emergency plans.

Employee Safety and Health Functions:
1. Understand the:
    a. Concepts of accident causation and prevention.
    b. Program objectives and personal applications.
    c. Need for implementation of safe practices.
    d. Communication system for safety and health items.
    e. Emergency plans and their roles.
    f. Employees' rights under legislation.
    g. Labor organizations' roles and rights.
    h. Environmental hazards of the workplace.

2. Comply with government regulations and company rules.
3. Use personal protective equipment as required.
4. Report accidents, injuries, illnesses, and hazards.
5. Take training as required.
6. Know how to contribute to the program.

Collateral Duty Safety and Health Personnel Functions:
1. Know the:
   a. Basics of accident investigation, analysis, and reporting.
   b. Policies and programs.
   c. Organizational levels and functions.
   d. Safety and health concepts and practices.
   e. Accident reporting and investigating systems.
   f. Line and staff roles in accident and injury prevention.
   g. Applicable governmental regulations.
2. Understand and make recommendations on programs.
3. Monitor adherence to governmental regulations and company rules.
4. Monitor and help staff in carrying out their functions.
5. Serve as a resource on safety and health matters.
6. Interpret safety and health information for senior management.
7. Monitor accident investigation and reporting activity.
8. Promote interest and participation in programs.
9. Inspect to identify hazards and unsafe practices.
10. Establish supplemental programs to attain objectives.
11. Make recommendations to line, staff, unions, and top management.
12. Use information that has been developed for accident prevention purposes.

Identifying training needs by referring to these functions is one way to approach the problem. Given its functions, what type of basic or extra training will any operational level need? This method of analysis has the advantage of pinpointing the safety and health functions of any particular position. Training can then be added for the new facility's occupation, changing procedures, new or modified equipment, and activities of committees or accident investigation teams.

INTERNAL SAFETY AND HEALTH TRAINING

Almost every organization has some provision for in-house training. However, when higher management contemplates a major revision in its safety and health training program, two questions arise: "How much training should the company attempt in-house?" and "What kinds of training can be handled better through an outside firm or consultant?" To help the manager make these decisions, here are some points to keep in mind about in-house training.

Advantages of In-House Training

1. It allows management to select the most knowledgeable people to conduct the training.
2. It allows management to fulfill personal objectives for those who want to be involved in training.
3. It improves morale by using company personnel as opposed to outsiders.
4. It allows management to safeguard its proprietary processes.
5. The total cost may be lower.

Disadvantages of In-House Training

1. The instructor may know the subject but be unfamiliar with training techniques (a poor teacher).
2. The instructor may be a good teacher but be unfamiliar with the subject.
3. Staff must spend time compiling knowledge and training materials.
4. Supervisors may resent having their best people taken for training positions.
5. Those who have interest and volunteer may not be the most capable teachers.
6. Ready-made courses available from outside sources may be cheaper.

Safety and Health Staff as Training Resources

The organization may already have one very good resource for in-house safety and health training: the safety and health manager or staff. They should always take part in any in-house training scheme, but how much is not certain. We question the wisdom of having the safety and health personnel do the actual training unless they were clearly hired for that purpose. (Some firms with regular safety and health staffs may include one or two staff members hired specifically as trainers.) Training can become so time-consuming that it prevents safety and health personnel from doing more urgent tasks. Also, all the disadvantages of in-house training apply to training via the safety and health staff. The safety and health person rarely knows actual work processes firsthand. Safety and health activities should not be separated from the regular daily activities in the workplace.

The safety and health manager provides extensive advisory services for all activities. However, being a training director dilutes the manager's effectiveness and that of the training staff. This is particularly true for routine areas like first aid, fire fighting, emergency rescue, and specific tasks. Someone else can better handle these training segments. The maximum benefit is achieved when the safety and health manager is an advisor to whoever gives the actual training. He or she can help select the right training methods and bring the right materials together. The manager with a sound knowledge of training methods and an overview of the objectives has greater leverage in an advisory role than as an instructor.

Training Strategy

A sound training strategy starts with the company policy itself. The policy should make a clear reference to safety and health training. Next, the training objectives should be clearly stated, and the responsibilities for different kinds of training should be explained. This does not relieve top management of its responsibility, but properly delegates part of it to the management levels that can best direct it. The overall responsibility should be assumed by a senior manager.

In terms of training objectives, the strategy is determined by required and expected changes in present performance levels. Some objectives will stem from statutory requirements. The strategy should always be based on:

1. The main safety and health problems of each functional area.
2. Problem areas needing improved performance.
3. Realistic and obtainable goals.

To make intelligent decisions about training strategies and aspects of the safety and health training program, the manager should ask several questions:

1. What are the expected benefits of training?
2. Who does the company need to train?
3. What levels of performance are required?
4. How long will training take? How much time can we allocate to training?
5. How urgent is the need for training? Does someone's life depend on it?
6. How many people need training?
7. Who will be expected to conduct the actual training?

Since line management implements the program, steps are needed to ensure that it is acceptable to them. The senior manager must present the training objectives, content, methods, and responsibilities to line management, making sure that they are understood. Line management's feedback may point out needed revisions. With these taken care of, the senior manager secures a firm commitment from the involved parties.

Arrangements should also be made to monitor the training and assess its value when completed. The training should be periodically reviewed for pertinence, quality, and timeliness, and changes made as required.

Special Safety and Health Training: The New Employee

Older, more seasoned employees may require "spot" training. They may already be familiar with safe work practices, the use of personal protective equipment, and how to report a workplace hazard. The new employee, however, represents a special case for training. He or she still has to learn about the organization. In training new employees, it is not enough to show them how to do the specific task for which they were hired. The training should include a much broader range of safety and health information, including:

Company safety and health philosophy.
Safety and health requirements as a condition of employment.
Employee safety and health responsibilities.
Safety rules and regulations.
Reporting injuries, accidents, and hazardous conditions.
Location of emergency exits, first aid kits, etc.

These subjects require the understanding and backing of top management and should be presented to all new employees. This early training period is an ideal time for top management to show its sincerity about safety and health. They should discuss it with new employees as part of their induction briefing, before or during the first work week. A few words on the subject from top management add viability and credibility to the program. A welcome and a few words of encouragement by the top executive (who they may not see again for weeks, months, or years) is a great way to start a new job. It gets things off on the right foot and gives management a chance to emphasize programs they feel strongly about. Certainly this emphasis means more when it comes from the top executive than from safety staff or a personnel clerk.

This is not enough, however, and there is a time to discuss safety and health for the specific job. This emphasis becomes the responsibility of the immediate supervisor, preferably in the workplace.

EXTERNAL SAFETY AND HEALTH TRAINING

Internal training is not always possible or appropriate, and often a decision must be made between internal and external training resources. Proper evaluation of the choices may be very difficult unless the manager has enough information on which to base the decision.

There are many external problems available. These are usually well known to the safety and health staff via safety and health publications. Thus, while we will review a few of the organizations and their offerings, the manager should check with his or her own safety and health staff before proceeding. If there is no staff safety and health professional, the manager should seek outside assistance. How the manager can secure such help is covered in this section. There are programs to meet nearly any need, whether to establish an overall training program or a specific training course. There are so many that we cannot mention them all. Lack of discussion here does not imply that they are not all equally good or equally capable of meeting the manager's training needs.

The National Safety Council

The NSC is the largest safety organization in the world. It is a nonprofit group with independent chapters throughout the United States. It covers nearly all areas of the safety and health professions. There are equivalent organizations throughout the Americas, as well as in England, Australasia, and many other areas. The NSC has a Training Institute at its Chicago headquarters, which present a variety of courses each year. These courses range from a few days to a series covering several weeks.

While national in scope, the NSC also offers safety and health training through local, independent chapters. These programs usually run from 1 to 3 days in length. While the courses vary widely to meet customer needs, adherence to national standards ensures quality control. The chapters normally give training in their own facilities, but also conduct programs at client companies.

Each year the NSC sponsors the National Safety Congress. For 1 week, there are a series of challenging seminars on every aspect of safety and health. Smaller, but similar, regional congresses throughout the United States are sources of valuable information and direct training for managers.

The NSC also offers a complete selection of safety and health training literature for most functional areas. These materials include films, slide-cassette packages, books, pamphlets, and packaged training classes, all suitable for group or individual use. The NSC can be contacted at 444 North Michigan Ave., Chicago, Il 60611 or through a local office. Services are available to both members and nonmembers.

Private Organizations

One private U.S. organization stands out in providing safety and health training in much the same manner as the NSC. The International Loss Control Institute (ILCI) has excellent home-based facilities, but also takes to the road with its offerings. Information on ILCI courses and services can be obtained from the International Loss Control Institute, P.O. Box 78, Loganville, GA 30249.

ILCI is only one firm out of hundreds that offer safety and health training and services. They range from small local groups to large professional organizations. A little research will give the senior manager a clear idea of who can be most helpful in developing a company's particular training program. There are also hundreds of capable safety and health professionals available on a consultant basis to provide tailor-made services.

The U.S. Government

Besides the private sector and professional safety and health organizations, the government can also be a source of training materials and information. The U.S. Department of Health and Human Services (HHS), through its National Institute of Occupational Safety and Health, has many health-oriented courses consistent with the OSHA act. Through sponsorship of NIOSH Education Resource Centers, opportunities for safety and health training are presented throughout the United States and its possessions. Regional HHS offices can give the manager more specific information on available programs and materials.

Colleges and Universities

Many colleges and universities throughout the United States offer short courses in safety and health subjects that could be classified as training. Some of these are presented under the auspices of NIOSH-sponsored Educational Resource Centers. Most, however, are the community's answer to the needs of the particular locality. Some offerings can be found through an NSC publication, *College and University Safety Courses*. Specialized short courses lasting for a few days or weeks are not always listed but are available in great number. Local inquiry seems to be the best way to locate these courses.

Professional Societies

Because safety and health is a multidisciplinary activity, there are many associated professional societies. Many of them offer safety and health training on a local or national level. Most also have excellent annual seminars or professional conferences. It is not possible to give a complete listing here, but an in-house safety and health professional or an outside consultant can furnish information on these offerings.

Professional Education

Education, by our definition, is directed at professional development and growth. A special word follows about college and university programs in this area. Training courses are only part of an organization's needs. Providing higher-level opportunities for safety and health personnel to develop skills and knowledge can be just as vital. Not only does this produce a greater degree of professional competence, which is useful to the organization, but it also boosts the interest and morale of the safety and health staff. The continual input of new information, more effective methods, and broader perspectives that safety and health people gain from professional education naturally benefits the company. Wherever possible, top management should encourage safety and health personnel to continue their professional education. They can do this by offering incentives such as paying for courses and textbooks, bonuses for course completion, and so forth. As part of this professional upgrading, management should know what professional educational opportunities exist locally and regionally. The NSC booklet mentioned earlier, *College and University Safety Courses,* can serve as a guide. The safety and health staff will surely bring other programs to management's attention once it has management's active support and encouragement to upgrade and develop professionally.

Safety and Health Consultants

Nothing is more wasteful of management's time and resources than developing a new internal safety and health program, only to find that a suitable program already exists and is readily available at reasonable cost. Similarly, the company may invest heavily in an in-house safety and health staff, only to learn later than an outside consultant could do the same thing at a more reasonable price. Many safety and health consultants can offer or design specialized programs and present them on site. Others can work on particular safety and health problems but do not provide direct training. There are also consultants who work full-time for a business (such as an insurance company), providing services for the business's clients. Any of these may prove suitable for the company's particular needs.

If management seeks help from an outside consultant, there is still the problem of finding a reliable professional to provide the exact services desired. Where does one start? Word of mouth may provide a few good names. Contacting the nearest professional society for the names and address of members who act as consultants may be a more reliable method.

One organization lists professional safety and health consultants. This information is available from the American Society of Safety Engineers, 1800 East Oakton Street, Des Plaines, IL 60018-2187. Ask for the *National Directory of Safety Consultants.*

Having obtained several names and made initial contact with a few consultants, how does the manager select the person best suited for the company's needs? Here are some points to consider:

1. What is the alternative if you don't use this consultant?
2. Why pick this particular consultant? (Do not use a low fee as a reason; find a better reason.)
3. What is the consultant's reputation? Make inquiries, as you would for any consultant.

4. How did you get in touch with the consultant? Is the source reliable?
5. Who are the consultant's references? Verify them and ask for an opinion.
6. Who are the consultant's associates or instructors?
7. Does the consultant have previous experience in your type of business? Is this kind of experience necessary?
8. How does the consultant evaluate his or her own effectiveness?
9. What can the consultant provide in the way of handouts, slides, viewgraphs, etc.? Will it cost extra?
10. Will the consultant furnish training outlines, facilities, and equipment?
11. Is the consultant willing to present training courses at your site?
12. Is the fee schedule suitable?
13. What kind of contract arrangements will the consultant make? (Be sure to use a contract if the training is complex or long-term.)
14. What is the consultant's attitude about working with and through your existing safety and health staff?

If you like the answers to these questions, you probably have a good consultant. Keep in mind, however, that even the best consultants cannot do good work if they lack the backing of top management. Besides finding out what can be done for you, find out what the consultant needs from top management to create and implement programs. Remember that management must decide whether in-house training is the best answer. This question may have to be reviewed from time to time as training needs change, as staff is added, or as programs become ineffective.

Training is a continuing need. It cannot be done on a one-time basis. For that reason, senior managers must keep informed of the ongoing success of training programs, keep them under continuous review, and make sure they are run according to plan.

TRAINING AND EDUCATION REQUIREMENTS FOR SAFETY PROFESSIONALS

There will always be safety and health personnel who require at least minimal training to do their inspections for compliance with regulations and standards. And there will always be a need for specialized short courses to prepare personnel for changing legal and operational requirements in the workplace. So too, will there always be a need for training to upgrade the level of safety and health professionals.

It is management's responsibility to see to the training and education of their professionals. If they want true professionals for the safety and health function, they should encourage continuing training and education through motivation, planning, and funding. The safety person's increased capability should be a prime consideration in career counseling and development.

Professional development is more than formal classroom work. It also involves travel to seminars and meetings to exchange and gather information, and includes acquiring an adequate safety and health library. Although many firms see the need for a truly professional approach to safety and health, many others still fail to provide the encouragement or funds to participate in training, educations, professional society participation, or even for literature to keep abreast of current practice.

It is not always clear what education and training meets the need of practitioners working with complex systems and sophisticated industries. For example, the complexities of advanced processes and techniques in the future will demand a type of safety professionalism that is beyond that available today. Now is the time to meet the demands of new, complex, technologically advanced workplaces, products, and processes. Not only are today's professionals unprepared to meet the safety requirements of such areas, but often the techniques to prepare them do not exist. From this we deduce that new safety and health requirements must be developed to meet tomorrows needs.

Today's more advanced managers see that quality control and reliability in products and processes are not only needed but must be nearly perfect to ensure economic survival. The message that it is right and practical to demand 100 percent quality and reliability is slowly sinking in as the United States struggles to meet the higher quality and reliability of foreign products. Still in the infant stage is the realization that safety and health also influence workplace and marketplace positions in a competitive world. Few have dared to admit that, just as we might expect no error in a product and process, we can expect no accident or injury in the workplace. Too many of the old clichés are still around, such as:

"There will always be human beings to cause accidents."
"You have to expect losses."
"You are never going to cut out all accidents."
"Zero accidents are not a realistic goal."
"Hazardous materials introduce hazards faster than we can recognize them."

This seems to mean that we willingly accept less than perfection; that it is okay to make mistakes, that it is acceptable to have accidents or not keep up with new workplace hazards; or that we can always expect workplace injuries and illnesses. No one claims to be perfect, but there is no reason to expect accidental losses, operating errors, substandard performance and products, or occupational illnesses. New safety and health viewpoints and philosophies are in order.

A MATTER OF EDUCATION

The contributions to products and processes made by the safety and health-associated sciences and disciplines are not found in the curricula of business schools (the source of most of our managers) or engineering schools. Most managers have never been academically exposed to the safety and health functions they control. They know little of the contributions that safety and health can make to their personal and operating success. They need safety and health education of a special kind. Safety and health in a product or process begins in the concept and design phases and continues throughout its lifetime and into the troublesome area of disposal. Yet the education of engineers about safety and health as resources to apply to product and process is almost ignored.

Safety and health education in the United States is largely subject to marketplace demands and this is fluid. If there is little demand for education, little is provided, except through the determination and foresight of a few educators or government agencies. Government-funded education programs tend to be given at well-established schools. Education in the field, however, tends to be a function of one faculty member's particular interests. If he or she leaves the university or has a change in priorities, the programs may disappear. In days of economic stress, unjustified and unsupported programs are soon dropped.

Safety and health as a discipline is now receiving higher visibility and greater academic stature. Government funding of safety and health training in colleges and universities has made even some of the top 20 universities aware of the field, particularly at the graduate level. With efforts to insert safety and health into business curricula, some university business schools have formally added safety and health curriculums. The complexity of modern safety and health practices, and organizational concern over safety and health, have resulted in much better-educated safety and health professionals. The percentage of safety practitioners with graduate degrees is roughly the same as the number holding undergraduate degrees 15 years ago. Recent surveys by the American Society of Safety Engineers reveal that the percentage of their members with undergraduate degrees is more than 75 percent and growing yearly.

Public interest in safety and health issues has increased the demand for college-educated specialists, particularly in environmental areas. There is a trend for safety and health specialists to assume environmental tasks, and this is sometimes reflected in their job titles. This trend can be traced to a shortage of environmental engineers and the closeness of some environmental concerns to safety and health, particularly in industrial hygiene.

While few of the top universities offer degrees in safety and health-related areas, most schools do offer some courses. Most programs are found in the approximately 2,500 undergraduate and community colleges. The variety and quantity of schooling is outstanding. Many schools offer associate, baccalaureate, or graduate degrees in safety, health, fire protection, and environmental health areas. Many of these also offer short, intensive courses, which are generally considered training rather than education.

The sheer number of schools involved precludes a detailed discussion of the degrees and courses offered or the exact department or person to contact at each school. Such reliable organizations as the NSC and the American Society of Safety Engineers seek to keep up with changes in offerings but find it difficult. The task of keeping such material up-to-date is periodically the subject of graduate theses. These are, unfortunately, out-of-date by the time they are available through University Microfilms in Ann Arbor, Michigan. At this writing, the best general guidance comes from the NSC's *College and University Safety Courses.* Although meant to be published each year, the NSC finds it hard to keep up with the task, and although the publication has a reasonable price, the returns make it a money loser.

It is nearly impossible to research and publish a current directory of colleges and universities with programs that would be valid by the time the reader sees it. We suggest four approaches to a person seeking information on local safety and health programs:

1. Contact the NSC for the latest issue of their *College and University Safety Course.*
2. Contact the student advisement office of the closest college and ask their help in finding safety and health programs. They have annual catalogs and databases to help advise students on career choices. They may be limited to helping you only with those colleges you name, since it is not practical to search thousands of catalogs for such a request.
3. Contact the closest chapter or national headquarters of the American Society of Safety Engineers, or talk to safety and health professionals.
4. Contact the NIOSH in Cincinnati, Ohio, where one small office periodically keeps track of this matter. They, in turn, can often tell you what other advice government offices may be able to offer.

Training

We have said that it is not possible to identify the many institutions and organizations that offer safety and health training. Most are entrepreneural in nature and cannot survive without the support of governmental agencies or paying students. Some offer safety and health training only at their headquarters, others present seminars at centrally located sites, and still others provide on-site special programs to meet requests. Our suggestion for those seeking training services is to look in a recent issue of one of these publications:

NSC's *Safety and Health* (monthly)
Occupational Health and Safety (monthly)
Best's Safety Directory (annual)

All three publications carry advertisements on training organizations and special seminars. The last one lists training organizations for nearly every aspect of safety and health.

REFERENCES AND BIBLIOGRAPHY

Anton, Thomas J., *Occupational Safety and Health Management,* New York: McGraw-Hill, 1979.
College and University Safety Courses, Chicago: National Safety Council, annual.
Dictionary of Terms Used in the Safety Profession, 3rd edition, ed. Stanley A. Abercrombie, Des Plaines, IL: American Society of Safety Engineers, 1988.
Education and Training in Occupational Safety and Health and Ergonomics, Geneva: International Labour Office, 1982.
Ferry, Ted S., *Safety Program Administration for Engineers and Managers,* Springfield, IL: Charles C. Thomas, Publishers, 1984.
Findlay, James V., *Safety and the Executive,* Loganville, GA: Institute Press, 1979.
General Materials Catalog, Chicago: National Safety Catalog, annual.
Good, Carter V., *Dictionary of Education,* New York: McGraw-Hill, 1983.
Grimaldi, John V., *Safety Management,* 4th edition, Homewood, IL: Irwin, 1984.
Hammer, Willie, *Product Safety Management and Engineering,* Englewood Cliffs, NJ: Prentice-Hall, 1980.
Hazard Control Information Handbook, Rockville, MD: International Institute of Safety and Health, 1983.
Heath, Earl D., and Ted Ferry, *Training for Workplace Safety: Improved Performance Through Training and Education,* Goshen, NY: Aloray, 1990.
King, Ralph W., *Industrial Hazards and Safety Handbook,* London: Newens-Butterworth, 1979.
ReVelle, Jack B., *Safety Training Methods,* New York: Wiley, 1980.
Supervisor's Safety Manual, Chicago: National Safety Council, 1985.

Chapter 5 / Part 2

Employer and Employee Training and Education in Occupational Safety and Health

INTRODUCTION

Definitions

The terms "training" and "education" are used to denote the information and instruction on occupational safety and health which employees must have to work at minimal risk to themselves, their fellow workers, and the public. In most countries in the industrialized Western world, the terms "information" and "instruction" are considered synonymous with the terms "education" and "training" when the subject of occupational safety and health is discussed. Accordingly, these terms are used interchangeably throughout this chapter.

Information and Instructions as Countermeasures

In the "war" on occupational injury and illness, safety and health practitioners like to speak of the "three E's": engineering, enforcement, and education. These have an interdependent relationship. None of them alone can do the job. In the past, engineering and enforcement were looked to for solutions to occupational safety and health problems more frequently than education and training. Recently, however, policy makers and legislators, as well as managers and safety and health practitioners, have begun to look more carefully at information and instruction and their role as occupational injury and illness countermeasures. Engineering is unquestionably the ideal solution, but it isn't always possible to design out all hazards. Enforcement, too, is limited by the number and quality of enforcers. Thus, other methods are needed to motivate and assist employers and employees in complying with occupational safety and health standards and with other recommended safe and healthy practices.

The French author and playwright Victor Hugo (1802-1885) once observed that "No cause can succeed without first making education its ally." Those whose cause is occupational safety and health know that, in the past, information and instruction were steadfast allies in the effort to secure a safe and healthy American workplace. Today the number of potentially hazardous substances and processes used in industry multiplies each year, making the worker's need to be informed more urgent than ever. Likewise, new governmental regulations, as well as the sophisticated safeguarding systems often required to control occupational hazards, must be fully understood by workers if they are to take advantage of the protection that these measures offer. It can easily be seen that if the occupational safety and health problem is to be solved, information and instruction must play an even greater role than they have in the past. As Cohen, Smith, and Anger (1979) concluded in their examination of techniques for influencing workers' self-protective behavior:

> Training remains the fundamental method for effecting self-protection against workplace hazards. Success depends on (1) positive approaches that stress the learning of safe behavior (not the avoidance of unsafe acts), (2) suitable conditions for practice that ensure the transferability of these learned behaviors to real settings and their resistance to stress or other interferences, and (3) the inclusion of means for evaluating their effectiveness in reaching specified protection goals with frequent feedback to mark progress.

Recently, a number of studies have focused on the importance of employee training as a countermeasure to injury and illness. For example, the National Academy of Sciences reported that 90 percent of grain elevator explosions could be avoided if employees were better trained and if the amount of grain dust in elevators was reduced. Another study (this one of coal mine accidents), released within the past 2 years by the National Academy of Sciences, recommended that the federal government double the current training requirements for coal miners in order to reduce accidents. In still another report of a rapid rail accident involving multiple fatalities in Washington, DC, investigators criticized management's failure to develop an adequate program of initial and recurrent training for the system's operating personnel.

In addition to these reports, which discuss the importance of training as an occupational injury and illness countermeasure, several research studies have examined industrial safety practices in general and safety training practices in particular. These studies incorporated opinion polls, analyses of factors common to companies with outstanding safety performances, and comparisons of safety program practices in companies with high versus low work injury rates. Typically, it was found that the companies with better records (1) included information on job risks and what to do about them in new employee orientation programs; (2) gave both initial and follow-up training in safe job procedures to all employees; (3) gave special safety training to supervisors; and (4) used a variety of safety training techniques, including lectures, films, group discussions, demonstrations, and simulations, in their training programs.

From data such as those generated by the Bureau of Labor Statistics (BLS) through its Worker Injury Report (WIR) surveys (these are surveys of individuals who have been injured on the job and who were then asked what types of training, if any, they had received in occupational safety and health risks and what to do about them), and from the types of inquiries which employers direct to OSHA concerning the content and timing of training programs for employees, it is apparent that additional guidance is needed by many employers.

WHAT OSHA REQUIRES IN TRAINING AND EDUCATION

One of the sections of the OSHA Act of 1970 (Public Law 91-596) that states or implies the responsibility of the employer to provide training and education to the worker is Section 6.(b)(7), which requires that employees be apprised of all hazards to which they are exposed, relevant symptoms and appropriate emergency treatment, and proper conditions and precautions of safe use or exposure. While OSHA has promulgated a large body of occupational safety and health standards with which employers must comply, none of these is a "training standard," as the term is normally understood and used. However, well over 100 current OSHA occupational safety and health standards, including many of the most recent ones, contain training requirements. And since these requirements are integral parts of the standards, failure to comply with them may result in a citation and penalty for the employer.

The training requirements for OSHA safety standards normally do not spell out precisely what is required of the employer. For example, the standard 29CFR1910.178(1), which relates to the use of powered industrial trucks in materials handling operations, states:

> Operator training. Only trained and authorized operators shall be permitted to operate powered industrial trucks. Methods shall be devised to train operators in the safe operation of powered industrial trucks.

In contrast to the often sketchy information included in a training requirement for a safety standard, as in the above quotation, the training requirement in a health standard typically covers these points:

1. Who is to be trained.
2. The date by which initial training is to be completed.
3. How often training is to be conducted.

4. What the training program should include, as a minimum.
5. Access to training materials.

Using as an example the standard "Occupational Exposure to Acrylonitrile (Vinyl Cyanide)" (29CFR1910.1045), the OSHA standard requires the employer to provide a comprehensive and specific training program for employees exposed to acrylonitrile ("AN"). The training requirement will be examined in light of the five questions listed above.

1. Who is to be trained? There are three such groups of employees: (a) All employees exposed to AN above the action level. An "action level" is an exposure level equal to one-half of the permissable exposure limit, above which certain precautionary measures, such as periodic monitoring and medical surveillance programs, must be conducted and below which only a limited number of the standard's requirements will apply (in this case, the action level is a concentration of 1 part per million as an 8-hour time-weighted average). (b) All employees whose exposures are maintained below the action level by engineering and work practice controls. (c) All employees subject to potential skin or eye contact with liquid AN.
2. By what date must initial training be completed? By January 2, 1979, the employer was required to institute a training program for, and ensure the participation of, the three groups of employees cited above. Since the standard for AN was published in the *Federal Register* on October 3, 1978, employers had 3 months to present the initial training to affected employees.
3. How often must training be conducted? Training is required at the time of initial assignment or upon institution of the training program, and at least annually thereafter.
4. What must the training program contain? The employer must ensure that each employee is informed of the following: (a) The information contained in Appendices A and B of the standard. Appendix A is a "Substances Safety Data Sheet for Acrylonitrile" and covers substance identification; health hazard data; emergency first aid procedures; respirators and protective clothing; precautions for safe use, handling, and storage; and access to information. Appendix B provides "Substance Technical Guidelines for Acrylonitrile" and covers physical and chemical data, spill, leak and disposal procedures, monitoring and measuring procedures, protective clothing, housekeeping and hygiene facilities, and miscellaneous precautions. (b) The quantity, location, manner of use, release or storage of AN, and the specific nature of operations which could result in exposure to AN, as well as any neces-

sary protective steps. (c) The purpose, proper use, and limitations of respirators and protective clothing. (d) The purpose and a description of the medical surveillance program required by the standard. (e) The emergency procedures required by the standard. (f) Engineering and work practice controls, their function, and the employee's relationship to these controls. (g) A review of the standard.
5. What about the workers' access to training materials? The employer must make a copy of Standard (29CFR 1910.1045) and its appendices available to all affected employees. In addition, upon request, the employer is required to provide to the Assistant Secretary of Labor for OSHA and the Director of NIOSH copies of all materials relating to the employee information and training program.

This, then, is the type of information that one may expect to find in the training requirements in standards such as those for dibromochloropropane (DBCP), benzene, inorganic arsenic, coke oven emissions and others.

State Requirements

In addition to OSHA, some states have enacted their own legislation or regulations concerning safety and health information and instruction to be provided to the worker. Following are excerpts from the General Industry Safety Orders and the Construction Safety Orders, respectively, for California, which has an OSHA-approved occupational safety and health plan:

General Industry Safety Orders. Section 3203, Accident Prevention Program. Every employer shall inaugurate and maintain an accident prevention program which shall include, but not be limited to, the following: 1) A training program designed to instruct employees in general safe work practices and specific instructions with respect to hazards unique to the employee's job asssignment. . . .

Construction Safety Orders. Section 1510. Safety Instructions for Employees. a) When a worker is first employed, he shall be given instructions regarding the hazards and safety precautions applicable to the type of work in question and directed to read the Code of Safe Practices. b) The employer shall permit only those employees qualified by training or experience to operate equipment and machinery; c) Where employees may be subject to known job site hazards, such as flammable liquids and gases, poisons, caustics, harmful plants and animals, confined spaces, etc., they shall be instructed in the recognition of the hazard, in the procedures for protecting themselves from injury, and in the first aid procedure in the event of injury.

What OSHA Compliance Officers Look For

Since many of the training requirements in the occupational safety and health standards are stated in broad terms, employers often are uncertain as to what documentation they should have available if their workplace is scheduled for an inspection. Accordingly, OSHA has taught its compliance officers to look for information such as the following when specific information is not given in the standard:

1. Powered industrial truck operators are required to be trained; thus, the employer should have evidence of a structured training program, including an outline of the knowledge and skills to be taught to the operator; a copy of the manufacturer's operator's manual for the equipment; a record of which personnel operating the equipment have been trained and when; evidence of retraining when new equipment is introduced; and the like.
2. Employees who are required to wear respirators must be trained in their proper fit and use, as well as their limitations. Thus, the employer should be able to demonstrate that the correct respirator is specified and available for each job where one is required; have written procedures covering the safe use of respirators in dangerous atmospheres that might be encountered in normal operations or in emergencies; and have evidence that the user has been properly instructed in respirator selection, use, and maintenance. Employers should have provided employees with an opportunity to handle the respirator, have it fitted properly, test its facepiece-to-face seal, wear it in normal air to achieve familiarity with it, and wear it in a test atmosphere before wearing it on their regular job.

To assist employers in developing training course outlines to comply with the training requirements of the occupational safety and health standards, OSHA has published "mini-course" outlines for those requirements found in the general industry, construction industry, and the maritime industry standards.

Penalties for Not Providing Information and Education

In addition to citing an employer for failure to provide the training and education identified in the OSHA occupational safety and health standards, OSHA compliance safety and health officers have invoked Section 5.(a) (1) of the Act, which states: "Each employer (I) shall furnish to each of his employees a place of employment which is free from recognized hazards that are causing or are likely to cause death or serious physical harm to employees." An example of the application of this authority is a case where OSHA charged an employer with failing to advise and supervise new employees adequately on the hazards of pressure vessels and hazardous materials associated with the petrochemical industry. Evidence revealed that a new employee was killed when the pthalic anhydride melt tank he was operating exploded. The employer contested the citation and penalty, but the Occupational Safety and Health Review Commission (OSAHRC) affirmed the citation and the proposed penalty.

Aside from the citations and penalties from OSHA which may result from failing to provide the information and instruction that an employee needs to work at minimal risk, there is the ever-present cost of having an employee who represents an increased threat to the self, to fellow workers, and to the public. If a worker has not been properly trained to perform a task, he or she is likely to be less alert to environmental hazards, and thereby at greater risk of incurring an injury or illness or of causing such injury or illness to another worker.

Reasons for Documenting Training

Few of the pieces of information or instruction that a worker should receive are spelled out in OSHA occupational safety and health standards. Thus, an employer should not keep training records solely because they may be required by federal or state legislation. Rather, an employer should regard such records as tools to be used in the design and implementation of effective training programs. Such records should be maintained for every employee who is scheduled for training or who has received training. Such records may be used:

1. To help establish whether there is a gap between the employee's current capacity and ability and the demands of the job he or she is doing or will be doing.
2. To provide a better picture of the personnel resources and the level of training of an organization's work force.
3. As a basis for future planning of courses.
4. As an aid in the evaluation and validation of training, which can be carried out only if information is available on the training received by the work force.
5. As evidence that the employer acted prudently in providing the worker with the information and instruction that the employer felt would increase the worker's skill and knowledge of the job and its attendant hazards, thus reducing the risk to which the employee is exposed.

All forms of training should be recorded in the simplest manner which meets both the employer's and external requirements. The employer's need to know what training which employees have received, and when, has extra significance during initial job training. At this stage, managers should not subject employees to risk by asking them to operate machinery or to carry out other tasks for which they have not yet been given appropriate training. Training records will also be valuable in conjunction with other safety records during the safety and health audit, and in reviewing establishment safety and health performance.

Right-to-Know: A Sound Business Practice

Former Assistant Secretary of Labor for Occupational Safety and Health Morton Corn, in a guest editorial in the *Journal of the American Industrial Hygiene Association,* observed: "The Occupational Safety and Health Act of 1970 adds to the Bill of Rights an additional freedom or liberty; the right of each worker to a safe and healthful environment."

Based on a study called the National Occupational Hazard Survey, NIOSH has estimated that 7 million workers in the United States are exposed to trade name products containing an OSHA-regulated toxic substance. More than 300,000 of these workers are exposed to trade name products containing one of the cancer-causing substances (carcinogens) regulated by OSHA. The adage that "What you don't know can't hurt you" was never more false than it is today. Blum (1978) observes:

> The law is moving toward an independent duty on the part of the employer to supply workers with information about adverse health conditions, separate from a cause of action for a given injury. From the general duty requirement of the common law emanates a legal duty to warn about harmful or dangerous conditions in the workplace and to provide the results of adverse physical exams given under the employer auspices. Presently, this does not stand as an independent cause of action; it can be raised only where the omission has led to a worker injury.

Tepper (1980) adds the following regarding the worker's right to know:

> Complete disclosure of risks is an essential element in offering employment since assumption of risk on other than a voluntary basis is fundamentally wrong. It is both morally correct and equitably necessary that workers understand and voluntarily accept the risks they assume as a condition of employment.

The National Academy of Sciences has said that it is unethical and unrealistic to assume that the right to know requires employers to inform workers about only the few chemicals that nearly everyone agrees are human carcinogens. The Academy believes that not only should the employee be informed about hazards at the workplace, but that the employer should make a reasonable effort to ensure that workers comprehend the relevant information.

Finally, NIOSH has suggested that workers ask themselves the following questions about their work environments:

1. Are there any substances used in this plant that are known to be harmful to your health?
2. Are trade name products used without full knowledge of what is in them?
3. Does the job require contact with mists, vapors, dusts, gases, or fumes that are potentially harmful? (The form of a substance or the way it is used can make it harmful. For example, wood dust may cause nasal cancer.)

NIOSH reports that if workers answer "yes" to any of these questions, they may be exposed to harmful conditions. NIOSH also suggests that if workers don't know the answers to these questions, they should find out if they are exposed to an occupational health or safety hazard and, if necessary, how to improve the situation. Employees are advised to check with their supervisor and, if they believe that they have been in contact with toxic (poisonous) or otherwise hazardous materials or situations, they should see their own physician, as well as the company's health staff if there is one.

The Hazard Communication Standard

The final rule on hazard communication was published in the *Federal Register* on November 25, 1983. All employers covered by the standard (Title 29, Code of Federal Regulations, Part 1910.1200) were to be in compliance with it by May 25, 1986. The purpose of this standard is to ensure that the hazards of all chemicals produced or imported by chemical manufacturers or importers are evaluated, and that information concerning these hazards is transmitted to affected employers and employees within the manufacturing sector. This information is to be transmitted by means of comprehensive hazard communication programs, which are to include container labeling and other forms of warning, MSDSs, and employee training. This occupational safety and health standard is intended to address comprehensively the issue of evaluating and communicating chemical hazards to employees in the manufacturing sector, and to preempt any state law pertaining to this subject. Any state which desires to assume responsibility in this area may do so under the provisions of Section 18 of the OSHA Act of 1970 (Public Law 91-596), which deals with state jurisdiction and state plans.

OSHA's Voluntary Training Guidelines

OSHA has developed and published a set of voluntary training guidelines to assist employers in providing the safety and health information and instruction needed for their employees to work at minimal risk to themselves, to fellow employees, and to the public. The guidelines are designed to help employers:

1. Determine whether a work site problem can be solved by training.
2. Determine what training, if any, is needed.
3. Identify goals and objectives for the training.
4. Design learning activities.
5. Conduct training.
6. Determine the effectiveness of the training.
7. Revise the training program based on feedback from employees, supervisors, and others.

The development of these guidelines is part of an agencywide objective to encourage cooperative, voluntary safety and health activities among Osha, the business community, and workers. Parts III and IV of this chapter are based on these voluntary training guidelines.

THE TRAINING PROCESS
Determining Whether Training Is Needed

The first step in the training process is a basic one: to determine whether the problem can be solved by training. When workers are not performing their jobs properly, it is often assumed that training will solve the problem. The assumption is that if workers knew better, they would perform better. This is not always the case. Gilbert (1967) states that more than half of the deficiencies that workers exhibit are deficiencies in execution, not knowledge. To distinguish between the two, he suggests asking a simple question: "Could this person perform correctly if his or her life depended on it?" If the answer is "yes," then the person has the necessary competence and the problem is a deficiency in execution. Faulty execution is due not to training but to other causes, such as inadequate feedback, where the worker lacks an understanding of the results or the value of the work; task interference, where other demands, such as personal problems, daydreaming, or telephone calls, compete for the worker's attention; punishment, where the desired performance involves tasks that are grueling, arduous, oppressive, or otherwise punishing; or lack of motivation, where ambition was never developed or was discouraged. All of these deficiencies of execution must be handled with remedies other than training. Train only to correct deficiencies in knowledge.

Problems that can be addressed effectively by training include those that arise from lack of knowledge or skills relating to a work process, unfamiliarity with equipment, or incorrect execution of a task. Whatever its purpose, training is most effective when designed in relation to the goals of the employer's total safety and health program. Such training covers both general safety and health rules and work procedures and is repeated, with modifications, following an accident or near-miss incident, or when observations of employee performance indicate a need for reinforcement of selected knowledge and skills.

Identifying Training Needs

If the problem is one that can be solved, in whole or in part, by training, then the next step is to determine what training is needed. For this it is necessary to identify what the employee is expected to do and in what ways, if any, the employee's performance is deficient. This information can be obtained by conducting a job analysis which pinpoints what an employee needs to know in order to perform a job correctly.

SAMPLE JOB HAZARD ANALYSIS
CLEANING INSIDE SURFACE OF CHEMICAL TANK — TOP MANHOLE ENTRY

STEP	HAZARD	NEW PROCEDURE OR PROTECTION
1. Select and train operators.	Operator with respiratory or heart problem; other physical limitation.	• Examination by industrial physician for suitability to work.
	Untrained operator — failure to perform task.	• Train operators.
		• Dry run.
		[Reference: National Institute for Occupational Safety and Health (NIOSH) Doc. #80-406]
2. Determine what is in the tank, what process is going on in the tank, and what hazards this can pose.	Explosive gas.	• Obtain work permit signed by safety, maintenance and supervisors.
	Improper oxygen level.	• Test air by qualified person.
	Chemical exposure — Gas, dust, vapor: irritant toxic Liquid: irritant toxic corrosive Solid: irritant corrosive	• Ventilate to 19-21% oxygen and less than 25% LEL of any flammable gas. Steaming inside of tank, flushing and draining, then ventilating, as previously described, may be required.
		• Provide appropriate respiratory equipment — SCBA of air line respirator.
		• Provide protective clothing for head, eyes, body and feet.
		• Provide parachute harness and lifeline.
		[Reference: OSHA standards 1910.106, 1926.100. 1926.21(b)(6); NIOSH Doc. #80-406]
		• Tanks should be cleaned from outside, if possible.
3. Set up equipment	Hoses, cord, equipment — tripping hazards	• Arrange hoses, cords, lines and equipment in orderly fashion, with room to maneuver safely.
	Electrical — voltage too high, exposed conductors.	• Use ground-fault circuit interrupter.
	Motors not locked out and tagged.	• Lockout and tag mixing motor, if present.

Figure 5-1.

When designing a new program or preparing to instruct an employee in an unfamiliar procedure or system, a job analysis can be developed by examining engineering data on new equipment or the MSDSs on unfamiliar substances. The content of the specific federal or state OSHA standards applicable to a business can also provide direction in developing training content. Another option is to conduct a job hazard analysis (See Figure 5-1). This is a procedure for studying and recording each step of a job, identifying existing or potential hazards, and determining the best way to perform the job in order to reduce or eliminate the risk. Information obtained from a job hazard analysis can be used as the content for the training activity.

If an employee's learning needs can be met by revising an existing training program rather than developing a new one, or if the employee already has

SAMPLE JOB HAZARD ANALYSIS
CLEANING INSIDE SURFACE OF CHEMICAL TANK — TOP MANHOLE ENTRY
(Continued)

STEP	HAZARD	NEW PROCEDURE OR PROTECTION
4. Install ladder in tank.	Ladder slipping.	• Secure to manhole top or rigid structure.
5. Prepare to enter tank.	Gas or liquid in tank.	• Empty tank through existing piping. • Review emergency procedures. • Open tank. • Check of job site by industrial hygienist or safety professional. • Install blanks in flanges in piping to tank. (Isolate tank.) • Test atmosphere in tank by qualifed person (long probe).
6. Place equipment at tank-entry position.	Trip or fall.	• Use mechanical-handling equipment. • Provide guardrails around work positions at tank top.
7. Enter tank.	Ladder — tripping hazard. Exposure to hazardous atmosphere.	• Provide personal protective equipment for conditions found. [Reference: NIOSH Doc. #80-406; OSHA CFR 1910.34] • Provide outside helper to watch, instruct and guide operator entering tank, with capability to lift operator from tank in emergency.
8. Cleaning tank.	Reaction of chemicals, causing mist or expulsion of air contaminant.	• Provide protective clothing and equipment for all operators and helpers. • Provide lighting for tank (Class I, Div. 1). • Provide exhaust ventilation. • Provide air supply to interior of tank. • Frequent monitoring of air in tank. • Replace operator or provide rest periods. • Provide means of communication to get help, if needed. • Provide two-man standby for any emergency.
9. Cleaning up.	Handling of equipment, causing injury.	• Dry run. • Use material-handling equipment.

Figure 5-1. (continued)

some knowledge of the process or system to be used, appropriate training content can be developed though such means as:

1. Using company accident and injury records to identify how accidents occur and what can be done to prevent them from recurring.
2. Requesting employees to provide, in writing and in their own words, descriptions of their jobs. These should include the tasks performed and the tools, materials, and equipment used.
3. Observing employees at the work site as they perform tasks, asking about the work, and recording their answers.
4. Examining similar training programs offered by other companies in the same industry, or obtaining suggestions from organizations such as the NSC, BLS, OSHA-approved state programs, OSHA full-service area offices, OSHA-funded state consultation programs, or the OSHA Office of Training and Education.

The employees themselves can provide valuable information on the training they need. Safety and health hazards can be identified through employees' responses to such questions as whether anything about their jobs frightens them, if they have had any near-miss incidents, if they feel they are taking risks, or if they believe that their jobs involve hazardous operations or substances.

Research has shown that current training programs on occupational safety and health for employees focus largely on physical or safety hazards at the work site and not on health hazards. Studies show that as much as 90 percent of the content of such training programs is related to physical hazards and only 10 percent to health hazards and what can be done about them. Thus, employers should ensure that the information and training programs they develop adequately address the health risks of a job.

Once the kind of training that is needed has been determined, it is equally important to determine what kind of training is not needed. Employees should be made aware of all the steps involved in a task or procedure, but training should focus on those steps on which improved performance is needed. This avoids unnecessary training and enables the training to be tailored to the needs of employees.

Identifying Goals and Objectives

Once the employer's training needs have been identified, employers can prepare objectives. Instructional objectives, if clearly stated, will tell employers what they want their employees to do, do better, or stop doing.

Instructional objectives do not necessarily have to be written, but in order for the training to be as successful as possible, clear, measurable objectives should be thought out before the training begins. For an objective to be effective, it should identify as precisely as possible what the individual will be able to do, to demonstrate that he or she has learned, or to demonstrate that the objective has been achieved. Instructional objectives should also describe the conditions under which the individual will demonstrate competence and define what constitutes "preferred" and "acceptable" performance.

Using specific, action-oriented language, the instructional objectives should describe the preferred practice or skill and its observable behavior. For example, rather than stating "The employee will understand how to use a respirator," it would be better to say "The employee will be able to describe how the respirator with which he has been provided works, how it is to be correctly fitted, and when it should be used." Objectives are most effective when worded in sufficient detail that other qualified persons can recognize when the desired behavior (performance) is exhibited.

Developing Learning Activities

Once employers have stated precisely what the objective(s) for the training program are, learning activities can be identified and described. Learning activities enable employees to demonstrate that they have acquired the desired skills and knowledge. To ensure that employees transfer the skills or knowledge from the learning activity to the job, the learning situation should simulate the actual job as closely as possible. Thus, employers may want to arrange the objectives and activities in a sequence which corresponds to the order in which the tasks are performed on the job if a specific process is to be learned. For instance, if an employee must learn the beginning processes of using a machine, the sequence might be:

1. To check that the power source is connected;
2. To ensure that the safety device(s) are in place and are operative;
3. To know when and how to throw the POWER-ON switch; and so on.

A few factors will help to determine the type of learning activity to be incorporated into the training. One aspect is the training resources available to the employer. Can a group training program that uses an outside trainer and film be organized, or should the employer personally train the employees on a one-on-one basis? Another factor is the kind of skills or knowledge to be learned. Is the learning oriented to physical skills (such as the use of special tools or manual materials handling) or to mental processes and attitudes? Such factors will influence the type of learning activity designed by employers. The training activity can be group oriented, with lectures, role play, and demonstrations, for example, or designed for the individual, as with self-paced instruction. The determination of methods and materials for the learning activity can be as varied as the employer's imagination and available resources will allow. The employer may want to use charts, diagrams, manuals, slides, films, videotapes, viewgraphs (overhead transparencies), audiotapes, or simply a blackboard and chalk, or any combination of these and other instructional aids. Whatever the method of instruction, the learning activities should be developed in such a way that the employees can clearly demonstrate that they have acquired the desired skills or knowledge.

Conducting the Training

With the completion of the steps outlined above, the employer is ready to begin conducting the training. To the extent possible, the training should be presented so that its organization and meaning are clear to the employees. To do so, employers or supervisors should:

1. Provide overviews of the material to be learned.
2. Relate the new information or skills to the employee's goals, interests, or experience wherever possible.
3. Reinforce what the employees learned by summarizing the program's objectives and the key points of the information covered.

These steps will assist employers in presenting the training in a clear, unambiguous manner. The above information has been summarized as follows:

1. Tell the employees what you're going to tell them.
2. Telling (and, where appropriate, showing them though demonstration).
3. Telling the employees what you've told them or demonstrated for them.

In addition to organizing the content, employers should develop the structure and format for the training. The content developed for the program, the nature of the workplace or other training site, and the resources available for training will help employers determine for themselves the frequency of training activities, the length of the sessions, the instructional techniques, and the individual(s) best qualified to present the information or demonstrate the skills. Figure 5-2 is an example of a lesson plan format used by some employers to identify what is to be taught. This plan is thought through long before scheduling the training session.

In order to be motivated to pay attention and learn the material and skills that the employer or supervisor is presenting, employees must be convinced of their importance and relevance. Among the ways of developing such motivation are:

1. Explaining the goals and objectives of the instruction.
2. Relating the training to the interests, skills, and experiences of the employees.

3. Outlining the main points to be presented during the training session(s).
4. Pointing out the benefits of the training (e.g., the employee will be better informed, have more skills, and thus be more valuable both on the job and on the labor market; or the employee, by applying the skills and knowledge learned, will be able to work at reduced risk).

An effective training program allows employees to participate in the training process and to practice their skills or knowledge. This will help to ensure that they are learning the material and permit correction if necessary. Employees can become involved in the training process by participating in discussions, asking questions, contributing their knowledge and expertise, and learning through hands-on experiences and role-playing exercises.

Assessing Program Effectiveness

To make sure that the training program is accomplishing its goals, assessment can be valuable. One critical component of training should be a method of measuring its effectiveness. A plan for assessing the training session(s), written or otherwise, should be developed when the course's objectives and content are developed. It should not be delayed until the training has been completed. Evaluation will help employers and supervisors determine the amount of learning achieved and whether the employee's performance has improved on the job. Among the methods of assessing training program effectiveness are:

1. Students' opinions. Questionnaires or informal discussions with employees can help employers determine the relevance and appropriateness of the training program.
2. Supervisors' observations. Supervisors are in a good position to observe an employee's performance both before and after the training and to note improvements or changes.
3. Workplace improvements. The ultimate success of a training program consists of the changes throughout the workplace that result in reduced error, injury, or accident rates, improved quality, and increased productivity.

However it is conducted, an evaluation of training can give employers the information necessary to decide whether or not the employees achieved the desired results and whether the training should be offered again at some future date.

Improving the Training Program

If, after evaluation, it is clear that the training did not give the employees the expected knowledge and skills, it may be necessary to revise the training program or provide periodic retraining. At this point, asking the questions of

	LESSON PLAN	LESSON NUMBER:
COURSE TITLE:		**LESSON LENGTH:**
LESSON TITLE:		
LESSON OBJECTIVES:		
SCOPE OF LESSON:		
INSTRUCTIONAL AIDS:		
METHOD(S) OF INSTRUCTION:		
REFERENCES:		
READING ASSIGNMENTS:		
LESSON PREPARED BY:		
LESSON APPROVED BY:		
DATE LESSON APPROVED:		

Figure 5-2. Sample lesson plan format.

Safety Management Planning

COURSE TITLE:	
LESSON TITLE:	
INSTRUCTORS NOTES	LESSON SCRIPT

Figure 5-2. (continued)

employees and of those who conducted the training may be of help. Among the questions that could be asked are:

1. Were parts of the content already known and, therefore, unnecessary?
2. What material was confusing or distracting?
3. Was anything missing from the program?
4. What did the employees learn, and what did they fail to learn?

It may be necessary to repeat steps in the training process, that is, to return to the first steps and retrace one's way through the training process. As the training program is evaluated, the employer should ask:

1. If a job analysis was conducted, was it accurate?
2. Was any critical feature of the job overlooked?
3. Were the important gaps in knowledge and skill filled?
4. Was material already known by the employees intentionally omitted?
5. Were the instructional objectives presented clearly and concretely?
6. Did the objectives state the level of acceptable performance that was expected of employees?
7. Did the learning activity simulate the actual job?
8. Was the learning activity appropriate for the knowledge and skills required on the job?
9. When the training was presented, was the organization of the material and its meaning made clear?
10. Were the employees motivated to learn?
11. Were the employees allowed to participate actively in the training process?
12. Was the employer's evaluation of the program thorough?

Establishing a Training Policy

A policy is a statement, either expressed or implied, of principles and their consequent rules of action, set up by executive leadership as guides and constraints for the organization's thought and action. The primary purpose of a policy is to enable the manager to relate the organization's work to its objectives.

Each organization should have a written policy with regard to the information and instruction that each employee is to receive during pre- and postplacement with the organization. The policy should spell out who, when, and under what conditions employees should receive orientation and initial, refresher, and remedial training. If employees are to receive special instruction when new equipment, materials, or processes are introduced, this should be spelled out in the policy; it should not be left to chance.

Likewise, if employees are to receive additional information and instruction following selected accidents or other incidents, this, too, should be spelled out. Some organizations require refresher training for employees who have been off work for 3 or more weeks for any reason (e.g., illness, vacation, training). Thus, this practice should be identified in the organization's training policy. Questions which may be helpful when formulating or revising a training policy include the following:

1. Is there a system in place for the identification of training needs?
2. Is the responsibility for training properly allocated?
3. Does the training policy cover all levels of employees, from new trainees to senior managers?
4. Should all operations involving known levels of risk be analyzed for their training requirements?
5. Is the requirement for periodic refresher or remedial training identified?
6. Is there a mechanism for assessing the training program to be used in improving the program? Have those who are to provide the feedback been identified, and have they been taught what to look for?

Establishing a Budget

A budget is a plan with personnel and dollar resources indicated for each line item (project or activity). A training plan and budget are no different in format from any plan and budget. A plan addresses:

1. Method (what is to be done and how).
2. Scheduling (when it is to be done and where).
3. Assignment of responsibility (by whom it is to be done).

Court decisions have held that an employer is just as responsible to employees for personal injuries caused by a defective system of using machinery as for injuries caused by a defect in the machinery itself. Thus, an employer should ensure that the employees receive the information and instruction required to alert them to workplace hazards and what to do about them when they are trained in job skills. Such information and instruction should be totally integrated into regular job training. Employees should not be considered qualified or competent to perform any operation until they have demonstrated the skills and knowledge required to perform the specific task(s) for which they have been hired. Thus, the plan and budget for training, including the occupational safety and health components, should be developed and submitted for approval at the time other plans and budgets are submitted.

As an integral part of the company's operation, training can be justified only if the benefits outweigh the costs; in other words, profits with training should be greater than profits without training if the training is to be considered worthwhile. This can be achieved by (1) decreasing training costs by designing, developing, implementing, and evaluating training programs more effectively or by (2) increasing profits through greater output (and reduced loss due to injuries and illnesses) by better-trained employees.

The fixed costs to be considered when developing a training budget include those following:

1. Salary of the training staff.
2. Rental rates for the training facility(ies).
3. Depreciation of instructional aids and other equipment.
4. Heat, light, and power for equipment used during demonstrations.
5. Phone.
6. Purchase of books, subscriptions to professional journals.
7. Administrative services and support costs provided by the company.

The variable costs to be considered when developing a training budget include the following. Variable costs, unlike fixed costs, may rise or fall in relation to the volume of training activity. Variable costs are based on forecasts of the type and amount of training projected and the numbers of trainees likely to be involved.

1. Wages of the trainees.
2. Fees for external courses (i.e., courses conducted away from the organization's training facility).
3. Fees for consultants and guest instructors.
4. Handout material for students (e.g., books, pamphlets, notebooks, paper, pencils, reproductions of selected materials).
5. Rental of special equipment.
6. Traveling expenses, accommodations, and meals when training is conducted off-site.
7. Wages for replacements during the time employees are in training.
8. Downtime and loss of production as a result of skilled workers being in a nonproductive assignment during training.

From the above, it is obvious that the training budget cannot be developed without a thorough knowledge of what the organization plans to do in terms of new product lines, increased production, relocation of selected activities, hiring plans, subcontracting for selected products and operations, and the like. Training, as a support activity, must be able to support fully the goals and objectives of the organization.

Matching Training to Employees

While all employees are entitled to know as much as possible about the safety and health hazards to which they are or may be exposed, and although employers should attempt to provide all relevant information and instruction to all employees, the resources for such an effort frequently are not, or are not believed to be, available. Thus, employers are often faced with the problem of deciding who has the greatest need for information and instruction.

One way to make this decision is to identify employee groups that are at higher risk. The nature of the work indicates that such groups should receive priority for information and instruction on occupational safety and health risks.

A second method of identifying employee groups at high risk is to examine the incidence of accidents and injuries, both within the company and within the industry. If employees in certain occupational categories have higher accident and injury rates than others, training may be one way to reduce these rates. In addition, a thorough accident investigation can identify not only specific employees who could benefit from training but also companywide training needs.

A method used by OSHA in designing its training and education effort for employers and employees and their representatives is shown in Figures 5-3 and 5-4. A branching model for the selection of workers for training in occupational safety and health is shown in Figure 5-3, and a worker selection matrix, utilizing the logic of the branching model, is shown in Figure 5-4. The selection of workers for occupational safety and health information, education, and training is based on the selection of the Standard Industrial

Model for the Selection of Workers for Occupational Safety and Health Training

Worker Population	Training Requirement Covered By Regulations	Employed In High Risk Industry (SIC #) Operation	Substantial Number Of Workers At Risk	Recognized Appreciable Reduction In Injury/Illness At Risk	Effective Training Available From Other Sources	Employed In Special Emphasis Industry/Occupation Hazard Category	Training Priority
							I
							II
							III
							IV Out Of Consideration

Figure 5-3. Model for the selection of workers for occupational safety and health training.

Illustrative Worker Selection Matrix for Training in Occupational Safety and Health

Industrial Classification (SIC #)	Covered By Regulations	High Risk	Relative Number of Workers At Risk	Is Training An Effective Countermeasure	Is There An Effective Employer (OSHA) Training Program	Is There Other Effective Training Available	Is Industry Involved With Special Emphasis Program	Training Priority
0001	YES	YES	HIGH	YES	NO	NO	YES	I
0006	YES	NO	LOW	NO	NO	YES	NO	II
0009	NO	NO	LOW	NO	NO	YES	NO	III
								IV Out Of Consideration

Figure 5-4. Illustrative worker selection matrix for training in occupational safety and health.

Classification (SIC) as the operative element; is limited to problems where training is an appropriate countermeasure (i.e., areas where a lack of skill and knowledge is a causative factor in the accident and the accident population is found to be deficient in skill and knowledge); and is limited to areas where existing training has not been effective. Emphasis is on high levels of safety and health hazards, a substantial number of workers at risk, small businesses, and special emphasis/special hazards occupational safety and health programs. This method may be of use to larger corporations when mapping out their training and education strategies and tactics.

Research has identified the following variables as being related to a disproportionate share of employee injuries and illnesses at the work site:

1. The age of the employee (younger employees have higher incidence rates than older ones).
2. The length of time on the job (new employees have higher incidence rates).
3. The size of the firm (in general, medium-sized firms have higher incidence rates than smaller or larger firms);
4. The type of work performed (incidence and severity rates vary significantly by SIC Code).
5. The use of hazardous substances (by SIC Code).

These variables should be considered when identifying employee groups for training in occupational safety and health.

In summary, information is readily available to help employers identify which employees should receive safety and health information and instruction, and who should receive it before others. Employers can request assistance in obtaining information by contacting such organizations as OSHA area offices, the BLS, OSHA-approved state programs, state on-site consultation programs, the OSHA Office of Training and Education, or local safety councils.

Training Employees at Risk

Determining the content of training programs for employee populations at higher risk is similar to determining what any employee needs to know, but more emphasis is placed on the requirements of the job and the possibility of injury or illness. One useful tool for determining training content from job requirements is the job safety analysis, described earlier. This procedure examines each step of a job, identifies existing or potential hazards, and determines the best way to perform the job in order to reduce or eliminate the hazards. Its key elements are (1) job description; (2) job location; (3) key job steps (preferably in the order in which they are performed); (4) tools, machines, and materials used; (5) actual and potential safety and health hazards associated with these key job steps; and (6) safe and healthy practices, clothing, and equipment required for each job step.

MSDSs can also provide information for training employees in the safe use of materials. These data sheets, developed by chemical manufacturers and importers, among others, contain lists of manufacturing or construction materials and describe the ingredients of a product, its hazards, protective equipment to be used, safe handling procedures, and emergency first aid responses. The information contained in these sheets can assist employers to identify employees in need of training (i.e., workers who handle substances described in the sheets) and to train employees in the safe use of the substances. MSDSs are generally available from suppliers, manufacturers of the substance, or large employers who use the substance on a regular basis; they can also be developed by employers or trade associations. They are particularly useful for those employers who are developing training programs in chemical use, as required by OSHA's Hazard Communication Standard (29CFR1910.1200).

APPENDIX: RESOURCES TO AID THE TRAINER

In addition to the federal government, several organizations publish and distribute occupational safety and health literature and produce training programs. However, there is no single clearinghouse for identifying, describing, assembling, and distributing occupational safety and health literature and training programs, even those produced by departments and agencies of the federal government. As a result, despite the existence of a huge amount of material, most of it factual and well written, much of it remains unknown to employers and employees alike.

OSHA and NIOSH Information Sources

Both OSHA and NIOSH are doing a creditable job of alerting the public to the materials that they produce and have had produced for them under contract. NIOSH and OSHA have mailed millions of copies of publications on health and safety work practices and guides. They also answer tens of thousands of requests from people in all kinds of jobs with all kinds of problems. Some requests are for NIOSH and OSHA publications; others are for answers to specific technical information questions. Because much of NIOSH's and OSHA's information is contained in publications, a question from an employer or employee can often be answered by referring to such publications. NIOSH and OSHA send catalogs of the publications that are available or provide specific information from their libraries or computers to those making inquiries.

Inquiries on NIOSH publications and training programs should be sent to NIOSH, 4676 Columbia Parkway, Cincinnati, OH 45226. Inquiries to OSHA for such information should be sent to the Office of Information, OSHA, U.S. Department of Labor, Washington, DC 20210.

Other Government Information Sources

There are three major information sources within the federal government from which the employer can obtain information on training program materials in occupation safety and health. These sources, along with their mailing addresses, are:

1. U.S. Government Printing Office, Washington, DC 20402. (Ask for a listing of publications in occupational safety and health.)
2. National Audiovisual Center, National Archives and Records Service, Washington, DC 20409. (Ask for a copy of the publication *Media for Safety and Health.*)
3. National Technical Information Service, U.S. Department of Commerce, Springfield, VA 22151. (Ask for a listing of publications and other reports on occupational safety and health.)

Nongovernment Sources of Information

Many quasi-public and private (both profit and nonprofit) organizations are involved in the production and dissemination of occupational safety and health training and educational materials. Some of the more prominent ones, along with the material and services they provide, are:

1. National Safety Council, 444 North Michigan Avenue, Chicago, IL 60611. The NSC is a public service organization which gathers, analyzes, and distributes safety information. Membership in the NSC provides organizations with a central source for developing safety programs for on-the-job, home, road, and recreational safety programs. Services and products of the NSC include technical assistance, statistical aid, access to its safety library, periodicals, safety manuals, MSDSs, incentives and awards, films and audiovisual programs, and safety aids and publications. The NSC also has a Safety Training Institute which offer home study courses, on-site industrial training programs, and workshops at the Institute for both supervisory and nonsupervisory personnel.
2. E. I. duPont deNemours and Company Inc., Applied Technology Division, Education Services. Clayton Building, Concord Plaza, Wilmington, DE 19898. DuPont's curriculum of over 25 safety training programs includes NEST (New Employee Safety Training program), STOP (Safety Training Observation Program) for both supervisory and nonsupervisory personnel, and the Safe Practices Series for nonsupervisory personnel, which covers hazards, equipment, tools, and other aspects of plant safety. DuPont also offers over 130 self-instructional vocational training programs covering such areas as carpentry, metal work, and instrumentation. DuPont's education consultation services include on-site visits to identify training needs, providing information packets on DuPont training programs, and tailoring programmed instruction courses to meet specific company needs.
3. International Loss Control Institute, Highway 78, P.O. Box 345, Loganville, GA 30249. The ILCI is an educational facility which provides safety and loss control training program audits and consultant services. It specializes in short conferences and seminars, continuing advisory services to conferees, ongoing loss control research, complete program audits (on-site evaluations and tailor-made programs), and the development of educational products. These educational products include books, training programs for supervisors, safety films and tapes, home study courses, and accident investigation manuals.
4. International Safety Academy, P.O. Box 8527, 1600 Arch Street, 12 Tower, Philadelphia, PA 19101. The Academy, with offices in Philadelphia, Houston, Atlanta, and Mission Viejo, California, offers seminars, safety training resources and films for in-house training programs, and consultations to identify and address specific loss control needs. Its film rental library includes over 800 titles on such topics as safety program management, property protection, transportation safety, occupational safety and health, specialized safety applications, life management skills, and off-the-job safety. Through its consultation program, it

can provide on-site loss control surveys and seminars, basic management review for loss control policies and procedures, safety manual development, and personally tailored seminars and training conferences.

Books

Most current textbooks in the field of occupational safety and health contain one or more chapters on training and education and their role in combating occupational injuries and illnesses. In addition to these sources, the following publications should prove helpful to the employer in developing training and education programs for employees:

> C. Richard Anderson, *OSHA and Accident Control Through Training.* New York: Industrial Press, 1975.
> I. L. Goldstein et al., *Employer and Employee Training and Education in Occupational Behavioral Action Intervention Strategies Training,* Industrial Publications Series BSC-101. Columbia, MD: Westinghouse Behavioral Services Center, 1975.
> Heath, Earl D. and Ted Ferry, *Training for Workplace Safety, Improved Performance Through Training and Education,* Goshen NY: Aloray, 1990.
> *International Correspondence Schools, Effective Training: A Guide for the Company Instructor*, Scranton, PA: International Correspondence Schools, 1968.
> Dugan Laird, *Approaches to Training and Development,* Reading, MA: Addison-Wesley, 1983.
> Robert F. Mager, *Preparing Instructional Objectives,* San Francisco: Fearon, 1962.
> Jack B. ReVelle, *Safety Training Methods,* New York: Wiley, 1980.
> W. Wayne Worick, *Safety Education: Man, His Machines, and His Environment,* Englewood Cliffs, NJ: Prentice-Hall, 1975.

Periodicals

The following periodicals will prove useful to the employer when developing training programs, especially those designed to provide information and instruction on occupational safety and health to the employee.

> *American Industrial Hygiene Association Journal.* Published monthly by the American Industrial Hygiene Association, 475 Wolf Ledges Parkway, Akron, OH 44311-1087.
> *National Safety News.* Published monthly by the National Safety Council, 444 North Michigan Avenue, Chicago, IL 60611.

Occupational Hazards. Published monthly by Penton/IPC Publications, Inc., 1111 Chester Avenue, Cleveland, OH 44114.

Occupational Health and Safety. Published monthly by Medical Publications, Inc., 3700 West Waco Drive, Waco, TX 76710.

Professional Safety. Published monthly by the American Society of Safety Engineers, 1800 E. Oakton Ave., Des Plaines, IL 60018.

Training and Development Journal. Published monthly by the American Society for Training and Development, 600 Maryland Avenue, S.W., Washington, DC 20024.

Training: The Magazine of Human Resources Development. Published monthly by Lakewood Publications, Inc., 731 Hennepin Avenue, Minneapolis, MN 55403.

REFERENCES AND BIBLIOGRAPHY

Blum, John D., "Revealing the Invisible Tort: The Employer's Duty to Warn," *Health Services and Health Hazards: The Employee's Need to Know,* ed. Richard H. Egdahl and Diana Chapman Walsh, New York: Springer-Verlag, 1978, p. 179.

Cohen, Alexander, Michael I. Smith, and W. Kent Anger, "Self Protective Measures Against Workplace Hazards," *Journal of Safety Research,* Fall 1979, p. 129.

Corn, Morton, Guest Editorial in the *Journal of the American Industrial Hygiene Association,* March 1976, p.

Federal Register, July 27, 1984, pp. 30290-30294.

Gilbert, Thomas F., "Taxonomy: A Scientific Approach to Identifying Training Needs," *Management of Personnel Quarterly,* Fall 1967, pp.

National Institute for Occupational Safety and Health, *A Worker's Guide to NIOSH,* Cincinnati, OH: NIOSH, October 1978.

Occupational Safety and Health Administration, *Job Hazard Analysis,* Washington, DC: OSHA, 1981.

Tepper, Lloyd B., "The Right to Know: The Duty to Inform," *Journal of Occupational Medicine,* July 1980, pp. 433-434.

U.S. Department of Commerce, National Technical Information Service, *OSHA Safety and Health Training Guidelines for General Industry (PB 239-310),* Washington, DC,

U.S. Department of Commerce, National Technical Information Service, *OSHA Safety and Health Guidelines for Maritime Operations (PS 239-311),* Washington, DC,

U.S. Department of Commerce, National Technical Information Service, *OSHA Safety and Health Guidelines for Construction (PS 239-312),* Washington, DC,

U.S. Department of Labor, *Occupational Safety and Health Administration, Informing Workers and Employers About Occupational Cancer,* Washington, DC: OSHA, 1978.

Chapter 6

New Installations

When a firm prepares a new facility or modifies an existing one, many elements and countless details must be coordinated to make it ready for safe and healthy use and occupancy. This chapter discusses which parties are usually involved, their primary responsibilities, and when those responsibilities should be carried out in relation to each other. The responsibilities related to safety and health are emphasized, but the information is applicable to all parties involved in readying the installation for use.

Because so many parties have responsibility, it is appropriate that one top management person be responsible for ensuring that all of the elements are considered and taken care of by someone. This is very important, since some elements overlap and others do not join, leaving gaps which will show up only when the installation is functional. Naturally, this is always undesirable from management's point of view, but where safety and health is concerned, even a small gap of this kind can have costly and tragic consequences.

The material in this chapter comes largely from the work of the Department of Energy's System Safety Development Center in the form of procedures to check on the work of contractors and subcontractors. It is applicable to any type of operation or installation, although very complex operations may require more in-depth consideration.

BASIC READINESS

"Basic readiness" refers to the components that are basic to all new or modified installations, not just to the safety and health functions. Our concern centers on three areas:

1. Structures (including services and hardware).
2. Appropriate procedural material (which translates into managerial control systems).
3. Personnel training.

Structures

Structures include associated services and hardware but concentrate primarily on buildings and grounds, since this is the most extensive area or component. It includes parking lots, roadways, yards, walkways, office buildings, laboratories, shops, warehouses, storage facilities, special buildings, and enclosures. The nature of the site must also be considered, including atoll tables (which affect floods and high water), soil settlement (which may involve drainage systems, former refuse dumps, back-filled soil, and so on), ground disturbance (which may involve mines, subterranean excavations, or geological formations), and acidity and sulfate-reducing bacteria (which may corrode or attack cables, pipes, tanks, conduits, etc.). Site considerations also include railroad spurs or hookups, water moorings, maneuvering space, and air corridors.

The senior manager can keep track of the pertinent factors involved in the building and grounds area if he or she considers their functions in terms of:

1. Basic structures and facilities.
2. Internal storage and transport.
3. Heating and cooling systems.
4. Electrical power.
5. Cleaning and waste.
6. Utilities.
7. Fire protection.
8. Security and exclusion from the area.

The new or unguarded installation should also be considered in view of the neighboring community. Sources of noise and vibration from neighboring industries should be reviewed, as well as the effect of the installation's own noise and vibration. The interaction of emissions from adjacent sites may be a factor, since some emissions could combine into toxic or explosive mixtures.

Managerial Control Systems

Structures considerations are tied directly to managerial control systems. Indeed, the managerial control system cannot work well unless all is well with the structures and grounds.

Managerial control systems should be designed to deal with both normal operations and abnormal events. The same managerial control system that applies to normal operations should provide some means of control over abnormal events as well. Both can be facilitated by the use of:

1. Written procedures.
2. Placards, signs, notification systems, alarms.
3. Training programs.
4. Supervisory personnel.

Of course, there is much more to managerial control than these four items, but they are basic to normal and abnormal operations control in the area of safety and health.

Responsibility and Accountability

Clear areas of responsibility and accountability should be established in the three basic areas: structures, managerial control, and personnel. To have the necessary procedures ready,

1. Top management. Overall responsibility rests with top management, which is most active in resolving discrepancies.
2. Occupant-user. Accepting the building or process as ready for use or implementation rests with the occupant-user.
3. Construction engineering. The premises must be built or designed to conform to codes, standards, or regulations, as well as the user's requirements. Only when this is achieved should the premises be turned over to the occupant-user.
4. Landlord. Most often the landlord is also the occupant-user, but sometimes he is simply the middleman who handles the entire construction and design process as a turnkey operation.
5. Quality assurance. The landlord, occupant-user, or constructor will all need a quality assurance system.
6. Safety organization. Safety organization must be part of the process long before the occupant-user is ready to take over the premises. It must be incorporated into the building and construction phase itself. As safety advisor Charles Critchfield points out in speaking of the safety expert's role, "He is the one person who can look at a layout design and point [to] the unsuspected design features which, while not technically unsafe, may tempt people to take shortcuts and place themselves at risk."

Under the typical arrangement, the occupant-user is responsible for several key functions:

1. Establishing basic functional specifications.
2. Reviewing the design against these specifications.
3. Resolving discrepancies via revision, an agreement to fix, or reference to higher management. For the occupant-user to agree to that readiness, responsibility usually falls into six groups:
4. Reviewing and testing the facility as built.
5. Resolving further discrepancies that may appear in the review.

6. Recommending release of the building for use.
7. Accepting the building for occupation.

Where the occupant-user is the corporation itself, these functions frequently fall on top management. Naturally, the manager will work closely with other members of the team, such as the designers and engineers, but ultimately top management bears the final responsibility for all of these steps. In fact, ultimately, management is responsible for ensuring that all areas—structural, management control systems, and personnel—are considered. Here, in brief, are the major responsibilities involved in each of these areas.

Structural Application:
1. Establishing specifications.
2. Constructing or modifying to specifications.
3. Inspecting and testing to specifications.

Management Control Systems Application:
1. Making certain that procedural requirements have been established.
2. Ensuring that the procedures thus created actually conform to the requirements.
3. Reviewing the newly-developed procedures in full.

Personnel Application:
1. Establishing the need for training requirements.
2. Determining the actual training requirements.
3. Reviewing the training procedures for adequacy.

A special word is needed about personnel training applications. in order to prepare personnel for the safe use of a new or modified installation, there are four factors that must be included in training:

1. General facility orientation.
2. Knowledge of written procedures.
3. Familiarity with placards, signs, notifications, and alarm systems.
4. Training of supervisory controls.

The manager must give these matters particular attention, since older employees, well trained and skilled though they may be on existing equipment and operations, may require retraining for both normal operations and abnormal events in the new facility. This may require a complete review and revision of existing training programs, with an eye to the special needs created by the new facility.

SAFETY AND HEALTH RESPONSIBILITIES

Let us now consider specific issues related to safety and health in the new installation. Higher management will probably need a comprehensive overview of safety and health in the new installation rather than a minutely detailed safety analysis. For this reason, we present a broad range of the safety and health areas involved in new installations, including:

Industrial safety
Fire protection
Industrial hygiene
Human factors
Vehicular safety
Emergency Preparedness
Criticality safety
Radiation protection
Environmental protection

While not comprehensive by any means, this list will help to pinpoint areas that the manager should be aware of.

Industrial Safety

Industrial safety begins with conformance to codes, standards, and regulations. Management must consider at least seven major areas in which these apply:

1. Posting requirements. This refers to OSHA forms, floor loading limit signs, notices on the use of personal protective equipment, notices of workers' compensation details, fire and evacuation instructions, warning signs, color coding of hazards, and any other posting or labeling requirements that apply to the specific operation.
2. Working surfaces. This covers trip hazards, guarding of floor openings, spill handling, steps/ladders/stools, guard rails, stairs, sanitation, housekeeping, and slippery areas.
3. Area access and egress. This refers particularly to adequate doorways and walkways, and to ensuring that means of exit are in compliance with both OSHA and fire codes.
4. Personal protective equipment. This covers coordination with industrial hygiene and refers to all required personal protective equipment, including hard hats, respirators, safety shoes, and glasses, as well as special equipment for fire brigades or emergency crews.
5. Material handling. This covers special equipment, operator training, scheduling, storage safeguards, rollover protection, and clearly designated loading areas.

6. Functional operations. This includes such items as power transmission, point of operation, chemicals and vapors, and machinery guarding.
7. Electrical. This considers supply, distribution, appliance load, extensions, interlock guards and warning signs.

Senior management does not need to know the specific details of the items listed, such as where to obtain protective glasses or the exact requirements for rollover protection. This list is simply for reference so that management can verify that all of these items have been taken into account by the responsible parties.

Fire Protection

Even if the facility is located next door to a city fire station, proper fire protection must be given serious consideration. The size of the problem, and the possible need for a professional fire safety consultant, can be seen just from the brief overview that follows. All of these areas must be considered:

1. Watchman service. The need for such service, personnel required, training needed, and frequency of service should be reviewed.
2. Building. The structure, occupancy, and lease requirements will all affect fire protection needs.
   a. Structure. Fire areas, wind, construction materials used, seismic problems, exposures, exits and access, electrical systems, interior finish, and alterations should be taken into account. The risk of a fire spreading may be drastically affected by alterations made after the building is completed.
   b. Occupancy. This refers to common hazards involved in operations, such as computers, radioactive materials, heating and cooling systems, cutting and welding operations, and use of flammable or explosive materials. Included are insulating or protecting hot parts, providing safe storage of flammable items, and use of approved equipment.
   c. Lease requirements. The contract or lease agreement should be carefully checked for specifications, funding restrictions, owner's responsibility, government restrictions, and inspection or insurance requirements.
3. Loss potential evaluation. This requires a careful review of all operations, equipment, and materials to determine the potential amount of loss in the event of a fire. Chemical plants, operations involving explosives, and oil and petrochemical operations require special study, probably by expert outside consultants. The Dow Fire and Explosion Hazard Index can be of great use as a management reference.

164 SAFETY AND HEALTH MANAGEMENT

As you can see, much of the information needed may be beyond the scope of the senior manager and may require the services of a special consultant. Considerable assistance can be obtained, however, from local government agencies, fire protection associations, and the organization's insurance carrier. In addition, there may be in-house expertise or adequate professional advice from outside.

Industrial Hygiene

Most senior managers do not have the expertise required for proper consideration of the industrial hygiene area, nor, for the most part, do they need it. However, the area is important enough, in terms of applicable codes and regulations, worker health and safety, to need a high degree of executive awareness. These are the main areas to be considered:

1. Lighting. Adequate lighting in terms of foot-candles, glare, shadows, special work needs, and an adequate power supply must all be considered.
2. Ventilation. Special attention should be paid to obtaining an adequate fresh air supply and an effective exhaust system, especially where operations may involve dust, fumes, or vapors.
3. Noise. Volume in decibels, special frequencies, source control, source enclosure, and personal protective equipment should all be considered.
4. Sanitation. This includes restroom adequacy, eating areas, trash removal, insect and rodent control, and general housekeeping.
5. Heat. This refers to wet and dry bulb temperatures, air movement, control of excess heat, and adequacy of heat when needed.
6. Toxic agents. This critical area includes the types of controls and monitors needed, measurement of toxic concentrations, personal protective equipment, employee exposure records, handling of overexposure emergencies, and disposal of toxic wastes.

Human Factors

We are coming to realize that many accidents, injuries, and operating errors are based on human limitations. These influences are gradually being documented, and there is evidence to suggest that they have a strong impact on almost every type of business and industrial operation. Safety considerations of human factors may involve reviewing to evaluate specific human errors in ordinary operations and devising programs or techniques to reduce these to a minimum or acceptable level. Among the influences involved in human error factors are the types of machinery in use and the types of operator errors possible; the selection and proper training of employees; medical follow-up to determine employee fitness; and studies of the strength, eyesight, reach, and coordination required in specific operations. In considering a new or modified installation, the entire installation must be reviewed to determine the points at which human error is most likely and what measures can be taken to reduce it (automatic vs. manual controls, safety shutoff systems, etc.).

Vehicular Safety

All vehicle and traffic safety programs involve certain codes, standards, and regulations, as well as ordinary vehicle inspection and maintenance programs. The following should also be included in any vehicle and traffic safety program:

1. Loading and unloading. A key consideration is having clearly designated loading and unloading zones for both passengers and materials. Any layout should minimize cross-traffic and traffic disturbance. All explosives, flammables, and toxic materials should be stored at an ample distance from these loading zones.
2. Bus stops. These should be adequate to handle normal passenger loads and banned from other use. All bus zones should be clearly marked, with designated passenger waiting areas. Bus stops should be located to minimize interference with ordinary traffic flow, and schedules should be prominently posted.
3. Company and personnel cars. Determine the need for parking areas or the adequacy of existing parking. Spaces should be clearly marked, all regulations posted, and employees informed about them. Crossflow should be kept to a minimum. Where possible, car parks should be located at the periphery of the premises and well away from hazardous operations or storage areas.
4. Traffic flow patterns. This may require considerable advance study and planning before marking spaces, aisles, and crosswalks. Procedures for review and improvement should be clearly established. Pedestrian walkways and aisles should be clearly marked, and traffic should be banned or limited in production areas. Fire and emergency lanes should be clearly marked, preferably with the advice and approval of local fire authorities.
5. Coordination with city, county, and police agencies. It is best to put one person in charge of this, to act as liaison and coordinator and to take responsibility for working with local traffic safety agencies, as well as with adjacent users and property owners.
6. Bicycles and motorbikes. The increasing popularity of bicycles, motorbikes, mopeds, and motorcycles has made it more imperative than ever that adequate facilities be set aside. Racks and bays should be located in a special area that is clearly marked and banned from use by regular

auto traffic and parking. Regulations should be clearly posted and disseminated to all employees.
7. Snow removal plan. A regular plan should be prepared well in advance of need, covering such things as special equipment or facility access, notices, times of clearance (preferably when employees have not yet arrived at the facility), disposal or dumping areas, sanding of walkways, steps and turns, and overall responsibility for carrying the plan out.

Criticality Safety

For those who must store, use, or ship items termed critical, special expertise is definitely needed, and should either be developed in-house or sought from outside sources. Of particular importance are the three areas of shipments, intra-area transfer, and storage.

A full discussion of critical items is beyond the scope of this book. We assume that companies that habitually use critical items are already well aware of the safety and health hazards and controls. The senior manager should be concerned with this only if the company is actively involved in the use of critical items.

Radiation Protection

The scope of the radiation problem generally requires expert knowledge. Because of its critical nature, radiation is thoroughly covered by governmental codes, regulations, and standards. A good deal of radiation safety and health consists simply of conforming to the regulations, since in many areas these are so strongly mandated that protection is more a matter of compliance then executive decision.

The senior manager, however, should be well aware of existing radiation hazards and their control, as they affect personnel, property, and the public. This area does not leave room for vague or sloppy operations and definitely calls for professional expertise. However, in working with the radiation expert, the senior manager may want to go over the following checklist in order to make sure that he or she understands both the hazards and the control measures involved:

1. Direct rdiation
2. Contamination
3. Air activity
4. Effluents
5. Internal transfer and storage
6. Shipping and receiving
7. Work process facilities and equipment
8. External sources and causes
9. Monitoring and communications systems
10. Building utilities and services

11. Satellite processes, facilities, and equipment
12. Emergency plans
13. Medical programs

Environmental Protection

Most of the existing regulations and standards for environmental protection deal primarily with the conservation of resources and the control of environmental impacts. The safety and health aspects are not always immediately apparent. However, recent developments—particularly newsworthy incidents or accidents—have made both the public and the government more aware of the possible safety and health consequences of environmental damage. Business itself in increasingly aware of the cost of wasting irreplacable or valuable resources or of rectifying environmental damage. Moreover, there is an increasing recognition of the need to present an environment-conscious image to the public.

Effective management must take the position that prevention or environmental protection is far less costly in the long run, avoiding costly waste before it becomes critical, or preventing environmental damage that may later cost the company millions to repair and millions more in lost goodwill. The effects on worker health and safety involve both the loss of valuable human resources and, often, millions in medical and compensation costs.

The senior manager, while he or she may not be an environmental expert, must ensure that expert guidance is available to consider:

1. Conservation of resources
    a. Energy. How much electrical, fossil fuel, labor, and other energy resources are needed, and how much is needlessly used.
    b. Land. Consider the use and location, local flora and fauna, recreational and other uses.
    c. Minerals. Available supply in the form of raw materials, and the possibilities of recycling and reclaiming must all be considered.
    d. Solar power. Blockage or reflection of light must be taken into account, as well as glare.
2. Impact and interactions
    a. Sociological commitments. Congestion, aesthetics, convenience, public benefits, and public opinion must all be considered.
    b. Material effluents and contaminants. This considers the control of liquid, solid, airborne, and other contaminants and effluents, as well as their probable effects on the environment, both immediate and long-term.
    c. Energy emissions. This involves not only radiation, thermal, electromagnetic, and mechanical energy, but also the problems of noise, heat and vibration.

Here, too, expertise is required, either from in-house specialists or outside consultants. The ramifications may involve engineering, public relations, and legal staffs as well, and the senior manager may need to work closely with the management of these departments.

SAFETY AND HEALTH SPECIFICS

Maintaining a safe and healthy working environment is an integral part of good management. It is not enough to add a paragraph on accident prevention to the company's policy manual, expecting that it will have some effect on accident and injury rates, or confine safety efforts to periodic walk-arounds or inspections. A good management system will enable the organization to identify clearly specific safety and health problems and develop measures to control them. In planning a new installation, management should determine that safety and health considerations have been included from the beginning in of design, construction, and maintenance. Naturally, an effective system will be far from simple. The manager can simplify his or her own task by breaking the safety and health picture down into its essential components. The simplest and most comprehensive breakdown should include:

1. Materials
2. People
3. Practices
4. Environment
5. Protective systems

Materials

Determine if the materials normally used in operations are, or should be, chemically analyzed. If this analysis reveals that some materials are hazardous, management should ensure that specific controls are included as part of the proper job procedures. Determine if the storage, handling, and disposal of waste materials are safe for workers. Investigate whether the disposal of environmentally hazardous waste is in accordance with regulations. Check to see if the approved standard procedures are used for corrosive, toxic, or radioactive materials. What safeguards are included in the system to make sure that these procedures are followed? What files, manuals, and records should be developed and kept to cover their properties and use?

People

Check to see if work station design, display, and location are ergonomically acceptable. Determine if a job analysis for safe operation has been undertaken when needed. Review training schemes for current adequacy, and check to see that they are periodically reviewed and updated to meet changing or anticipated requirements. Determine if the personnel department is aware of safety and health considerations in its hiring criteria for specific tasks (i.e., eyesight, strength, coordination) and update training needs. Confirm that training is being provided for new workers, including safe working practices, and for those being transferred to new jobs or new facilities. Find out if high-risk areas need to be examined by appropriate experts to identify and eliminate hazards and threats to safety and health. Are there adequate safety and health guidelines to cover visitors as well as workers? Finally, is there an effective communication and feedback system between workers and management?

Environment

Is proper attention being paid to pollution in order to maintain acceptable standards? Is there a plan of action in the event of a disaster or accident? Have critical areas been mapped for noise, radiation, dispersion of gases, and so on? Are the conditions of working surfaces such as floors, scaffolds, and benches safe? Have contaminants which affect health in the workplace been identified and proper handling procedures established? Is there an adequate system of medical monitoring, where needed, to deal with potential health hazards? What existing environmental conditions may have an impact on the facility (noise, emissions, and vibration for neighboring installations)?

Protective Systems

First, determine if safety procedures and equipment exist for energy and power sources and if they provide adequate protection. Do not overlook moving plant equipment, as well as radioactive, corrosive, or toxic materials which may also require protective equipment and procedures. Do the protective measures meet standards or mandatory regulations? Are consultants available when needed? Are employees fully aware of their responsibilities for using protective equipment and procedures? Does the company provide proper protective equipment, and are there standards for its selection, use, maintenance, and replacement? What controls exist to ensure that workers use such equipment when required? Is adequate training provided in the use of protective equipment and practices?

THE ROLE OF CHANGE

New or modified installations necessarily bring about changes in operation. If the operation is completely new, everything is changed for the personnel involved. If it is a modification of an existing operation, it may require learning new ways of doing things, giving up old habits and previous experience. What was formerly a good procedure may now be a potentially disastrous way of doing things: what was formerly a simple operation may present new and unexpected complications. All these considerations may require extensive retraining, and especially a readjustment of attitudes and expectations.

The role of change in accidents has become more visible in the last 15 years. The change or revision record on engineering drawings has been used as a source of clues in uncovering the causes of accidents. A thorough review method is needed if unwanted side effects are to be controlled.

Ironically, changing over to new, supposedly safer facilities is usually accompanied by an increased accident rate. A beneficial change in one part of an operation can have a detrimental affect on a subunit. This may call for systematic review of all subunits or subsystems, beginning with the lowest-level subunit and moving up in the system, looking for unwanted changes or hazards. The problems in each subunit should be resolved before moving higher in the system.

Although we are talking primarily about new installations, the process of change is present in all installations, even the most stable and entrenched. Change can be of any type—new employees, transfers, new work orders, new work projects, or a new supervisor.

Where new processes or installations are involved, there is an additional danger in the use of older procedures or equipment. A good example of this is the case of a 3,000-psi compressor. In an effort to make the operation safer, management decided to replace the compressor with a new, lower-psi device. The new compressor seemed to be the answer to the problem, and several years went by without any notable difficulties. Then, however, production demands stepped up, and the new compressor was not adequate to handle them. Why not bring the old compressor back into use as well? The solution seemed very simple. However, the old compressor had been sitting idle for several years, and nothing had been done to overhaul or maintain it. Workers had either gotten used to operating the new compressor or else had had no experience with the old one. Far from solving the production problem, the old compressor caused two nearly fatal accidents before management was alerted to the danger.

The lesson here should be clear: before putting old equipment back into use or installing it in a new facility, management must carefully consider its impact. Will it need retooling or maintenance? Will it require retraining or reacquainting workers with it? Will it place unanticipated strain on any other part of the system (a machine that cycles at a much faster or slower speed than the rest of the system, for instance)? Can it be thoroughly integrated into the new system? What adaptations should be made before putting the machine into use? This is not to say that older equipment cannot and should not be integrated into new processes or installations where possible. It is merely to point out that extra care should be exercised in doing so, and the safety and health impacts thoroughly considered. Changing back to the old is as potentially hazardous as changing over to the new.

Sensitivity to change (either impending or probable) is a key component in the makeup of an effective manager. In fact, it is sometimes thought of as a nearly intuitive capacity. Fortunately, even the manager without this intuitive flair can make use of techniques which will help identify, anticipate, and direct change. Most managers, in fact, now recognize that one of management's key tasks is to guide the company through necessary changes and to help plan and orchestrate anticipated changes. A good deal of current safety and health literature, in fact, now stresses the need for management to be sensitive to change and the fact that change has become an everpresent factor in safety and health programs.

Systems that are in the process of changing generate new or additional hazards. The manager must be alert for changes in every part of the operation: processes, transfers, new equipment, startups, new standards or legislation, materials, even changes in the management system itself. The alert manager, in fact, must learn to differentiate important changes from trivial or unimportant ones and concentrate on dealing with the significant ones. This sensitivity to change and the ability to distinguish between important and trivial changes is the mark of an exceptional manager. It is important in any area of business, but in the area of safety and health it can be crucial.

Changes have two major characteristics that the manager should be aware of. The first one is simple to grasp: once change has started, it tends to keep on going. It does not usually stop by itself. This is especially important to keep in mind when the change is an undesirable one. The manager must clearly understand that, without intervention, the undesirable change will continue and may even grow more pronounced. An escalating accident rate, for instance, is an undesirable change: it will not decrease if the manager does not become concerned with it. The appearance of an undesirable change in any part of the operation is a signal that some action is required.

The second characteristic is a chain-reaction effect: changes interact, compounding the effect. That is, a small change tends to cause other changes. These changes, in turn, cause still other changes. The result is an ever-increasing chain reaction causing more and more change. The manager must anticipate change and establish countermeasures before the change occurs. Countermeasures are most effective if applied early in the change sequence. Later, a countermeasure that is addressed to the original change may have little or no effect on the subsidiary changes already produced, and each one of these changes may still have to be dealt with.

EMERGENCY PREPAREDNESS

Although Chapters 17 and 18 discuss this subject in detail, emergency preparedness cannot be left out in discussing new installations or modified activities. Codes, standards, and regulations chart most of our course in emergency preparedness, but at least six other areas call for our attention: (1) plans and procedures, (2) training, and testing, and practice, (3) communications, (4) equipment, (5) utilities, and (6) response and amelioration. Let us consider these in a bit more detail:

1. Plans and procedures. There should be plans for operational problems, nuclear/radiological situations, civil disorders, sabotage, natural disasters, and national emergencies.
2. Training, testing, and practice. This involves employee procedures, notifications, monitors/sensors, alarms, response systems (equipment for fire, emergencies, etc.), personnel for response teams (fire, medical and health), and external assistance. Advance training, testing of equipment and procedures, and emergency drills are vital ingredients.
3. Communications
   a. Equipment. This includes signs, directions, monitors/sensors, PA or broadcast systems, alarms, two-way communication devices, and emergency manuals or directions.
   b. Personnel. Establish who is in charge of response teams; what the individual responsibilities are; and who will provide emergency information, deal with the public, and provide emergency care for employees and members of the public. Exact personnel needs should be projected in advance, and all emergency team members should have thorough training, refresher courses, and updating at regular intervals.
4. Equipment. Provide for providing the maximum emergency and personnel protection. Two broad areas are involved:
   a. Fixed equipment. This includes a rescue station, blankets, stretchers, ropes, poles, emergency operating stations and center, shelters, supplies, fire fighting and first aid equipment, eyewashes, showers, treatment rooms, and cots.
   b. Mobile equipment. This is vital for the fire, medical, rescue, health, and other teams. It includes emergency vehicles such as fire trucks, ambulances, portable generators, portable communications systems, and other mobile equipment.
5. Utilities. Automatic and manual cutoffs should be provided, and backup and redundant systems installed and periodically tested. Emergency power supplies, water, and telephone or other communications systems should not be overlooked.
6. Response and amelioration. The intent here is to limit further damage and injury. A fire department, emergency medical services, emergency cadre and rescue team, and emergency security teams may all be involved. Preventing the spread of fire, fumes, or water, shutting down

equipment, evacuating buildings, hosing down dangerous spills, and thoroughly checking all equipment and facilities before they are put back into use are integral parts of this operation. Subsidiary damage can also be prevented by prompt follow-up action. This includes such things as posting extra security guards in a damage area (to prevent scavenging or break-ins), boarding up or blocking broken windows or entrances, and closing or controlling public access to damaged areas.

THE MANAGER'S SAFETY AND HEALTH ROLE

Not every manager is faced with the design and construction of new installations, but almost every one is faced with periodic modifications of existing facilities, introduction of new processes or equipment, and basic changes in operations. New and modified installations present unique hazards but also an opportunity to prevent future accidents. The manager should consider it an unparalleled opportunity to do things right before the building, modification, or process is put into use.

The checklist that follows consists of questions designed to help the manager who is faced with such new or modified operations. By reviewing these questions, the senior manager can detect whether all the proper actions have been taken in the conceptual, design, and preparatory stages. This is far cheaper and more beneficial than costly corrective action and retrofit later.

THE MANAGER'S SAFETY AND HEALTH CHECKLIST

1. Do I set specifications for the operability and maintainability of equipment? Is a human factors review provided to ensure operability, maintainability, and reliability?
2. Do I require maintenance programs consistent with the system's functional operation?
3. Do I require adequate system inspection?
4. Do I insist on safe work arrangements of hardware and personnel?
5. Do I provide for consideration of the total work and the external environment?
6. Do I provide adequate operational specifications?
   a. Equipment testing and qualification programs?
   b. Supervisory training and qualifications?
   c. Testing and evaluation procedures?
   d. Personnel selection and evaluation?
   e. Personnel training and qualification?
7. Do I specify monitoring points and establish monitoring programs?

8. Do I provide emergency plans and provisions?
9. Do I establish disposal plans for operational residue and waste? For final disposition of materials and hardware?
10. Do I provide for independent quality and safety reviews?

REFERENCES AND BIBLIOGRAPHY

Nertney, R. J., *Occupancy Use Readiness Manual,* Springfield, VA: National Technical Information Service (NTIS), 1976.

Nertney, R. J., and Bullock, M. G., *Human Factors in Design,* Springfield, VA: NTIS, 1976.

Chapter 7

Maintenance Programs and Safety

The operation of any system is no safer or more efficient than the system's maintenance program. In addition, even an excellent system can develop problems over time. Continual maintenance is needed to keep it operating safely, as it was designed to do. Establishing and evaluating a maintenance program is a necessary part of safety and health management. This chapter provides criteria for evaluating the maintenance program from both a functional and a managerial viewpoint, and demonstrates how safety and health are involved at every step.

The *Maintenance Managers Guide* used by the Department of Energy specifies 18 areas of consideration, shown in Figure 7-1. In relating these areas to safety, we must consider three primary factors:

1. The maintenance activity and its role in safety and health.
2. The operational safety and the system being maintained.
3. Providing safety for the maintenance system itself.

PROGRAM COVERAGE

The ideal maintenance program should consider all of the elements shown in Figure 7-1 but still allow the manager to concentrate on those that apply most directly to the company's particular operations. Complex or sophisticated operations will require all the elements mentioned. For small or relatively straightforward operations, some elements may not be needed. For the most part, however, all areas will be applicable to some degree, if only as checkoff items, to make certain that coverage of the maintenance program is complete.

For each of these checklist items, the manager must include the proper allocation of personnel in terms of their skills and the organization of their functions. Further, he or she must ascertain that financial and material resources have been made available to support the program.

Program Policies

The manager should ensure that policies are adequate in terms of safety. The general maintenance policy statement should be supported by all relevant safety-related criteria. Such support policies should spell out the criteria for

Figure 7-1. Elements of a maintenance program.

maintainability in the design and modification process; it should also cover situations in which the process is running or is shut down. Adequate maintenance criteria and instructions should be included for new and modified equipment, and there should be provisions to cover repair vs. replacement of parts and equipment. The policy should clearly spell out the relationship between the maintenance program and other staff and operating functions, including quality assurance and reliability. This description should specifically mention and include safety considerations.

The policy should also include adequate statements about job and process safety in the maintenance program and should assign safety responsibilities for the same. Along with this, there should be a statement of authority relating to safety. What happens when a maintenance person detects an unsafe operation that does not involve his or her direct maintenance? Suppose that it is a matter of imminent danger? Suppose that it is a matter of interfacing operations when no one on hand is clearly in control of the situation? Definite lines of authority should be established and spelled out in this policy to cover such instances.

Work Order System

The manager should determine if the maintenance program is backed up by an adequate work order system. Note that we refer to it as a *system*. It includes numerous elements that must work together to be successful—

hence, a system. In other words, the work order itself should be designed to include several items needed to make the entire system work. From the safety viewpoint, the work order should have a space that describes the reasons for performing the work. It should include target start and completion times, with the actual times entered later. It should state who has authorized the work, who will coordinate or approve it, and any special functional requirements needed.

In order for the work order system to function properly, the work order should also state exactly which items and processes will be affected by the work. It should state who will provide the required job, safety, and process supervision. This last ensures quality control of the maintenance activity.

The work order should also provide a checklist of appropriate categories and priorities, such as "preventive maintenance," "repair," "installation," "must complete before startup," "urgent attention required," and so on. All have their proper place in maintenance scheduling and performance. Preoperational testing should be included in these categories as part of the maintenance program.

Obviously, the complete work order system we have described embraces far more than is usually included for routine maintenance. The manager will have to consider the company's specific systems and decide what part of the work order system can be simplified without endangering the system or the operation.

Materials Control

"Materials control" refers to every stage of materials handling, from selection through storage and use and, finally, disposal of waste. Planning controls must provide for safety considerations in the selection of materials, required protective equipment for handling the materials, and the possible use of alternative materials to reduce excessive hazards. This calls for coordination between safety and purchasing personnel, and possibly even approval of certain purchases for safety.

Storage of materials is a growing problem, since the growth of new, hazardous combinations of materials far exceeds our knowledge of them. New combinations are being formed at the rate of about 2000 per year. This calls for careful issue, use, and return of such materials, including restrictions on authorized personnel. There must be careful control of sensitive and hazardous materials in temporary as well as permanent storage. Testing must be done under carefully controlled conditions in proper areas. Where hazardous materials are involved, maintenance activities may be more complex and hazardous than the actual functional process. They may require more planning and preparation skill than a regular operation might take. Precautions must be taken in case something goes wrong. It is vital to provide for spills, unsealed containers, waste, mixed materials, and surplus materials.

Equipment Records

Adequate equipment records are an integral part of maintenance safety program coverage. There should be properly maintained records on equipment startup, failures, misuse, routine maintenance, and maintenance failures. This information should be fed back to the design, engineering, and accident programs. There should also be records of calibration, rebuilding, modification, and use. In addition, descriptive files should be maintained, including operator-service manuals, drawings, parts lists, specification sheets, warrantees, and quality assurance requirements.

Preventive Maintenance

Defining and meeting safety-related reliability requirements and specifications are essential to preventive maintenance. Preventive maintenance records may reveal maintenance needs in areas not already part of the program and can help determine what is required; this may mean including other areas under preventive maintenance. Similarly, records may show that preventive maintenance is being overused and that some other solution would be more cost effective. There needs to be an adequate balance between preventive maintenance and breakdown (on-call) maintenance.

Preventive maintenance is usually justified economically for continuous and/or highly automated processes, where the cost of lost production is high and the failure of one piece of equipment can bring the entire process to a halt.

The efficient organization of preventive maintenance is a complex and skilled activity requiring systematic record keeping, inspection procedures, job analysis, and work programming and planning. The safety staff cannot be expert in all of these activities, but they should know the subjects well enough to be able to make substantial contributions to the program from a safety and health viewpoint. Such safety knowledge requires thorough understanding of the hazards that may arise:

1. Through failure to carry out preventive maintenance at the right time.
2. In the course of maintenance operations themselves.
3. Through faulty maintenance, which leaves the equipment in a more dangerous condition than it was before.

Mandated Preventive Maintenance

On safety grounds, preventive maintenance is required where failure might be dangerous (e.g., in boiler and pressure vessels). In these cases, standards and regulations will be the guide, such as the OSHA regulations or the English Factories Act. The manager may have little choice but to comply with required inspection and preventive maintenance programs.

Evaluation and inspection of items required by law form a base on which a more thorough system of preventive maintenance can be built. Without reference to a particular act or country, we can generally expect mandated preventive maintenance to be required in cases that involve:

1. Factory cleanliness and housekeeping.
2. Hoists and lifts.
3. Chains, ropes, and lifting tackle.
4. Breathing apparatus.
5. Safety belts and ropes.
6. Reviving apparatus.
7. Steam boilers.
8. Steam receivers and containers.
9. Water-sealed gas holders.
10. Air receivers.
11. Fire warning devices.
12. Electricity insulating stands.
13. Ventilation for metal grinding.
14. Power presses.
15. Radiation sources.

Certain operations, such as those involving chemicals or health facilities, will find special requirements mandated for their operation. Further information on the details of this maintenance may be found in the Merritt Company's *OSHA Reference Manual.*

On-Call Maintenance

"On-call" maintenance, sometimes called "unplanned," "as-needed," or "breakdown" maintenance, refers to the response of maintenance personnel to a breakdown in equipment and processes. On-call maintenance cannot be predicted or planned in advance, as preventive maintenance can, and it may be difficult to project the time and costs involved. However, provisions must be made to meet the requirement and avoid costly shutdowns or equipment damage. How to make such provisions without knowing in advance how much on-call maintenance will be required can be a major problem for management, especially at budget time. However, there are some general guides that may be quite helpful.

On-call maintenance is often the cheapest and most flexible type of maintenance to perform. It means waiting until a fault appears or an item breaks down and then deciding whether to repair it, scrap it, sell it, or replace it with a similar item. This policy assumes that replacement or repair facilities are readily available, so that spare parts need not be carried, and that there will be minimum delays in starting the process or equipment again. It can be extremely useful when this actually happens, as when the operation involves the use of only standard, and readily available equipment or tools.

Equipment which operates as an individual unit, separate from an operational process, and whose breakdown does not constitute a safety hazard is often best run until it breaks down. It is then overhauled or replaced by reconditioned equipment. Even where on-call maintenance is preferred to systematic preventive maintenance, however, it is important to establish that the item is not operating in a dangerous condition before it finally breaks down. Some sort of inspection or preventive maintenance may be in order to ensure that the limits of safe operation are not exceeded.

The manager will quickly see that some operations can be safely and efficiently dealt with on an on-call maintenance basis, while others will require a more elaborate and costly system of preventive maintenance.

Job Planning

Safe operation will require an adequate communication system between equipment users and maintenance groups. Feedback from the users is sometimes the only way of determining maintenance needs or effectiveness. Where possible, maintenance supervisors can conduct work site visits and inspections to gather information directly, but inspection by users, technical or engineering personnel, and safety persons may also be needed to provide information to the maintenance staff.

Performing a maintenance task calls for adequate engineering and technical backup before, during, and after the job is completed. Such assistance may involve in-house personnel, vendor representatives, or consultants. It may also require keeping and checking a safety literature file, especially where new hardware or materials are involved, or where special repair and operation manuals, manufacturer maintenance recommendations, parts lists, or maintenance checklists may be needed.

Maintenance tasks should be properly sequenced to ensure that adequate parts, materials, and equipment have been ordered and are on hand before work begins. There should be an approval system for starting a job and an inspection system for signing it off as completed.

Backing up a maintenance program calls for a lot of planning to ensure that:

1. Proper tools and equipment are on hand.
2. Materials and parts are available.
3. Tools, parts, materials, and equipment have been properly tested and inspected.
4. Job safety analysis is used. Chapter 8 presents complete details on performing and recording the job safety analysis.

MAINTENANCE PROGRAMS AND SAFETY   181

5. Safe work permits are used.
6. Human stress factors are considered.
7. Personnel are operationally ready to carry out the job (i.e., they have received and understand the work order, have been properly briefed on the task, have all necessary parts and equipment, and are prepared to perform the work required).

There must be a clear understanding between the user supervisor and the maintenance supervisor so that each one's responsibilities are clearly understood. Clear-cut procedures are especially needed to close out the finished work. Who, for instance, will ultimately accept the product or job? Inspection by both the maintenance supervisor and the user supervisor may be desirable before putting a piece of equipment back into use; either or both may have the authority to reject a particular piece of maintenance work or to require rework or additional work. Definite guidelines should be written for these cases, clearly spelling out where the final responsibility rests.

Work Sampling

It is necessary to have a system to sample the maintenance task for effectiveness and completeness. This may be achieved by inspections by the maintenance supervisor or maintenance staff, the safety staff, quality assurance staff, operations-project groups, or technical and engineering specialists. Records may be kept and periodically checked to see, for instance, how often the same maintenance task must be redone, whether second or third maintenance calls were needed before a problem was solved, or how much time elapsed between a request for maintenance and its completion. Parts of this task may be included in the regular safety audits; other parts may require special attention.

Maintenance Scheduling

Proper scheduling of work must take certain factors into account. In maintenance scheduling, the task will be easier if certain things are kept in mind:

1. Jobs should be reviewed for priority before scheduling, and the schedule itself should be continually reviewed in light of new priorities.
2. There should be a definite system for handling backlogs when they occur.
3. Active work should be scheduled so that several maintenance groups or crews will not need the same equipment, facilities, or tools at the same time; so that sequential processes are not repeatedly interrupted; and so on.

4. There should be job-to-job, job-to-process, intergroup, and worker-to-supervisor coordination.
 5. Formal job inspection, closeout, cleanup, and record keeping should be completed before removing a job from the schedule.

Backlog Control

Even with the most adequate staff and the best planning, backlogs do happen from time to time. They can occur at any stage of the maintenance cycle. The backlog may develop with work orders received, but not yet planned or scheduled. It may develop where jobs are already planned and scheduled, but must wait until equipment, materials, or personnel are available. Again, the maintenance tasks may be quickly and efficiently performed, but a backlog can develop in the inspection and approval processes, so that completed jobs have to be held. (This can be crucial in operations in which equipment must be inspected before being returned to use.)

Management should have definite plans and procedures for dealing with backlogs. They should cover:

 1. Extra personnel resources, whether in-house or outside.
 2. Procedures for securing extra equipment, materials, or tools.
 3. Redistribution of the workload where possible.
 4. Streamlining for "bare minimum" requirements (i.e., determining which steps can be telescoped, postponed, or eliminated to get the job done, and which cannot).

When backlogs occur too often or too regularly, it is a sign that management should review the entire maintenance system to see if procedures are too cumbersome, personnel insufficient, equipment or materials not readily available, and so on. Pinpointing the cause of the backlog is essential, as it may point to other operational or management failings (e.g., the purchase of substandard equipment that requires too frequent repair, inadequate training of workers in the proper care and handling of equipment, and so forth).

Performance Measurement and Improvement

Like any part of operations, maintenance should be periodically evaluated, measured against standards, and, where needed, improved to meet those standards. The specific parts of the maintenance system that should be evaluated and reviewed are:

 1. Work order system.
 2. Materials control.
 3. Equipment records.
 4. Preventive maintenance.

MAINTENANCE PROGRAMS AND SAFETY

5. Job planning.
6. Work sampling.
7. Scheduling.
8. Backlog control.
9. Productivity control.
10. Control vs. actual performance.
11. Prioritization.
12. Engineering standards.
13. Equipment failure analyses.

Control vs. Actual Performance

Maintenance program control functions comprise the responsibilities of management/supervision/staff. They include scheduling, prioritizing, inspecting, budgeting, allocating personnel, developing maintenance system standards, record keeping, and personnel training. These should all be subjected to independent evaluation and review, separately from the actual performance of maintenance functions.

The "actual performance" of maintenance function refers to actual field performance in accomplishing the maintenance task. It depends largely on having workers with proper skill and training, establishing a proper personnel selection process, and providing the proper equipment, tools, and materials to complete the job. None of these performance functions can be properly performed, however, if the control function has not provided properly for it. There is a close interrelationship between the effectiveness of the control function and the effectiveness of the actual performance, and that interrelationship must be maintained. Personnel on either level can become overly concerned with their immediate tasks and forget their obligation to the other.

Prioritization

This refers to setting priorities for elements of the maintenance program, giving the higher priority to those likely to prevent, ameliorate, or solve the most problems if they are worked on first.

Engineered Time Standards

Adequate engineered time standards can have an important direct impact on the total effectiveness of the safety and health program in terms of change, corner cutting, job phasing, job coordination, work deferments, and the like. Time standards should be applied to productivity measurement, cost estimates, and schedule maintenance. Both the safety program and the maintenance program should involve the use of engineered time standards; management, however, must be alert to see that safety and health are not sacrificed for the sake of keeping on schedule.

Cost Identification and Control

In order to allocate resources properly, adequate cost identification and cost controls must exist. In identifying and controlling maintenance costs, management must be sure to include safety-related work performed by the maintenance organization. In many instances, it will be the maintenance people who actually perform tasks for the safety organization or offer support resources for certain safety-related tasks. These should be treated as an integral part of the maintenance organization's work, and proper allocations should be made for them.

Repair vs. Replacement Criteria

Eventually, even in the newest facility, management is faced with the task of deciding when it is no longer reasonable to repair or maintain a piece of equipment and when to replace it. Criteria for making this decision should be established well ahead of time, instead of allowing the exigencies of the situation to make the decision by default. Many items not directly associated with maintenance can influence the decision and should be taken into account in establishing the criteria. These should include the time and cost required to repair the equipment, the age of the equipment, productivity of old vs. new equipment, write-off status, expected productivity activity, and hazards presented, among other things.

Rather than establishing an arbitrary set of criteria, the senior manager should seek input from several key departments or functions in helping to define reasonable standards. These include:

1. *Accounting,* which can provide depreciation and write-off figures, as well as cost factors.
2. *Engineering,* which may be aware of new developments in equipment or process design, as well as safety engineering factors.
3. *Production,* particularly where the decision may be influenced by expected or scheduled production activity.
4. *Safety and health,* particularly where management should know about expected or pending regulatory changes affecting equipment or processes.
5. *Quality control,* which may be able to furnish relevant data on waste, rework, substandard work, and so on.
6. *Maintenance,* which may have the most accurate information on when a piece of equipment is no longer repairable or operable.

Whoever ultimately makes the decision on whether to repair or replace should be familiar with these criteria and should be able to justify his or her decision in light of them.

Replacement and sale of worn items are particularly common among operators of car fleets or similar equipment. Only basic maintenance such as lubrication, servicing, adjustment, and the like is carried out. The item is sold when its operating efficiency begins to fall, or when the item becomes dangerous or is likely to require lengthy or costly overhaul. The automobile is probably the most common example, but the principle can be applied to many items.

Scrapping is also common, particularly with inexpensive, easily replaceable items such as drill bits, saw blades, ropes, slings, lights, and so on. If the item is liable to become dangerous before it breaks or wears out, this should be recognized and a system or inspection of preventive maintenance employed so that the item is scrapped or replaced before it becomes hazardous.

Personnel Training

Maintenance is so important to safety that maintenance personnel should be selected with special care for their experience, ability, alertness, and training in accident prevention. They are faced with a complex and ever-changing set of hazards, unlike the regular patterns of activity usual for most of the work force. They should receive special training in the use of ladders, slings, ropes, and protective equipment, as well as with the ordinary tools of their trade. Because of their movement around the installation and their exposure to a wide variety of situations, they should be well trained in first aid, rescue, and fire-fighting activities as well.

Their training should encompass recognition and control of hazards, and they should be notified of the introduction of any new hazard or material into the areas they cover. For particularly hazardous jobs, the safety staff should have a representative to arrange and agree on safe procedures, including a detailed outline of the various steps to be taken. When maintenance management makes a change in equipment, the user of the equipment should be fully informed, and management should see to it that this is done.

MAINTENANCE AND ACCIDENTS

It is easy to see that many accidents are caused by a lack of adequate maintenance. It is just as important for safety and health-minded management to recognize that accidents can also be caused by the maintenance activity itself. The first sort of accident can be prevented by providing adequate maintenance. But what kind of preventive actions can be taken to control the maintenance-related accident? Answering the following questions can help the manager pinpoint areas that need special attention to prevent accidents.

1. In connection with the maintenance activity itself:
    a. Are the maintenance requirements spelled out in plain, easy-to-understand language?
    b. Are maintenance standards thoroughly analyzed and spelled out?

c. Has on-call maintenance caused by equipment failures been analyzed for cause, such as reasonable wear and tear, misuse, poor maintenance, substandard equipment, or fatigue?
   d. Is the maintenance schedule adequate?
   e. Is the maintainability of the equipment adequate? The design may preclude adequate or convenient maintenance.
   f. Have proper training and equipment been provided for maintenance personnel?
2. In executing the maintenance plan:
   a. Is the maintenance task adequately performed? If not, what area seems to be at fault—task assignment, task safety analysis, pretask briefing, job allocation, failure to use proper safety controls, or personnel deficiencies?
   b. Are logs, labels, or color coding used at the point of operation to show the maintenance status of equipment?
   c. Are periodic preventive maintenance inspections properly performed? Are jobs properly inspected after completion?
   d. Are lockout procedures applicable, and are they properly followed where needed? Are work permits required and properly used where needed?
   e. Are work orders properly filled out and followed through, including special instructions and authorizations?
   f. Are adequate time and personnel allocated for the maintenance task?

BUILDING MAINTENANCE

So far, we have spoken of maintenance mainly as it pertains to equipment and processes. But buildings also require maintenance to prevent dangerous conditions from developing. Checking and correcting faults on a regular basis is less costly and safer than waiting for damage to reach serious proportions. The maintenance program should provide for regular inspections, preventive maintenance schedules, and prompt attention to building hazards.

The safety specialist is usually not an expert on building maintenance, but he or she should be sufficiently knowledgeable to recognize hazards and distinguish between good and bad practices. Special attention and knowledge are required to look after:

1. Foundations, footings, and column bases.
2. Structural members.
3. Walls.
4. Ceilings and floors.
5. Roofs, gutters, and roof-mounted items.
6. Tanks, tank towers, and stacks.

7. Loading platforms.
8. Underground maintenance.

There are particular items to check on in each of these areas, and some of them (such as checking structural members) call for special expertise. Sometimes the engineering staff will prove to be a valuable resource in checking the soundness and safety of the building, but often outside expertise is needed as well. Professional associations, insurance companies, and governmental agencies can also provide helpful guidelines or inspection services. The manager should not underestimate the importance of establishing reasonable standards and effective working systems to maintain building safety.

Maintenance of Lighting Systems

One area of building maintenance that requires special mention is lighting systems. Generally, lighting maintenance is on-call, and safety problems arise in installing, replacing, handling, and disposing of lighting accessories.

However, lighting can drastically affect the safety of operations and even the health of workers. Maintenance personnel should be thoroughly briefed where lighting is critical and where special care must be given to the intensity or color of lighting, to matching existing lamps, to glare, or to reflection problems that may develop if replacement procedures are careless or incorrect. Special follow-up may be required in these situations, with final inspection being made by a member of the safety or engineering staff to ensure that the criteria have been met and the proper type of lighting has been used.

Pipe Maintenance

In speaking of pipe maintenance, we are dealing with two different systems: the ordinary piping systems that are an integral part of the building structure itself (such as water and sprinkler pipes, conduits, heating pipes, and so on) and those systems that may be part of special process equipment. Both require careful attention, and maintenance must be meticulously planned and performed.

The most important safety and health factor in connection with piping is the proper identification of piping systems. The stories of mishaps in this area are legendary, possibly because the consequences can be so extreme. Opening the wrong valve or connecting the wrong pipes can cause serious accidents. Such mistakes as connecting steam pipes to hand rails or linking toilet drains to oil-water separators can create dangerous situations. They can also have fatal results, as recently happened in a hospital where a worker accidentally connected operating room lines designed to carry oxygen and anesthetic gases. Two patients died before the error was discovered. Make no mistake, this is a very dangerous area of operation which calls for every possible type of precaution. No shortcuts should be allowed.

Equally important in this area is the safety and health of the maintenance worker, particularly when the work involves hot pipes or those used to carry flammable, toxic, or hazardous substances. Adequate protective equipment, ranging from gloves to breathing apparatus and full body suits, should be provided and their use insisted upon.

Lubrication

This should be considered part of the regular preventive maintenance program, though it may also be required in the course of on-call maintenance as well.

A full survey of lubrication requirements should be made and the information made a matter of record. These records should be kept with the regular machine records. All lubrication operations should be carried on without danger from moving parts. Special features and fixtures may be required to permit this, or equipment may need to be shut down while lubrication is performed. Above all, lubrication should be carried out on a carefully programmed schedule. An adequate supply of lubricating guns, cans, oils, and greases should be provided to ensure easy compliance with the schedule. Care should be taken to ensure that recommended lubricating compounds are available, including special lubricants, and that manufacturers' recommendations are complied with.

Follow-up is an essential part of lubricating operations from a safety and health standpoint. Excess and spills should be promptly cleaned up, as they pose a particular hazard on walking surfaces or on machine parts, where they may come into contact with the worker (such as handles, dials, or controls). Maintenance personnel should also be well briefed and alert to report any parts or processes they note during lubrication, which may involve other maintenance or safety problems (i.e., excessive wear on machine parts, missing guards, evidence of misuse, and so on). In this way, even the routine job of lubrication can be incorporated into the larger plan of periodic inspection and maintenance.

CONCLUSION

Safety and health considerations in the maintenance program are often too little appreciated but have tremendous importance. They can be compared to the essential maintenance activities required on a ship or aircraft, where proper functioning can be a life-or-death matter. They are just as important to the well-being of an industrial firm. The catastrophic possibilities are equally great, though possibly not as spectacular or as publicly visible.

This chapter has discussed the most essential areas of safety and maintenance interactions. For the chief engineering or maintenance supervisor, the treatment is undoubtedly oversimplified. But for the senior manager, this overview should provide an adequate checklist for review of the organization's maintenance program from a safety and health standpoint. In addition, it should point up how closely related the two are: safety and health programs are no more effective than the maintenance programs that help support them.

REFERENCES AND BIBLIOGRAPHY

Buys, J. R., *Events and Causal Factors Charting,* Idaho Falls, ID, System Safety Development Center, 1978.

Grimaldi, John V., *Safety Management,* 4th edition, Homewood, IL, Irwin, 1984.

Johnson, W. G., *MORT Safety Assurances Systems,* New York, Marcel Dekker, 1980.

King, Ralph W., *Industrial Hazards and Safety Handbook,* London, Newens-Butterworth, 1979.

Nertney, R. J., *Safety Considerations in Evaluation of Maintenance Programs,* Idaho Falls, ID, System Safety Development Center, 1978.

Raheja, Dev, "System Safety: Concepts and Applications," *Hazard Prevention,* March-April 1980, pp. 21-22.

Chapter 8

Job Process Control and Safety

INTRODUCTION

All too often, when things go wrong, we try to treat the symptoms rather than the causes. This is especially true when a workplace problem arises or a safety and health event has adverse effects. When production falls off in a department, we put pressure on the supervisor. If the accident level increases, we may lecture the workers. If a worker makes a costly error with a machine, we simply dock his or her pay. We overlook the fact that there may be good reasons for the drop in production or an obvious cause for the accidents or the worker's error. It is the symptom which catches our eye.

Sound management, however, seeks to discover the cause of the trouble, trace it to its source, and correct it. This is the only way to keep the event from being repeated. Tackle the cause, not the symptom. As we suggested earlier, the system should be examined for failures which allowed the event to happen. The drop in production may be due to material shortages or critical equipment breakdowns. Too many accidents point to sloppy housekeeping, poor supervision, or an inadequate training program. Perhaps the worker's error was due to poor machine design, inadequate instruction, or poor maintenance. All can indicate a management deficiency at some level of operation, such as an inefficient purchasing system, an inadequate training program, or failure to establish proper job procedures.

It is up to the principals of an organization to review and approve its long-range safety and health objectives. In a small corporation, this responsibility may rest on one or two persons who determine the broad policies. More commonly, however, a committee representing top management and the heads of functional departments approves the plans and programs, which include safety and health planning and programming. These top management people, more than anyone else in the organization, should be deeply concerned with discovering the causes of workplace problems and developing sound systems and processes to recognize, diagnose, and correct them.

THE MAIN ELEMENTS OF JOB PROCESS CONTROL

Safety and health considerations are not extraneous to day-to-day operations. They are an integral part of every process and procedure in the organization. Establishing job processes and procedures that effectively incorporate safety and health considerations is management's responsibility.

The astute manager knows that this requires more than a properly worded policy statement in the corporate manual or a formal list of objectives. The area of job processes and procedures deals not just with abstractions but with actual workers, the things they do, and how they go about doing them. We might call this the "real-life" arena of safety and health, where every facet of the planned safety program will be proved out.

Basically, job process and procedure control involves three main elements:

1. People
2. Hardware
3. Procedures

These three factors cannot be separated; they are interdependent, each affecting the others and being affected by them. In order to judge the adequacy or effectiveness of job processes and procedures, the manager must ask three basic questions:

1. Do the people match the hardware?
2. Does the hardware match the procedures?
3. Do the procedures match the people?

The interdependent relationships thus spelled out are illustrated in Figure 8-1. These represent the main element of job process control.

PEOPLE REQUIREMENTS

A large hydraulic press is a powerful, dangerous implement involving several types of energy in its operation. It cannot be made 100 percent safe. Its safe operation does not depend solely on the hardware. It depends on three factors: the hardware, the use of safe procedures, and the proper actions of personnel. In addition, it requires a sound evaluation of the entire process to do the job safely and properly. We will begin by considering the personnel requirements, and what management should do to ensure the best and safest use of people.

Providing adequate control of personnel involved in the job process system has four basic requirements. As shown in Figure 8-2, these are:

1. Establishing an appropriate employee selection process.
2. Providing adequate training.
3. Establishing an adequate testing and qualification program.
4. Evaluating the current status of each employee.

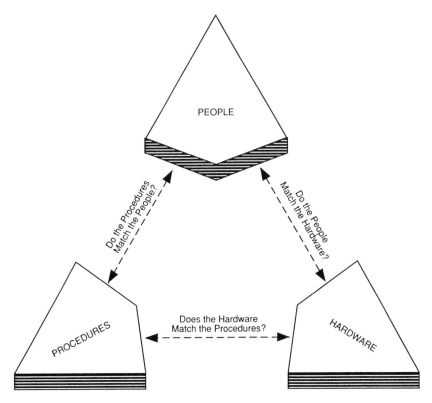

Figure 8-1. The main elements of job process control.

Figure 8-2. Personnel requirements involved in job process control.

Selection

The steps in establishing a well-rounded personnel selection and placement program consist of:

1. Adopting a clear selection policy and making it known to all supervisory personnel.
2. Determining the actual job requirements, with special regard for the psychological and physiological demands.
3. Knowing the applicant's interests, experience, abilities, and attitudes.
4. Determining the applicant's physical capabilities through medical examination.
5. Placing the applicant by matching job requirements with his or her abilities, interests, motivation, background, and physical capabilities.
6. Acquainting the worker with the job, using a planned procedure.
7. Regularly following up the placement.

It is necessary to establish criteria for the selection of personnel before determining the methods of selection. The safety-related job requirements must be defined in order to determine the physical and mental capabilities needed for the job. Outside assistance, such as that of someone who knows the job and/or safety requirements extremely well, may be needed, and this person should be consulted before establishing criteria. Employees should be selected for their ability to match the work environment and interface with the procedural and managerial controls. For example, a hearing-impaired person should not be placed in a situation where keen hearing is necessary to carry out the job safely. However, if the task calls for good eyesight and no particular use of hearing, we could conclude that a hearing-impaired person with 20-20 vision matches the requirements very well.

Keep in mind that we are discussing only the safety and health aspects of personnel selection (though some of these steps apply to all personnel selection criteria).

There are a number of standard tests which can help us select the right personnel. These are designed to evaluate such traits as:

Emotional stability
Self-sufficiency
Extroversion
Dominance
Confidence
Sociability

These evaluations may be very useful when job safety depends heavily on such things as ability to work as part of a team, ability to perform calmly under pressure, ability to make quick decisions, and so on.

Similar tests for occupational interests and aptitudes build a useful profile of the person's real innate abilities. In some cases, aptitude tests to measure mental ability, spatial perception, and motor ability may be more pertinent to the job involved.

Tests of motor ability are generally psychomotor tests, that is, combined tests of sensory and muscular abilities, which may be grouped as follows:

1. Control precision
2. Multilimb coordination
3. Response orientation
4. Reaction time
5. Speed of arm movement
6. Aiming
7. Rate of control
8. Manual dexterity
9. Arm-hand steadiness
10. Wrist-finger speed

In certain types of work, safe operation depends much more heavily on these skills than it does on intelligence, personality traits, or interest.

The point should be obvious, however: merely administering tests is not the complete answer. Workers should be tested for specific traits or aptitudes needed on the job. Elaborate and costly testing of manual dexterity will do little good in selecting personnel for jobs that rest heavily on an ability to get along with others, general intelligence, or emotional stability. Also, skilled personnel may be needed to administer and grade the tests, which can represent a rather costly investment.

In this area, the manager is best advised to consult a professional, who can give expert advice on available testing instruments and practices and, where needed, design testing programs tailored to the organization's needs.

Training

At the time of selection, personnel should be screened to see if they are already suitably trained and skilled or if they will need a training program. The training program should be adequate in scope, depth, and detail, as described in Chapter 5.

It is not our aim to repeat all the points made in Chapter 5. However, applying personnel training needs to the question of job process control raises three questions:

1. Is the training kept current with changes in hardware, procedures, and management control? How is this done?
2. Does the personnel department know which risks it can accept and which ones it must refer to a higher management level?
3. Is training directly related to the procedures to be followed on the job?

Making training work, even when it is well designed and presented, calls for the active support of supervisors and managers. Supervisory and management personnel should be well trained in safety and health, should understand the principles of risk management, should be able to identify hazards, and should know which risks are acceptable and which are not. Mere training of the new employee is not adequate unless thorough training of supervisors and management has been accomplished first.

Testing and Qualification

A testing and qualification process that is regularly monitored and verified will ensure that proper training is being conducted. This might call for simulators, tests, or examinations. A system of verifying the person's current qualification status is needed, particularly where workers are hired as already trained or skilled. Testing and qualifying should be performed not just at the time of hiring, but on a continuous and periodic basis as well. If task criteria are changed, new testing and qualification are needed for all employees. The new examination should focus on the new criteria, and procedures should be established for retraining and requalifying the existing work force. The manager should make sure that this testing is conducted on the work site, wherever possible, so that individuals can demonstrate their skills under realistic conditions.

Evaluation

It is necessary to establish training processes to evaluate not only the current status of personnel training but also the support functions of supervisors, staff, and management in the training process. For example, supervisors must understand their responsibilities in order to assess their employees. They should know what performance indicators might reveal problems. In some cases, the program may involve supervisory backup for medical programs and employee assistance programs for drug and alcohol problems. How effectively does the supervisor give on-the-job training? How alert is he or she to the existence of training deficiencies? How closely does the supervisor follow established and approved training materials or procedures?

HARDWARE REQUIREMENTS

The next basic of job process control is hardware or, more properly, hardware and plant (sometimes referred to as "equipment").

The elements of the total hardware and plant picture are complex. They may consist of hundreds or even thousands of separate machines, tools, support systems, and buildings. For the sake of simplicity, it is easiest if the manager views hardware in terms of its "life cycle." That is, the manager can make a thorough review of the hardware by considering each stage of its life cycle separately. Basically, there are eight stages in the hardware's life cycle,

196 SAFETY AND HEALTH MANAGEMENT

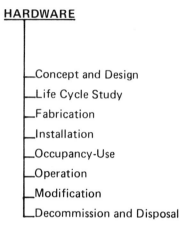

Figure 8-3. Stages of the life cycle in plant and hardware control.

as illustrated in Figure 8-3. Each of these stages will be discussed separately, with special attention to the control systems needed.

From the simplest standpoint, effective hardware control of the job process system involves four main tasks:

1. Determining the need and design requirements; making a life cycle study and safety analysis of the product.
2. Constructing and acquiring the product, inspecting it as a quality control function, and testing it for suitability.
3. Operating, maintaining, and inspecting the equipment.
4. Decommissioning and disposing of it at the end of its useful life.

For a more detailed look, let us consider each of the eight life cycle stages involved and the controls necessary for each, particularly as they involve safety and health.

Concept and Design

The concept and design process is not a simple matter of engineering. We should not consider it complete until it is clear that the process will be safe for the workers directly involved and will not be a hazard to other employees. A process is not safe just because a careful, alert, safety-minded worker can carry it out without injury. The process must first be engineered for safety; that is, it should be designed to preclude the possibility of injury as much as possible. Human factors should definitely be considered in this phase of the process.

There are few problems involving human engineering that cannot be solved, though they may take time and money. The manager must clearly understand the importance of taking the extra time and cost involved in this stage of the hardware's life cycle. If necessary, he or she should insist on the services of a human engineering consultant. The manager must understand that it is essential to remove as many hazards as possible before placing a process or piece of equipment in operation.

Thus, from the time the hardware is first conceived of and throughout the design phase, safety must be a prime concern. There are five steps involved, given here in order of their importance:

1. Design for *minimum hazard* and *maximum safety* throughout the design phase.
2. Reduce hazards through the use of *safety devices*. These can help reduce hazards which cannot be controlled by design to an acceptable level.
3. Incorporate *warning devices* to detect and alert the worker of hazards. When hazards cannot be eliminated or reduced to an acceptable level, use devices that will warn of hazardous conditions.
4. Develop *procedures to reduce hazards.* When design and warning devices still leave residual hazards, develop special procedures to deal with the problem. (Examples of this might be redundant safety devices on a press, two sets of brakes on a car, or special emergency procedures in case all brakes fail on a vehicle.)
5. *Identify the residual hazards.* Hazards remaining after all the above steps have been taken should be communicated to management, so that they can decide if the remaining risks are acceptable or not.

Design solutions should accurately reflect the criteria, specifications, and requirements involved, instead of being an afterthought to solve problems. From the outset, management should firmly establish guidelines for safety and health. For instance, it should specify that emergency procedures must be easy to perform and should provide for safe shutdown during anticipated emergencies. It should stipulate that the least hazardous energy forms should be sought, and limited to those needed to perform the operation. It should, for instance, require design engineers not to use 440 volts when 110 will do the job. If management presents clear safety and health guidelines, engineering will follow suit.

It is during the design phase that human factors should be considered and built into the hardware. This calls for an expert and is definitely not the place for management to stint on expenses. It provides the best opportunity to "design out" problems of hardware-human interface, allowing safer operation throughout the entire life cycle of the equipment or hardware.

Life Cycle Study

A thorough safety analysis must involve a life cycle study. Plans must ensure that this study will be initiated during the planning phase. The study should review each stage of the hardware's life, from planning and design through fabrication, operation, maintenance, and disposal. It may (and often does) include an analysis of the environmental impact of the hardware as well. These reviews should be carried out at predetermined milestones during the hardware's life, preferably by independent reviews rather than by the staff directly associated with the use of the hardware.

Fabrication

Once the design is completed, the hardware must be fabricated. Here, too, proper controls must be established and rigorously followed.

One aspect of fabrication control involves following well-established guidelines, such as accepted engineering practices, quality assurance practices, engineering and design codes, standards, regulations, and the use of standardized parts or components. The other aspect deals with the introduction and use of the fabricated parts or components in the system itself. Fabrication controls should be placed on both in-house and outside work, with special attention to in-house and outside work, with special attention to in-house operations. These may require separate identification, quality assurance, and documentation controls. Specifications should be kept up-to-date for in-house fabrications and checked at every stage.

Installation

The installation of hardware calls for considerable advance preparation. First, care must be taken to ensure that there is no damage or degradation of the hardware between fabrication and installation. Next, an installation plan and description should be prepared and reviewed to ensure that it provides clear and precise information for installation personnel. Postinstallation inspection should be performed before the equipment is tested or put into use to ensure that all directions and specifications were complied with. This should include checking all clearances between the new equipment and other parts of the system, use of proper grounding and bracing, adequate electrical wiring and shielding where specified, and particularly to see if any damage was done to the hardware during installation.

Occupancy-Use

The preparation of the hardware for facility use is a complex business, as we saw in Chapter 6. It involves applying stringent criteria to ensure operability and maintainability, particularly to meet safety and health needs. Determining readiness for occupancy or use may involve inspecting for operational readiness as well as startup of functional operation. A suitable plan to determine occupancy-use readiness should be developed well ahead of time.

Operation

Before hardware is placed in operation, management must make sure that all safety requirements have been specified, made available, and are ready for use. These should include operating instructions, manuals and maintenance documents, emergency instructions, and records that will be kept with the hardware. Documentation of the requirements should be up-to-date and available to the user. If the new hardware replaces old hardware, the latter should be removed from the system, along with its documentation.

The work force should be given an adequate pretask briefing to consider the effect of changes, maintenance, and any new or potential hazards. Preoperational training should cover all these factors, as well as operating instructions for the new equipment. Adequate technical support should be provided at the work site, along with careful scrutiny of the interface between operations and maintenance personnel.

An analysis should be made for each task, including the potential for error, injury, damage, or unwanted energy flow. Task safety should be determined by a job safety analysis. (This will be covered in more detail in the next section).

Operational control requires adequate maintenance support and inspection of equipment, as well as a plan broad enough to cover most aspects of operation. Gaps in the coverage should be called to management's attention. The plan should also provide for failure analysis, particularly to determine the causes of the failure. As covered in the previous chapter, the plan should also provide for minimizing maintenance-related problems.

Finally, dry runs or tests should be conducted to prove out all hardware and procedures, adjustments of oversights, and any final adjustments prior to the first hands-on operation.

Modification

We have already spoken at some length about the hazards involved in changing or modifying hardware or processes. There should be a formal program to ensure that modified procedures and hardware do not present new hazards of which we are unaware. An adequate review process should be undertaken whenever modifications are made in the system. Chapter 6 presents an in-depth look at the problems involved and the necessary steps to be taken in dealing with them.

Decommission and Disposal

We have already touched on the need to have well-established policies and criteria for determining when hardware or equipment should be discarded or replaced (see "Repair vs. Replacement Criteria" in Chapter 7). When hardware reaches this stage of its life cycle and the decision is made to discard it, safety considerations do not cease to matter. Removing or disposing of hardware can involve actual and potential hazards to safety and health. Management should be aware of these long before the need to dispose of the hardware arises.

Definite plans should be made to minimize disposal problems and hazards. These plans should consider the hazards involved in dismantling and moving equipment, safe disposal areas or processes, potential environmental impacts when machinery is simply junked or discarded, scrapping operations, and so on.

If the hardware is involved in complex operations or is very large, plans should be made for sectional removal, without completely redesigning or shutting down during the replacement process. This should take into account the impact of changed or modified operations, temporary arrangements, and so forth. These detailed disposal plans should also be subjected to a thorough job safety analysis to detect and eliminate hazards.

JOB SAFETY ANALYSIS

We have mentioned job safety analysis in several earlier chapters. This analysis is so often used as a safety and health tool that a few words on the technique are appropriate. In a well-thought-out safety and health program, emphasis is on the way the system, rather than the individual, operates. For that reason, we will examine the job safety analysis in terms of the system. Both job and system safety analyses are based on three facts:

1. A specific task or operation can be broken down into a series of relatively simple steps.
2. The hazards associated with each step can be identified.
3. Solutions can be developed to control or eliminate these hazards.

Job or system safety analysis can be used to prevent accidents at any level of operation. To simplify this for the senior manager, we will consider three levels: (1) policy level, (2) procedure level, and (3) performance level. The policy level involves the work of top management; the procedural level is that of middle management and staff; and the performance level is where the work is actually carried out.

The job safety analysis, for instance, can point out the need for changes in safety and health policies, thereby involving the top management or policy level. It also allows management to detect problems in safety and health procedures, involving middle management and staff in procedural corrections. Finally, at the working or performance level, the job safety analysis can call attention to the need for protective equipment for workers who handle hazardous materials. (This level of job safety analysis is often used as a basis for developing emergency plans and procedures.)

Job safety analysis is a simple process that involves four steps. In their proper sequence, these are:

1. *Job selection.* The first step is to select the task to be analyzed. The analysis is done for one task at a time; thus, several analyses may be required to cover a complex task or operation.
2. *Job breakdown.* Next, the task is broken down into its basic steps. For example, if the task is to load a cart with waste material, the steps might be: determine the amount of waste, select an appropriate cart, bring loading tools, position the cart, and so on. While this may seem very detailed and time-consuming, most tasks can be broken down into 10 to 15 steps.
3. *Hazard identification.* Then the actual or potential hazards associated with each step are identified. There is a definite procedure to follow in doing this. Typically, we might first ask if worker can slip, trip, or fall, or overexert himself by lifting, pushing or pulling. We then ask if the worker may come in contact with energy sources, chemicals, excessive heat, and so on.
4. *Hazard control.* Finally, we stipulate procedures or equipment to reduce or eliminate each hazard listed in step 3. If step 3, for instance, indicates that the worker might be exposed to toxic fumes, we could specify that he or she must wear breathing apparatus. If one of the hazards is the need to lift heavy objects, we might specify that two workers perform the task, that a forklift be used, and so on.

As the job safety analysis is performed, a chart should be developed, as shown in Figure 8-4. This will list the basic job steps, the corresponding hazards, and the recommended safe procedures and equipment for each step. The completed chart can then be used as a training guide for new personnel or as a safety review guide for experienced workers. It also makes an excellent guide for reviewing new or infrequently performed tasks. The job safety analysis can also be included in the supervisor's safety manual and can be extremely useful in day-to-day operations or during the safety inspection or audit. The analysis, at least initially, is done in the workplace with employee participation, perhaps by the supervisor or a committee, with subsequent management review.

MANAGERIAL REQUIREMENTS

Essential to the development of sound job processes is the establishment and implementation of adequate managerial control systems. As shown in Figure 8-5, managerial control involves four main areas:

1. Policy
2. Responsibility

JOB SAFETY ANALYSIS

Job Title/Description	Location/Department	Prepared By: R.G. Williams Date: 2/9/82
Drilling Machine Operations	Machine Shop "B" Building 4	Supervisor: Harley J. Peters

KEY JOB PROCESS STEPS	TOOLS/ MATERIALS USED	POTENTIAL HAZARDS Conditions or Actions Which Could Cause an Injury / Affect Health	RECOMMENDED SAFE PROCEDURES Personal Protective Devices Special Clothing, Procedures
1. OBTAIN MATERIAL	Tote boxes or containers	Strains from lifting or handling parts;	Exercise correct lifting practices;
		Injuries to fingers or hands caught in pinch points, or cuts caused by sharp edges	Wear hand protection and/or use mechanical lifting aids.
2. MOVING MATERIALS TO THE JOB	Cart, truck or mechanical lift	Strains from pushing or pulling, falling, slipping, tripping; hand or body caught between objects	Get help to move heavy or bulky parts. Keep work area clear and work surfaces smooth. Keep aisles free of obstructions;
3. SET-UP	Drill press, drills, related equipment	Unguarded belt or pulleys;	Provide guards for belts, pulleys, exposed moving parts;
		Unsecured machines or parts;	Secure machine to floor or base;
		Lighting poor, area around machine cluttered; machine close to traffic	Provide adequate lighting. Clear area near machine. Set up barrier if needed.
		Controls awkward to reach;	Move controls for easy reach;
4. DRILLING OPERATIONS		Loose fittings, flying chips, oil spray, falling parts;	Be sure chuck key is removed and all parts and fittings are secure. Eye and face protection to be worn;
		Clothing/hair caught in drill	Keep hair in place with cap or snood; ties or loose clothing should not be worn — shop coat or apron to fit.
		Spinning parts caught in drill	Back-up bars or hold-down devices to be used to secure material;
		Fumes generated by drilling action on certain types of material or cutting compounds	Make sure there is adequate ventilation and/or devices to remove fumes from worker area.
5. FINISHED WORK REMOVAL	See steps 1 and 2.	See steps 1 and 2.	See steps 1 and 2.

Figure 8-4. A typical job safety analysis. (Reprinted from *The OSHA Reference Manual*, copyright 1982 by The Merritt Company)

3. Authority
4. Accountability

A brief look at each of these should be helpful to the manager.

Policy

A current and effective policy should have enough scope to address all the major problems likely to occur in the course of operations. It should include the major concerns of human outlook, cost-efficient legal compliance, and

JOB PROCESS CONTROL AND SAFETY

Figure 8-5. Managerial control system requirements.

so on. For this and other reasons, it is essential that the policy be stated in writing. Moreover, the policy should be written in a lucid and simple style so that it can be implemented without conflict or confusion.

Without a sound policy, the remaining functions of managerial control responsibility, authority, and accountability cannot operate effectively, either separately or as an interacting triangle. (Chapter 3 presents a thorough discussion of policy and should be referred to when framing and developing a policy.)

Responsibility

The policy should clearly define who is responsible for what. This will ensure that responsibility is accepted at the proper level, particularly so that line management can accept risks. Guidelines and rules should ensure that the needs, capabilities, and responsibilities of supervisors are provided for as well. Top management must make it absolutely clear where responsibilities are located in the organization if those responsibilities are to be met.

Authority

If the overall program is to fulfill the intention of the policy statement, authority must be properly assigned. Safety and health programs and functions cannot be implemented without authority. Authority to correct system failures, for example, must be clearly delegated, and published in a form and style that is easily grasped. Special care should be taken not to leave interface gaps when authority is delegated.

Through proper delegation, top management can ensure the effective control of lower-level programs. While this delegation does not, of course, relieve top management of its overall responsibility, it ensures that top-level policy is actually carried out at the operating level. Moreover, vesting authority at the proper level increases the program's credibility, as well as its effectiveness.

204 SAFETY AND HEALTH MANAGEMENT

Accountability

Responsibility and authority are only two of the necessary elements that allow top-level programs to be carried out at the procedural and performance levels. There is a third indispensable element: accountability. Simply stated, top and middle management must always be accountable for the operations under their purview. It is true that they can delegate some authority and hold subordinates accountable for their particular portion of the program. But ultimately, top management remains accountable for the overall safety and health program.

PROCEDURAL REQUIREMENTS

The next major element in the job process control system is the development of procedures. By now it should be clear that, as carefully as we may select personnel and hardware, and as effectively as we may frame our policy and delegate authority, our operation will only be as safe and efficient as the procedures we use.

To some extent, we have already looked at some of the procedural ramifications, particularly in speaking of the role of safe procedures in the job safety analysis technique. However, we need to take a closer look at what is specifically required in the development of safe procedures. There are three essential factors that must be included in developing safe procedures. As illustrated in Figure 8-6, these are:

1. Participation
2. Testing
3. Feedback

Let us look briefly at each of these three requirements.

Participation

Safe procedures cannot be established in a vacuum or even on the drawing board. Engineers, designers, and managers may write the safety and health procedures, but in the end, these are not the people who will carry them out.

Figure 8-6. Procedural requirements involved in job process control.

Their involvement is important, but the participation of the on-site worker is absolutely essential.

The worker is the person who is actively involved in performing a task. Workers know what the job requires in terms of actions, and if they are seasoned, they have established their own routines, perhaps have discovered the most efficient motions or arrangement of tools, are likely to have developed shortcuts (often at the expense of safety), and know exactly what difficulties the procedure involves. Actual work-site studies may be extremely helpful, as in developing the job safety analysis or reviewing existing analyses against actual workplace practices. Where possible, workers should be asked for their input, both verbally and in demonstrating work practices, as well as making suggestions based on actual experience.

Not only will this participation enable the manager to develop a more realistic picture of actual workplace needs and practices, but it can also make the resultant official procedures more acceptable to the work force. Particularly where safety and health practices are concerned, this morale factor can be a tremendous asset.

Testing

Even when procedures have been developed and committed to paper, they should be considered tentative. The only way to determine the validity of procedures, check the flexibility of plans, and verify the clarity of instructions is through testing. Both regular and emergency procedures should be tested under actual workplace conditions by the workers who will actually be using them. This may require a period of monitored activity or the use of emergency drills or simulations, with direct observation by management or by an appropriate committee of engineering, safety, and management personnel.

Findings should be thoroughly reviewed and, where needed, corrections should be made. The resultant revised procedures should then be tested in exactly the same way, with further review and revision. Only after thorough testing and review should the procedures be regarded as firm and incorporated in the permanent job process system.

Feedback

Even when adequate procedures have been developed, management must not fall into the trap of thinking that they are fixed and permanent. Change is a constant factor in the workplace, and procedures may require periodic adjustment. A vital part of this adjustment process is the existence of an adequate feedback system running all the way from top management to the worker and back.

Feedback which originates with the worker (such as an employee suggestion) must find its way to management. Management must indicate that the message was received and will be fully considered. If the suggestion will be acted on, this too should be acknowledged. The feedback chain is not complete until the worker confirms that he or she has received management's acknowledgment. Likewise, feedback which originates with management (such as a revised procedure) must find its way to the worker. A system should be established to verify that the affected workers have received the message (perhaps initialing a copy of the new instructions, signing an acknowledgment form, or simple supervisor verification).

Failure to give feedback can destroy an effective safety and health program, as when employees become convinced that management has not even seen their suggestions or when management assumes that everyone has received the new directive when in fact they have not. Feedback may be the last of the main elements of the job process, but it is by no means the least important. In many ways, it is the indispensable link that makes the whole system work.

INTERFACING THE ELEMENTS

At the outset, we stated that there are three major elements involved in the job process control system: people, hardware, and procedures. Each of these interfaces with the others. As shown in Figure 8-7, these interfaces are:

1. People to hardware
2. Hardware to procedures
3. Procedures to people

Each of these interfaces, the point at which one element of the system meets and affects another, represents a critical point. For instance, we cannot know how safe our hardware is by considering the hardware alone; we must look at what happens when it interfaces with people and procedures.

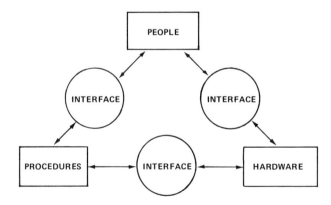

Figure 8-7. Job process interfaces.

System control involves considering each of these interfaces and asking some pertinent questions.

People to Hardware

Are the people basically compatible with the hardware? (This includes aptitudes, skills, and physical and mental traits.)
Is personnel training designed for existing hardware? Is it adequate?
Has hardware been thoroughly tested to determine its safety for personnel?
Are personnel requirements reviewed when hardware is changed, and is retraining provided?

Hardware to Procedures

Were procedures developed with existing hardware in mind?
Was hardware designed or selected to conform to existing procedures?
Have procedures been tested on the actual hardware?
Has new hardware been introduced since the procedures were written?

Procedures to People

Are the contents and language of the procedures appropriate to the people? (Highly technical instructions may be beyond the grasp of some workers, bilingual manuals may be needed, and so on.)
Do existing procedures match what workers have actually been trained to do?
When new procedures are developed, is training modified to match them?
Do the procedures match the capabilities of current personnel?

These questions should be used as a checklist, particularly prior to the start of a new or modified job process or the incorporation of new procedures in an existing job process. They can greatly aid the manager, both in pinpointing potential trouble spots and in maintaining an adequate job process control system.

The heart of the job process control system is these relationships between worker, machine, and environment or, as we have termed them, people, hardware, and procedures. While each involves distinct, complex questions that must be asked and requirements that must be met, no part of the system can be considered entirely separate from the others. Effective job process control must include continued attention to the interfaces between these elements.

Senior management, more than any other level or function, must perform this task, using its overview to develop job processes that are not only adequate and efficient, but that include sound safety and health practices from the planning stages on.

REFERENCES AND BIBLIOGRAPHY

Bird, Frank E., Jr., and Robert Loftus, *Loss Control Management,* Loganville, GA, Institute Press, 1976.

Briscoe, G. J., J. H. Lofthouse, and R. J. Nertney, *Job Safety Analysis,* Idaho Falls, ID, System Safety Development Center, 1979.

Bullock, Milton G., *Work Process Control Guide,* Idaho Falls, ID, System Safety Development Center, 1979.

Grimaldi, John V., *Safety Management,* 4th edition, Homewood, IL, Irwin, 1975.

Hendrick, Kingsley, USCG, "Job and Systems Analysis." USAF Study Kit, June 1979.

King, Ralph W., *Industrial Hazards and Safety Handbook,* London, Newens-Butterworth, 1979.

Nertney, R. J., *Occupancy Use Readiness Manual,* Springfield, VA, National Technical Information Service, 1976.

———, *Management Factors in Accident and Incident Investigation,* Idaho Falls, ID, System Safety Development Center, 1978.

Pavlov, Paul V., *Job Safety Analysis.* Idaho Falls, ID, System Safety Development Center, 1979.

Pope, William C., *Managing a Housekeeping Activity,* Alexandria, VA, Safety Management Information Systems, 1978.

Chapter 9 / Part 1

Engineering Controls: A Comprehensive Overview

INTRODUCTION

More than ever before, the executive, whatever his or her other duties, must be aware of and concerned with employee safety and health. This is no longer a job solely for the safety expert. It is important to take a broad overview here. Too often, the executive thinks strictly in terms of OSHA or other regulations. True, the passage of OSHA in 1970 highlighted the need for safety and health controls in the workplace. But it is in the manager's own best interest to recognize that much of what has occurred since that time would have happened whether or not OSHA had become law.

As workplace environments become more technologically complex, the necessity of protecting the work force from safety and health hazards becomes more pressing. A glance at Figure 9-1 proves this: these are just some of the typical workplace operations in which workers should be protected. Whether they should be protected via personal protective equipment, engineering controls, administrative controls, or a combination of these is an open question. However, one thing is clear: ensuring that they receive the most cost-effective protection available is good business in the long run.

What constitutes "the most cost-effective protection available," however, is a very complex issue. Increasingly, senior executives are faced with choices and decisions that require a broad background and great knowledge. Should they urge the company to adopt elaborate engineering controls or trust to the effectiveness of personal protective gear? Should they recommend complete process or hardware redesign, or simple modification of existing systems? In either case, what are the likely costs and benefits, and, just as important, the limitations or risks of each approach?

This chapter attempts to answer some of these questions for the executive. To provide the broadest possible grasp of the issue, four areas will be examined: (1) the broad question of engineering controls vs. personal protective equipment; (2) the basis of cost/benefit analysis that is usually applied to the question; (3) government regulations most relevant to the issue of protection; and (4) specific areas of application, such as hardware and process design and redesign, energy control, and cost savings.

OPERATIONS REQUIRING ENGINEERING CONTROLS AND/OR PERSONAL PROTECTIVE EQUIPMENT

Acidic/Basic Processes and Treatments	Grinding
Biological Agent Processes and Treatments	Hoisting
Blasting	Jointing
Boiler/Pressure Vessel Usage	Machinery (Mills, Lathes, Presses)
Burning	Mixing
Casting	Painting
Chemical Agent Processes and Treatments	Radioactive Source Processes and Treatments
Climbing	Sanding
Compressed Air/Gas Usage	Sawing
Cutting	Shearing
Digging	Soldering
Drilling	Spraying
Electrical/Electronic Assembly and Fabrication	Toxic Vapor, Gas, Mists and Dust Exposure
Electrical Tool Usage	Welding
Flammable/Combustible/Toxic Liquid Usage	Woodworking

Figure 9-1. Operations requiring engineering controls and/or personal protective equipment.

ENGINEERING CONTROLS VS. PERSONAL PROTECTIVE EQUIPMENT

The Basic Question

Efforts to ensure that workers perform their jobs in safe environments have been a source of increasing controversy during the past few years. Faced with increasing regulatory control, industry complains that the rules are prohibitively costly and unnecessarily inflexible. Adequate personal safety equipment, many managers feel, provides a reasonable level of protection without costly system modification or redesign.

On the other hand, such safety equipment has never inspired the confidence of labor. Personal protective gear is never a substitute for engineering controls, they maintain. Indeed, attempts to change the worker rather than the workplace have failed to prove that personal protective equipment provides any advantage over the engineering controls approach.

Manufacturers of such gear, as well as union and government officials, agree that even the best equipment will not work without comprehensive worker training and medical surveillance. However, even companies with extensive worker education programs can encounter stiff resistance to the use of safety devices by employees.

Advocates of engineering controls take supporters of personal protective equipment to task for overlooking the spirit of the OSHA law. The employer should provide a workplace free of hazards, they maintain. Emphasizing the use of personal protective gear over engineering controls puts the onus on workers to protective themselves.

Even while agreeing that, in the best of all possible worlds, every workplace would be completely hazard free, the business executive is faced with the common sense of the real-life situation: What can the company reasonably afford? At what point does the cost of engineering controls cut dangerously close to the profit margin of corporate survival? At the same time, the executive is faced with issues that are just as pertinent: What are unsafe conditions really costing the company, both in production and in human terms? What factors need to be considered as costs? As benefits? What, finally, is a human life worth?

These questions are not the concern of business alone. The same issues have recently precipitated extensive cost benefit analyses by OSHA itself, "against the goals envisioned by the Congress when it passed environmental, safety and health laws," according to Basil Whiting, former Deputy Assistant Secretary of Labor for OSHA.

In addition, three landmark cases involving cost/benefit analysis, OSHA and safety and health standards have recently made their way through the United States judiciary to the Supreme Court. It is notable that when the Court ruled to uphold the OSHA cotton dust standard in June 1981, it accepted the principle that OSHA standards are "feasible," provided that they stop short of endangering an industry's financial viability. Specifically, the Court limited its review to three economic questions:

1. How must OSHA demonstrate the economic feasibility of its standards?
2. Does OSHA's test of economic feasibility (that a standard does not put an industry out of business) satisfy the act's requirements?
3. Does the act require OSHA to demonstrate that a standard is reasonably necessary and that its cost bears a reasonable relationship to its benefits?

The Basic Decisions

Are engineering controls an alternative to personal protective equipment, or is it the other way around? This chicken-and-egg debate has become an emotionally charged issue. Proponents on both sides argue their beliefs heatedly, but there is little well-founded evidence to support their cases. The reason for this unfortunate situation is simple: there is no single solution to all the hazards found in industry. The only realistic answer to the question of which method is better is, "It depends." Each situation requires an independent analysis, considering all the known factors, so that a truly unbiased decision can be reached.

In short, the debate becomes meaningful only when we take it out of the theoretical realm and look at actual workplace operations. What specific alternatives is the executive looking at? To illustrate this, Figure 9-2a compare engineering controls and personal protective equipment for a variety of

212 SAFETY AND HEALTH MANAGEMENT

SUMMARY OF EMPLOYEE SAFEGUARDS		
TO PROTECT	PERSONAL PROTECTIVE EQUIPMENT TO USE	ENGINEERING CONTROLS TO USE
Breathing	Self contained breathing apparatus, gas masks, respirators, alarm systems	Ventilation, air filtration systems, critical level warning systems, electrostatic precipitators
Eyes/Face	Safety glasses, filtered lenses, safety goggles, face shield, welding goggles/helmets, hoods	Spark deflectors, machine guards
Hands/Arms/Body	Gloves, finger cots, jackets, sleeves, aprons, barrier creams	Machine guards, lockout devices, feeding and ejection methods
Head/Neck	Bump caps, hard hats, hair nets	Toe boards
Hearing	Ear muffs, ear plugs, ear valves	Noise reduction/isolation by equipment modification/substitution, equipment lubrication/maintenance programs, eliminate/dampen noise souces, reduce compressed air pressure, change operations *
Excessively High/Low Temperatures	Reflective clothing, Temperature controlled clothing	Fan, air conditioning, heating, ventilation, screens, shields, curtains
Overall	Safety belts, lifelines, grounding mats, slap bars	Electrical circuit grounding, polarized plugs/outlets, safety nets

Figure 9-2a. Summary of employee safeguards.

EXAMPLES OF THE TYPES OF CHANGES THAT SHOULD BE CONSIDERED INCLUDE:	
• Grinding instead of chipping	• Electric tools in place of pneumatic tools
• Pressing in lieu of forging	• Welding instead of riveting
• Compression riveting over pneumatic riveting	• Mechanical ejection in place of air-blast ejection
• Wood or plastic tote boxes in lieu of metal tote boxes	• Wheels with rubber or composite tires on plant trucks and cars instead of all-metal wheels
• Use of an undercoating on machinery covers	• Wood in place of all-metal work benches

Figure 9-2b. Operational changes as employee safeguards.

workplace activities. Figure 9-2b presents operational changes that may achieve the same ends.

There are numerous textbooks and other references that describe the quantitative methodologies already available for making the "best" decision, as we will see later in this chapter. Unfortunately, none of these analytical approaches addresses the big question: "How much is one human life worth?" Recognizing that they will never be able to factor this social variable into their equations, managers must learn to make decisions regarding employee

ALLOCATED COSTS*	OVERHEAD COSTS	INTANGIBLE FACTORS
Raw materials	Building lights and power	Compatibility with existing equipment
In-process materials	General utilities (electricity, gas, water)	Flexibility for future change
Equipment investment	Plant supplies	Improved customer service
Acquisition cost	Janitor supplies	Pilferage losses/savings
Operator costs	Property taxes	Reduced manual effort
Fixed costs	Fire insurance	Operator morale
(depreciation, interest	Payroll taxes	Pride in facility
expense, preventive	Rent	Good will
maintenance)	Space utilization	Community relations
Variable costs	Inventory value of spare parts	
(fuel cost, replacement	Worker's compensation insurance	
parts, demand maintenance)		
In-process labor		
Maintenance department labor		
Storeroom labor		
Foremen's salaries		
Night watchman's salary		

NOTE: Cost categories vary among firms. In some organizations attempts are made to assign dollar values to some items listed here as intangible.

*Items shown are those that are readily measured and allocated to a specific operation, product, service, or project. They can be associated with both personnel and equipment. Overhead costs, shown in the second column, pertain to overall operations, and cannot readily be assigned to a single product, service or operation.

Figure 9-3. Typical cost factors to be considered.

safety and health with known facts and the best information available. Generally speaking, most managers must begin by examining the usual costs to be considered when justifying any new project. Figure 9-3 presents typical cost factors which should be considered when a safety or health project proposal is evaluated.

Some Examples of Typical Cases

Perhaps the most effective means of demonstrating the recurring dilemma of choosing between engineering controls and personal protective equipment is to provide several typical examples. While these examples may not have a direct bearing on the reader's own company operations or safety and health problems, they should be useful to study. For one thing, they will give the manager an idea of what specific factors may have to be weighed in the decision-making process. For another, they provide a good starting place or point of approach for tackling problems in the manager's own company.

Case 1: Excessive Noise Levels. At the beginning of every noise control project, precise noise reduction objectives must be established. This may mean compliance with the latest OSHA standards, or the company itself may want to improve acoustical comfort in the workplace. Suppose that a new machine shop facility is being constructed. Previous experience with noise confusion, impulse-noise reverberation, and inaudible public address systems has convinced the owners that the new building should be acoustically treated to provide a better, more efficient working environment.

Taking into account the volume and materials involved, the untreated acoustical properties of the facility are first calculated and compared to acceptable noise levels. Recognizing that the use of baffles is the most cost-effective engineering control to achieve these levels, comparisons are then made, using a dollars-per-decibel-reduction formula. Three major variables influence the cost effectiveness of noise reduction by baffles:

1. Cost per sabin added (a sabin being a measure of noise absorption)
2. Installation costs
3. Maintenance of replacement costs

Other factors which may contribute to costs are investment tax credits, changes in productivity (either increases or decreases), and fire insurance rate revisions. All of these variables must be considered when trying to determine the most cost-effective engineering controls of minimizing noise.

The results of these analyses should then be compared to a similar analysis which considers the use and cost of such personal protective equipment as earplugs. ear valves, or ear muffs. Naturally, the resultant decision will vary, depending on the facility and equipment, the personnel to be exposed, and the costs of each choice.

Case 2: Excess Airborne Particulate Levels. Dust collection systems can be divided into two groups: elevated-temperature systems and ambient-temperature systems. In general, elevated-temperature systems, such as those serving boilers, dryers, ovens, or melting furnaces, do not lend themselves to plant air recirculation, so the question of excessive airborne particulate levels may not be as crucial. However, the exhaust of ambient-temperature systems is often considered a source of recirculated air, and the particulate levels are naturally of greater concern. To begin with, the exhaust of ambient-temperature systems results in an excessive waste of energy during the heating season. While these systems do not allow economical indirect recovery methods because of the lower temperatures involved, the vast amounts of energy present in the exhaust air can be recovered at a reasonable cost by using a dust removal system capable of air recirculation.

It falls to the safety, industrial, or plant engineer to establish the cost effectiveness of engineering control systems vs. personal protective equipment such as respirators, masks, or self-contained breathing apparatus.

Case 3: Corrosive Chemicals Processing Area. The damaging fumes which emanate from acid baths can be easily isolated from human operators by means of remote control handling devices or robots. A less costly (but potentially more hazardous) engineering control is the use of local exhaust systems located so as to eliminate virtually all of the dangerous fumes. Both alternatives should be compared to the use of personal protective gear, such as self-contained breathing apparatus and impervious protective clothing designed to cover the operator completely.

Naturally, each situation will vary. In some simple cases, respirators, chemical goggles, aprons, and gloves may be sufficient personal protective equipment to afford the necessary coverage. In more complicated situations, even the most sophisticated equipment may not be enough, and engineering controls become mandatory. Safety, industrial, and plant engineers should be expected to provide the necessary analyses in order to ascertain the extent of the hazard to employees whose work exposes them to the corrosive fumes.

Engineering Controls for Machine Tools

Machine tools (such as mills, lathes, shearers, punch presses, grinders, drills, saws, and the like) are among the most commonly used types of industrial equipment. They also provide an example of common conditions in which only a few limited items of personal protective gear are available for use. In these cases, the problem to be solved is not personal protective equipment vs. engineering controls, but rather which engineering control(s) should be used to protect the machine operator.

Crushed hands and arms, severed fingers, blindness—the list of possible machine-related injuries is as long as it is horrifying. There seem to be as many hazards created by moving machine parts as there are types of machines. Safeguards are essential for protecting workers from needless and preventable injuries.

A good rule to remember is: Any machine part, function, or process which can cause injury must be safeguarded, Where the operation of a machine or accidental contact with it can injure the operator or others in the vicinity, the hazard must be either controlled or eliminated.

Safeguarding of dangerous moving parts is especially vital in three basic areas:

1. The point of operation. This is the point where work is actually performed on the material, such as cutting, shaping, boring, or forming of stock.
2. Power transmission apparatus. This includes all components of the mechanical system which transmit energy to the part of the machine performing the work. These components include flywheels, pulleys,

belts, connecting rods, couplings, cams, spindles, chains, cranks, and gears.
3. *Other moving parts.* These include all parts of the machine which move while the machine is working, such as reciprocating, rotating, and transverse moving parts, as well as feed mechanisms and auxiliary parts of the machine.

A wide variety of mechanical motions and actions may present hazards to the worker. These include the movement of rotating teeth and any parts that impact or shear. These different types of hazardous mechanical motions and actions are basic to nearly all machines, and recognizing them is the first step in protecting workers from the dangers they present.

The basic types of hazardous mechanical motions and actions are:

Motions
    Rotating (including in-running nip points)
    Reciprocating
    Transverse
Actions
    Cutting
    Punching
    Shearing
    Bending

What must a safeguard do to protect workers against mechanical hazards? Engineering controls must meet these minimum general requirements:

1. *Prevent contact.* The safeguard must prevent hands, arms, or any other part of a worker's body from making contact with dangerous moving parts. A good safeguarding system eliminates the possibility of the operator or another worker placing their hands near hazardous moving parts.
2. *Be firmly secured.* Workers should not be able to remove easily or tamper with the safeguard; a safeguard that can easily be made ineffective is no safeguard at all. Guards and safety devices should be made of durable material that will withstand the conditions of normal use. They must be firmly secured to the machine.
3. *Protect from falling objects.* The safeguard should ensure that no objects can fall into moving parts. A small tool which is dropped into a cycling machine can easily become a projectile that can strike and injure someone.
4. *Create no new hazards.* A safeguard defeats its own purpose if it creates a hazard of its own (such as a shear point, jagged edge, or unfinished surface) which can cause an injury. The edges of guards, for instance, should be rolled or bolted so that they present no sharp edges.

5. *Create no interference.* Any safeguard which prevents a worker from performing the job quickly might soon be overridden or disregarded. Proper safeguarding can actually enhance efficiency, since relieves the worker's apprehensions about injury.
6. *Allow safe lubrication.* It should be possible to lubricate the machine without removing the safeguards. Locating oil reservoirs outside the guard, with a line leading to the lubrication point, will reduce the need for the operator or maintenance worker to enter the hazardous area.

All power sources for machinery are potential sources of danger. When using electrically powered or controlled machines, for instance, the equipment as well as the electrical system itself must be properly grounded. Replacing frayed, exposed, or old wiring will also help to protect the operator and others from electrical shocks or electrocution. High-pressure systems, too, need careful inspection and maintenance to prevent possible failure from pulsation, vibration, or leaks. Such a failure could cause explosions or flying objects.

Machines often produce noise or unwanted sound which can result in a number of hazards to workers. Not only can this startle and disrupt concentration, but it can interfere with communications, thus hindering the worker's safe job performance. Research has also linked noise to a whole range of harmful health effects, ranging from hearing loss and aural pain to nausea, fatigue, reduced muscle control, and emotional disturbances. Engineering controls, such as the use of sound-dampening materials, as well as less sophisticated hearing protection, devices such as ear plugs and muffs, have been suggested as ways of controlling the harmful effects of noise. Vibration, a related hazard which can cause noise and thus result in worker fatigue and illness, may be avoided if machines are properly aligned, supported, and anchored.

Because some machines require the use of cutting fluids, coolants, and other potentially harmful substances, operators, maintenance workers, and others in the vicinity may need protection. These substances can cause ailments ranging from dermatitis to serious illnesses and disease. Specially constructed safeguards, ventilation, and protective equipment and clothing are possible temporary solutions to the problem until these hazards can be better controlled or eliminated entirely from the workplace.

Machine Safeguarding Methods

There are many ways to safeguard machinery. The type of operation, the size or shape of stock, the method of handling, the physical layout of the work area, the type of material, and the production requirements or limitations will help to determine the appropriate safeguarding methods for the individual machine.

As a general rule, power transmission apparatus is best protected by fixed guards that enclose the dangerous area. For hazards at the point of operation, where moving parts actually perform work on stock, several kinds of safeguards are possible. One must always choose the most effective and practical means available. Generally speaking, we can group safeguards under five general classifications:

I. Guards
   A. Fixed
   B. Interlocked
   C. Adjustable
   D. Self-adjusting
II. Devices
   A. Presence sensing
      1. Photoelectrical (optical)
      2. Radiofrequency (capacitance)
      3. Electromechanical
   B. Pullback
   C. Restraint
   D. Safety controls
      1. Safety trip control
         a. Pressure-sensitive body bar
         b. Safety tripod
         c. Safety tripwire cable
      2. Two-hand control
      3. Two-hand trip
   E. Gates
      1. Interlocked
      2. Other
III. Location/distance
IV. Feeding and ejection methods
   A. Automatic feed
   B. Semiautomatic feed
   C. Automatic ejection
   D. Semiautomatic ejection
   E. Robot
V. Miscellaneous aids
   A. Awareness barriers
   B. Miscellaneous protective shields
   C. Hand-feeding tools and holding fixtures

Figures 9-4 through 9-6 provide specific information regarding machine safeguarding.

METHOD	SAFEGUARDING ACTION	ADVANTAGES	LIMITATIONS
Fixed	Provides a barrier	Can be constructed to suit many specific applications In-plant construction is often possible Usually requires minimum maintenance Can be suitable to high production, repetitive operations	May interfere with visibility Can be limited to specific operations Machine adjustment and repair often requires its removal thereby necessitating other means of protection for maintenance personnel
Interlocked	Shuts off or disengages power and prevents starting of machine when guard is open; should require the machine to be stopped before the worker can reach into the danger area	Can provide maximum protection Allows access to machine for removing jams without time-consuming removal of fixed guards	Requires careful adjustment and maintenance May be easy to disengage
Adjustable	Provides a barrier which may be adjusted to facilitate a variety of production operations	Can be constructed to suit many specific applications Can be adjusted to admit varying sizes of stock	Hand may not enter danger area — protection may not be complete at all times May require frequent maintenance and/or adjustment The guard may be made ineffective by the operator May interfere with visibility
Self-Adjusting	Provides a barrier which moves according to the size of the stock entering danger area	Off-the-shelf guards are often commercially available	Does not always provide maximum protection May interfere with visibility May require frequent maintenance and adjustment

Figure 9-4. Machine safeguarding: guards.

Personal Protective Equipment as Engineering Control

All too often, the manager and even the safety expert is inclined to think of a fundamental dichotomy between engineering controls and personal protective equipment. Properly seen, these alternatives are not necessarily mutually exclusive. The seasoned expert who has a thorough grasp of the purpose of both of these approaches can take a broader view, understanding that personal protective equipment itself is often part of overall engineering control. That is, personal protective equipment is seen as one more engineering control device, to be combined with other devices.

Engineering controls, which eliminate the hazard at the source and do not rely on the worker's behavior for their effectiveness, offer the best and most reliable means of safeguarding. For that reason, engineering controls must be the manager's first choice for eliminating machinery hazards. But whenever an extra measure of protection is necessary, operators must wear protective clothing or personal protective equipment as well. As an example, guards or barriers should be installed on a machine. But if there is still a

METHOD	SAFEGUARDING ACTION	ADVANTAGES	LIMITATIONS
Photoelectric	Machine will not start cycling when the light field is interrupted When the light field is broken by any part of the operator's body during the cycling process, immediate machine braking is activated	Can allow freer movement for operator	Does not protect against mechanical failure May require frequent alignment and calibration Excessive vibration may cause lamp filament damage and premature burnout Limited to machines that can be stopped
Radiofrequency (capacitance)	Machine cycling will not start when the capacitance field is interrupted When the capacitance field is disturbed by any part of the operator's body during the cycling process, immediate machine braking is activated	Can allow freer movement for operator	Does not protect against mechanical failure Antennae sensitivity must be properly adjusted Limited to machines that can be stopped
Electromechanical	Contact bar or probe travels a predetermined distance between the operator and the danger area Interruption of this movement prevents the starting of the machine cycle	Can allow access at the point of operation	Contact bar or probe must be properly adjusted for each application; this adjustment must be maintained properly
Pullback	As the machine begins to cycle, the operator's hands are pulled out of the danger area	Eliminates the need for auxiliary barriers or other interference at the danger area	Limits movement of operator May obstruct work-space around operator Adjustments must be made for specific operations and for each individual Requires close supervision of the operator's use of the equipment
Safety trip controls: Pressure-sensitive body bar Safety tripod Safety tripwire	Stops machine when tripped	Simplicity of use	All controls must be manually activated May be difficult to activate controls because of their location Only protects the operator May require special fixtures to hold work May require a machine brake

(continued on next page)

Figure 9-5. Machine safeguarding: devices.

METHOD	SAFEGUARDING ACTION	ADVANTAGES	LIMITATIONS
Two-handed control	Concurrent use of both hands is required, preventing the operator from entering the danger area	Operator's hands are at a pre-determined location Operator's hands are free to pick up a new part after first half of cycle is completed	Requires a partial machine with a brake Some two-hand controls can be rendered unsafe by holding with arm or blocking, thereby permitting one-hand operation Protects only the operator
Two-hand trip	Concurrent use of two hands on separate controls prevents hands from being in danger area when machine cycle starts	Operator's hands are away from danger area Can be adapted to multiple operations No obstruction to hand feeding Does not require adjustment for each operation	Operator may try to reach into danger area after tripping machine Some trips can be rendered unsafe by holding with arm or blocking, thereby permitting one-hand operation Protects only the operator May require special fixtures
Gate	Provides a barrier between danger area and operator or other personnel	Can prevent reaching into or walking into the danger area	May require frequent inspection and regular maintenance May interfere with operator's ability to see the work

Figure 9-5. (continued)

METHOD	SAFEGUARDING ACTION	ADVANTAGES	LIMITATIONS
Automatic Feed	Stock is fed from rolls, indexed by machine mechanism, etc.	Eliminates the need for operator involvement in the danger area	Other guards are also required for operator protection — usually fixed barrier guards Requires frequent maintenance May not be adaptable to stock variation
Semiautomatic Feed	Stock is fed by chutes, moveable dies, dial feed, plungers, or sliding bolster		
Automatic Ejection	Work pieces are ejected by air or mechanical means		May create a hazard of blowing chips or debris Size of stock limits the use of this method Air ejection may present a noise hazard
Semiautomatic Ejection	Workpieces are ejected by mechanical means which are initiated by the operator	Operator does not have to enter danger area to remove finished work	Other guards are required for operator protection May not be adaptable to stock variation
Robots	They perform work usually done by operator	Operator does not have to enter danger area Are suitable for operations where high stress factors are present, such as heat and noise	Can create hazards themselves Require maximum maintenance Are suitable only for specific operations

Figure 9-6. Machine safeguarding: feeding and ejection methods.

chance of injury, the worker should also be required to wear safety goggles, gloves, or whatever personal protective equipment is appropriate.

To provide adequate protection, protective clothing and equipment must always be:

1. Appropriate for the particular hazards.
2. Maintained in good condition.
3. Properly stored when not in use to prevent damage or loss.
4. Kept clean and sanitary.

Protective clothing is, of course, available for different parts of the body. Hard hats protect the head from the impact of bumps and falling objects when the worker is handling stock; caps and hair nets can help keep the worker's hair from being caught in machinery. If machine coolants can splash or particles can fly into the operator's eyes or face, then face shields, safety goggles, glasses, or similar kinds of protection may be necessary. Hearing protection may be needed when workers operate noise machinery. To guard the trunk of the body from cuts or impacts from heavy or rough-edged stock, protective coveralls, jackets, vests, aprons, and full-body suits are available. Workers can protect their hands and arms from the same kinds of injury with special sleeves and gloves. Likewise, safety shoes and boots, or other acceptable foot guards, can shield the feet from injury in case the worker needs to handle heavy stock which might drop or fall.

It is important to note that protective clothing and equipment themselves can create hazards. A protective glove can become caught between rotating parts, or a respirator facepiece can hinder the wearer's vision. Not only are alertness and careful supervision required whenever protective clothing and equipment are used, but special care should be taken in selecting them. Both company-issued and employee-purchased protective gear should meet appropriate standards.

Other aspects of the worker's dress may present additional safety hazards. Loose-fitting clothing might become entangled in rotating spindles or other kinds of moving machinery. Jewelry, such as bracelets and rings, can catch on machine parts or stock and lead to serious injury by pulling a hand into the danger area.

Alternatives to Strict Engineering Controls

The competent executive wants to learn about any alternatives, hybrid systems, or midway strategies that can make use of both engineering control approaches and other, less expensive control strategies. Further, the manager wants to know about applicable decision-related criteria. Therefore, this section presents material that will be useful in the selection and application of company resources to industrial safety and health problems.

Safety and health engineering control principles are deceptively few:

1. Substitution
2. Isolation
3. Ventilation (both generalized and local)

In a technological sense, an appropriate combination of these strategic principles can be applied to any industrial safety or hygiene control problem to achieve a work environment of satisfactory quality. It is rarely necessary or appropriate to apply all these principles to any specific potential hazard. However, a thorough analysis of the control problem must be made to ensure that the proper method is chosen to produce control in a way that is compatible with the technical process. The method must also be acceptable to the workers in terms of the day-to-day operation, and it should be possible to accomplish it with an optimal balance of installation and operating expenses.

Substitution. Although it is one of the simplest engineering principles to apply, substitution is often overlooked as an appropriate solution to occupational safety and health problems. There is a tendency to concentrate on correcting a problem rather than eliminating it. For example, in considering a vapor exposure problem in a degreasing operation, the first inclination may be to provide ventilation. A better solution might be simply to substitute a solvent with a much lower degree of hazard. Substituting less hazardous substances for more hazardous ones, substituting one type of process equipment for another, or even substituting a safer process for a more dangerous one may provide the most effective hazard control at minimal expense.

This strategy is often used in conjunction with safety equipment: substituting safety glass for regular glass, replacing unguarded equipment with properly guarded machines, or replacing safety gloves or aprons with garments made of materials more impervious to the chemicals being handled. Since this sort of substitution is often done as an immediate response to an obvious problem, it is not always recognized as an engineering control, even though the end result is every bit as effective.

Substituting one process or operation for another is seldom considered, except in cases of major modification. In general, changing any process from a batch to a continuous type of operation reduces its potential hazard. This is true primarily because continuous operation involves less worker contact with the process materials compared to batch operation. The substitution of one process for another can be applied on a very basic level (e.g., substitution of airless spray for conventional spray equipment to reduce worker exposure to solvent vapors). Substituting a paint dipping operation for a paint spray operation can reduce the potential hazard even further. In any of these cases, automating the process can further reduce the potential hazard.

Isolation. The principle of isolation is frequently thought of as simply installing a physical barrier or guard between the hazardous operation and the worker. It is important to note, however, that isolation can often be provided without a physical barrier. The appropriate use of distance or time provides equally effective isolation in some instances.

Perhaps the most common example of isolation as a control strategy is associated with the storage and use of flammable solvents. Large tank farms with dikes around the tanks, underground storage of some solvents, detached solvent sheds, and fireproof solvent storage rooms within buildings are all common in American industry and are good examples of isolation. Frequently, applying the principle of isolation also yields other benefits, such as noise control, remote control materials handling (as with radioactive substances), and local exhaust ventilation.

Ventilation. Workplace air quality is largely determined by the design and performance of the exhaust system. An improperly designed hood, or one that evacuates air too slowly, can easily contaminate the atmosphere and affect workers in its vicinity. This simple example raises one of today's most urgent issues: the relationship between atmospheric emissions (regulated by the EPA) and occupational exposure (regulated by OSHA). The main question is: What is done with gases generated as a result of industrial operations or processes? Many of these emissions cannot be exhausted directly into the atmosphere because of their toxic or hazardous nature. Some can be exhausted indirectly into the atmosphere (through the general ventilation system) or recirculated to the workplace. Ventilation system design and operation must take into account the exact nature of these emissions, as well as the necessity and type of respiratory gear needed to protect the work force.

COST/BENEFIT ANALYSIS

By now, it should be apparent that many factors must be taken into account when deciding on the feasibility and necessity of certain engineering controls. In one frame of reference, the operative question is, "What type of engineering controls or protective equipment will provide the greatest safety and health benefits?" The second half of the question, for the conscientious manager, is, "What will it cost?" Weighing the costs of any particular engineering control device against its expected benefits is an essential part of the decision-making process, which we will examine in depth in this section. There are several approaches to cost/benefit analysis that are particularly appropriate to the question of engineering controls vs. personal protective gear.

There are many quantitatively oriented techniques for determining the desirability of a capital investment. However, even the most sophisticated methodologies are not capable of rationally considering the value of life itself. For this reason, the person who performs such analyses must provide sufficient documented input to make supportable choices possible. The manager must insist on having this information, particularly when faced with choices or decisions involving worker safety and health.

Cost/benefit analysis has been defined as a costless economic arrangement in which the gains can be distributed to make the entire organization better off. In a thorough cost/benefit analysis, both the present and future benefits and costs associated with a potential investment are determined and compared. (Future benefits and costs must be calculated to take the inflation rate into account and translated into present-day dollar values.)

For proper comparison, both benefits and costs must be expressed in common monetary units (e.g., 1990 dollars). They are usually compared either by (1) by computing the ratio of benefits to costs or (2) by subtracting costs from benefits to generate net benefits, if any. Benefit/cost ratios greater than 1, or positive net benefits, are normally considered a sufficient economic justification for the contemplated investment.

Serious questions can arise if the principal benefits anticipated are saving of lives and/or reduction of the frequency and severity of injuries. Obviously, these cannot be reasonably quantified in monetary units, so grave theoretical and conceptual difficulties are involved. The same difficulties also arise where costs involve the loss of life and/or increases in the frequency and severity of injuries.

Virtually all safety and health cost/benefit studies assign fixed monetary values to the loss of life or limb. Values for a life in such studies are typically obtained either by computing the discounted future income of individuals or by computing the discounted differences between future earnings and personal consumption. These concepts and approaches are useful but by no means universally accepted. They have been criticized on a number of grounds, including the charge that they are economically irrelevant.

Costs and Benefits

A complete cost/benefit analysis demands knowledge of all germane costs, as well as both the tangible and intangible benefits anticipated from a specific investment. To take just one common example, consider the direct and indirect costs involved in accidents and injuries. The direct costs are relatively simple to determine, since they involve just two main items: medical expenses and workers' compensation payments. The indirect or hidden costs are numerous and complex, as illustrated by Figure 9-7. These, according to Arthur J. Naquin, who compiled the list, are just some of the hidden or indirect costs involved in accidents or injuries.

The manager should try to develop an estimate of the average cost for each of the indirect and direct items. Working with the firm's industrial engineers, insurance company and cost accountants, it is possible to pro-

1. Wages or salaries paid to injured employees over and above the amounts required to be paid under the provisions of a state Worker's Compensation act. (Many companies find it advisable from an employee-relations point of view NOT to curtail a worker's take-home pay during the period that the employee is disabled).
2. Wages or salaries paid for time not worked on the day of the injury.
3. Cost of the inefficiency of an injured worker after he returns to work on either a "modified (light) duty" job or to his regular job.
4. Cost of vocational rehabilitation of an injured employee.
5. Costs of selecting and training a replacement employee if the injured's production schedule must continue. This may not amount to very much if an already qualified employee can be shifted to the injured's job.
6. Cost of the "start-up" inefficiency of a replacement employee. Again, this may not be a sizable item.
7. Costs of time lost by an injured's co-workers who stop work:
    - To assist the injured by rendering first aid or by transporting the injured to a medical facility.
    - Out of sympathy for the injured.
    - Because of emotional upset (shock).
    - Out of curiosity.
    - To "loaf" (while supervision is busy assisting injured worker).
    - To wait until work team can be reorganized.
    - To wait until an "all clear" signal is sounded following a general alarm.
    - To serve as a member of an accident investigation team or committee.
8. Cost of bringing medical treatment to an injured employee.
9. Costs arising out of the need to transport an injured worker(s) to a medical facility from a fairly faraway point such as an off-shore drilling platform, the transportation to be effected by helicopter, crew boat, or ambulance.
10. Costs of overtime brought on because of an accidental injury.
11. Costs of curtailed production efficiency of foremen, supervisors, and other management and staff personnel while they are engaged in:
    - Assisting injured employee(s). Getting him (or her) to a medical treatment facility.
    - Arranging for the injured's production to be continued.
    - Investigating the cause or causes of the accidental injury. This may involve the taking of one or more photographs of the scene of the injury; the production of scale drawings; the preparation of a written report.
    - Preparing the required accident-injury reports, including those required under the provisions of the Occupational Safety and Health Act.
    - Notifying the Regional Office of the U.S. Department of Labor-OSHA of the accidental injury if such notification is required.
    - Attending such hearings as may be called to investigate the accident.
    - Arranging for the selection and training of a replacement employee if the severity of the injury requires that this step be taken.
    - Arranging for the selection and training of new personnel caused by the resignation(s) of the injured's co-workers; the resignations being induced by the occurrence of one or more occupational injuries.
    - Calling on members of the injured's family, if this step is indicated.
    - Assisting the company's legal department, if a tort suit arises out of the accidental injury.
12. Cost of cleaning up following an accidental injury.
13. Cost of repairing or replacing damaged machines or equipment.
14. Cost of the lack of production of machines, tools, or equipment that are rendered inoperative because of an occupational injury.
15. Cost of spoiled raw materials and semi-finished or finished goods and products.
16. Cost due to failure to fill orders on time (penalties, forfeits).
17. Loss of bonuses that could have been earned for ahead-of-time delivery of goods, materials, equipment, etc.

(continued on next page)

Figure 9-7. Indirect, hidden, or uninsured accident-injury costs.

duce rough order-of-magnitude (ROM) estimates for most of these accident-related cost items. With this knowledge, the executive is better prepared to compare the anticipated costs associated with accidents or injuries with the expected costs of engineering controls, personal protective equipment, or both. One type of input needed is illustrated in Figure 9-8. What types of injuries are likely in the particular operation in question?

18. Costs that arise out of the loss of good will as, for example, the reluctance on the part of a purchaser to continue placing orders with a firm whose occupational injuries delay promised shipment.

19. Cost to an employer due to provisions of employee welfare and benefit programs.

20. Costs that arise out of a depression of workers' morale with an attendant loss of efficiency.

21. Costs that arise out of labor-management contract bargaining that arise out of a company's poor safety record.

22. Costs of attorney's fees and other legal expenses that are incurred because of tort suits that arise out of an occupational injury.

23. Uninsured medical costs not covered by a worker's compensation insurance policy.

24. Penalty costs assessed by OSHA Compliance Officers because of violations of provisions of the Occupational Safety & Health Act which are discovered as a result of the occupational injury.

25. Penalty costs of complying with OSHA standards that an occupational injury exposed as not being in compliance with a prescribed standard.

26. Cost of transporting OSHA Compliance Officers to the scene of an occupational injury; a cost voluntarily incurred by a company or corporation. This cost can be of considerable size if the Compliance Officer is transported to an off-shore drilling platform out in the Gulf of Mexico some 100 miles from shore.

27. Costs of near-miss accidents in which no employee is injured, but which do result in damages to machines.

28. All other hidden, indirect, or uninsured costs which result from an accidental employee injury.

Figure 9-7. (continued)

ACCIDENT CLASSIFICATION SYSTEM

AREA OF INJURY		TYPE OF INJURY	
Head	Abdomen	*Wounds:*	*Burns:*
Face	Hand	Laceration	Heat
Eye	Finger	Contusion	Chemical
Teeth	Thumb	Infection	Friction
Neck	Hip	Foreign body	*Skin:*
Collarbone	Rupture	Puncture	Dermatitis
Shoulder	Leg	*Eyes:*	Irritation-rash
Arm	Thigh	Foreign body	Fracture
Elbow	Knee	Burn corrosive	Strain
Forearm	Skin	Burn heat	Sprain
Wrist	Ankle	Burn flash	*Gases:*
Chest	Foot	Wound	Nausea
Ribs	Instep	Irritation	Dizziness
Back	Toe	*Pains*	Irritation

Figure 9-8. Accident classification system.

In addition, these same cost items are the benefits which can be realized as a result of installing, using, and maintaining safety-related controls or equipment. Some items, however, remain costs despite the controls or equipment, while others become benefits. Which items remain costs and which become benefits will vary, depending on the specific situation and the controls and equipment being analyzed.

Generally, the following formula is used to analyze the cost/benefits feasibility of a given investment. First, the expected number of accidents over the anticipated usable life of the controls or equipment is determined.

This is then multiplied by the ROM average cost per accident. If the product of these two figures is equal to or greater than the cost of the investment, the company is well advised to make the purchase. On the other hand, if the product is less than the investment cost, the decision to purchase the equipment or install the controls cannot be made solely on a financial basis.

Alternative Analyses

Cost/Effectiveness Analysis. Cost/effectiveness analysis provides an important alternative to cost/benefit analysis, since it compares the costs of alternative means for effectively achieving an agreed-upon goal. The means could be engineering controls, personal protective equipment, or a combination of approaches. The goals are often expressed in public policy as laws and standards. Much of the philosophy and methodology of cost/effectiveness analysis was derived from cost/benefit analysis. As a result, there are many similarities between the techniques, which causes many people to confuse them. To clarify this issue, let us look at the three basic requirements of a cost/effectiveness analysis:

1. The systems or projects being compared must have common goals or purposes.
2. Alternative means for meeting the goal must exist.
3. The capability of measuring the cost and effectiveness of each system must exist.

Requirement 1 spells out one of the fundamental differences between the cost/benefit and cost/effectiveness methods. In the cost/benefit approach, both costs and benefits must be measured and compared in common monetary units. In a cost/effectiveness analysis, the systems being compared must have common goals or purposes, which do not have to be expressed in monetary units.

Systems Analysis. Economic analyses of complex alternatives (such as a decision to purchase personal protective equipment or to install engineering controls) are most valuable when they can take into account all the relevant considerations facing the decision maker. This is often called the "systems analysis" approach.

Systems analysis is the comparison of alternative means of achieving a desired result. This is accomplished by systematically investigating proper objectives. The cost, effectiveness, and risks associated with alternative policies or strategies are compared quantitatively, and additional alternatives are formulated.

In systems analysis, the analyst can systematically vary the inputs to assess their effects on both costs and output. Given a specific resource, he or she can thus determine the best combination of inputs to maximize the output. This approach gives improved "visibility" to the cost/performance relationship.

These brief descriptions are by no means comprehensive. Many other techniques and approaches have been successfully applied, ranging from fault-free analysis to complex mathematical modeling. The point, however, should be clear to the manager: the analytical inputs needed for sound decision making often require expert knowledge. The manager cannot afford to make crucial safety and health decisions "blind" and will need all the input available.

GOVERNMENT REGULATORY DEMANDS

Company operations, accident rates, and costs are by no means the only pertinent inputs the manager has to consider. Increasingly, when making decisions that involve engineering controls, personal protective equipment, and general safety and health strategies, the executive must deal with constraints and requirements imposed by government regulators.

While these regulatory agencies are examined in greater depth in Chapter 21, we will look briefly at the three laws that have the most direct impact on engineering controls and personal protective equipment decisions: the Toxic Substances Control Act (TSCA), the Resource Conservation and Recovery Act (RCRA), and the Occupational Safety and Health Act (OSHA).

This section addresses the regulatory demands of these federal statutes from the perspective of managers who must decide whether to install engineering controls that will enable their companies to meet these standards, or simply to discontinue certain operations altogether, as when they cannot justify the associated costs of regulatory compliance.

Toxic Substances Control Act (TSCA)

The EPA has responsibility for administering both the TSCA and the RCRA, both initially enacted in 1976.

Until the TSCA, the federal government was not empowered to prevent chemical hazards to health and the environment by banning or limiting chemical substances at a germinal, premarket stage. The impetus behind the law was the idea that production workers, consumers—indeed, every American—would be protected by an equitably administered early warning system controlled by the EPA.

This broad law authorizes the EPA administrator to issue rules prohibiting or limiting the manufacturer, processing, or distribution of any chemical substance or mixture which "may present an unreasonable risk of injury to health or the environment." The EPA administrator may require testing (at a manufacturer's or processor's own expense) of a substance after finding that:

> The substance may present an unreasonable risk to health or the environment.

There may be a substantial human or environmental exposure to the substance.
Insufficient data and experience exist for judging a substance's health and environmental effects.
Testing is necessary to develop such data.

This legislation is designed to cope with hazardous chemicals such as kepone, vinyl chloride, asbestos, and fluorocarbon compounds (freons), as well as polychlorinated biphenyls (PCBs). The company that manufactures or processes these compounds must base its decisions re engineering controls, worker protection and process modifications, on TSCA requirements or test findings.

Resource Conservation and Recovery Act (RCRA)

Enacted in 1976 as an amendment to the Solid Waste Disposal Act, the RCRA sets up a cradle-to-grave regulatory mechanism or tracking system for such wastes from the moment they are generated to their final disposal in an environmentally safe manner. The act charges the EPA with the development of criteria for identifying hazardous wastes, creating a manifest system for tracking them through final disposal, and setting up a permit system based on performance and management standards. These affect the generators, transporters, owners, and operators of waste treatment, storage, and disposal facilities. It is expected that the RCRA will be a strong force for innovation and eventually lead business to a broad rethinking of chemical processes. That is, companies will have to look at hazardous waste disposal not just in terms of immediate costs, but rather with respect to full life-cycle costs. Here, too, many management decisions can be made only by consulting the specific requirements affecting company operations.

Occupational Safety and Health Act (OSHA)

Among the most comprehensive sets of regulations within OSHA is Subpart Z, Toxic and Hazardous Substances. Essentially, the standards specify employee exposure limits for a large number of substances, as well as providing computational formulas for measuring exposure. They also require that feasible engineering and administrative controls be used to comply with the standards. When no feasible controls exist, personal protective equipment must be worn or other protective measures taken.

Other sections of OSHA address specific standards for workplace safety relative to noise levels, air quality, and other related environmental criteria. All of these may radically affect management's need to reassess company practices and processes, install tighter control systems, and provide greater worker protection. In many cases, they may also involve the decision of whether to continue or discontinue a particular process or operation.

After numerous claims and counterclaims by industry, unions, governmental agencies and workers, various legal briefs have been argued all the way to the U.S. Supreme Court. The resultant conclusion is that business must find a way:

1. To produce safe chemical substances
2. To control unsafe waste products
3. To provide safe, uncontaminated workplaces and environments.

Superfund Amendments and Reauthorization Act (SARA)

The Superfund Amendments and Reauthorization Act (SARA), also known as the the Comprehensive Environmental Response, Compensation and Liability Act (CERCLA), was greatly revised in 1986 and signed as the Superfund Law. This law places many requirements on business and industry, as do the TOSCA, RCRA, and OSHA. It is described further in Chapter 18, Part 3.

AREAS OF APPLICATION

"No man [or woman] is an island." Nor can engineering control decisions be made apart from related considerations such as ergonomic/human factors input, process/operations design criteria, and energy-saving controls. This section examines how engineering controls relate to hardware and systems design functions, as well as to energy control systems.

Hardware Design and Redesign

In creating or modifying both mechanical and electrical equipment, machine designers must consider the performance characteristics of machine operators as a major constraint. To ignore these would be the same as ignoring the limitations of human capabilities. Equipment designers especially concerned with engineering controls must also understand the principles of human factors (ergonomics).

Equipment designers, like managers, are aware that there are certain tasks which people can perform with greater skill and dependability than machines, and vice versa. Some of these positive performance characteristics are noted in Figure 9-9.

In addition designers of equipment and engineering coontrols need to know human performance limitations, both physically and psychologically. They should know that the interaction between people and their operating environment presents a never-ending challenge in assessing safety vs. hazard or health vs. contamination. Figure 9-10 lists the six pertinent sciences which are most closely involved in the design of machines and engineering controls. When engineering controls are being considered, it is reasonable to expect

232 SAFETY AND HEALTH MANAGEMENT

POSITIVE PERFORMANCE CHARACTERISTICS — SOME THINGS DONE BETTER BY:

PEOPLE	MACHINES
Detect signals in high noise fields	Respond quickly to signals
Recognize objects under widely different conditions	Sense energies outside human range
Perceive patterns	Consistently perform precise, routine, repetitive operations
Sensitive to a wide variety of stimuli	Recall and process enormous amounts of data
Long-term memory	Monitor people or other machines
Handle unexpected or low-probability events	Reason deductively
Reason inductively	Exert enormous power
Profit from experience	Relatively uniform performance
Exercise judgement	Rapid transmission of signals
Flexibility, improvisation and creativity	Perform several tasks simultaneously
Select and perform under overload conditions	Expendable
Adapt to changing environment	Resistance to many environmental stresses
Appreciate and create beauty	
Perform fine manipulations	
Perform when partially impaired	
Relatively maintenance-free	

Figure 9-9. Positive performance characteristics—some things done better by people vs. machines.

PEOPLE PERFORMANCE SCIENCES

Anthropometry:	Pertains to the measurement of physical features and characteristics of the static human body.
Biomechanics:	A study of the range, strength, endurance and accuracy of movements of the human body.
Ergonomics:	Human factors engineering — especially Biomechanics aspects.
Human Factors Engineering:	Designing for human use.
Kinesiology:	A study of the principles of mechanics and anatomy of human movement.
Systems Safety Engineering:	Designing which considers the operator's qualities, the equipment and environment relative to successful task performance.

Figure 9-10. People performance sciences.

that designers will make full use of the principles established by specialists in these human performance sciences.

Process Design and Redesign

A stress (or stressor) is a physical or psychological feature of the environment that requires operators to exert themselves unduly to continue performing. Such exertion is termed "strain" or "stress and strain." Common physical stressors in industrial workplaces are poor illumination, excessive noise, vibration, heat, and the presence of harmful atmospheric contaminants. Much work has been done to determine the effects of each of these on the worker.

Unfortunately, much less is known about their effects when they occur simultaneously, in rapid sequence, or over extended periods of time. Research suggests that such effects are not simply additive, but synergistic, thus compounding their damage. In addition, when physical work environments are unfavorable to equipment operators, two or more stressors are generally present: high temperature and excessive noise, for example. The solution of process design and redesign is relatively easy to specify but costly to implement: design the physical environment so that all physical characteristics are within an acceptable range.

Marketed in the United States since the early 1960s, industrial robots offer both hardware and process designers a technology which can be used when excessively hazardous or uncomfortable working conditions are expected or already exist. Where a job situation poses potential dangers, a robot should be considered as a substitute for human operators. Hot forging, die casting, and spray painting fall into this category. If work parts or tools are awkward or heavy, an industrial robot may fill the job. Some robots are capable of lifting items weighing several hundred pounds.

An industrial robot is a general-purpose, programmable machine which possesses certain human-like capabilities. The most obvious characteristic is the robot's arm, which makes it suitable for a variety of uncomfortable or undesirable production tasks. Increasingly, hardware and process designers are suggesting the use of robots in their designs and redesigns. Their feasibility will be determined by the specific company and operation involved, and their initial cost will be subjected to the regular cost/benefit analysis ordinarily applied to any proposed investment.

Energy Control and Cost Savings

Energy control and cost savings have become major concerns for industry. No executive can be unaware of the escalating fuel and energy costs, the critical shortages that can drastically impede production, and the need to reclaim and recycle as much energy as possible. This section reviews energy-saving controls from the perspective of both potential safety and health advantages and expected cost savings.

In American industry, top management is concerned with the bottom line. In energy management, that bottom line is how efficiently a plant performed this month, compared to last month, compared to a year ago. Energy management performance can be calculated only if accurate energy monitoring and reporting systems exist.

Control charts, which were originally developed for statistical quality control, can be usefully applied to energy control as well. Despite their name, control charts do not control usage; rather, they help monitor process behavior in relation to the expected or desired behavior. It is recognized that the process being monitored will contain normal as well as abnormal variations. By looking for abnormal variations, management will be able to recognize a situation in need of remedial action.

If a firm is monitoring its energy consumption, the impact of hardware and/or process alterations on the use of energy should be immediately apparent. Even a computer simulation can graphically demonstrate the resultant impact of a modification on a company's energy utilization. It is thus possible to see both improved and curtailed energy management performance as a direct consequence of such changes. The application of a computer simulation should assist the decision maker in selecting from among two or more options. For example, incorporating a particular type of engineering control may result in an increase in energy consumption, while a different control would not. (This assumes, of course, that both alternatives will provide employees with adequate protection from a hazard or contaminant.)

Present and future energy sources, availability, and costs, as well as worker safety and health requirements, must be taken into account by the decision maker. Unfortunately, today's manager faces this challenge with very little historical guidance. This is a new area which must be explored carefully and intelligently. Growing energy costs, as well as expanding government regulatory requirements, constitute a new challenge to concerned, responsible executives. To date, there is no simple answer or pat formula that will yield a sure-fire answer.

CONCLUSION

This chapter has looked at engineering controls from a variety of perspectives. From any angle, it is apparent that this managerial challenge is unlikely to disappear or even diminish in the near future. If anything, it will continue to grow more complex as time goes by.

Faced with increasing energy costs, growing governmental expectations, greater union involvement, and stronger worker pressures, executives must continually analyze new requirements and update their knowledge of the alternatives available to them. Purpose of this, of course, is to select the control and protection systems that will yield the greatest benefits within the company's budget.

BIBLIOGRAPHY

"Accident: Related Losses Make Cost Soar," *Industrial Engineering,* May 1979, p. 26.

"Accident Prevention: Your Key to Controlling Surging Workers' Compensation Costs," Occupational Hazards, November 1979, p. 35.

"Analyzing a Plant Energy-Management Program: Part I—Measuring Performance," *Plant Engineering,* October 30, 1980, p. 59.

"Analyzing a Plant Energy-Management Program: Part II—Forecasting Consumption," *Plant Engineering,* November 13, 1980, p. 149.
"Anatomy of a Vigorous In-Plant Program," *Occupational Hazards,* July 1979, p. 32.
"A Shift Toward Protective Gear," *Business Week,* April 13, 1981, p. 56H.
"A Win for OSHA," *Business Week,* June 29, 1981, p. 62.
"Buyers Should Get Set for Tougher Safety Rules," *Purchasing,* May 25, 1976, p. 34.
"Complying with Toxic and Hazardous Substances Regulations—Part I," *Plant Engineering,* March 6, 1980. p. 283.
"Complying with Toxic and Hazardous Substances Regulations—Part II," *Plant Engineering,* April 17, 1980, p. 157.
"Computers Help Pinpoint Worker Exposure," *Chemecology,* May 1981, p. 11.
"Conserving Energy by Recirculating Air from Dust Collection Systems," *Plant Engineering,* April 17, 1980, p. 151.
"Control Charts Help Set Firm's Energy Management Goals," *Industrial Engineering,* December 1980, p. 56.
"Controlling Noise and Reverberation with Acoustical Baffles," *Plant Engineering,* April 17, 1980, p. 131.
"Controlling Plant Noise Levels," *Plant Engineering,* June 24, 1976, p. 127.
"Cost-Benefit Decision Jars OSHA Reform," *Industry Week,* June 29, 1981, p. 18.
"Cost Factors for Justifying Projects," *Plant Engineering,* October 16, 1980, p. 145.
"Costs, Benefits, Effectiveness, and Safety: Setting the Record Straight," *Professional Safety,* August 1975, p. 28.
"Costs Can Be Cut Through Safety," *Professional Safety,* October 1976, p. 34.
"Cutting Your Energy Costs," *Industry Week,* February 23, 1981, p. 43.
"Elements of Effective Hearing Protection," *Plant Engineering,* January 22, 1981, p. 203.
"Energy Constraints and Computer Power Will Greatly Impact Automated Factories in the Year 2000," *Industrial Engineering,* November 1980, p. 34.
"Energy Managers Gain Power," *Industry Week,* March 17, 1980, p. 62.
"Energy Perspective for the Future," *Industry Week,* May 26, 1980, p. 67.
"Engineering and Economic Considerations for Baling Plant Refuse," *Plant Engineering,* April 30, 1981, p. 34.
"Engineering Control Technology Assessment for the Plastics and Resins Industry," NIOSH Research Report, Publication No. 78-159.
"Engineering Project Planner, A Way to Engineer Out Unsafe Conditions," *Professional Safety,* November 1976, p. 16.
"EPA Gears Up to Control Toxic Substances," *Occupational Hazards,* May 1977, p. 68.
"Fume Incinerators for Air Pollution Control," *Plant Engineering,* November 13, 1980, p. 108.
"Groping for a Scientific Assessment of Risk," *Business Week,* October 20, 1980, p. 120J.
"Hand and Body Protection: Vital to Safety Success," *Occupational Hazards,* February 1979, p. 31.
"Hazardous Wastes: Coping with a National Health Menace," *Occupational Hazards,* October 1979, p. 56.
"Hearing Conservation Implementing an Effective Program," *Professional Safety,* October 1978, p. 21.
"How Do You Know Your Hazard Control Program Is Effective?" *Professional Safety,* June 1981, p. 18.
"How to Control Noise," *Plant Engineering,* October 5, 1972, p. 90.
"Human Factors Engineering—A Neglected Art," *Professional Safety,* March 1978, p. 40.
"IE Practices Need Reevaluation Due to Energy Trends," *Industrial Engineering,* December 1980, p. 52.
"Industrial Robots: A Primer on the Present Technology," *Industrial Engineering,* November 1980, p. 54.

"Job—Safety Equipment Comes Under Fire; Are Hard Hats a Solution or a Problem?" *The Wall Street Journal,* November 18, 1977, p. 40.
"New OSHA Focus Led to Noise-Rule Delay," *Industry Week,* June 15, 1981, p. 13.
"OSHA Communique," *Occupational Hazards,* June 1981, p. 27.
"OSHA Moves Health to Front Burner," *Purchasing,* September 26, 1979, p. 46.
"OSHA to Analyze Costs, Benefits of Lead Standard," *Occupational Health & Safety,* June 1981, p. 13.
Patty's Industrial Hygiene and Toxicology, Volume I, 3rd revised edition, New York, Wiley Interscience, 1978.
"Practical Applications of Biomechanics in the Workplace," *Professional Safety,* July, 1975, p. 34.
"Private Sector Steps Up War on Welding Hazards," *Occupational Hazards,* June 1981, p. 50.
"Putting Together a Cost Improvement Program," *Industrial Engineering,* December 1979, p. 16.
"Reduce Waste Energy with Loads Controls," *Industrial Engineering,* July 1979, p. 23.
"Reducing Noise Protects Employee Hearing," *Chemecology,* May 1981, p. 9.
"Regulatory Relief Has Its Pitfalls, Too," *Industry Week,* June 29, 1981, p. 31.
"R01 Analysis for Cost-Reduction Projects," *Plant Engineering,* May 15, 1980, p. 109.
"Safety and Profitability—Hand in Hand," *Professional Safety,* March 1978, p. 36.
"Safety Managers Must Relate to Top Management on Their Terms," *Professional Safety,* November 1976, p. 22.
"Superfund Law Spurs Cleanup of Abandoned Sites," *Occupational Hazards,* April 1981, p. 67.
"Taming Coal Dust Emissions," *Plant Engineering,* May 15, 1980, p. 123.
"The Cost-Benefit Argument: Is the Emphasis Shifting?" *Occupational Hazards,* February 1980, p. 55.
"The Cost/Benefit Factor in Safety Decisions," *Professional Safety,* November 1978, p. 17.
"The Economics of Safety: A Review of the Literature and Perspective," *Professional Safety,* December 1977, p. 31.
"The Design of Manual Handling Tasks," *Professional Safety,* March 1980, p. 18.
"The Hidden Cost of Accidents," *Professional Safety,* December 1975, p. 36.
"The Human Element in Safe Man-Machine Systems," *Professional Safety,* March 1981, p. 27.
"The Problem of Manual Materials Handling," *Professional Safety,* April 1976, p. 28.
"Time for Decisions on Hazardous Waste," *Industry Week,* June 15, 1981, p. 51.
"Tips for Gaining Acceptance of a Personal Protective Equipment Program," *Professional Safety,* March 1976, p. 20.
"Toxic Substances Control Act," *Professional Safety,* December 1976, p. 25.
"TSCA: Landmark Legislation for Control of Chemical Hazards," *Occupational Hazards,* May 1977, p. 79.
"Were Engineering Controls 'Economically Feasible'?" *Occupational Hazards,* January 1981, p. 27.
"Were Noise Controls 'Technologically Feasible'?" *Occupational Hazards,* January 1981, p. 37.
"What Are Accidents Really Costing You?" *Occupational Hazards,* March 1979, p. 41.
"What's Being Done About Hazardous Wastes?" *Occupational Hazards,* April 1981, p. 63.
"Where OSHA Stands on Cost-Benefit Analysis," *Occupational Hazazds,* November 1980, p. 49.

Chapter 9 / Part 2

Practical Applications of Engineering Controls

INTRODUCTION

So far, we have spoken mainly about alternative means of ensuring safety and health in the workplace: engineering controls vs. personal protective equipment. Let us assume that the manager has carefully studied and applied all the relevant data, from cost/benefit analyses to government regulatory standards. Sooner or later, the manager comes to the conclusion that the company must rely on engineering controls, rather than personal protective equipment, in certain phases of company operations.

The manager is now faced with choices and decisions of quite another sort. The debate between engineering controls and other means of worker protection is over. Attention now focuses on the specific question of applying engineering controls to the particular company's operations. As with any other decision, the manager is seeking to find the best, most efficient, and most effective means of using these techniques.

At this point, the manager requires information of a very different sort: less involved with the theories of engineering controls vs. personal protective equipment, and more involved with the practical application of specific engineering controls in the workplace.

This section will explore these practical questions in greater depth, with a view to giving the manager the broad base of knowledge needed to select specific alternatives. This will enable the manager to make the best use of the engineering and design recommendations available to achieve greater workplace safety and health.

ENERGY: A CAUSE OF ACCIDENTS

When the word "energy" is used today, particularly in regard to management, the manager is most likely to think in terms of saving energy, conserving energy, or simply controlling the costs of energy. There is another aspect of energy, however, that is far more central to the business of safety and health management: the control of energy, particularly through effective engineering controls.

Because this aspect of energy is unfamiliar to most senior managers, this chapter will attempt to present a thorough background and rationale in order to demonstrate how vital the control of energy is both to safety and health and to profitable, efficient operation.

"Energy" is usually defined as the physical capacity to do work, and obviously is essential to business and industry. Our main concern here is the control of energy. The design, use, and evaluation of engineering controls are based on the need to do something about unwanted transfers of energy. All accidental injuries and damage result from the application of specific forms of energy in amounts exceeding the resistance of materials, products, tissues, and so on. Managing unwanted energy is a fruitful approach to safety and efficiency that holds much promise.

The types of energy which cause injury and damage are limited, but the forms in which they occur and the variety of the energy carriers are innumerable. Moreover, we constantly compound this problem as we ingeniously develop more powerful types of energy and put them to work in new forms.

Figure 9-11 illustrates the point. The senior manager should be aware of these potential causes of accidents. Even if his or her organization is not

ENERGY SOURCES

Corrosive	Kinetic/Linear	Pressure Volume
Acids	Fork Lifts	Boilers
Caustics	Carts	Surge Tanks
Decons	Presses	Gas Bottles
	Crane Loads	Pressure Vessels
Electrical	Power Tools	Stressed Members
Diesel Units	Trucks	Gas Receivers
High Lines		
Transformers	**Kinetic/Rotational**	**Radiation**
Wiring	Motors	Storage Area
Motors	Pumps	Radioactive
Power Tools	Cafeteria Equipment	Contamination
	Laundry Equipment	Lasers
Explosive	Gears	Medical X-Ray
Dynamite	Centrifuges	Radiography
Dusts		Equipment Noise
Gases	**Mass, Gravity**	
Nitrates	Human Effort	**Thermal**
Primer Cord	Stairs	Furnaces
Caps	Elevators	Boilers
Squibbs	Scaffolds	Steam Lines
	Ladders	Sun
Flammables	Excavations	Fire Boxes
Packing Materials	Cranes	Lead Pots
Rags	Hoists	Convection
Gasoline		
Solvents	**Nuclear**	**Toxic**
Diesel Fuel	Receiving Area	Fluorides
Building Content	Shipping Area	Lead
Gases	Storage Tanks	Asbestos
	Shops	
	Assembly Area	
	Laboratories	

Figure 9-11. Energy sources and typical examples.

concerned with the most obvious uses of energy (dynamos, heavy machinery, and generators), even the most modest operation does involve many of these potential causes of injury, whether in the employee cafeteria, the elevator, or the stock room.

Let us begin by defining an "accident" as an event involving an unwanted transfer of energy. Energy produces injury and damage unless there are adequate controls or barriers. Different types of energy can work together to produce hazards. This interaction is common in accident causation. Some examples of this interaction are:

1. An electrical spark ignites a flammable vapor, resulting in an explosion. This, in turn, injures workers and damages the building.
2. An electrical shock causes a worker to fall from a ladder.
3. A liquid spill results in a fall which damages the equipment, as well as injuring the worker.

SPECIAL PROBLEMS

Designing safeguards which will reduce exposure to energy release and subsequent hazards needs special attention. Every operation involves several potential sources of unwanted energy transfer, many of them unrecognized. Consider, for instance, just a few examples of potential exposure involving equipment:

1. Direct contact with moving parts
2. Work in process (such as ejection, conveying, discharge, etc.)
3. Mechanical failure
4. Human failure arising out of curiosity, distraction, fatigue, chance taking, or fear

Some equipment is recognized as potentially hazardous and in need of special precautions and guarding devices. This includes power presses, flywheels, elevators, lifting equipment, boilers, and other pressure equipment. The hazards of unwanted energy release and its potential consequences are easily visualized with these items. Designers and safety engineers have developed numerous ways to approach all of these hazards. In many instances, there are standards and regulations requiring specific controls, including governmental requirements such as OSHA standards.

We are sometimes surprised to find so many ordinary places and situations where there is great potential for injury or damage due to energy release. Take the kitchen, for example. Unofficial figures indicate that catering and kitchen staffs have substantially higher rates of lost-time accidents than those in other industries or occupations. The main causes are falls (from slippery floors), burns and scalds, mishandling of materials, equipment misuse, cuts from hand tools, and improperly designed or used machinery.

Even in operations that do not involve large or complex machinery, day-to-day activities may include the use of numerous common hand tools ranging from the cafeteria worker's electric knife to the maintenance man's electric drill to the draftsman's electric eraser. The common and ordinary nature of these hand tools may lead us to disregard or underestimate their potential for injury. Common hand tools, however, present unusually great electrical hazards ranging from a frayed power cord or overloaded outlet to more serious hazards. These latter generally fall into four areas:

1. Working in contact with earthed metal
2. Working in awkward positions
3. Working in hot or damp sites
4. Working at heights above floor level

All of these situations require special provisions to eliminate or overcome their hazards.

Static Electricity

Static electricity is in a class by itself in presenting hazards in connection with energy exchange. Overcoming the problem of static electricity is largely an engineering control matter. Static electricity control must definitely be considered in the design phase. However, problems sometimes do not develop until well after the design phase and frequently require attention in postdesign situations.

In order to evaluate the possibility of static electricity hazards in industrial processes, we must first understand its effects. Static electricity is generated by the contact and separation of dissimilar materials. This generation cannot be prevented. If one of the materials is a nonconductor or an electrically insulated conductor, static electricity charges can accumulate on it. The amount of accumulation is limited by the electrical insulating valve of the material surrounding the collector.

Several processes which may produce static electricity are recognized as common to industry. They include:

1. The use of power belts and similar mechanisms in which poorly conductive materials move over or between pulleys or rollers.
2. The flow of fluids through pipes or conduits, or from openings into tanks or containers.
3. The production and movement of dust clouds.
4. The use of rubber-tired vehicles.
5. The flow of gases from openings.
6. The accumulation of charges on personnel.

Thus, we can find static electricity nearly anywhere—on textiles, paper, or plastics. In most cases, the generation of static energy (such as that coming from a moving belt) cannot be prevented. However, many things can be done to control it, mostly through various grounding processes. Likewise, static electricity is unavoidably generated when liquids move in contact with other materials. Mixing, pumping, filtering, agitating, or even pouring liquids from one container into another will generate static electricity. Some hydrocarbons, in fact, tend to accumulate very high static charges. But again, control is possible even when static cannot be completely prevented.

Similarly, electrostatic charges in moving dust clouds cannot be prevented; in fact, the dust retains its charge after it settles. The best solution is to keep the formation and dispersal of dust as close to the point of origin as possible.

Rubber-tired vehicles accumulate static electricity, as you may discover when you touch your car door. And surely everyone has had the experience of walking across a carpet and then feeling a spark jump when they touched some object.

The simple examples of what happens when we reach for the car door or shake hands with a person in a carpeted room may not seem to have much to do with accident-producing hazards, but they do point to the real problem involved. The major hazard created by static accumulations is the possibility of a spark. Even sparks of a relatively low voltage can ignite a flammable vapor and cause a large explosion. This may be an acute problem in areas where paint spraying, tank cleaning, petroleum transfer, grain handling, printing processes, and so on are occurring.

Static electricity is not difficult to test for, and it is usually controllable when detected. Again, much of the hazard can be handled by engineering controls, either in initial design and manufacture or after the fact.

Although the presence of static electricity is often demonstrated by shock, sparks, or the attraction to or repulsion of light particles from surfaces, it is best measured in industry by a variety of instruments built for the purpose.

Human Factors Considerations

Machine hazards are themselves often compounded or influenced by the human element and further by the environment. The total hazard reduction program must include human factors considerations if it is to be effective. Some examples of human factors considerations are:

1. The selection and training of operators.
2. The effects of long hours, shifts, and personal problems.
3. Machinery layout, particularly proper working space.
4. Regular inspections and maintenance.
5. Improved light and noise levels.
6. Good air quality and freedom from toxic fumes.

7. Temperatures conducive to alertness and good working conditions.
8. Suitable support or position for the operator.
9. Natural or reflex actions of the operator.
10. Comfortable protective clothing and devices.
11. Good housekeeping standards.

Engineering controls for safe operation, particularly control design, are closely linked to human factors. Some of the areas where human factors enter into the engineering design phase are:

1. The proper design of dial instrumentation to permit quick and easy interpretation, as well as accurate information.
2. The selection of the proper type of control to meet varied requirements, such as fast and precise operations, easy grasping distance, and minimum distance between controls.
3. The selection of a particular type of control to permit precision work involving little force.
4. The type of control needed for a particular task (such as a handwheel, level, or pedal).
5. Controls that require force over a wide range but little precision (such as cranks, hand wheels, or levers).
6. Proper relationships between controls and displays (such as the pointer moving in the same direction as the control).

The proper design of controls should never be left to expediency or intuition. Too many times, what has seemed to be right to the user or designer turned out to be entirely inadequate under the stress of operation. Ergonomic engineers should be brought into the design phase and should interact with the ultimate user or operator.

The human factor aspect of hazards and process control raises special questions for our consideration, such as these:

1. What sense organs are used to receive information? Will a buzzer, light, horn, dial, or verbal order be most effective?
2. Does the type of discrimination involved call for lights, colors, sounds, or dial readings?
3. What physical response is called for: rotating a handle, pushing a button, turning a wheel, or stepping on a pedal?
4. Is physical movement called for? Does the movement interfere with incoming information? What force is required?
5. What are the speed and accuracy requirements: reasonably close, hairline accuracy, slow and deliberate adjustment, or speedy response?

6. Will the process be influenced by environmental conditions such as high humidity, unusual temperatures, noise, light levels, the presence of chemicals, or other factors?

The following questions should be asked about machines, equipment, and surrounding areas:

1. Would a guard or other device eliminate or isolate the hazard?
2. Should the hazard be identified by a color, sign, alarm, or light?
3. Would interlocks protect the worker from making the wrong move?
4. Should the machine or circuit be designed to be fail-safe?
5. Is there a need for standardization?
6. Are easily identified and accessible emergency controls needed?
7. What unsafe condition would be created if the proper operating sequence were not followed?

POLICY AND PLAN

Safety begins with top management's commitment but is fully effective only if it is backed up by an equal commitment to and from engineering controls. Company policy should indicate that safety should be incorporated into the machine or operation at the design phase or before procurement when ready-made equipment is used. Once management is dedicated to hazard reduction, a plan is needed to accomplish this; the plan must incorporate the sound use of engineering input. This approach might include:

1. Consulting the standards for appropriate health and safety legislation.
2. Reviewing the loss experience of the industry to determine why some companies have good records and other do not. (It is not necessary to have your own accident experience to review, as might be the case with a new operation.)
3. Determining what procedures should be used to design out hazards that have resulted or could result in injury or damage.
4. Meeting with production and maintenance managers to discuss safety and reliability features.
5. Consulting local and state codes relative to safety and health, being careful not to make these the standard. (Since they are minimal requirements, the resulting program would also be minimal.)
6. Consulting the company's insurance underwriters. They can not only supply valuable information on the industry's accident experience, but loss prevention expertise as well.
7. Developing checklists and flow diagrams to highlight exposures and hazards which need to be overcome.

8. Establishing an order of priority and a timetable to accomplish the objectives.
9. Breaking the total plan into logical and definable segments. Assigning responsibilities and authority for task accomplishment in each segment to specific individuals.

Controls that result in safer equipment and processes should be established before the design phase and should consider the safe design, construction, operation, and maintenance of the process, facility, or equipment. The prime objective of an engineering group should be to design equipment and processes so that exposures to hazards are eliminated or controlled.

ENERGY BARRIERS

Since unwanted energy transfers can be harmful or wasteful, several concepts or strategies have been developed to provide barriers for systematic energy control. Proper energy management should include such preventive barriers, not only to prevent injuries but to reduce waste due to misplacement of energy. Effective energy barriers not only protect workers but lead to fewer equipment failures, less spoilage, fewer rejects, less rework, less damage in transit, and fewer production failures. Energy management thus serves efficiency and economy, as well as safety.

The Department of Energy cites 13 strategies for systematic energy control:

1. *Limit energy.* Use the minimum amount of energy needed for the task, including smaller containers of materials, less voltage, and reduced concentrations of chemicals.
2. *Use safer substitutes.* Use safer chemicals, designs requiring less maintenance, and safer materials-handling devices. In many instances, simpler devices and safer compounds may be better and cheaper as well.
3. *Prevent energy buildup.* Use regulators and governors to control or reduce speed; better housekeeping to reduce dust and spills; and so on.
4. *Prevent release of energy.* Use safety factors in design, including energy turnoffs, insulation on pipes, protective enclosures, or railings.
5. *Provide for slow release of energy.* This includes the use of safety valves, bleed-offs, and wider shoulders on roads.
6. *Channel energy release away.* Separate the energy sources in space or time by storing oxygen and explosives separately, having a safe distance between stored chemicals and work areas, and so on.
7. *Have barriers on the energy source.* Electrical insulation, fences, noise reduction, or baffles are examples.
8. *Have barriers between energy sources.* Walls, shields, and acoustical insulation should be used where necessary.

9. *Have barriers on human objects.* This includes preventive bumpers, protective equipment, and clothing.
 10. *Raise the injury threshold.* Physical preparation and damage-resistant materials can help achieve this.
 11. *Ameliorate once the energy is released.* Shutting off the power, removing injured personnel, stopping traffic, isolating the area, and putting out the fire are all ameliorative measures.
 12. *Rehabilitate.* Restore the system to a safe working condition or repair the equipment before using it again.
 13. *Control the wrong energy input.* Use a circuit breaker or other form of shutoff where none is provided.

While simple in concept, there are some very sophisticated techniques for thorough application of energy barriers. While these are beyond the average manager's need to know, he or she should still be aware that approaches to containing energy are available and ensure that the organization is making maximum use of them.

DEALING WITH RISK

Wherever energy is being used, there is always the risk of unwanted release and consequent injury or damage. The manager should be fully informed about the specific risks involved in the organization's operations. Once he or she becomes aware of a risk, the manager has four possible courses of action for dealing with it:

1. Terminate it.
2. Treat it.
3. Tolerate it.
4. Transfer it.

A company may, for instance, decide to discontinue or abandon a process that has become too risky or dangerous (terminating it). However, this is not usually possible in practice, as it might mean abandoning the very processes that are basic to a company's operations. So the usual solution is to treat the risk, that is, minimize it or safeguard against it, either before a system becomes operative or through corrective action later.

However, before any risk can be treated, it must first be identified. A review of the company's own lost-time injuries is a good place to start, along with a review of similar industry experience. Identifying the hazards peculiar to the industry before analyzing one's own operation is often useful.

As always, do whatever is possible in the design and concept phases. However, keep in mind that it is often necessary to accept risks that already exist and that were not designed out in the first place. In that case, redesigning the process to correct the hazards may be the best solution. If that is impractical (as it often is), try to reduce the hazards in the existing process to an acceptable level through one of the strategies already listed. If that is still inadequate, install warning devices to alert workers to the hazards. Procedures should then be developed to reduce and further control the hazards. If a high level of risk still remains, it needs to be clearly identified and accepted by the proper level of management.

No matter how carefully the manager thinks the company's risk situation has been studied, it should continually be reviewed. Every new chemical, process, or worker can constitute an unrecognized risk. And for the company's sake, every manager must take the attitude that the most dangerous risk is the one that is not recognized.

Process and operation flow charts are excellent devices for identifying risks. The hazard points (and their consequent control points) can easily be detected. Building diagrams and detailed process charts serve a similar purpose.

The task of identifying and detecting hazards should not be left to the safety department alone. Because risk identification and control should begin at the design phase, it is, properly speaking, a task for the engineering department as well, if not more so. The task of engineering safety in and hazards out is a task for the engineers, the suppliers, and the designers. The safety professional is merely a resource in helping them to detect the problem areas and offering possible solutions based on his or her experience and education. While the senior manager may not be an engineering expert, he or she can certainly set the standard or performance that the engineering department is expected to maintain. For many top managers, this can be assured by working on one simple assumption: the designer has not done the job correctly if it is necessary to add guards to machinery to protect against hazards.

CLASSES OF HAZARDS

In addition to identifying a hazard, it is extremely useful to classify it. The most obvious benefit of classifying hazards is its usefulness in establishing priorities. The hazards posing the greatest threat to the facility and personnel can then be given the highest priority for corrective action.

A common method of classifying hazards is:

Class A hazard. A condition or practice with the potential for causing the loss of life or limb, or the loss of a structure, equipment, and/or material.
Class B hazard. A condition or practice with the potential for causing serious injury or property damage, though less severe than Class A.

Class C hazard. A condition or practice with the potential for causing a nondisabling injury or nondisruptive property damage.

There are other methods of classifying hazards, but any sound system should immediately describe the *loss potential.* If such a classification system is known and applied, the busy senior manager can scan reports and pick out items requiring immediate attention (such as Class A hazards), while less hazardous conditions (Class C) can be dealt with in a more routine fashion by the supervisor, safety committee, or maintenance staff.

MACHINE GUARDING

So far, we have stressed the fact that safety and health on the job begin with sound engineering and design. The engineer and designer will be familiar with most of the common hazards to be dealt with in the design phase. For the senior manager, however, a brief overview will probably be useful, highlighting the most common hazards found in equipment and the ones requiring particular alertness.

The most common sources of mechanical hazards are unguarded shafting, shaft ends, belt drives, gear trains, projections on rotating parts, chain and sprocket drives, and any exposed components of rotating parts. Where a moving part passes a stationary part or another moving part, there can be a scissor-like effect on anything caught between the parts. A machine component which moves rapidly with power, or a point of operation where the machine performs its work, are also typical hazard sources.

There are probably over 2 million metalworking machines and half that many woodworking machines in use that are at least 10 years old. Most are poorly guarded, if at all. Even the newer ones may have substandard guards, in spite of OSHA requirements. Guarded machinery may be roughly categorized as:

1. Forming, punching, and shearing machines, including power and hand presses, forging hammers, power brakes, and shears.
2. Milling, drilling, and drinding machines, including drill presses, boring mills, lathes, shapers, and grinders.
3. Sawing, ripping, and planing machines, including circular and band saws, jointers, and planers.

There are several factors which affect the safety of machine design and make it an increasingly important matter:

User demands. Machine purchasers are demanding that the tools and machines they buy meet OSHA, local, state, and company safety standards, particularly machine guarding and noise standards.

Competition. Machine manufacturers are compelled by competition to develop safer machines or tools in order to market them to safety-conscious buyers.

Product liability. The growing threat of product liability claims is forcing machine builders to produce the safest product possible.

Safe machinery is not a concern of U.S. industry alone. For instance, the Factory Act in the United Kingdom for supplying safe machinery states: "It shall be the duty of any person who designs, manufactures, imports, or supplies any articles for use at work to insure, so far as is reasonably practical, that the article be so designed and so constructed as to be safe and without risks to health when properly used." This is equally appropriate to U.S. industry and is a sane guideline to follow.

All of this adds up to the fact that it is no longer acceptable merely to supply bare and unguarded machinery, leaving it to users to design and improvise their own safeguards later. The time and effort required to remove the hazard from the job are sometimes better spent in removing the human being from the job instead and automating a machine or process. Machine designers and manufacturers are increasingly aware that this is becoming more and more a part of management thinking, and most of them have responded by attempting to design equipment with maximum built-in safety.

Managers have an obligation to purchase and install machinery in compliance with safety and health laws. Machinery builders in general are aware of these laws, but they are also aware that their customers differ in their commitment to safe operation. Generally, a manufacturer will have several models available, ranging from those with subcompliance-level design and guarding to models with extra safety features designed in. Generally, the safer the model, the greater the initial cost. The decision to buy cheaper equipment, however, may prove to be a false economy. When compliance inspectors discover the lack of safeguards, pressure will be brought to fulfill the safety laws. The best guards are designed and built into the mechanical operation as a basic part of the equipment. A machine with guards added later is not only more costly, it is never as safe as a machine which was well designed in the first place and is as economical to operate.

OSHA legislation should not be taken as the highest or most desirable standard; in fact, its requirements are frequently minimal. Guarding fabricated in the plant may actually be better than that specified by the standards. The manager should not only be aware of this fact, but should also keep in mind that machine guarding is not the only way to keep the company hazard free. The manager should learn about other available techniques before making a decision.

When deciding whether to guard or not to guard, keep this in mind:

1. Never risk more damage than you can afford to pay for (possible loss of life or limb can never be fully measured or paid for).
2. Don't risk a lot for a little (that is, if cost is not a big factor, then use guards).
3. Always consider the odds. Consult the services of a safety professional, if need be, to find out how great the chances of an accident are.

Objectives of Machine Guarding

The basic objective of machine guarding is to prevent personnel from coming in contact with revolving or moving parts such as belts, chains, pulleys, gears, flywheels, shafts, spindles, and any working part that creates a shearing or crushing action or that may entangle the worker. Those actions or motions classified as the most dangerous are rotating, reciprocating, and transverse motions. These can result in in-running nip points, cutting actions, punching, shearing, and bending. Even a smooth, slowly rotating shaft can grasp clothing or hair; mere skin contact can force an arm or hand into moving machinery.

Machine guarding is visible evidence of management's interest in the worker and its commitment to a safe work environment. It is also to management's benefit, as unguarded machinery is a principal source of costly accidents, waste, compensation claims, and lost time.

Selection of Proper Machine Guards

Where guards are required, it is vital that they be properly designed, constructed from the right materials, and properly secured in place.

Selecting the right material for the guards is the first step. There are numerous guides for this, which are familiar to both designers and engineers. For the manager, however, two points are important:

1. Whenever possible, use fire-resistant material.
2. Ensure that the material is strong and durable.

Proper design is also essential. Here again, there are many guides available to the engineer or designer, such as tables that advise on mesh openings, minimum gages to be used, and heights above the floor or platform. The guard should be firm and secure in place, yet easy to remove. Where it is necessary to change belts, make adjustments, or lubricate, the guard should be provided with a hinged or movable section to allow for this.

The basic requirements for a mechanical guard are:

1. It should be sturdy and preferably constructed in one piece.
2. It should permit required maintenance without excessive disassembly labor.
3. It should have no detachable parts which, if removed, would affect its usefulness.
4. It should be properly mounted so as not to shift, rattle, or interfere with work.
5. It should be easy to inspect as part of the periodic checkup program.
6. It should, where possible, be part of the integral design of the machine, rather than something extra added on later.

Beyond these basics, a properly designed and installed guard should also:

1. Prevent unexpected tripping, slipping, and falling into or against a dangerous mechanical motion.
2. Protect the unexpected or reflex action of an employee who is not thinking.
3. Contain flying materials or particles, such as slivers, dust or debris, or broken parts or pieces of the machine itself.
4. Warn of potential danger if and when the guard is removed.
5. Protect the inexperienced worker who unintentionally places himself in a dangerous position.
6. Reduce the possibility of stock being unexpectedly thrust into the moving parts.

Types of Machine Guards

There are many types of machine guards, some with special features adapted to particular types of equipment. In general, the types of machine guards are:

Fixed-barrier guards. These prevent access to a danger zone but permit feeding and removal of stock. They can either be attached to the machine or an integral part of the tool or die.

Two-hand or remote controls. Where work requires access to the point of operation, the operator's hands are taken out of the danger zone by requiring both hands to be on controls or at a safe location to put the machine in operation.

Interlocked movable barriers. These prevent operation of the machine until the operator's hands are removed from the danger zone.

Presence-sensing devices. Basically, these are similar to movable barriers, except that electrical beams, magnetic fields, or other devices sense the operator or his or her hands and prohibit operation until the danger zone is clear.

Pull-out devices. Cables attached to the thumb, hand, or arm pull the operator's hands clear of the danger zone at the stroke of the machine.

Sweep devices. These are activated by the motion of the machine itself, and sweep across the point of operation just ahead of the closing of a nip or shear point to clear the danger zone.

Color Coding of Hazards

In addition to mechanical guards, color coding may be used to alert workers of hazards. Typically, there are standard colors which workers can learn to recognize. While there is no universal standard, and while systems may vary from company to company, bright and easily visible colors are generally effective. A typical coding system might be:

Black and yellow stripes: Dangerous hazards.
Red: Stop, danger, or fire (though it is sometimes also used to indicate a start button).
Green: Safety.

A common technique is to paint a bandsaw start button red to indicate danger when pushed. When the guard is removed, the inside might be painted yellow and black to warn of danger when open. A safety shutoff might be painted green. The combinations are many and are a useful supplement to the actual guards. But keep in mind that they are not an acceptable substitute for guards or other safety devices.

PURCHASING

The responsibility for providing safeguards should lie with the equipment vendor. Management should provide the vendor with guidelines, however, to indicate the standards expected. These should, as a minimum, conform to the standards of the American National Standards Institute (ANSI), as well as to applicable federal and state requirements. The only exception would be by written agreement with the purchaser specifying special, stronger, or better protection.

Obviously, where outside vendors are expected to provide materials and equipment to conform to certain safety standards, there must be an in-house source of information, a checkpoint for quality inspection, and an interface to ensure that these standards are actually met. In most cases, the purchasing department is the logical source of these functions. However, purchasing should work closely with both the engineering and safety staffs. These departments can help purchasing review all specifications, provide data or inspection services, and keep purchasing up-to-date on new developments in standards and equipment.

Purchasing, with this help from the safety and engineering staffs, can do much to ensure that proper safety standards are met. They should insist that machines be delivered already equipped with safety features. In cases where the safety equipment cannot be built to company standards, the company's safety and engineering staffs may have to take on the task of making the process or machine safe before it is used. Purchasing should inform these staffs when such a need arises. In all cases, purchasing should be aware that substandard materials and equipment may constitute a safety hazard, and should be guided by safety as well as cost considerations.

THE COST OF SAFE PROCESSES

Much has been said about the costs of accident prevention. In examining the area of engineering controls, we must specifically ask, "What is the cost of safer processes?" Three considerations involved are:

1. Design costs.
2. Operation costs.
3. Planning and consequence limiting costs.

We must point out, however, that even if these costs are changed to include safety considerations, they do not materially increase the price of a process or equipment operation (see Figure 9-12). That is, design costs, operations costs, and planning costs are not substantially increased when they include safety considerations.

Besides the three costs listed above, there are additional important design and process safety costs. These include:

1. All redundant control features for safety purposes.
2. All machine guards and protective devices installed on the machine.
3. The increased cost of additional metal thickness required by design codes to provide a margin of safety.
4. Systems installed to remove toxic and explosive materials produced by a process, or to maintain a safe and healthy working environment.
5. Fire protection measures, including escape routes, fireproof materials, and barriers.
6. Layout costs necessary for protecting the particular process.

These are only some of the financial commitments involved in engineering controls. However, it is nearly impossible to separate precisely the costs of safety from the costs that would be encountered in any sound and efficient

PRACTICAL APPLICATIONS OF ENGINEERING CONTROLS 253

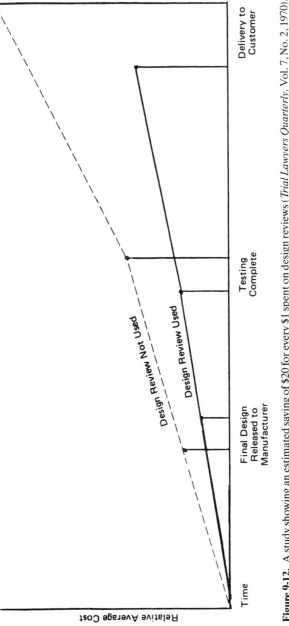

Figure 9-12. A study showing an estimated saving of $20 for every $1 spent on design reviews (*Trial Lawyers Quarterly*, Vol. 7, No. 2, 1970).

operation. This is particularly true when safety considerations are integrated in a process from the design phase on.

While recognizing that engineering is not usually the manager's field of expertise or strongest interest, we must emphasize that he cannot afford not to make it his business. Engineering controls are an integral and necessary part of any effective safety and health program, just as they are a vital part of designing and operating a profitable business.

BIBLIOGRAPHY

Greenberg, Leo, and Don O. Chaffin, *Workers and Their Tools,* Pernnell Publishing Company.
Johnson, W. G., *Management Oversight and Risk Tree,* Washington, D.C., U.S. Government Printing Office, 1973.
―――, *MORT Safety Assurance System,* New York, Marcel Dekker, 1980, Chapters 3 and 4.
"Static Electricity," *Safety in Industry,* Washington, D.C., U.S. Department of Labor.
Trial Lawyers Quarterly, Volume 9, No. 2, 1970.

Chapter 10

Management's Stake in Accident Investigation

So far, we have spoken mainly of management's interest in preventing accidents and devising systems that will reduce the number of accidents on the job. The fact remains, however, that accidents do happen, even in the best-managed systems. Management's actions following an accident, particularly in accident investigation and follow-up, can be crucial. With a properly conducted accident investigation, the manager can unearth the causes of accidents within the system and take remedial action that will not only increase workers' safety and health but strengthen the overall effectiveness and profitability of the organization.

THE IMPACT OF ACCIDENTS

There are few times when management needs help more than it does when there has been a significant (and perhaps costly) mishap on the premises. It may seem as if everyone and everything puts pressure on the principals of the company when an accident has taken place. A few of these pressures are as follows:

1. There is a disruptive event calling for some type of decisive action.
2. Chaos accompanies every accident, and order is needed.
3. Work processes are disturbed, sometimes shut down.
4. Profit margins may shrink or vanish.
5. Expensive machinery may be damaged or lost.
6. There may be dollar losses fay beyond insurance coverage.
7. Valuable members of the organization may be taken out of action.
8. The news media may play up the incident, and bad press coverage can damage the corporate image.
9. Families of the injured may need attention.
10. Public confidence in the organization may be shaken.
11. Stockholders and board members may demand explanations.
12. Governmental investigations may result, with attendant finger pointing and time-consuming demands for attention. Violations of regulations may surface.

13. Doubt will be cast on other operations in the organization. Operational, supervisory, and management changes may be required.
14. Insurance and workers' compensation costs will probably rise as a direct result of the mishap.
15. On-the-fence customers and clients may take their business to the competition.
16. Top and middle management may have to spend long, valuable hours engaged in investigating and taking corrective action.
17. The morale of the whole work force may be adversely affected.

These are only a few of the management concerns when a mishap occurs. They may not all be present, but they are all possible after an accident, and they point up management's need for involvement. A look at accident investigation itself, before the accident has happened, will ease many of these pressures that can later arise from an accident.

The Cost of Mishaps

Most people think of the cost of an accident solely in terms of themselves, that is, the direct cost of a mishap that involves them personally. This concern usually surfaces when serious injury or loss involves our family, our friends, or our property. We think of lost earnings, lost transportation or valuable goods, or the inability to perform some task or reach some goal. Think of any recent mishap that has occurred to you. Consider the cost in terms of lost salary, inconvenience, delays, pain, hospitalization, or any undesired consequence. Multiply this by millions of similar events each year. It adds up to a $143.4 billion loss each year, according to the National Safety Council—a fantastic drain upon our country and resources.

How much did that single undesired event that you considered cost? Consider just the hospital or medical costs and lost wages (generally known as "direct costs"). It is fairly easy to place an exact dollar amount on these direct costs. However, we are seldom able to assess all of the other (or "indirect") costs of the accident. Estimates of such costs range from 2 to 200 times as much as the direct costs. A conservative ratio is 3:1. That is, the indirect costs run an average of three times the direct costs.

In other words, if your accident involved hospital bills and lost wages of $5000, then we should consider at least $15,000 in indirect costs—for a total of $20,000, the average total cost of a disabling injury today. We tend to overlook these indirect costs because they do not come directly out of our pockets. But, if we consider what our accident cost in lost production, cost of temporary replacement, damaged property to replace, or lost sales or service, we can readily see that it costs someone a good deal more than $5000. The manager is all too aware that the "someone" is usually his or her organization.

The point is that no matter how small the cost may seem to the individual, it is invariably far greater when the indirect costs are considered. On a national scale, the costs of accidents have increased more than the costs of inflation, even in those years where accident rates and numbers of accidents have decreased. Moreover, accidents take approximately 100,000 Americans lives every year. Some businesses have actually been forced to close because of accidents. Their closures, in turn, have cost thousands of workers their jobs. It is estimated that roughly 75 million worker-days are lost to accidents each year due to work mishaps. Imagine what this means in terms of lost production alone—particularly to a society where the productivity of workers is a key factor in competitive workplaces that must emphasize production costs and quality workmanship to stay in business. The fact is that that we cannot afford accident, injury, and environmental mishap costs. When the cost is prorated among all workers, it is about $410 per person (In 1988). This grim picture calls attention to three more facts of interest to the manager:

1. The cost of accidents grows faster than the rate of inflation.
2. The cost of health and safety prevention programs goes up faster than the cost of accidents.
3. We may be able to control this vicious circle of spiraling costs through more effective mishap investigations.

Management Inefficiencies

Safety and health managers have long recognized that allowing mishaps to occur is a costly way to do business. There is little justification, on moral or sound business grounds, for not preventing mishaps and injuries as much as we can. Astute managers realize that mishaps signal that there is something wrong with the way their organization operates, and recognize that they are signals of management malfunctions at some level of operation. This strongly implies that management's inefficiencies are powerful contributors to mishaps.

It may take many such inefficiencies, even hundreds, before a recordable mishap occurs. Indeed, management inefficiencies may *never* result in a recordable event but they cost the company money all the same, day in and day out. Without a recordable event to bring them to management's attention, they continue to drain company assets and diminish profits for years. Mishaps are a sign of inefficient operation, bad business, and poor operating practices, and provide a sharp focal point for examination. Good mishap investigation uncovers, among other things, management inefficiencies which contribute to the event. Correcting them lessens the opportunity for similar mishaps and results in more efficient operations.

Morale

Have you ever noticed what happens after an accident at work or play? In a work situation, people generally stand around and talk for a while, returning to work only slowly and reluctantly, subdued and sobered by the event. Their work pace slows down, and they are wrapped up in their own thoughts and concerns. In some situations, the business may close early, people will simply go home early. Their morale is low, and they don't feel like working. Their interest and productivity drop.

This effect often results in a chain reaction that spreads to other departments or activities. When the activity does continue, another mishap may quickly occur, since people are distracted and don't have their minds on their tasks. Even if no second mishap occurs, production and activity drop noticeably. If the mishap also calls attention to a bad or hazardous work situation, there may be far-reaching results, well beyond lost work time or lowered production. Low morale may result in workers' complaints or refusal to work, decisive union action, or worker and union pressure to do something about the bad situation. Teamwork may fall to a low point as each individual becomes preoccupied with the accident.

While lowered morale resulting from mishaps cannot easily be measured, its presence is definitely felt, both on the job and in the profit picture. Good mishap investigation can be a vital part of management's efforts to restore morale. Where workers are aware that management is taking immediate steps to track an accident to its sources, many of the less obvious effects of lowered moral are eased. Rumors can be quickly dispelled, the record can be set straight, workers' fear or concern over hazardous conditions can be relieved, and even the process of answering specific questions about the accident can help workers refocus their attention and dismiss the accident from their minds, confident that something is being done about it. Properly conducted, an immediate accident investigation demonstrates management's interest, concern, and good intentions. Even more important, it can prevent a similar mishap from happening in the future, further damaging an already demoralized work force.

Public Concern

At any given time, we can expect mishaps of public concern to be in the news. It may be an aircraft crash, a school bus wreck, an oil spill, a fire in a senior citizens' home, a mishap in a nuclear facility, soil contamination from hazardous chemicals, or a broken dam. These events attract a high degree of public interest and bring adverse public opinion to bear on the company and the operation. The loss in travel, sales, or orders is well known. We hear of canceled aircraft sales, demands for new school bus designs, refusal to buy a refinery's output, a clamor for criminal neglect proceedings, replacement of high government officials, a call for indemnity payments, and tougher standards and inspection requirements. A mishap often results in loss of stockholder confidence, dropping stock prices, canceled expansion plans, and so on. This all comes as a result of public concern over mishaps.

This concern may be misplaced, may be proper, or may result in inappropriate corrective action. But it is a public concern, and it does get action. On a far smaller scale, thousands of times each year, there are local mishaps that generate the same type of local action as a result of loss of public confidence and the concern of the citizenry. In your town, there are places where people would rather not work because of a history of mishaps; this means less qualified workers going into the work force or good workers leaving. There are roads you avoid because so many mishaps occur on them; businesses in that area suffer as a result. We urge our children to avoid certain sports because of injuries to local players; the sport fails in the neighborhood. There is a lawn mower that you won't buy because a neighbor was hurt using one; there are fewer sales for that dealer.

Public concern is a strong weapon against those considered at fault in mishaps, particularly if they represent an establishment subject to public suspicion. Public alarm over a mishap will not only be reflected in loss of sales and goodwill; the public may become antagonistic as a result of their concern. Once adverse public reaction builds because of a mishap, it may take years to overcome. It may never be overcome. Events, or even possible events that generate public concern, are extremely expensive for a company. They require the company's most costly resources to counteract them, in the form of many years of high officials' times and the use of expensive law firms. It is small wonder, then, that most companies immediately respond to public concern when a mishap occurs. Taking corrective action is far more costly than preventing the mishap in the first place. Some organizations spend much of their resources preventing and investigating mishaps that might be of public concern, even though the dollar loss might be low. For example, we might spend a lot more to investigate a minor mishap with dangerous chemicals and no injuries than we would on a fatal traffic accident.

WHO ARE THE STAKEHOLDERS?

Who has a stake in mishaps? The most concerned stakeholders are the victims, whether individuals or the corporation. On the corporate level, the prime stakeholders are the senior executives, who must cope with the consequences a mishap has on the corporation as a whole. Often the senior executive has the most direct stake in uncovering the causes of a mishap. This section is directed to those with a personal stake in investigation because they must direct, conduct, review, or apply their findings of the investigation in some way. The stakeholders may be those who carry out investigations, such as the supervisor/foreman, line manager, safety professional, staff manager, special committee, or outside consultant. This includes the specialized investigator who, dues to his or her special skills and training, spends a great deal of time investigating mishaps. Each of these stakeholders may have special problems and special investigative strengths, which should be reviewed.

Senior Management

We tend to think of mishaps, chemical spills, and health hazards as being concerns of the operating level. However, recent developments spurred by public concerns and regulatory actions have forced senior management and staff into the picture. They have become aware, often painfully, that they are being held liable, sometimes personally by name, for anything that affects their employees, community, profits, and operations. This pressure is heightened by public and regulatory concerns about mishaps that involve the environment and disease. Senior management's increased visibility in this area has mandated a more personal stake in mishap investigation.

Senior management and staff that are involved in mishaps include, for example:

1. *Chief executive officer (CEO)*—overall corporate performance.
2. *Chief financial officer (CFO)*—bottom-line performance.
3. *Risk manager*—resource management to cover losses.
4. *Division vice president*—direct responsibility for operations and performance deficiencies.
5. *Plant manager*—his or her workers and operation.
6. *Personnel manager*—employees and workers' compensation.
7. *Corporate safety and health manager*—hazard and mishap avoidance resource.
8. *Purchasing/logistics manager*—safe products and machinery.
9. *Medical director*—safe and healthy work force and workplace.

These are only some of the senior staff who have responsibilities, and a sample of their stake in mishap prevention and investigation.

Supervisor/Foreman

The supervisor/foreman is the person closest to the action. Since the accident takes place in his domain, he often investigates the mishap. When his operation is so large as to experience numerous mishaps each year, someone will probably help with the investigative duties. In large corporations, the supervisor or foreman may conduct only those investigations involving minor injuries or small losses of resources. However, whether or not he actually conducts the investigation, the supervisor of the operation will always be involved in furnishing much data and information. He may prepare all or some of the reporting forms.

The supervisor/foreman seldom has specialized investigative training. While there may be reasons for this lack of training, there is no real excuse for it. If it is the supervisor's job to investigate, he has every right to expect management to prepare him for the task. Training is available from several sources, and management would do well to secure it.

It is reasonable to expect the supervisor to investigate mishaps that occur in his department. His personnel, equipment, and operations are involved, and his operations were being carried out when the mishap occurred. He knows more about the operations than anyone else. Moreover, he is considered responsible for the mishap in this area. He may, in fact, face reprimand or demotion if mishaps occur too frequently. He definitely has a stake in clearing up the causes of accidents.

The very reasons that make the supervisor/foreman the logical person to conduct the investigation are also the reasons why he should not be involved. His reputation is on the line. The causes uncovered may reflect adversely on his effectiveness or method of operation. His closeness to the situation may preclude an open and unbiased approach. The more thorough the investigation, the more likely he is to be implicated as contributing to the event. While it should never be the objective of an investigation, some stigma may invariably be involved. For that reason, it requires all the integrity and objectivity a supervisor can muster to carry out an investigation of his own department.

Regardless of who actually performs the investigation, the supervisor/foreman should be informed of the findings. The investigator must turn in a thorough report, with firm findings and recommendations for specific corrective actions to keep the event from being repeated. The supervisor will have to carry out most of the recommendations. He can do more to prevent a recurrence than anyone else, since he controls the day-to-day operation. He can communicate directly with department personnel to ensure corrective action or compliance with rules. He knows how to get information from his own group. He speaks their language and generally has their trust. In short, he is in a better position than anyone else to prevent a recurrence.

Finally, the supervisor/foreman is a major stakeholder in the smooth and efficient operation of his department. Correcting hazardous conditions and operational errors uncovered by a thorough investigation is essential to the welfare of his immediate operation. It ensures increased production and reduced operating costs, and shows that he is in control of his operation.

Line Manager

The line manager is the person in the management chain of command who has authority at least one level above the first-line supervisor. He deals with the supervisor and foreman, rather than with the workers, and can be considered middle management. His involvement, however, stems from the same background as the supervisor's. At a higher level, his people and his operation are involved; he has overall authority to take corrective actions on deficiencies uncovered by an investigation. It is to his advantage to see that mishaps are promptly investigated and corrective actions are quickly implemented.

He may merely review the investigations carried out by his supervisors or foremen and must be able to recognize when a thorough job has been done, as well as being able to tell when the findings are credible. Like the supervisor, his reputation is riding on a good investigation, so the same drawbacks apply to him that apply to the supervisor. The deficiencies that appear as the cause of an accident may be his own, rather than the supervisor's.

Safety Professional

In an organization large enough to have a full-time safety professional, that person should be considered the organizational expert on accident investigations. While the supervisor may actually conduct most of the investigations, the safety professional will have a major role in reviewing accident reports for adequacy and thoroughness and, most important, to see that the appropriate corrective actions are suggested and carried out. His role in evaluating investigations permits him to function as an advisor to top management. However, desirable it is to keep the safety manager in the role of evaluator, there are appropriate times for him to conduct the investigation as well.

The safety professional is more likely than anyone else to have developed investigative skills, particularly as a result of the special training that has been part of his professional education. If the supervisor receives special training in investigation, the safety professional will often be the instructor, or at least arrange for the training. This requires a higher degree of skill than that needed by the other members of the organization. This skill will certainly not be wasted. Where the supervisor does not have adequate time for investigative duty, the safety professional may have to do it.

There are also instances in which it is preferable to have the safety professional perform the investigation rather than the supervisor or line manager. The safety professional should make the investigation when:

There are implications of high public interest.
There have been a series of similar mishaps.
A complex case calls for investigative skills beyond those of the supervisor.
There is considerable potential for liability actions.
There has been loss of life or of large resources.
There is likely to be an investigation by governmental agencies.

Another reason for having the safety professional investigate accidents is that he is not normally a stakeholder; that is, he has no vested interest in the mishap or the findings. With no special interest in the department concerned, he can assess facts and information free from bias, prejudice, and the pressures of responsibility for operation. This advantage may be reduced when workers or supervisors fear that he is too management oriented or when the findings of an investigation may reflect on his own competence.

Staff Managers

Staff managers are those persons at the middle management level who serve in an advisory capacity. Examples include the personnel director, medical director, labor relations manager, and supply/logistics manager. These persons are seldom involved in line operations, but their functions are often enmeshed in the causal factors of accidents, so they should be familiar with the investigative process. For example, if mishaps are occurring because workers lack the physical ability to handle the task, it might be necessary to review the worker selection process with the personnel and medical directors. Should it turn out that faulty materials or equipment are involved, the purchasing manager may be a stakeholder.

Since staff managers are far removed from the front line of first-level supervision, there may be a tendency to overlook them. They can, however, play a critical role in uncovering the causes of mishaps, and they may serve as valuable experts on whose knowledge the investigator can draw.

Committee Investigations

Investigating committees can fall into two classifications:

1. General or standing safety committees that investigate mishaps as one of their normal functions.
2. Special committees appointed for a particular investigation.

The organization's general safety committee is often charged with investigating accidents when there is no one else trained or available to do the job. In a smaller organization, mishap investigation falls within its scope. This can work to the organization's advantage when committee members serve for extended periods of time. It allows members to develop broad experience and knowledge of plantwide activity, as well as experience in applying investigative methods. In many organizations where the safety committee performs the investigative functions, special training is provided in investigative techniques and strategies.

Since they operate on an organizationwide basis, members will conduct more investigations than if they were investigating only their own departments, which further sharpens their investigative skills and experience. In addition, the committee approach has the advantage of bringing people with diverse backgrounds and knowledge together to examine a specific problem associated with a mishap. Since they do not represent just the department involved, they have no vested interest, and can be more objective and unbiased. For this reason, they can often make better investigations and recommendations. Committee findings also tend to be better received than those coming from representatives of the management/staff function, particularly where committee members are drawn from the supervisory and worker levels.

Even in organizations where investigations are routinely handled by other persons, special investigative committees are sometimes appointed after unusual or very serious mishaps. The very nature of a particular incident may call for special expertise or knowledge (as in the case of fires or of mishaps involving radiation or explosives) in order for an in-depth investigation to be made. This is particularly true where public concern, liability, workers' compensation, or regulatory violations may be an issue. Where the consequences of a mishap are particularly grave, only a special committee may be considered objective enough to perform a thorough and unsparing investigation.

As with general safety committees, a special committee's findings may be better received, particularly when fellow workers or supervisors appear to be involved as causal factors in the accident. Committee findings may also receive a better reception when the investigation results are made public or reported to the work force at large.

For two reasons, care should be taken in choosing the members of a special committee. First, it is essential to secure members who have the required expertise or knowledge if that is a major factor. Second, they should be respected by workers and management alike, and should not be regarded as representing one viewpoint exclusively (i.e., a committee composed only of members of the top management, supervisory, or worker levels).

Committee action does have several disadvantages, however. First, it takes people away from their normal routines to do what some consider the supervisor's job. Occasionally, committee members will resent the system, feeling that it makes them perform the unpleasant task of finding fault with their peers. There is also the danger of unwise compromise to reach consensus, which may simply obscure essential points.

Other Investigations

It may not be necessary for the organization to perform all its own investigations. An insurance carrier may investigate certain mishaps, using its own investigators, as in the case of incidents involving boilers, vehicles, or fires. In some cases, governmental agencies will conduct an investigation. In either of these situations, facts needed for other purposes can be taken from their reports. However, the manager who is truly interested in using investigations as an accident prevention technique may find that the insurance company or governmental agency investigation is not adequate, or that it concentrates on issues other than management inefficiencies or control systems. An in-house investigation may still be desired, if only to supplement the outside work. Results of investigations by insurers and governmental agencies are not always available. In any case, these investigations seldom reveal all the causes and costs needed to develop a comprehensive plan of mishap prevention.

PRIORITIES FOR INVESTIGATION

There is no question that management should investigate matters of great public concern, serious injury, high property damage, or fatalities. This, in fact, may be required by law. Where managers fail in this regard is in making a sufficiently detailed investigation to aid them in preventing similar incidents or in applying the findings to correct the underlying causes.

The term "mishap" refers to all of those unexpected, undesired events (many of which are called "accidents") that cause loss of resources. There are various definitions of "accident." For example, OSHA defines it in terms of injury or illness. The firm's auto insurance carrier may use a broader definition. Some governmental and private entities may require an accident report based strictly on dollar costs. When this happens, some mishaps (accidents) costing hundreds of thousands of dollars in lost resources may not even be recorded because there was no injury or illness. Many firms, in fact, conduct investigations only if required by law. We have made it plain that any unwanted, unexpected loss of resources is considered a mishap, even if not officially recorded.

This is a shortsighted attitude on management's part and may end up costing the organization enormous amounts of money in the future. Suppose, for instance, that corporate regulations define an event as reportable if it causes $250 worth of loss or damage, without an injury or illness. If the loss is only $245, the company reasons that an investigation is not required. This reasoning contradicts the whole purpose of the investigation. All of the same causal factors present might later cause a $10,000 accident. Had it not been so intent on that $5 difference, management might have conducted an investigation that would have prevented the second, costlier event.

A dollar value may not be the proper basis for determining if an investigation is required, but by the same token, neither is an injury. For each recorded injury, there may be hundreds of similar mishaps that caused no injury at all. The same causal factors are still present. The same mistakes of management and supervision are there, as are the same operator errors and design deficiencies. The ideal approach is to investigate all undesired, out-of-the-ordinary events that keep workers from doing the job efficiently. They deserve the same careful scrutiny that a loss of life or a serious injury would bring. In fact, investigating these apparently minor mishaps is the most efficient and profitable way to prevent the very costly accidents that may eventually result from the identical causes.

Management should, therefore, investigate any mishap that may reveal causal factors which could also cause reportable events. Where resources do not permit an investigation of all mishaps, a selective sampling process may be useful, as may the investigation of mishaps with the greatest potential for damage. Another alternative is to investigate all incidents that the law requires, plus every fifth or tenth mishap, whether it appears to be significant or not. This regular sampling may yield extremely valuable results or may pinpoint areas where a thorough review of operations would prevent potential losses.

Senior managers sometimes question the wisdom of investing resources in the prevention of insignificant events, rather than concentrating on costly or catastrophic events. Since investigation is a preventive technique, they reason, inquiring into minor mishaps is not a profitable use of resources. If even the minor mishaps require thorough inquiry, the manager may be uncertain about what constitutes a "thorough" or "proper" inquiry.

When should there be a detailed investigation? Consider the following:

1. If the operating environment involves high hazards, the prevention efforts (investigation included) deserve a correspondingly high investment of resources. Highly hazardous operations have the same causal factors as less hazardous operations; only the size of the probable loss or the likelihood of mishap is different. Even the less hazardous operations can involve very costly resources and deserve careful attention.
2. A situation with supposedly minimum loss potential is often deceptive, ultimately involving very high losses. The relatively common back or eye injury is a case in point. This may be given scant attention because of its very simplicity and the fact that extended absence from the job is either rare or is fully covered by workers' compensation. In reality, these are two of the most expensive injuries and involve large hidden expenses for management. The total cost of lost production, increased insurance costs, training replacement, worker rehabilitation, and other benefits for a single worker disabled by a back injury may be equal to that of several deaths.
3. Sophisticated investigators and managers realize that even the common, minor events that seem to cause only inconvenience, irritation, or delay are symptoms of inefficient management and operations. They consider the investigation of each as simply good business. Not only can the elimination of these minor problems bring major safety and health benefits, but they also allow management to achieve greater overall efficiency and profitability.

Astute managers recognize all these factors, and frame investigative policies that will take them into account. They recognize that they, as senior executives, are top stakeholders in accident investigation, and that it is in the best interests of themselves and the company to make accident investigation a key part of the overall safety management program.

HOW TO ENSURE A GOOD INVESTIGATION

Most people think that a good investigation is made when competent persons examine every possible factor that may have been involved in a mishap and include it or discount it in the final report. Far from it! The most important things in any investigation, regardless of the resources, are preparation and action.

The senior manager is the key to a complete, quality mishap investigation, not by doing it personally, but by ensuring that certain preparations and actions take place. Ensuring a good investigation involves three steps:

1. *Prepare* for a mishap investigation before the mishap.
2. Use the right resources and techniques to gather *information,*
3. Take the proper *action* after investigating the causal factors.

These three steps—preparation, information gathering, and action—can be further simplified by highlighting the two most important factors in excellent investigations:

1. *Preparing* properly for mishap investigation.
2. Taking the right *action* after the report forms are completed.

To summarize, the most important factors ensuring a good investigation are what is done *before* the mishap happens and what is done *after* gathering all of the information and completing an investigation form. These two critical steps can only take place with the direction, guidance, and support of senior management. The best possible investigation of facts surrounding a mishap can be made only if the organization is prepared to investigate. All of the facts and findings in the world will not do any good if no action is taken on them. Thus preparation beforehand and action afterward are the keys to the best investigation. Further, being prepared to investigate and taking action on the findings ensures that the most information is uncovered and acted on, regardless of the resources available.

BIBLIOGRAPHY

Accident Facts, 1989 Edition, Chicago, National Safety Council, 1989.
Accident/Incident Investigation Manual, Washington, DC, National Technical Information Service (NTIS), 1976.
Benner, Ludwig E., Jr., *Four Accident Investigation Games,* Oakton, VA, Lufred Industries, 1979.
Ferry, Ted S., *Elements of Accident Investigation,* Springfield, IL, Charles Thomas, 1978.
——— "Accident Investigation and Analysis: A Dozen Steps for the Safety Professional," *Professional Safety,* Park Ridge, IL, American Society of Safety Engineers, January 1981.

——— *Modern Accident Investigation; An Executive Guide,* New York, Wiley Interscience, 1981.
——— *Accident Investigation for Supervisors,* Des Plaines, IL, American Society of Safety Engineers, 1988.
——— *Modern Accident Investigation and Analysis,* 2nd edition, New York, Wiley Interscience, 1988.
Jefferies, J. W., "Accident-Injury Loss Sequence," *Journal of the National Safety Management Society,* n.d.
Nertney, R. J., and J. R. Fielding, *A Contractor Guide to Advance Preparation for Accident Investigation,* Washington, DC, National Technical Information Service (NTIS), 1976.
Pope, William C., "Manualizing Your Safety Function," Alexandria, VA, Safety Management Information Systems, 1975.

Chapter 11

Workers' Compensation and the Cost of Safety

THE DEVELOPMENT OF WORKERS' COMPENSATION

Workers' compensation, as everyone knows, is a system of insurance designed to protect workers who are injured on the job or who incur job-related health impairments. Among its benefits, it ensures prompt payment of medical costs and provides a portion of the injured employee's lost income. Generally, we think of the system's benefits to the employee rather than to the employer. The latter may not be so obvious, even to managers who are extremely familiar with workers' compensation. As far as they are concerned, workers' compensation laws require them (or their insurance carriers) to pay for the benefits due to the injured workers. However, under the workers' compensation system, liability actions against the employer are drastically reduced. The worker can collect without the employer or employee having to undertake costly legal action and court fees.

Liability and Compensation

Workers' compensation (comp) developed largely as a result of the problems that emerged during the Industrial Revolution. As countries evolved from agrarian-based craft economies into fully industrialized nations, it became evident that one significant price for this industrialization would be a greater number of accidents, often resulting in serious injury or fatality to workers. As industrial technology grew more complex and labor intensified, the number of severe job-related accidents grew.

Prior to the development of workers' comp laws, an injured worker's only recourse was to initiate a tort liability action against the employer, based on the employer's alleged negligence as the cause of the injury. Our common law system is deeply rooted in the principle of negligence, which holds that a person has a legal responsibility to exercise reasonable or due care in dealing with others. This principle asserts that a prudent employer has the responsibility to provide a reasonably safe working environment, with suitable warnings of hazards and reasonable safeguards on machinery and processes. If an employer fails to exercise due care, and this results in injury or damage to a worker, the injured worker can sue to recover the loss.

However, in order to win in court, the worker had to prove that the employer was negligent, and that there was a causal connection between this negligence and the injury. The employer, in turn, had several common law defenses available, posing significant obstacles to a successful action by an employee. For example, the employer could claim "contributory negligence" on the part of the employee. If an employee contributed to the accident through his or her own failure to exercise due care the employee could be barred from collecting any judgment.

Other defenses available to the employer were developed from the English law case of *Priestly v. Fowler* in 1837. This case established the doctrines of "assumption of risk" and "common employment," which were significant defenses for the employer.

Assumption of risk means that an employee who voluntarily accepts employment knows or should have known about the risks of the job. By accepting the employment, the employee also accepts the risks. It is also emphasized that the employee is being paid sufficiently to compensate for the risks of the job.

The doctrine of common employment means that the best judges of an employee's competence to perform an assigned task in a safe manner are the employee's fellow workers, since they are in a far better position to know the basics as well as the hazards of the job. Furthermore, these fellow employees, should they discover unsafe actions on the part of others, have a responsibility to notify the employer. Their failure to notify the employer is the key point. If the employer did not know about the unsafe actions, how could he be held responsible for them or their consequences?

Besides these defenses, the employer had the advantage of superior economic power, which provided him with far better legal counsel. These factors, as well as a court system generally favorable to the employer, made it evident that injured employees would frequently be unsuccessful in tort actions. Even when the injured employee was successful, the wheels of justice moved so slowly that it was extremely difficult to obtain the award promptly. In the meantime, the injured worker was faced with the prospect of having no income, plus medical bills and ordinary living expenses for himself and his family.

It soon became evident that some accidents, however unfortunate, could not be attributed to any one person, but were the result of a new and complex technology. Existing laws simply did not apply to these new conditions. Also, injured and maimed employees became an increasing burden to society, since many of them could no longer obtain gainful employment. Inevitably, society began to demand that something be done to aid the victims of industrialization, both for humane reasons and to keep them off charity and welfare rolls.

Early Developments

The concept of workers' compensation originated in Germany during the late 1830s. In 1838, Otto von Bismark passed the Prussian Employer's Liability Law, which held railways liable to both the passengers and the employees in the event of train wrecks. In effect, this law established the worker's right to compensation. Later, the Unified Prussian States passed a Sickness Act, to which was later added an Accident Insurance Law. Under this law, the injured employee was given assistance without regard to fault or liability. The employee was required to help pay for the cost of this insurance, but it is important to note that the employer paid most of the cost. These actions established a far-reaching precedent: that of providing compensation without a determination of fault. Moreover, it laid the groundwork which ultimately required the employer to pay the cost of these benefits.

Numerous workers' compensation laws were passed in industrialized Europe during the latter part of the nineteenth century. These laws usually required an employer to provide prompt and reasonable payments for work-related injuries, regardless of fault. In return, the employee had to give up the right to sue the employer. The cost of workers' comp, it was reasoned, should be passed on through the chain of commerce as part of the cost of production.

Workers' Compensation in the United States

The development of workers' compensation in the United States lagged behind that of industrialized Europe. However, our own experience with industrialization was taking its toll, and the problem soon made itself felt. Severe work accidents in textile mills, mines, railroads, and other new industries created a public outcry which, inevitably, moved many states to take action. This action first took the form of employer liability laws, which were passed during the period from 1900 to 1910. These laws sought to hold the employer liable for work-related accidents; most of them, however, proved either unsatisfactory or unconstitutional. In 1910 the State of New York did pass a workers' comp law, but it too was later held unconstitutional.

In 1911, however, several states passed the first lasting workers' comp laws. These, like their European counterparts, were based on the no-fault concept of compensation for injuries without the need to establish fault or liability. By 1913 over one-quarter of the states had workers' comp laws to protect employees, and by 1948 all states had them.

Today, virtually all employment in the United States is covered by workers' compensation. The insurance for workers' comp can be compulsory or elective, depending on the specific state's laws. In states where it is elective, the employer may choose to accept or reject coverage. However, if he or she rejects it, the employer also loses the three common law defenses. For this reason, it is highly unlikely that an employer will reject coverage.

State requirements and regulations vary widely. Some states require coverage for private employment, while others may exempt those with a stipulated number of employees. Many states also exclude certain types of employment because of the nature of the work. For example, in some states, domestic servants, gardeners, and other casual employees are specifically excluded. (However, most jurisdictions permit employers voluntarily to cover employees in an exempted class.) Minors are covered by workers' comp in all states. In some states, additional compensation is also provided and penalties are assessed against the employer. Also, many states provide minors with special legal benefit provisions which may override "sole remedy" provisions.

Workers' compensation insurance was developed with these basic objectives in mind:

> To provide injured workers with reasonable and prompt income in order to cover most of the economic loss due to the injury, to take care of the medical costs associated with it, and to provide reasonable income benefits to dependents.
> To provide the employee with a "sole remedy" for work-related injuries, thus reducing the burden on courts caused by the delays, extra work, and long trials which occur in private liability litigation.
> To relieve the financial burden on public and private charities, which would otherwise have to care for and support uncompensated workers.
> To reduce the need for extensive legal fees to obtain attorneys and witnesses, as well as other legal costs of both the employee and the employer.
> To encourage employers to implement accident prevention programs (since continued accidents will inevitably increase the cost of compensation and, hence, the cost of production).

THE BENEFITS OF WORKERS' COMPENSATION INSURANCE

Benefits provided to injured workers through the workers' compensation system are usually in the form of:

1. Medical benefits.
2. Cash benefits.
3. Rehabilitation.

Medical benefits pay all of the costs associated with medical treatment as a result of an injury and are usually provided without a dollar or time limit. Most work-related injuries involve only medical benefits, since substantial physical impairment or wage loss does not occur. These medical benefits represent approximately 32 percent of the total dollars paid out in the compensation system. Medical benefits are unlimited in all states.

The purpose of cash benefits under compensation is to replace the loss of income or earning capacity of the injured worker due to the occupational injury or disease. The amount of the benefit and the length of time over which it will be paid are based on the type of disability involved. There are four types of disability:

Temporary total disability
Permanent total disability
Temporary partial disability
Permanent partial disability

By far the largest number of cases involve temporary total disability. This type of disability implies that, although the employee is totally disabled during the period when the benefit is payable, he or she is expected to make a full recovery and return to work, with no lasting impairment. (Permanent total disability generally indicates that the injured worker is totally and permanently unable to be gainfully employed.)

Income benefits for temporary and permanent total disability are expressed as a percentage of the employee's normal wage. Most states use a formula to determine the amount of benefits the person is entitled to, as well as to calculate the maximum and minimum benefits. Some states also limit the total number of weeks and the total dollar amounts of benefit eligibility. The current maximum weekly benefit in most states is 66⅔ percent of the injured worker's normal wage. Where there is permanent total disability, most states provide payments for life.

Most awards and most of the money paid out as income are for temporary total or permanent partial disability. Permanent partial injuries are generally divided into "scheduled" and "nonscheduled" types for the purpose of determining benefits. Scheduled injuries involve the loss of, or loss of use of, specific body members, where wage loss based on the nature of the impairment is presumed. Loss of a hand, eye, or fingers is compensated according to a predetermined schedule, hence the name. In most states, the actual amount payable is a specific number of weeks of benefits multiplied by the weekly benefit involved. Since disabilities involve current earnings or wage-earning ability, in many states weekly benefit payments for temporary or permanent partial disability of the nonscheduled types are based on the concept of "wage loss" replacement. That is, compensation pays the difference between the wages earned before and after the injury. In some states, nonscheduled permanent partial disabilities are compensated as a percentage of the total disability.

In the event of fatal injuries, the employee's spouse receives a fixed amount or a percentage of the deceased worker's annual income. Usually benefits cease if the spouse remarries. Dependent children also receive death benefits until they reach a specific age (though if the children are handicapped or unable to become independent, benefits may continue throughout their lives). An established amount is also provided to cover funeral costs for fatally injured workers.

In order to reduce the number of small claims, most states impose a waiting period during which income benefits are not payable. This waiting period does not affect medical payments, which begin immediately after the accident. If disability continues for a certain number of days or weeks, most laws provide for income benefits retroactive to the date of the injury. Thus, if a worker suffers only minor injury, causing him or her to miss only a few days of work, the worker is eligible only for medical payments. This, in fact, is usually the case: workers' comp pays for the medical costs, and the employee's lost work time is covered by the normal sick leave benefit provided by the company.

Workers' comp programs generally include rehabilitation as an integral part of the medical benefits. In most cases, this merely involves therapy to return the worker to a condition in which he or she can return to work. However, where permanent disabilities are involved, this rehabilitation may extend beyond medical treatment in the form of vocational rehabilitation and retraining to prepare the worker for a new kind of work suitable for the physical impairment. Rehabilitation benefits also include a subsistence allowance and special fund sources to finance the rehabilitation. The Federal Vocational Rehabilitation Act is now in effect in all states, providing federal funds too aid states in vocational rehabilitation for the occupationally disabled. Rehabilitation is considered mandatory, and an employee can lose up to 50 percent of benefits by refusing to undergo it. The employer or insurance carrier is responsible for providing up to 26 weeks of rehabilitation services, and up to 26 additional weeks if necessary. In most states the employee receives temporary total benefits of $66\frac{2}{3}$ percent of the weekly wage during rehabilitation.

Second-injury funds are another feature of most state systems. These were developed to meet the problems created when a preexisting injury is complicated or worsened by a second injury, producing a disability greater than that caused by the second injury alone. The purpose of these funds is to encourage the employer to hire the physically handicapped worker; in effect, they limit the impact of such compound losses on the employer, as well as equitably allocating the cost of providing benefits to such employees.

Under the second-injury fund program, the employer need only pay compensation related to the disability caused by the second injury alone. The employee, however, will receive a total benefit appropriate to the combined disability. The difference between what the employer pays and what the employee receives is paid out of the second-injury fund.

If there were no second-injury funds created by law, an employer might be held liable for compensation due for the total resulting disability, rather than just the injury that occurred while the worker was actually in his or her employ. (For instance, a worker who had lost one eye while working for Company A could lose the remaining eye while working for Company B, thus becoming totally blind, even though, technically, he lost only the sight of one eye while employed by the latter company.) Under those circumstances, an employer might be inclined to refuse to hire the handicapped because of the increased risk. For this reason, second-injury funds are advocated by most jurisdictions.

Compensable Injuries

Workers' compensation was designed specifically to cover the cost of injuries arising out of the work environment, rather than providing social assistance for accidents that occur on the outside. For this reason, certain basic requirements were established to define a "compensable injury," one for which the worker can collect compensation. The primary requirements accepted by all jurisdictions are that the injury must:

Arise out of employment (AOE).
Arise during the course of employment (COE).

These requirements suggest, first, that the worker's employment must be the source of the accident or injury (AOE). If no employment relationship exists, the injury is not compensable. Second, the accident must occur during the time, place, and circumstance of employment (COE). The intent of COE is to restrict compensation to those injuries that are directly tied to job activities. There has been considerable conflict over the precise limits of AOE and COE, and judicial or administrative rulings have tended to broaden these definitions over the years.

Some occupational diseases and cumulative traumas have been disputed, based on these AOE or COE requirements. It is generally agreed that these illnesses, to be compensable, must arise out of employment and be due to causes or conditions characteristic of and peculiar to the particular trade, occupation, process, or employment. For example, a respiratory ailment that can be traced directly to the worker's exposure to harmful dust in the workplace could be treated as an occupational disease, and deafness resulting from repeated exposure to excessive workplace noise could be considered a cumulative trauma. Ordinary diseases to which the general public is exposed (such as flu, or pneumonia) are specifically excluded. Highly subjective health impairments, such as a stress-induced emotional illness, are often questioned. It is debatable just how much of the stress is due to the work environment and how much to marital, financial, or other personal problems. These subjective illnesses pose a critical concern to workers' compensation administrators. First, the incidence of cumulative trauma cases and awards has been increasing significantly in recent years. Second, medical research has suggested certain links between workplace conditions and medical problems that may not be apparent for many years (as with certain carcinogenic substances, which were not recognized as cancer-causing agents during the worker's period of employment). The disease may not actually develop until years later, when the worker is no longer in the same company or industry.

Employer's Liability

The "sole remedy" or "exclusive remedy" concept in workers' compensation insurance protects the employer against worker liability suits in most states. That is, workers' comp is the employee's sole remedy, and the employer cannot be sued for additional compensation. However, employers' liability for work-related injuries may still exist in certain circumstances under most state laws. For example, one form of liability in most states can increase an injured worker's benefits to some specified maximum amount if the injuries are caused by "serious and willful misconduct" by the employer. These additional benefits, assessed as a penalty, are uninsurable and must be paid by the employer. A typical case might involve an employer with a long history of similar accidents which are controllable through reasonable action by the employer (such as refusal to supply exposed workers with guards or protective equipment).

On the other hand, in some states the employee's benefits can be reduced by a specified amount if he or she is found guilty of serious and willful misconduct which is the primary cause of the accident. However, these actions are rare exceptions, and very few cases are ever filed.

Some states also permit an employee to sue a fellow employee if his or her injuries were caused by an intentional assault by that person or if the fellow employee was intoxicated.

In many states, if the compensation law fails to provide a remedy, the injured employee may seek justice in court. These cases may occur, for example, if the employee is ineligible for workers' compensation benefits. This would include cases involving exempt employments, illegal employments, or noncompensable injuries or diseases. It might also include third-party actions, such as are common in contract construction. For example, a general contractor who hires a subcontractor to work on a job site may be sued by the subcontractor's employee if the employee sustains an injury due to the general contractor's negligence. In some states, even though an injured employee receives state compensation benefits, his or her spouse may sue the negligent employer for the loss of the partner's services.

An important area for third-party employee actions is product liability. Several recent decisions have supported work-related injury actions by employees in this area.

In one case, *Douglas v. Gallo Winery* (1977), the employer manufactured processing equipment, which was sold to other companies as well as being used by the company in its own operations. An employee who was injured by this equipment filed suit for product liability against the manufacturer, who was also his employer. This case did not really extend the scope of workers' compensation, but it does shift the worker from the category of "employee" to the category of "general public." Thus, the worker has the same right to sue that he or she would have had by purchasing the machinery for personal use, rather than using it in the course of employment.

In a more recent case, *Johns-Manville Products Corp. v. Superior Court* (1980), the California Supreme Court held that the employer may be subject to common law liability where the employer's intentional misconduct goes beyond his failure to ensure that the tools or substances used by the employee or the physical environment of the workplace are safe. Such cases would include an employer's physical assault of an employee; fraud; or an employee's injury related to, though separate from, an initial work-related injury.

In this case, the plaintiff alleged that the employer was aware that the worker had contracted a serious disease from the work environment but concealed this knowledge from the worker. This allegedly fraudulent concealment, according to the plaintiff, aggravated the illness.

WORKERS' COMPENSATION ADMINISTRATION

Workers' compensation laws are administered either through the state court system, by a special commission or board established for this purpose, or by a combination of both. The primary concern in administering a worker's comp program is to ensure the prompt payment and effective disposition of cases. Without an effective benefit delivery system, many of the problems associated with common law liability actions would remain.

An effective workers' comp administration system should include:

1. Supervision or monitoring of statutory compliance by employers, employees, insurance carriers, and the legal and medical professionals involved in the program.
2. Investigation and adjudication of disputed claims.
3. Supervision of medical and vocational rehabilitation funds.
4. Management and disbursement of second-injury funds.

5. Collection and analysis of occupational claims data, and evaluation of the program's effectiveness and efficiency.

The more important statutory provisions of administration involve the time limits in which employers must be advised of the employee's injury, the time limits in which the injury claim must be filed, the claim settlement conditions, and regulation of attorneys' fees in litigated cases. The administering agencies are generally given power to impose sanctions or penalties for violations of statutory requirements.

The Cost of Workers' Compensation

Employers are required either to obtain workers' compensation insurance or, if financially able, to self-insure. A few states require employers to insure through a monopolistic state fund; other states permit the employer to purchase insurance from a competitive state fund or private carrier. Most states permit the option of self-insuring as well.

The premium rates for workers' compensation are developed by the National Council on Compensation Insurance, which was established specifically for this purpose. The Council grew out of a 1915 conference which agreed that rate making for compensation programs could not be handled by each state separately. The Council was established as an independent agency whose primary purpose is to promulgate the premium rates charged by insurance companies.

As established, the Council collects data on workers' comp loss experience incurred by member companies. Future rates are developed based on an analysis of this prior accident cost data. The rates established by this agency serve as the standard for a nationwide rate-making procedure approved by the National Association of Insurance Commissioners.

The Council's basic manual is standard for all private insurance carriers. It establishes the Council's rules, procedures, and rates applicable to workers' compensation insurance. Since all states now have statutes regulating workers' comp rates, the Council must file annually all proposed rate revisions and all supporting data with these state authorities. In most states, public hearings must be held before the rates can be revised.

The workers' compensation premium "rate" is the amount charged per $100 of payroll. A base rate is established for each major work classification. The work classification system was established to group the work operations of all employers who have exposures common to that type of employment. Injury frequency and severity, as you might imagine, differ considerably between, say, retail grocery stores and heavy construction firms, so that a single rate would not be equitable for both. On the other hand, one construction firm's injury experience will be similar to those of other construction firms. Thus, it is necessary to establish rates according to work classification.

For example, the work classification code for a retail grocery store is 8006, and this code is assigned to all grocery stores. If the rate for this classification is $3.50, the insured will have to pay $3.50 for each $100 of payroll; the resultant figure will be the premium.

Most states have "minimum rate" laws, which make it illegal to charge less than the minimum rate assigned to the work classification. However, a few states have advocated "open rate" laws, which permit insurers to charge lower rates for some classifications whose loss experience does not warrant the collection of higher premiums.

Premiums can also be modified to reflect the individual employer's loss experience. This modification is calculated by comparing the employer's actual loss experience with the loss experience of all members in the same classification. (The loss experience developed for each classiciation is used to predict future losses. Thus, the insured is paying a premium based on his or her expected losses. If the actual losses are considerably different, an adjustment can be made.)

For example, if the employer's losses are consistently very low, he or she might pay 20 percent less than the average employer in this classification. On the other hand, if his or her losses are consistently much higher than the average, the employer might have to pay 20 percent more. Besides being an equitable way of allocating actual losses paid out, this system provides a powerful incentive for the employer to practice accident prevention and create a safer work environment.

Rising Compensation Costs

The explosion of workers' compensation insurance rates during recent years has caused great concern to employers and to the insurance industry. Some of this inflationary trend is tied directly to the spiraling cost of health care in general (since most benefits are paid for medical expenses). However, much of it is apparently the result of substantial improvements in the benefits available to injured workers.

Most insurance executives agree that the primary factors which have caused workers' compensation costs to increase sharply are:

1. High benefit levels have made the benefits more attractive to injured workers (in other words, they are worth striving to keep).
2. Liberal interpretations of workers' comp benefits by workers' compensation boards have increased the utilization of these benefits.
3. Attorneys and unions have assisted employees in gaining the maximum benefits possible from the system.

4. The number of recognized occupational diseases (as opposed to specific accidents) is growing and will continue to do so.
5. Many states are experiencing longer durations of claims for disabled workers. This may be an indication of a contentment with benefits and a loss of the incentive to return to work.

Some insurance executives have suggested that the sharp rise in workers' compensation rates may also be due to state efforts to increase benefits without taking corresponding action to tighten controls on the administration of the benefit delivery system.

A National Commission on Workers' Compensation was formed in 1972 to evaluate the workers' comp system nationwide. The Commission's basic conclusion was that state laws were not living up to their potential in providing benefits to employees. As a result, the Commission made eighty-four recommendations to improve the system, of which nineteen were considered essential. Almost all of these recommendations were related to the improvement of benefits, although little effort was made to pinpoint and attack administrative abuses of the system. Yet experience shows that the system can be needlessly wasteful, costly, and subject to abuse. Definite action will eventually be needed if compensation costs are to be brought back into line.

Workers' comp benefits are tax free. Therefore, as benefit levels approach wage levels, the incentive to return to work is bound to diminish, as the worker's benefit income may be equal to his or her regular wages after taxes. In addition, more than one member of the employee's family may be working. If one wage earner is drawing compensation benefits and several other members of the household are employed, the family may be as comfortable as or even more comfortable than it was before.

Thus, many insurance executives believe that both the higher benefits and the higher taxes deducted from regular payroll wages act together as an incentive to keep workers on disability rather than returning to work.

FUTURE TRENDS IN WORKERS' COMPENSATION

Many states are now attempting to reform their compensation laws to eliminate administrative abuses so that benefits will go only to the industrially disabled, as originally intended. One such measure is Florida's "wage-loss" system, which was established in 1979. Under this program, permanent partial disability benefits are provided for wages lost as a result of the disability, rather than for the disability alone. Thus, an injured employee will have to prove that the injury caused an earnings loss in any postinjury work; otherwise, the permanent partial disability is restricted to time away from work.

In order to renew supplemental wage loss benefits, workers receiving permanent total disability benefits must prove that they cannot earn 85 percent of what they earned before the injury. This program does, however, provide lump sum payments for the loss of body members, such as eyes, limbs, or fingers, or for death. The supplemental benefits under the program have been substantially increased as well.

There has been some congressional support for a national compensation program as a means of standardizing the benefits available to injured workers and complying with the recommendations of the National Commission. However, the National Commission itself was opposed to a federal compensation system and advocated maintaining the state programs.

The movement for a national compensation system has recently lost support, perhaps because of administrative problems in the existing federal programs. These, for instance, have resulted in skyrocketing costs for workers covered by the U.S. Longshoremen and Harbor Workers' Act and the Federal Compensation Act. Both of these federal programs have received sharp criticism, as they provide high benefits under excessively liberal guidelines, which bend over backward to resolve all doubts in the worker's favor.

Given current economic pressures and soaring costs, there will undoubtedly be more efforts to reform the administrative agencies which oversee compensation benefits. Ideally, the benefits should be provided only to those who have actually sustained work-related injuries or illnesses. The benefits received by an injured worker should be limited to a level that will encourage a return to work as soon as it is physically possible to do so. The level, in short, should neither create undue hardship nor provide a hidden incentive to remain "disabled."

CONTROLLING WORKERS' COMPENSATION COSTS

One of the most important objectives of the workers' compensation system is to prevent or minimize occupational accidents. It is generally agreed that our no-fault system should encourage an honest study of the causes of accidents, rather than the concealment of fault which tends to prevail under the tort or negligence system. Most compensation insurance carriers offer excellent safety engineering or loss control services to their policy holders, which can greatly aid the employer in improving workplace safety.

It is generally recognized that employers can control their accident costs through an effective loss control program, emphasizing accident prevention and efficient claims administration. Accidents, whether they result in personal injury, property damage, or both, are a major drain on corporate resources. In fact, the "direct" or insurable workers' comp costs are minuscule compared with the "indirect" or uninsured costs. These uninsured costs include the time wasted by supervisors and other employees, the cost of damaged products or equipment, downtime in processing operations, the cost of first aid, the cost of training a replacement worker, and so on. They represent a far greater cost to the employer and have enormous impact on the cost of production.

A good loss control effort should involve a partnership between the employer and the insurance carrier. The insurer's loss control representative should be asked to review and analyze existing loss control programs and offer recommendations for improvement. The employer should also take advantage of the supervisory training, industrial hygiene, and other loss prevention services that are available from the insurance carrier. If used properly, these services can be extremely valuable in helping to control losses before they occur and the cost of losses after they occur.

Insurance carriers can also provide valuable claims administration assistance to help employers control the cost of the accidents that do occur. This assistance can include developing systems to improve the efficiency of claims reported, providing methods for meaningful accident investigations, and performing timely reviews of the injured worker's case so that management is well informed of his or her condition and situation. Another important service is rehabilitation, which requires a close working relationship between the employer and the insurance carrier.

Implicit in these services is the partnership between the employer and the insurance carrier. Both have an interest in preventing and controlling losses. The time and money spent by the employer to improve his or her loss prevention and control efforts can result in much lower workers' compensation costs, as well as a more efficient operation and a safer and healthier work environment.

BIBLIOGRAPHY

Chamber of Commerce of the United States, Analysis of Workers' Compensation Laws, Washington, D.C., U.S. Chamber Publication No. 6383, 1981.

Malecki, Donald S., James H. Donaldson, and Ronald C. Horn, *Commercial Liability Risk Management and Insurance,* Vol. 1, Malvern, PA, American Institute for Property and Liability Underwriters, 1978.

Mehr, Ronald I., and Emerson Cammack, *Principles of Insurance,* Homewood, IL, Irwin, 1980.

Sheridan, Peter, J., "The Evolution of America's Industrial Safety Movement," *Occupational Hazards,* September 1975.

Chapter 12 / Part 1

Design and Purchasing

Effective safety and health management begins long before the worker appears in the shop or a machine process is put into use. This is a critical fact which the astute manager must remember. It is less costly and more efficient to correct safety and health hazards before they become part of the workplace. The aim of effective management should be to see that safety and health hazards are, to all intents and purposes, eliminated before they exist. This may appear to be a very high ideal, even an impossible one. The method for achieving it, however, is relatively simple: proper attention to safety and health in the design and purchasing phases.

Unfortunately, safety and health have too often been omitted from the goals of the designer and the purchasing manager. Too often a product is placed in use or on the market without adequate thought being given to safety and health. While the machine or product may do its job very well, it may also be too dangerous to use. Only after field use, resultant accidents, and expensive suits are the safety and health hazards considered. This is far too late. Safety and health management begins with design and purchasing.

DESIGN

The philosophy we recommend is both sound and simple: accident prevention begins during the product design phase. This rightly implies that, to some extent, the designer is responsible for many accidents. Particularly where sophisticated equipment and products are involved, many failures occur as a result of inadequate design. Even where the designer does not actually commit errors in design, he or she may still permit the omission of factors which would have prevented accidents or protected users. If a designer cannot eliminate hazards completely, he or she should at least endeavor to minimize the chance of accidents by reducing the possibility of errors in use. Plants and stores are full of accidents waiting to happen because these simple design principles are overlooked.

This does not mean that safety and health rest solely on the efforts of the designer. Safe design is a complex multi- and interdisciplinary effort, involving many people and functions on whose input and cooperation the designer must rely. Design defects involve a host of considerations, including:

- Material stress factors
- Human stress factors
- Construction materials
- Safety guard devices
- User needs
- User evaluation
- Foreseeable emergencies
- Accident or crash worthiness
- Proper operating instructions
- Codes, standards, and accepted practices
- Manufacturing specifications
- Product service experiences and facilities
- Properly trained users
- Design envelopes
- Financial constraints

This short list presents only a few of the inputs the designer must take into account. It is plain that the designer cannot know everything; a coordinated, supportive effort involving many types of expertise is needed.

Supporting Roles in the Design Process

There are too many supporting roles, functions, and disciplines in good design to be described extensively. A few of them, however, are so essential that some discussion is required. In particular, the senior manager should be aware of these:

1. Designers
2. Managers
3. Supervisors
4. Safety personnel
5. Maintenance personnel
6. Committees

Even a brief examination of each of these should illustrate how complex and far-reaching design issues are.

Designers. The design of equipment and products is essentially evolutionary. That is, it is always changing as experience, new materials, and new applications evolve. Designers must constantly acquire wider experience and knowledge to keep abreast of their trade's requirements. Their position is critical, since it is a key to eliminating problems and forestalling unsafe situations. In addition to doing design work, they must assess and design safety and health hazards, be knowledgeable about safety and health regulations, know the environmental demands of the workplace, and be aware of the various engineering controls that can solve their problems.

A few examples of the things an engineering designer might do to approach safety and health problems include these:

Design containers to minimize vibration and impact effects.
Design for maximum tolerances and clearances to reduce criticality and to ease maintenance requirements.
Provide panels for easy access to internal parts and to reduce errors from inaccessibility. (This creates the need for a new set of design standards to ensure that panels will not be removed or left off during operation.)
Include safety requirements on production drawings.
Minimize the need for special tools. (This guards against improper or infrequent maintenance).

These few concepts are merely a preface to the detailed design considerations that we will examine later in this chapter. However, they illustrate just a few of the areas in which the designer can minimize later safety and health hazards.

Managers. The management role in design starts with a corporate policy that clearly indicates that safety is to be built into the job and the product. Each firm has its own management philosophy, its own way of viewing safety and health, consistent with its underlying objectives. No single organizational plan would be useful or appropriate for all firms. One sound approach is to establish a corporate committee to develop the design requirements of the firm. This approach also meets some of the requirements of the Consumer Product Safety Commission and helps senior management make decisions regarding the equipment or products that the firm produces. (This approach will be discussed at greater length later in the chapter.)

It is easy to see that the safe production of equipment is closely related to safe processes, as outlined in Chapter 8. Any change in process that results from changes in equipment or product design must be considered as a source of possible hazards. Even a change in a gasket or sealant can have catastrophic results in certain circumstances. Management must ensure that any changes are carefully considered by knowledgeable and responsible parties, with an eye to possible safety and health consequences.

Typical management functions and inputs relating to design are:

Changing the training criteria and programs to coincide with new designs.
Initiating medical studies on suspected origins of occupational illness.

Coordinating the personnel-design interface by informing industrial relations personnel of environmental safety and health hazards.

Senior management's demonstrated interest will have a great impact on design personnel, keeping safe design in the forefront of their thinking. Management's emphasis will also ensure that the proper material is purchased and that the finished product is safely used. If management makes it clear that they want to eliminate hazardous environments and equipment, rather than just resorting to protective devices and equipment, then that is the way design personnel will operate. Where protective and rescue equipment is still deemed necessary, management should back the purchase of equipment which meets the current standards and the approval of various agencies or industrial associations.

Supervisors. The supervisor is closer to the firing line than the manager, though his or her role involves carrying out the desires of management. The supervisory role can be illustrated by looking at two primary safety and health tasks: instructing and ensuring.

The supervisor, who has daily contact with workers, must continually *instruct* the employees in:

Proper use of equipment and proper actions in the operating environment.
Proper use of safety equipment and devices.
Precautionary procedures to be followed.
Procedures to be followed in case of error or emergency.

Even when he or she knows that all workers have been properly instructed, the supervisor still has the ongoing task of *ensuring* a safe operation. Among other things, it is important to ensure that:

Employees understand the safety and health instructions.
Employees can carry out the instructions.
Equipment is being operated in accordance with instructions.
Proper tools are being used for the job at hand.
Guards and other safety equipment are installed and used.
Equipment is functioning as it was designed to do, and in a manner that operators can manage.
Instruments are labeled and easy to read, and controls and dials are operating as they should.

Safety Personnel. While the safety professional cannot be expected to make competent judgments on all design functions, he or she is still in a position to carry out many of the coordinating functions of management. For this person can ensure that:

Proper job descriptions and performance standards have been established and regularly reviewed for design and engineering personnel.
All work is properly checked by a superior before release. (The safety professional does not necessarily do this personally, but ensures that it is done.)
Design and engineering people are properly trained in safe practices.
Specifications are written and followed for all designs.
Changes in materials and processes are suitable from a safety and health standpoint.
Equipment and product failures receive proper analysis.

The safety specialist can function as a member of the design and review teams and can be a valuable source of input for the designer. Based on experience and training, this person should be able to point out design features which may be technically correct but unsafe in actual use. Even more than the engineer or designer, the safety specialist is aware of the hazards that may arise in the workplace or in the day-to-day operation of equipment, and can bring these to the designer's attention. Regular contact with the safety professional will keep engineers and designers aware that safety and health are primary considerations in proper design.

Maintenance Personnel. Managing a safe and healthy working environment must include good maintenance. A machine or process that looks hazard free on the drawing board may develop serious problems in the course of normal use, wear and tear, or as a result of misuse. Generally, it is the maintenance staff who must deal with such potential problems. Involving maintenance people in the development stage can help ensure that the design includes sound and proper maintainability. This involvement is particularly critical in terms of anticipated maintenance requirements on newly designed products.

Maintenance personnel should definitely be consulted about problems involving the location of access panels, lubrication points, the need for special tools or equipment, clearances, and tolerances and recommended intervals for routine maintenance or upkeep. In turn, when a problem relating to any of these areas develops, maintenance should be able to channel the information back to design so that similar errors can be avoided in the future.

Maintenance personnel must be alert to any special safety and health problems posed by new products or equipment. They should be encouraged to detect and report any deterioration of the equipment, either in function or in maintainability.

Committees. It was mentioned earlier that one of the soundest approaches to safety and health in design was to establish a corporate committee to check on the design of new equipment and products throughout various stages. There is no absolute rule about the membership of such a committee, but a high-level design review panel might include:

1. Top management
2. Legal counsel
3. Marketing manager
4. Engineering manager
5. Human factors expert
6. Safety manager
7. Loss control manager
8. Quality and reliability control manager
9. Maintenance manager
10. Plant manager
11. Supervisor

Not all of these people are necessary or appropriate in every instance, but this list should serve as a checklist for top management, to make sure that every angle of design review has been considered and that no key inputs are missing.

Such a committee should have a regular secretary so that committee decisions are implemented and follow-up is carried out. These suggestions, directives, and corrective actions must be documented and the documents placed in the permanent files or records. The secretary should keep track of all recommendations and suggestions, from initial input through final follow-up, as well as keeping communications flowing in both directions.

Besides the roles specifically mentioned so far, there are many other functions involved in a complete consideration of the safety aspects of design. For example, the equipment or product supplier interacts with engineering, manufacturing and purchasing; the distributor interacts with the retailer, local distributor, and lessor. These are only a few of the many persons who might be involved.

An Idealized Procedure

Several effective procedures have been established for product and equipment design. These usually center on (1) establishing the functional objectives of the product, (2) selecting or inventing the appropriate means of reaching these objectives, and (3) designing the product. Too often, only the last step is properly considered or performed.

The National Aeronautics and Space Administration has developed an idealized flow of safety functions for new products which is readily adaptable to most organizations. The safety-conscious manager will recognize that this idealized system can have immediate practical effects. Briefly, the steps involved are:

1. Perform safety analysis of components and subsystems.
2. Identify energy sources and the necessary control devices.
3. Identify areas of inadequate energy control. At this point, one of two approaches may be followed:
    a. Recommend corrective action to management, participate in further design review, and prepare further safety analyses.
    or
    b. Develop new safety criteria, receive management/engineering approval, evaluate all safety controls for adequacy, incorporate the changes into Step 2 for further review, and carry out Steps 3 and 3a.

PHASES

In order to ensure that safety and health considerations are given adequate attention at every stage of the design process, the manager needs a clear grasp of the exact sequence of phases involved and the problems associated with each one. The typical life cycle of a product or process involves the following steps:

1. Product specification
2. Initial product design evaluation
3. Prototype review and testing
4. Final product tooling and design
5. Quality control and reliability
6. Field service and feedback
7. Development of technical and sales documents (operating)
8. User/operator training
9. Export marketability (if applicable).

The last step is necessary if the product is to be exported, since many countries, concerned about the safety of the products they import, have special standards. This is especially important in dealing with Great Britain, Sweden, Canada, France, Germany, Australia, and New Zealand. This step can become very complicated, as the standards of the various countries not only differ but often conflict with each other. Part of this intercountry conflict will be eliminated when most European countries adapt to the single set of standards of the European Community by 1992.

On the other hand, second- and third-level countries change their standards as they become more sophisticated and gain experience in global manufacturing. Witness the recent example of China, which used failure to meet its standards as one reason to call off a vehicle manufacturing deal that was 6 years in the making. The global trend to assemble and market products with components coming from many countries means that all suppliers must conform to the set of standards of the marketer, who may sell to many countries. More recently, the easing of international tensions between Western and Iron Curtain countries signals the advent of new sets of standards for multinational firms to meet.

Safety in design, it should be obvious, is a very complex matter, and the average manager may find it extremely difficult to achieve a comprehensive overview. In the author's experience, such an overview can be readily achieved by breaking the entire process down into its component phases and spelling out the problems and questions of each. One especially useful approach, then, is to break design safety down into five phases:

1. Concept phase
2. Design phase
3. Production phase
4. Operations phase
5. Disposal phase

While other approaches may be equally useful, the senior manager should at least consider this method. Let us take a look at what is involved in each of these five phases.

Concept Phase

Between 70 and 90 percent of the design decisions affecting safety and health are made in the conceptual phase. For that reason, this is the time to look at the widest possible range of consequences. In fact, most design errors are generated during this phase.

The philosophy that accident prevention begins during the design conceptual phase rightly implies that the designer is a party to many subsequent accidents. Even when the designer has made no actual errors, there may still be errors of omission, such as failing to include safeguards that would have prevented accidents or protected personnel. When a designer cannot completely eliminate a hazard or the possibility of an accident, he or she must try to minimize the possibility of user errors that could generate mishaps.

The concept phase involves the use of systems analysis, considering both product function and operating environment. At this stage, alternative design studies are often conducted. One tool, known as the "product evaluation checklist," may be especially helpful. A typical checklist would ask:

1. What is the product and the intended use?
2. How many will be produced?
3. What are the critical environments?
4. What are the critical manufacturing steps?
5. What are the failure modes and their consequences?

During the concept phase, the designer should also consider the human engineering aspects of the product. These include use, misuse, abuse, assembly, installation, and maintenance.

The design review is another part of the concept phase, involving such techniques as failure modes and effects analysis, and fault tree analysis. (These techniques, however, do not apply solely to the concept phase; they may overlap with or be used in other phases as well.) During the concept phase, safety devices should also be reviewed for compliance with both industry and governmental standards or other applicable codes.

Design Phase

Design errors are not always obvious. Indeed, they are often very subtle, and are more than simple or easily detectable mistakes. A design, for example, may be technically sound and practical but inadequate for the actual operating conditions. Both the design and its specifications should be verified during this phase. A wide variety of tools can be applied, including failure modes and effects analysis, fault tree analysis, and other types of hazards analysis. These are generally within the domain of the system safety engineer. Such verification would also include criticality analysis, reliability analysis, and qualification testing. The design phase is also the time to verify document control (more on that shortly), production plan prototype safety, and reliability testing.

A formal design review is more than just a scheduled, systematic review and evaluation of the design. For one thing, it should be conducted by personnel who are not usually directly associated with the development of the process or product. However, they should be knowledgeable about the product or process and should have responsibility for various elements of the product throughout its life cycle. Although the review is discussed here as part of the design phase, exactly when it should be performed is a debatable matter. If it is performed too early, there is little to discuss. If it takes place too late, remedying errors and oversights may be too costly or difficult. The most formal and complete design review may be best held just after prototype testing; other, less formal or extensive reviews may be held starting with the conceptual phase. Repeated design reviews are not unreasonable.

Appropriate techniques, including failure mode and effects analysis, fault tree analysis, and human factors analysis, should be a normal part of the design review.

The manager should definitely consider retaining such a specialist as a product safety engineer, both for this and for other phases, to ensure that safety principles are adhered to in the design. This specialist should be brought in for reviews at later stages as well. Retaining expert help should ensure the thorough application of fail-safe features, fault-tolerant designs, warning devices, redundant controls, proper energy paths, derating, environmental insensitivity, isolation principles, shielding, redundancy for operation, and emergency features.

The technical safety of all equipment should be checked in the design phase. This is the best time to consider the equipment in relation to both plant design and work environment. This should include protection against heat, water, chemicals, and unauthorized use or access. In some cases, actual plant redesign may be needed in order to control unusual hazards arising out of the product or process. The manager should keep in mind that this redesign process is far less costly if it is done before new equipment is installed, rather than waiting until an accident happens to demonstrate the inadequacy of the existing plant design.

This is also the time to consider guards of all types, interlocking devices, machine feeding, work environment, noise, access to work points, electrical bonding, and product spillage. Keep in mind that this list is not complete, and is only meant to suggest some of the items to be considered. Still, a glance at the number of items involved should make it obvious that the designer will need help. Management support is essential, and a variety of talents must be made available for the designer to draw on.

The complexity and scope of the design phase pose special problems for senior managers. How are they to direct their efforts properly and oversee the design process? One useful method is to divide the design phase into three areas: (1) hazards identification, (2) human considerations, and (3) experience and feedback. By considering each of these separately, the manager can achieve a detailed overview.

Hazards Identification. One of the most essential early steps of design involves identifying exactly the hazards of the process or product. The designer must first recognize a hazard before he or she can successfully deal with it. There are several commonly used design analysis techniques to ensure systematic identification of all probable hazards. The two major approaches are generally classified as (1) fundamental and (2) technical.

The fundamental approach consists of a study of all possible hazards that may exist. They are first examined qualitatively, to be sure that they are actually hazards, and then quantitatively, to calculate the overall probability of an accident or event arising from them. This approach is sound, whatever the new product or its use.

The technical approach involves a careful study of as many accidents as possible in order to identify the hazards that had a role in them. The findings are then applied to the design of new equipment and processes or the redesign of existing systems. The technical approach nearly always involves some version of failure mode and effects analysis and fault tree analysis. Since the average person has little occasion to deal with either of these techniques, a few words of explanation may be appropriate here.

Failure modes and effects analysis looks at the possible failures of each component of a product or system and determines the effect of each failure. It is equally useful when a system involves many parts, several failure modes, and moderately complex designs.

Fault tree analysis organizes the cause-and-effect relationships of an accident or event, deducing the various combinations of failures and errors that could result. By using statistical information, it can also determine the probability of an accident. It works well when there are many parts, many modes of failure, a complex design, and a sophisticated operation. It can also give information for making decisions, choosing alternatives, and allocating resources. Figures 12-1 and 12-2 illustrate some of the qualitative information that the techniques can yield.

The objectives of the hazard analysis process are as follows:

1. To detect and correct potential accident situations during the design phases.
2. To perform such analysis systematically until every activity has been covered by a hazard evaluation.
3. To perform such analysis whenever a process or equipment is changed.

CLASSIFICATION	EFFECT
I. Catastrophic	Death or complete system loss a possibility
II. Critical	May cause injury, severe occupational illness or major system damage
III. Marginal	May cause minor injury or illness, or minor system damage
IV. Negligible	Will not result in injury, illness or system damage

Figure 12-1. Qualitative severity index.

DESCRIPTION	INDIVIDUAL ITEM INCIDENCE
Frequent	Likely to occur frequently
Reasonably probable	Will occur several times in item's life
Occasional ($> 10^{-4}$)	Likely to occur in life of item
Remote (10^{-4} to 10^{-6})	Assumed this hazard will not exist
Extremely improbable ($< 10^{-6}$)	Probability cannot be distinguished from zero
Impossible	Physically impossible

Figure 12-2. Qualitative probability index.

The rigorous application of such analytic techniques may be outside the expertise of most managers. However, they should insist that knowledgeable personnel perform these analyses and should study the results themselves.

Human Considerations. No matter how meticulously a design is studied from a technical or engineering standpoint, the manager must remember one very important fact: sooner or later, every product or process will involve people as well as parts. Human considerations can radically affect the safety, efficiency, and reliabililty of the best-designed process or system.

All too often, we blandly assume that an efficient worker is automatically a safe worker. However, this frequently is not true; efficiency is not enough to guarantee safe performance. The worker who tries to achieve efficiency by taking shortcuts or eliminating safeguards may be less safe than the slower, less efficient worker who takes time, follows instructions, and pays attention to the work. In many cases, the worker needs to be informed of the primacy of safety, to be warned of the consequences of following improper procedures, or to be alerted to his or her own carelessness or inattentiveness.

Almost every mishap can ultimately be traced to a human error, although it may not be an operator error. It may have been an error on the part of the designer, the production worker, the maintenance worker, the supervisor, or anyone even remotely connected with the process, including the worker 2000 miles away who carelessly put the wrong label on a container.

It is true that certain tasks involve inherently hazardous human-machine interactions, not all of which can be eliminated. However, there are things that can be done to reduce those hazards to a minimum. The designer can not only design out technical hazards but also minimize the possibility of human error. (By "human error," we mean anything that is inconsistent with normal behavior patterns, or anything that deviates from prescribed procedures.)

Too often the human worker's limitations receive little attention during the design phase. Managers are left in the position of having to fit the worker to the machine later on. They are urged to instruct, train, or motivate the worker to avoid the hazard when the hazard should have been eliminated from the equipment in the first place. Fitting the worker to the task is a poor approach, particularly in light of the fact that minor changes in equipment can greatly improve performance and reduce accidents.

Fortunately, a good deal of human factors research focuses on developing design criteria which can be directly applied to product design. Applying these criteria can be a straightforward task which may even be performed by computers. In other instances, complex judgments enter the picture, involving all the experience and ingenuity the designer has. In some cases, the vital data simply does not exist and must be developed or compiled. Consider, for example, some of the questions that arise from the increasing role of women in the workplace. What type of safety harness does a woman window washer need? Does the female anatomy lend itself to frequent raising of the head to check an overhead guage? Does the length of a woman's arm in relation to her torso permit her to operate a console with the same ease as the male worker for whom it was originally designed? These are real problems, as the differences in male and female configuration can lead to serious accidents in the workplace. Here the equipment designer may find that studies must be undertaken to develop the needed information and criteria.

In a complex system, applying design data may be expensive and time-consuming. Some decisions may involve trade-offs based on qualitative severity and probability indexes (see Figures 12-1 and 12-2). A hazard classified as "negligible" on the Severity Index but "reasonably probable" on the Probability Index might be acceptable to the designer, while one classified as "critical" on the Severity Index and "occasional" on the Probability Index would not.

Although the design of physical components is largely an engineering chore, human factors input should be considered essential. Tracking down relevant data sources and evaluating their applicability, however, may overwhelm the designer, and outside expertise may be needed. The manager should make it clear that, where needed, such expertise should be sought, and should give his or her full support to securing it.

What are some of the considerations in approaching a design problem from the human engineering viewpoint? The specialist will sometimes use a series of questions, such as:

1. Which functions must be performed by humans?
2. When the human must exercise control, what type of device is appropriate?
3. Can each control device be easily identified?
4. Is there appropriate redundancy in critical systems?

These questions are only illustrations of the types used. The actual list may run to fifty or more items. Keep in mind, too, that this is only a first step in identifying the problems. Once they have been identified, further detailed human factors knowledge is needed to solve them.

In considering the human-machine interface, it is important to stress that there is really no such thing as "an average person." In actual use, a system may be operated by highly skilled persons and relatively unskilled ones, workers with great dexterity and workers with less, attentive operators and inattentive ones. The system must operate safely and effectively for all of them.

Closely associated with human factors is the problem of equipment misuse. This has gained much attention due to recent liability judgments, making management aware that misuse must be designed out and guarded against so as to prevent such damaging liability suits.

Human factors can play a key role in the various techniques for error reduction that the designer may use. The human factors specialist, for instance, can design a reliable system even though it involves a variable human component. Knowing the characteristics of the intended users make it possible to design equipment to fit their capacities and limitations. Analyzing their errors will also give direction for redesign, automation, or plain "goof-proofing." This type of improvement in the design is generally more effective than trying to achieve improvement by selection, training, supervision, and motivational campaigns.

Human factors engineering is far too complex for a thorough discussion here. However, even this brief glimpse should make it clear to the manager that human engineering approaches can help solve many design problems. While common sense will be of great assistance, human engineering expertise is the best way to ensure adequate consideration. The senior manager may not possess this knowledge, but he or she must insist that an expert be sought and consulted as an integral part of the design process.

The overall benefits that human factors engineering can offer include:

1. Greater system effectiveness
2. Fewer performance errors
3. Fewer accidents
4. Minimization of redesign and retrofit after the process is operational
5. Reduced training time and cost
6. More effective use of personnel, with fewer restrictions on qualifications.

Experience and Feedback. One of the manager's main concerns should be to make the greatest possible use of existing information so that it does not first have to be developed or experienced. Many other businesses and industries have recorded their experiences with design problems, and several government studies summarize research in the area. This information is readily available as a base, so it is not necessary to start from scratch.

Surprisingly, one of the most often overlooked sources of experience may be found within the organization itself. Accident records may indicate causal factors involving design. Various staff and line personnel may have direct knowledge of design hazards, some of which may already be on record as complaints, suggestions, safety committee reports, and so on. Such user complaints and comments should be reviewed with an open mind, as they can provide an experience base and feedback that cannot be purchased for any price. The experience of trainers and medical personnel can also reveal trouble spots, as can a search of on-call maintenance records.

Production Phase (Manufacturing)

After all the theoretical considerations of the conceptual and design phases, the manager must be concerned with the practical problems of the production or manufacturing phase.

Actual production will call for careful configuration control. At this stage, engineering changes should not be permitted unless they have been approved by all parties and been fed back into the design for analysis and evaluation. Material should be used exactly as specified in the design recommendations. Even changes in vendors should be taken into account to ensure that there is no loss of quality or reliability. During this phase, careful documentation is vital. It should include both the ingoing raw materials and subcomponents and the outgoing product.

Experience has shown that product malfunctions can result from faulty design, manufacturing defects, exceeding specified limits, and environmental effects. While there are other ways for failures to occur, these particular failures are design associated and have a decided bearing on most problems. In the past, designers tended to overdesign a product to maximize safety. With increased sophistication and cost consciousness, this approach is now seldom used. More frequently, a margin of safety to handle expected operations is set, and a margin above that for the unexpected is allowed.

The failures stemming from construction error often involve both production and design. Sometimes a bolt or other component is substituted without approval of the design group and subsequently fails. A construction error may also result from a fault in assembly, and from a subsequent inspection error which overlooks or passes such omissions, substitutions, or other errors.

Production personnel have several responsibilities related directly to equipment design. It is their task to:

1. Engage in research to find new production techniques.
2. Analyze new production models and recommend design changes.

3. Coordinate and schedule new methods in order to minimize costs while maximizing production.
4. Pinpoint hazards that require design revisions.
5. Conduct hazard analyses of processes to locate possible failures.
6. Recommend tool and equipment designs that will cope better with the hazards of the process.
7. Keep manufacturing engineers informed of new hazard recognition procedures.
8. Ensure that purchased production machinery does not present unacceptable hazard potentials.

The quality control function is essential to the manufacturing or production phase. It is responsible for inspection, testing, and careful control of receiving, both in the process and in the final product. Specifically, quality control's tasks include the following:

Examining, testing, and inspecting all materials and final products.
Advising the purchasing department when a vendor supplies unacceptable products.
Conducting studies to determine if alternative designs, materials, and process methods could improve the product.
Analyzing new engineering models and prototypes to establish quality standards and to determine potential quality problems.
Ensuring that manufactured products meet product liability standards.
Ensuring that products are delivered with fail-safe devices and are free from any hazard that is not properly controlled.
Offering ideas for improvement of the product's quality based on hazard studies.
Providing failure rates on materials.

Still another discipline related to production is reliability. This involves both failure and failure reduction. Reliability, safety, and design do overlap, but they are not necessarily synonymous. "Reliability" refers to malfunctions, while "safety" is usually concerned with injury or property damage. Their relationship is obvious, but bear in mind that reliability does not ensure safety. It is possible to have high reliability and low safety. A classic example is the old-fashioned hand axe: almost nothing can go wrong with the implement, but its potential for doing injury or damage is almost unlimited.

Operations Phase (User)

This phase may not begin until long after the design phase is completed, but it is still an integral part of the design process. The operations phase, after all, provides the actual feedback on product successes or failures, which in turn must be evaluated and integrated by design, either in redesigning the product or in designing future products. This phase also involves labeling, instructions, modifications, notifications, and recall activities, some of which are discussed more thoroughly later in the chapter.

Disposal Phase

Finally, the operations phase terminates with the disposal of equipment and process by-products. (A complete discussion of handling obsolete and worn-out equipment is presented in Chapter 13.) Special consideration should be given to the use of equipment or products when they are beginning to wear out or are not performing up to standards. Part of the design process itself should consist of establishing criteria for identifying when equipment should no longer be used. Disposal criteria should also consider the materials used in a product or process and its by-products. Some of these can pose hazards long after an item is removed from use and require special disposal techniques. Health and safety regulations must be considered in connection with these hazardous by-products or materials, as there may be mandated standards for their disposal.

COSTS

The costs of design for safety and health cannot be stated with precision. They are probably best expressed in terms of a percentage of engineering costs. A thorough system safety analysis, for example, has been estimated to be about 5 percent of the engineering costs.

The money invested in safe design can be a good investment, paying off handsomely in many ways. A safety feature may simultaneously effect a reduction in weight, volume cost, accident rate, lost time, workers' compensation costs, insurance premiums, and so on. Even where it increases costs, the safety feature may pay for itself many times over in reduced accident and insurance rates. Safety through a reduction in power may offer great cost advantages in energy savings, especially for electrically operated devices. The operating efficiency of an electrical system is often directly related to the safety of the design. Greater efficiency means less wasted energy and, hence, a lower overall operating cost.

The decision to make design changes may not be based on direct costs but on the change's potential for reducing accidents. Various scaling techniques may provide a valid basis for such decisions, as they allow management to calculate the payoff for costly changes. The scaling of accident potential can be based on criteria developed for this or a similar purpose. One such scaling device for accident potential and decision making is shown in Figure 12-3.

The senior manager should carefully review Figures 12-1, 12-2, and 12-3 and consider using or modifying them to suit the organization's needs and

	ACCIDENT POTENTIAL CLASS	RESULTANT COSTS	
		PERSONS	PROPERTY
I.	Safe	Scratches	$10
II.	Marginal	Medical	$100
III.	Moderate	Lost time < 10 days	$1,000
IV.	Hazardous	> 10 days Permanent	$10,000
V.	Critical	Death, 5 injuries	$100,000
VI.	Catastrophic	Multi-death	$1,000,000
VII.	Super	Many deaths	$10,000,000

Figure 12-3. Typical scaling device for accident potential and cost analysis.

criteria. Properly used, they allow the manager to make judgments regarding the seriousness of design deficiencies and the value of making costly design investments. While they are subjective and qualitative in nature, the charts do present a general overview on which the manager can base his decision.

GUIDELINES

The design documents themselves usually provide important guidelines, often referring to many additional sources for information. This can amount to a stack of information several feet high, not counting the reams of documentation spelling out local and government requirements. Published safety requirements and practices are contained in a wide variety of documents, including the *Federal Register,* OSHA, state codes, and industry standards. When no codes exist, design is governed by the basic principles of engineering, the materials to be used, current guidebooks, accepted practices, and general safety factors.

So complex and detailed are these various guidelines that most designers and managers require a checklist to make sure that all relevant ones have been taken into account. Many companies and associations have developed their own checklists based on the needs and standards of their operations or industry.

Checklists are created for self-checking by the designer, the engineer, the assembler, the packer, and so on. A design checklist, for example, should include the following checkoffs as an absolute minimum:

1. Completeness and accuracy of product or system specifications.
2. Completeness and accuracy of block diagrams for the system or product.
3. Applicable safety standards followed.

4. Applicable human engineering standards followed. (These should cover removal of devices, failure of safety devices, and provisions for safety devices that remain operable during the life of the product.)
5. Adequate design review for fail-safe operation.
6. Thoroughness of failure reports and documentation bearing on the final design.
7. Accuracy of computations.
8. Accuracy of drawings, including material and construction specifications.

A number of checklists are available for the designer to use in evaluating his or her efforts. In considering the human element in equipment design, we recommend turning to a human factors consultant, who can provide useful and relevant checklists. One such checklist, giving measures to be taken by the designer or methods engineer to prevent accidents and injuries, can be found in Hammer's *Handbook of System and Product Safety,* published by Prentice-Hall.

PRODUCT LIABILITY AND DOCUMENTATION

In recent years, the area of product liability has become increasingly prominent in the headlines and in the thinking of management. While any product liability case is a cause for concern, those that involve a product's basic design are often the most crucial. Since all products are subject to the same design process, manufacturers tend to fight design cases the hardest. A case involving just one product could raise questions in the public's (or the manufacturer's) mind about the safety of all their products' designs.

Generally speaking, the manufacturer must exercise "reasonable care" in designing the product for the purpose for which it is intended. Unfortunately, reasonable care may not be enough, even when defined by custom or by national standards. The plaintiff may simply attack the standards for not meeting acceptable practices. In some notable recent cases, manufacturers who thought that they were exercising reasonable care in the design process nevertheless found themselves liable for extraordinary judgments. (The concept of reasonable care is further discussed in Chapter 20.)

Claims based on product liability have become a serious threat to corporate products and a major concern to management. Liability claims are extremely costly, whether settled in or out of court, and even when the defendant wins the case or the costs are covered by insurance. In addition, these claims represent only a portion of the acual losses sustained when the safety and reliability of a manufacturer's products are called into question. These losses include such things as the cost of warranty programs when the product does not perform properly, damage to the corporation's reputation, loss of sales, and the expense of recall and modification programs. The auto and drug industries provide highly visible examples of this trend and its attendant costs. Growing public awareness of the liability remedies available to them makes it likely that this trend will not only continue but grow even more pronounced.

Apart from the fear of liability suits, manufacturers and suppliers have an obligation to ensure that their products are safe. This means that the manufacturer (and more specifically, the designer) must:

Ensure that products are designed and constructed to be safe in use.
Examine and test finished products to ensure that this is so.
Establish the risks involved in the use of the product, and clearly warn of the dangers and how to avoid them.
Provide clear, detailed instructions on the safe use of the product.

A management decision to skimp on procedures that will guarantee the safety and reliability of the product may be enormously expensive. "Meeting the cost of the competition" is not a viable defense in a court of law.

Many attorneys argue that errors in design are simple things which should not and would not have happened if the manufacturer had used a good design review process. In other cases, the manufacturer has been held liable for developing new safety devices and not notifying existing users of their availability.

The rapid changes in court attitudes and precedents prevent us from giving a final or even up-to-date word on the real-world situation. The good manager makes it a point to keep informed of the latest cases, precedents, and developments that affect his or her business, often relying on the legal department or insurance department to report on them. Moreover, it is the manager's business to see that the information does not simply stop at his or her desk, but is passed on and applied in the design and design review processes. This can often eliminate costly liability actions before they happen.

Safe design is a dynamic affair. Court decisions, the state of the art, and public opinion make it necessary to reevaluate product safety more frequently than in the past. The standard of yesterday may not be acceptable today. Added to this is the manufacturer's responsibility to design for the foreseeable use of the product. A brief checklist of the things which must constantly be reevaluated by the design team includes:

Structural weakness
Contamination of or by the product
Corrosion or deterioration of the product
Toxicity in any form
Component and product incompatibility
Difficulty in using the product due to its design
Changing state of the product's material (such as brittleness as the material ages)

DESIGN AND PURCHASING   303

Possible fatigue failures
Maintenance and service hazards
Lack of control of the product
Lack of proper guarding of moving parts

These few examples of design considerations require full documentation when carried out.

"Documentation" refers to all of the written and audiovisual materials relating to a product, including sales, maintenance, and training materials. Too often, documentation does not begin until the product is ready for the market, when there may not be enough time for the task. It is important that formal documentation begin at the earliest part of the conceptual process. The manufacturer establishes his concern for a safe product by such careful documentation. To carry this out, a documentation coordinator should be appointed, preferably under the direction of the design review committee. This coordinator should specify the documentation file's contents and see to it that there are no gaps in correspondence or activity. The news media of today constantly carry references to missing documentation in court cases, which is a persuasive reason for documenting everything. Among other things, detailed records should be kept on:

Materials control
Purchasing
Process engineering
Reliability
Packaging information
Quality control
Customer feedback
Maintenance feedback
Customer complaints

Records of these events and processes can play a vital part in economic and legal decisions. Management's role in documentation includes active participation in a safety plan, a quality control plan, and a recall plan. Additional management participation and awareness are involved in design records, safety analyses, inspection and testing results, compliance certifications, and defect notices. Careful documentation should make it possible to trace a raw material forward to the components in which it was used or back to its original supplier. Such records should be readily accessible and kept for the life of the product. Five years of retention has often been recommended, but this is too little. A product manufactured 10 to 15 years ago can be the subject of a liability suit tomorrow.

As a general rule, it is better to have one documentation package too many than one too few where product liability is concerned. Newer and better techniques of information storage can make such long-term document retention more practical as well.

PURCHASING

The traditional role of the purchasing department is to obtain supplies that meet the specifications set up by those with the authority to order them. Purchasing seeks to provide proper quality, proper quantity, and timely delivery at the lowest cost. Today the purchasing department has one other function: it must serve as a vital link in the company's safety and health program.

The purchasing agent can do much to smooth the way for safety and health by insisting that all machines and other equipment be delivered with safety equipment in place, thus precluding expensive design retrofits. Demand is forcing manufacturers to safeguard their equipment properly. This applies, of course, to the supplier as well as to the company's own products. In some cases, the emphasis on price makes it necessary for the supplier to furnish equipment without safeguards as an option. This is an extremely dangerous procedure; there is little defense in court against it for either the user or the supplier. The OSHA emphasis on safeguards is helpful in this regard, but it furnishes no guarantee of more than minimum protection.

Senior management can assist by discouraging the purchase of equipment which does not meet or exceed standards and which later has to be expensively retrofitted or pulled out of service. This involves setting clear and adequate guidelines for purchasing to follow, clearly spelling out that safety, not cost alone, must be a determining factor in equipment purchase. With this kind of management backing, purchasing can do a much more effective job of securing safe equipment and products. Management can also insist that other departments and functions provide purchasing with needed input, feedback, and support. For example, as a sound practice, the safety manager should check equipment specifications before ordering.

Actually, purchasing plays a more active role in safety and health than most of us realize. Some examples of purchasing responsibilities that affect safety and health are:

Purchasing required protective equipment.
Ensuring that the supplier provides proper servicing instructions.
Having the supplier provide training where needed.
Ensuring that the materials purchased meet established standards.
Ensuring that hazard information is obtained from suppliers.
Providing for economical sources of supply that meet both standards and quality requirements.
Ensuring that vendors provide proper labeling, including cautions, warnings, and hazard labels.
Ensuring that vendors of toxic or hazardous substances provide proper containers for shipment and storage.

Purchasing should be oriented to consider safety and health in making all purchases. Particular caution is needed in the purchase of personal protective equipment, where defects could cause injury or illness. Equipment for moving heavy loads is no place to cut corners. In the purchase of fluids, suitable storage containers should be considered. Unsuspected hazards can often be found in ordinary items such as hand tools, cleaning rags, paint cabinets, and the like. Purchasing should consider the life of the product, sharp edges or points, maintenance requirements, and hazards to a handler's health, including excess fatigue or stress. Purchasing can reasonably expect help in determining the safety of materials from engineering, safety, medical, and other personnel.

Safety experts can devise and put into writing the standards to guide the purchasing department, but purchasing still has the responsibility to ask for and consider inputs from various sections. Engineering, for example, can specify necessary built-in safeguards, with the safety specialist checking all such plans and specification. Where orders are repetitive, safety can furnish or develop standard lists of repetitive items. This will ensure that uniform quality and specifications are maintained.

In some organizations, the responsibility for specifications covering safety and health rests entirely with the department that is to use the product. This works well only where there is a full-time purchasing agent and a full-time safety professional.

The purchasing agent should be able to rely on the safety manager as someone who can:

1. Provide specific information about process hazards that can be eliminated by design or guarding features.
2. Supply information about safe equipment, tools, and materials.
3. Provide specific information about workplace, fire, and health hazards.
4. Provide information about federal and state safety and health requirements or standards.
5. Provide information on current industry or association safety and health standards.
6. Supply accident experience information on equipment or materials for use in future ordering.

Safety can also help purchasing by securing support from other departments (such as financial backing) for safety- and health-related purposes. While this is usually a purchasing or management function, it is also the type of activity that safety can effectively coordinate or promote. In turn, the safety manager can expect the purchasing department to be familiar with plant and process hazards, to ask for safety information when in doubt, to have a rough idea of how and when materials and equipment will be used, and to participate in accident investigations where equipment or materials purchased from outside vendors may have been involved. Purchasing can also be expected to confer with requesters before ordering or purchasing, and to confer with department heads to ensure that the products they order comply with accepted safety and health standards. Here again, purchasing and the safety professional should work closely together.

CONCLUSION

Design and purchasing departments may seem like strange bedfellows. However, while their functions are often unrelated, they are joined where the materials and components for a new design are to be purchased, and where company-designed products are to be used in-house and supported by outside suppliers. This relationship, in turn, is supported by line users and staff, such as the safety manager. Whether a piece of equipment is manufactured or purchased, design and purchasing are both affected by mandated compliance and safety and health regulations.

The philosophy that carefully designed-out hazards on new equipment and products is better than added safeguards and retrofits applies equally to design and purchasing. Design must operate according to that philosophy in designing new equipment and products; purchasing must keep it in mind when selecting products or equipment designed by outside suppliers or manufacturers. Moreover, design must rely on purchasing to secure the materials needed to manufacture new equipment or products, and purchasing must look to design for consultation and input on acquisitions for nondesign uses.

BIBLIOGRAPHY

Accident Prevention Manual for Industrial Operations, 9th edition. Chicago: National Safety Council, 1988.
Bird, Frank E., Jr., and Robert E. Loftus, *Loss Control Management,* Loganville, GA, Institute Press, 1976.
Boyle, William G., *Designing Production Safety Systems,* Tulsa, OK, Petroleum Publishing, 1979.
Bullock, Milton G., "Appendix B, An Alternative Hardware Maintenance Program Tree," *Safety Considerations in Evaluation of Maintenance Programs,* Idaho Falls, ID, System Safety Development Center, undated change.
Clarke, Richard, "Safety Priorities for New Transport Aircraft," *SAFE Journal,* Fall 1980, pp. 16-19.
Ferry, Ted S., and D. A. Weaver, *Directions in Safety,* Springfield, IL, Charles C. Thomas, 1976.
Findlay, James V., *Safety and the Executive,* Loganville, GA, Institute Press, 1979.
Firenze, Robert J., *The Process of Hazard Control,* Dubuque, Kendall-Hunt, 1978.
Gray, Irwin, *Product Liability,* New York, American Management Association, 1975.
Grandjean, E., *Fitting the Task to the Man,* London, Taylor and Francis, 1980.
Greenberg, Leo, and Don O. Chaffin, *Workers and Their Tools,* Midland, MI, Pendell Publishing, n.d.
Grimaldi, John V., and Rollin H. Simonds, *Safety Management,* 4th edition, Homewood, IL, Irwin, 1984.

Hammer, Willie, *Handbook of System and Product Safety,* Englewood Cliffs, NJ, Prentice-Hall, 1972.
———, *Occupational Safety Management and Engineering,* Englewood Cliffs, NJ, Prentice-Hall, 1976.
———, *Safety Management and Engineering,* Englewood Cliffs, NJ, Prentice-Hall, 1980.
Hulet, Michael W., "The Great Black Hole," *Professional Safety,* December 1986, pp. 37-41.
Johnson William G., *MORT Safety Assurance Systems,* New York, Marcel Dekker, 1980.
King, Ralph W., *Industrial Hazards and Safety Handbook,* London, Newens-Butterworth, 1979.
Mann, Jim, *Beijing Jeep,* New York, Simon & Schuster, 1989.
McCormick, Ernest J., *Human Factors in Engineering and Design,* New York, McGraw Hill, 1976.
Nertney, Robert J., and M. G. Bullock, *Human Factors in Design, SSDC-2,* Springfield, VA, National Technical Information Service (NTIS), 1976.
Peters, George A., "What to Do About Product Liability," *Professional Safety,* September 1980, pp. 19-21.
Petersen, Dan, *Techniques of Safety Management,* 2nd edition, New York, McGraw-Hill, 1978.
Product Safety and Liability: A Loss Control Handbook, Chicago: Alliance of American Insurers, 1979.
Wynholds, Hans, and Lewis Bass, *Product Liability,* Princeton, NJ, ECON, Inc., 1977.

Chapter 12 / Part 2

Purchasing Controls

PURCHASING CONTROLS: WHAT'S IT ALL ABOUT?

Managing a purchasing controls program can involve any or all of the following:

Researching the standards for equipment and supplies.
Finding reliable suppliers.
Specifying safety features in purchase orders.
Verifying that goods meet specifications.
Evaluating substitutions.
Identifying loss potentials in routes or modes of shipment.
Communicating on materials and equipment hazards.
Protection in storage.
Salvage of unusable items.
Disposal of hazardous items.

The organization's purchasing staff can do a lot for the safety program. To realize the potential, active cooperation of purchasing and safety personnel is necessary.

FUNDAMENTAL PURCHASING CONCEPTS

The role of the purchasing department is to obtain the items and services that are needed. A secondary role is to obtain these at a favorable price, thus contributing to the organization's profit and effectiveness. These are common objectives of purchasing personnel, and many of them can relate directly to safety objectives.

1. *Reliability and continuity of the supply.* The manufacturing of products or the providing of services requires a continued supply of operating materials. In purchasing, these are called "MRO" items. If the supply is interrupted, substitutes may be made to keep operating. Substitutes often don't meet all of the specifications or standards. A substitute can have qualities which make it less safe than the normal item. If these qualities are not identified, accident controls may unknowingly be weakened.
2. *High inventory turnover.* This is desired to have the lowest practical quantity on hand. The purpose is to minimize the storage costs and aging of inventory items. Excess stock can create numerous safety problems,

such as improper stacking, blocking of aisles and exits, and increased fire or health hazard exposure if the materials are hazardous substances. Aged inventory may be inferior. If it consists of chemicals, these may be unstable. Aged inventory may also not be up to current standards if improvements have been made in the newer inventory. The improvements may be followed by removal of previous accident controls as unnecessary, in which case the aged inventory reintroduces a hazard.

3. *Economy of quantity.* Purchasing personnel often find quantity discounts. If they are not aware of hazards resulting from excess stocks, they will make the quantity buy. This causes a conflict with the objective of high turnover, including the hazards of excess materials such as congestion, fire potential, and deterioration from the environment.
4. *High value-to-cost ratio.* Getting the most of what is wanted leads to efficiency. When only process needs are known, only those needs may be met. The best cost item may not meet safety standards unless those standards are an integral part of the specifications.
5. *Low inventory waste ratio.* Waste is accompanied by disposal costs. It can also create accident, illness, and liability potentials. Excess waste is a symptom of many problems in engineering, training, and motivation, and these problems also cause accidents.
6. *Low operating cost.* This show efficiency, which is directly related to safety. An exception occurs when there is false economy—that is, the purchase of cheap substitutes. These prove more costly in the long run when the quality, waste, and accident costs are added in.
7. *High compliance with specifications.* Industry standards and methods are the result of extensive study and experience. Oversights can be costly in terms of productivity and safety, and replacement or retrofit multiplies the costs and losses. A prime objective in establishing purchasing controls is to identify the appropriate standards.
8. *Good vendor relations.* The vendors who look on your problems as their problems are valuable. Vendors who appreciate business won't sell customers items that don't meet their needs unless they are unaware of those needs. Vendors' resources can be used to help identify both process and safety specifications.

Policies and Instruments

In order to meet these objectives, purchasing departments have policies and procedures for organization personnel to follow. Important among these guides are:

1. Control of purchases through central offices so that standards can be enforced. Many hazards can slip in if numerous people are given buying power.

2. Finding of sources of supply by purchasing to control the use of vendors. Purchasing personnel can better identify reliable, responsible suppliers. They can also control problems introduced by the small local purchase or the salesperson's free samples.
3. Stating of specifications with written purchase requests to prevent substandard equipment and materials. The user should know, or make an effort to learn, the possible safety requirements and then make these known to the purchasing department before the order is issued.
4. Approval for purchases of materials with significant fire or health hazards. A control committee, or at least the safety staff, should review and approve the initial purchase of an item to make sure that the risks are known and can be controlled before the item is brought in.

Common Instruments

Several purchasing instruments have been developed to control problems or make purchasing more efficient. Understanding these instruments is the first step in managing purchasing controls.

The purchase order is the instrument used to ensure that the organization gets what it asks for. It is an extremely valuable business document. Under law, it has the full status of a contract. It obligates the vendor to provide and the purchaser to pay. The specifications given in the purchase order must be met by the vendor who accepts it. Consequently, critical safety and health features must be stated in the purchase order. Many order forms have standard statements of terms. These should be read carefully before assuming that they will ensure the desired safety features. Any additional safety requirements must be spelled out on the order form.

The purchase requisition is the organizational form that customarily tells the purchasing staff what the user needs. When there are safety standards to be met, and to be stated in the purchase order, they should be described on the purchase request. Then purchasing personnel will be aware of the needs and can seek a proper source.

A purchase order change notice is used to amend an order. It amends the contract and is legally needed. The change notice form is used to keep the revised order from being interpreted as a new or additional order. Change is a high-risk process, and changes should be handled carefully to avert voiding safety specifications.

A receiving report is used to note any variation between the items ordered and those received. When substitutions are made, it is important to note, evaluate, and communicate the fact that the item received was not the one specified. This should prompt a review of the potential change in risk and an evaluation of the effectiveness of accident control when coupled with the substitute item.

SELECTING SOURCES OF SUPPLY

Sourcing is a critical responsibility of the purchasing department. The ultimate success of a new product or service, the profitability of an existing product or service, and the organization's competitiveness in the field depend largely on the quality of the supplier selected. Logically, then, all selection of suppliers should be made, or at least influenced, by those qualified in supplier selection methods. However, the safety staff can also supply valuable information on the safety record of the potential supplier or on the reputation of its safety products. Without a regular dialogue between safety and purchasing personnel, the organization is the real loser.

Several steps are common to effective sourcing for suppliers. These are:

1. Know precisely what you plan to purchase. If you don't know at the start, use several potential suppliers to learn.
2. Identify safety criteria to the suppliers.
3. Establish a list of prospective suppliers.
4. Obtain relevant information about prospective suppliers and secure bids from those who are judged suitable.
5. Evaluate the bids to select the supplier.
6. Negotiate the final terms, award the contract, and, if appropriate, notify all other bidders.
7. Develop a relationship with the prime supplier and other potential suppliers for future supplies.

Supplier sourcing is the essence of purchasing. The next few sections give some guidelines on the safety aspects of sourcing.

Knowledge of Users' Needs

Inadequate materials or equipment are often traced to vague purchase requests. If you don't know what you want, chances are that you are not going to get it. Complete knowledge of the specifications for the material or equipment is essential to effective purchasing. To obtain this information, the starting point is the prospective user. He or she can be the designer of a new system or modification; or, at other times, the manager following through on an inspection, investigation, or safety analysis. The completeness of the information and the effectiveness of purchasing control depend on the research done. It is chancy to assume that the person who originated a purchase request has done the full research related to safety unless he or she has had safety management training in purchasing controls.

A complete study of operating conditions, of the hazardous properties of materials and equipment, and of the synergistic effects of other materials and equipment, as well as of the work environment, is often needed. The originator of the purchase request may know some of the safety needs and some possible sources for the items requested. But there may be additional needs and better sources. The originator, the purchasing manager, and the safety manager have to work together until they know everything necessary about the item. The depth of the study will depend on the item's significance in the system and the degree of risk involved.

There should be additional inputs on loss exposure precedents identified through accident investigations, safety and health program activities, and new safety research. When safety information is considered at the prepurchase stage, total costs are reduced and system reliability is increased.

Possible suppliers can be used to help do the preliminary research. Purchasing staff can contact them and arrange to observe the equipment and materials in use elsewhere, or get the vendor's inputs on safety requirements.

Knowledge of Specifications

The possible specifications for an item have many facets. There are legislated standards. There are judicial standards set by applicable precedents and decisions at law, and there are the consensus standards of the industry. Often there are manufacturers' standards and even the users' preferences established through brand names.

Internally, there are design standards and codes, parts specifications, variation tolerances, and manufacturing process and testing specifications.

Other internal or external specifications may include packaging, labeling, shipping, storage, handling, service, maintenance, and warranty specifications. Still other user specifications might be based on the interface with other materials or processes where a known hazardous condition must be encountered or might be created. Failure to give all applicable criteria to the supplier could be disastrous.

Effect of Loss Exposure

The energy contact and transfer concept of accident causation is an invaluable aid to loss exposure identification. Whenever there is a potential for energy transfer in a material, equipment, or facility to be purchased, there is a possibility of controlling the exposure through a purchasing specification. References such as *Best's Safety Directory* or the National Safety Council's *Accident Prevention Manual for Industrial Operation* identify standards-setting organizations which can be sources for safety specifications. The amount of knowledge needed about the item varies with the economic, production, quality, and safety considerations of the material or equipment. The greater the potential for energy transfer, the more precise knowledge of safety requirements is required for a proper decision on a purchase.

Supplier Criteria Development

Determining the criteria for supplier selection is an important part of purchasing management. If the criteria on which suppliers are selected are not properly developed, the organization may purchase inferior materials or equipment, may pay excessive prices for what is specified, may wait for custom manufacture and delivery, or may buy an unknown, potentially costly loss exposure.

Some common questions used to develop criteria for supplier selection are:

1. What will be the average quantity of a purchase?
2. What will be the approximate quantity variation as the economy rises and falls, as industry and consumer practices change, and as internal conditions (growth, other product or service lines) change?
3. What will be the smallest quantity or size item purchased? What will be the largest?
4. What will be the frequency of repeated orders?
5. What assurance of stability in supply is needed?
6. What changes in the material or equipment are anticipated?
7. What routine technical services are needed by the user?
8. What technical services are needed by the user in emergency or critical situations?
9. What routine maintenance technical services are needed? What is needed in emergencies or critical situations?
10. What installation or initial-use technical services are needed?
11. What are the transportation options from the vendor to the point of delivery?
12. What loss and liability exposures may occur during transportation?
13. What storage will be required at the organization's locations?
14. What storage will be required at the vendor's locations?
15. What loss exposure is possible during storage?
16. What special storage or handling facilities will be needed as a consequence of the hazard exposure control or the quantity purchased?
17. What handling is required after receipt?
18. What handling is required during transportation?
19. What loss exposures exist during handling?
20. What facilities are needed to handle the equipment or material between receipt, storage, and use?
21. What design technical services are needed?
22. What skill training or job instruction is required before the workers can handle the equipment or materials to be purchased? Must this be supplied by a vendor?

23. What protective equipment is required?
24. Will there be a requirement for additional purchases of equipment or of repair and cleaning materials?

Listing Prospective Suppliers

Establishing lists of prospective suppliers is a two-stage process. First is a general search for possible sources of a material, an item of equipment, or a service. Second, is an inquiry about the qualifications of each possible supplier, who should then be evaluated on the appropriate factors. The factors should be ranked in order of importance. Some possible ones to consider are:

Quality of work
Ability to deliver as desired
Performance history; satisfaction of customers
Warranties and claims history
Production facilities
Production capacity
Commitments to other customers
Technical capability, including patents
Packaging capability
Reputation in the industry
Financial position
Loss control management
Communication system
Management and organization
Operating controls
Attitude of management
Labor relations record
Technical services
Maintenance services
Training services
Geographic location
Desire for business

The appropriate length of listings and comparative ratings of prospective vendors are quite variable. For some items, a choice of two suppliers may be satisfactory. For other items, one may need very long lists and choose multiple suppliers at any given time. In some cases, a supply source may even have to be developed. In other words, a potential supplier may have to be induced to produce an item or service.

At times, lists are restricted by an organization or governmental policy to buy within a local area, state, or country. Some governmental contracts specify sourcing from small businesses or from firms owned by minority groups. Other constraints can come from social, economic, or political pressures and limit the possible lists of suppliers.

The purchasing department that has been operating for some time has lists of suppliers for various items. These lists have been validated through purchasing experience.

This is another reason why all sourcing should be done through purchasing; all the unreliable suppliers have probably already been weeded out. Even so, safety offices can aid the judgment of supplier reliability. Safety staff can evaluate the prospective supplier's accident potential, as well as the effect on the reliability of deliveries and on the quality and cost of a supplied item. Accidents both delay deliveries and increase the cost of materials and services supplied. The vendor with a higher accident rate will have higher costs or may use inferior materials to stay competitive on price.

Obtaining Information About Vendors

When dealing with suppliers not used previously, there are risks of substandard performance in terms of quality, quantity delivered, delivery time, safety, cost overruns, and services provided. There are also the dangers of legal imperfections or outright fraud. More often, the consequences will be wide variations in quality, due to purchasing the marginal product of a supplier who is operating near full capacity, or excessive cost due to buying from a company whose products exceed the quality needed. These risks can be reduced with each bit of information gathered about the supplier.

The goal is to determine which suppliers can deliver the needed item, in satisfactory quantity and quality, at the time needed, and at a competitive price. This may also include the capability and willingness to supply for as long a period as needed.

In most cases, for purchases with little significance other than quantity and price, the inquiry about a supplier will be limited to a request for a price quotation. General safety and quality standards should, of course, be included in even such routine inquiries. These standards may range from compliance with guarding on machines and tools, to packaging and labeling for hazardous substances, and even to safe practices for contract workers on site to perform services.

For more critical purchases, organizations need extensive information about potential suppliers. Purchasing offices often use a capabilities survey form which asks the prospective vendor for appropriate information, such as:

    Name, address, and parent or subsidiary organizations that are involved in the order
    Principal officers and their titles
    Key personnel concerned with providing the subject item or service

Regular products or services
Size of facilities: office, research, plant, storage
Number of personnel in each of the above functions
Number of shifts and hours worked
Union affiliations and contract expiration dates
Customer references on items to be supplied
Credit references from suppliers
Financial statement
Work capabilities and quality tolerances
Equipment owned or leased relative to the item or service
Special skills related to the item or service
Patents and copyrights owned related to the item
Unused production capacity over existing commitments
Accident frequency, severity, and cost summaries

Using the Vendor's Information

The accident record is rarely required. This is unfortunate, because there is as much descriptive information in safety data as there is in production data. It has been well established that safety is a function of management effectiveness. Safety data correlate well with productivity, quality, and cost control. When sourcing, you can compare the prospective suppliers' safety data with that of the industry and get an indication of management's effectiveness. Industry averages are readily available in the National Safety Council's *Accident Facts* booklet (address: "National Safety Council, 444 N. Michigan Avenue, Chicago, IL 60611).

When accident frequency and severity are used in supplier evaluations, the rates for a period of several years should be examined carefully. One year's record may be significantly affected by unusual circumstances and may be disproportionately high or low. However, a sustained high loss record may well be an indication of unreliability or of excessive cost for product quality.

Potential unreliability can be determined through a study of fire incidents. Businesses experiencing major fires traditionally have patterns of small fires or repeated fire hazards discovered on inspections. A 10-year study of business fires in the United States and Canada conducted by *Occupational Hazards Magazine* revealed that, of the businesses which had major fires, 43 percent failed immediately and an additional 28 percent failed within 2 years.

A supplier subject to a major fire due to inadequate safety management has a high potential for sudden interruption of deliveries. Other types of accidents pose similar threats. The safety staff can help the purchasing department by conducting safety evaluations of potential vendors' loss records and safety programs.

Suppliers' Safety Management Performance

An even better measure of potential suppliers' reliability is obtained through a safety management audit. The modern concept of safety management uses measurements of performance of program activities rather than relying on the traditional accident rates. The true correlation between the causes of past accidents and present management's performance is questionable.

A British health and safety study in 1976 concluded that information more meaningful than accident statistics can be obtained through systematic auditing of physical safeguards, systems of work, rules and procedures, and training methods. Objective measurements of performance have been shown to be both practical and instrumental in effective management. Several audit systems have been developed and are available for this use. One of these, the International Safety Rating System (by the International Safety Rating Council, P.O. Box 345, Loganville, GA 30249) comes in six different versions—for general industry, mining and maritime operations, offshore operations, hotel management, and retail trades. Another one is available from the International Safety Academy (P.O. Box 19600, Houston, TX 77024). Several large firms, such as General Motors, Owens-Corning Glass, and others also have their own systems.

On-Site Surveys

Suppliers are frequently eager to contract for orders which they are not presently equipped to fill. Small businesses seek such orders and use them as the basis for expansion. There is evidence that some suppliers, after receiving an order, attempt to renegotiate the terms of the contract based on their real capability. Renegotiation may mask an attempt to pass substandard goods or services. Safety standards can be among the areas designed for cuts by suppliers once they have obtained the contract.

On-Site Survey Teams

An on-site survey of the prospective supplier's facility and personnel is in order when the purchase is critical. Such surveys need to be planned and organized. Procurement survey teams often include purchasing, production, engineering, and quality control personnel. A safety and health professional on the survey team can evaluate the myriad loss exposures and judge which ones will affect the reliability of the vendor's performance. Through objective safety management assessments or audits, the safety professional can also aid in the appraisal of the vendor's management. Quality of management, as evidenced by the quality of involvement in loss control, will be the ultimate control of the supplier's performance.

The on-site survey is especially important in the case of potential suppliers of hazardous materials, of critical safety equipment or items with a high potential for liability, and of waste disposal services.

Supplier Selection

Some questions which can help define a potential supplier's strong and weak points are as follows:

1. Does the supplier have the equipment, people, materials, and other resources to handle the order? If not, are the plans to acquire the resources credible? Would they allow adequate testing of facilities and training of people? Is there a possibility that safety factors will cause the supplier to try to renegotiate after the buyer is committed and has lost other potential sources of supply?
2. Are other orders being negotiated which would use the same people or property resources, creating problems similar to those stated above?
3. Are the equipment and other resources capable of consistently producing the quality and quantity needed from the supplier?
4. Does the supplier's history of equipment, condition, and repair after accidents indicate abuse or abnormal use that could forewarn of substandard performance? Older equipment or excessive shutdowns for repairs can indicate potentially costly acceptance inspections for the buyer.
5. Are the facilities and equipment well maintained and orderly? If a supplier takes shortcuts on preventive maintenance and housekeeping to maintain production, it will ultimately affect cost and reliability.
6. Are overhead costs comparable to those of the industry and of other potential suppliers? High overhead can indicate disorder and potential poor item quality.
7. What are the supplier's controls for fire, flood, windstorm, and other catastrophic loss potentials? Prevention is vital, and emergency preparedness is another measure of a vendor's quality of management.
8. Are working conditions in compliance with safety standards?
9. Is human factors engineering employed to enhance workers' well-being and productivity?
10. Is the lighting good in general areas and particularly good in areas where fine detail work is to be done?
11. Are walking surfaces, machine guarding, environmental controls, and so on in compliance with legislated standards? The degree of compliance in a supplier's facility is a strong indicator of attention to safety standards on items or services supplied.
12. Is the workplace efficient but not rushed? A well-managed organization has neither idle nor harried people. Confusion or aimless activity

indicates inadequate job training and instruction. The buyer will pay the price in terms of quality, consistency, delivery, and reliability.
13. Is there effective material and equipment control? Are in-process materials produced in excess and piled haphazardly throughout the work area?
14. Are tools not in use left lying about? This may indicate inadequate production scheduling, bottlenecks, waste, and variations in quality which can lead to accidents, losses, and inadequate deliveries.
15. Are materials and processes in use that could have an adverse effect on the prospective order? Dusts, fumes, mists, and vapors can contaminate wide areas. They are a source of later health hazards or of abnormal wear and loss exposures.
16. Are receiving and shipping facilities well managed? Delays and damages to raw materials and finished goods can cause serious losses to you later.
17. Are all materials-handling equipment operators trained and qualified?
18. Do materials-handling equipment, warehouse facilities, and stores show signs of abuse or damage?
19. Do shipping people know how to label, pack, and load the items to be ordered to prevent damage? This could be significant in terms of shipping damages, or in liability for shipments en route that are free on board (FOB) the vendor's shipping dock.
20. Is the supplier currently making or providing items or services like those to be ordered? Are the quality and handling of the orders up to standard? What are the differences in specifications that will require changes? Can the vendor make changes? Inappropriate substitutions can often arise under such conditions. Substitutions can lead to violations of safety standards and to accidents.
21. If the supplier is a foreign company, or if items are to be sent to a foreign division or subsidiary, does the supplier have the knowledge and ability to comply with standards for the intended place of use?

The answers to these questions indicate the potential contribution that the safety professional can make or should be prepared to make to the sourcing process.

Self-Sourcing

The "make-or-buy" question is one that the purchasing staff often considers. The company itself is a potential supplier of its own goods and services. The possible advantages in terms of cost, reliability, control, and convenience are matters for the buyers to consider, but safety aspects are a concern that is sometimes overlooked.

Safety resources can be valuable to the purchasing manager studying a make-or-buy decision. Using an item and making it can be quite different. The industry's safety experience may give some preliminary indications of the risks that might be involved. Contacts through professional organizations may help determine the experiences in the industry.

Known precedents and identified hazards are the basis of safety standards. Part of the task of standards research is shifted to the supplier. If an organization decides on self-sourcing, it takes on the full responsibility for compliance with safety standards on the items it makes for itself and in the production of those items. The organization weighing a make-or-buy decision may be well versed in the safety and health hazards in the final process of using the goods or services. However, it may face entirely different exposures in the earlier stages if it becomes its own supplier.

Raw materials processing, such as crushing, can involve extreme physical hazards in materials handling, in the crushing itself, and in the by-products generated, such as dust. Chemical processes often produce by-products or intermediate products which are hazardous only in this intermediate form. The control of processes, the storate and handling of intermediate products, and the disposal of wastes have safety considerations which might affect the feasibility of a make-or-buy decision.

One final word on safety considerations in self-sourcing. That word is "liability." Product liability is firmly attached to the manufacture of components and materials. The courts have given many large compensation awards to injured users when a product did not conform to design and manufacturing standards. The capability to make a component, a material, or a service safer often mandates this action. An employee who is injured when using equipment or materials made by the organization can, in addition to claiming workers' compensation, sue the organization, as a third party under the dual capacity rule, as the manufacturer of the item. At times, make-or-buy decisions are made without recognizing them as another source of potential danger and liability.

Sourcing from Experience

Within the records of an organization is a wealth of information about potential suppliers. The people who operate, repair, service, and supply are untapped reservoirs of data. Interviews with these people can help identify the safety problems of certain products or suppliers. These people know when the problems arose because standards were not met. This information can be obtained through the same interviewing methods used with accident witnesses.

TRANSPORTATION OF GOODS

Purchasing often includes transportation of equipment and materials. Frequently, vendors will ship FOB origin. In this situation, the buyer owns the goods as soon as they leave the vendor's loading dock. The buyer becomes responsible and liable for what happens during shipment. Another potential for loss arises during interplant transfers. Many processors and manufacturers do parts of the processing or assembly at different plants. Items are transported between plants. In any mode, transportation increases the chance of losses.

When goods are moved from one place to another, they can be exposed to various hazards. Heat, sunlight, water, or reactive chemicals can damage the goods themselves. In addition, the reactions can expose third parties to fire, smoke, vapors, and many other forms of injury. After an accident, many organizations have learned that they were moving hazardous materials through densely populated areas at peak time periods. Others simply did not foresee the need to protect goods against possible weather elements.

An important part of purchasing controls is determining exactly where the organization has ownership and responsibility, ensuring adequate protection during transportation, and controlling travel routes to minimize the loss exposure.

DISPOSAL EXPOSURES AND PROBLEMS

Disposing of waste, scrap, by-products, outdated materials, worn-out tools, or obsolete equipment can create numerous loss exposures. People can be injured in handling these items. There can be pollution of the environment or harm to public health. People can be injured while attempting to use the salvaged items or in a chance encounter with them in a dump.

Toxic Waste

Waste is inevitable. Chemical processes produce by-products of the desired reactions. Additional waste is generated by inadequate engineering controls of the process, as a result of employee errors, or through accidents.

The following are critical steps in managing a waste disposal program:

1. Identify and record the chemical content of all wastes and outdated materials.
2. Label all containers of wastes as to type, hazard, and chemical content.
3. Select a proper method for disposal. If a disposal contractor is used, verify that proper methods will be employed.
4. Document each disposal or shipment.

Salvage Equipment

Obsolete, worn-out, or unneeded equipment may be disposed of as waste or scrap material or sold for reuse. In each form, it can create loss potentials.

When disposed of as waste, to be put in a dump area, the item must be in a zero-energy state. All springs, accumulators, flywheels, and other devices to store energy need to be deenergized. Any parts that can move and inflict injury need to be secured. Enclosed spaces where people could be trapped should be sealed.

When equipment is sold for reuse, it is important to make the buyer aware of any deficiencies that could affect the safety of use. Often obsolete equipment has been idle for some time. It may not have had safety features which are currently required. If it is worn out, certain repairs may need to be made by the buyer before the item is safe for use. Salvaged equipment should be tagged to indicate deficiencies or hazards. The bill of sale should indicate the condition and release the seller from liability.

A Disposal Program

Several purchasing authorities contend that there is no such thing as waste; that there is a buyer for everything at the right price. The seller has a moral and, in most cases, a legal responsibility to promote the buyer's safety. Modern purchasing guidelines suggest that a disposal program include:

1. A top management policy statement.
2. A description of duties and responsibilities.
3. Definitions of what and when items are to be disposed of.
4. Procedures for identifying, listing, reporting, transferring, refurbishing, and labeling items for disposal and making them safe.

HAZARDOUS MATERIALS MANAGEMENT

Many potentially hazardous materials are used in almost every workplace. These include flammables, combustibles, toxics, corrosives, cryogenics, and reactives. Purchasing is the critical point of control for identifying these potential hazards, limiting the exposure, and communicating the hazards to employees. This chapter is concerned with managing the purchasing of these items, rather than the related physical sciences and engineering controls.

Identifying Hazards

Effective control of exposure depends on proper identification of the hazards. It is preferable to identify these hazards before the material enters the workplace. The material may be so hazardous that the intended product or service is not worth the risk. Special storage areas may need to be built to isolate the risk or to segregate reactives. Personal protective equipment may have to be obtained and fitted so that employees can handle the material safely. Medical supplies and training may be required for emergency care. Employee education programs may be needed to acquaint people with the hazards of the material and the procedures for handling it safely. It is best to do these things before the hazard is on site.

Material Safety Data Sheets

The starting point for identifying hazards is the material safety data sheet. When use of a substance is planned, purchasing can obtain preliminary information from prospective suppliers. In addition, generic safety data sheets are available from several safety reference books.

The material safety data sheet contains several standard items of information (Figure 12-4).

Chemicals Control Group

Many organizations use a chemicals or materials control group to study potential risks and to approve materials before an initial purchase is made. This group researches the benefits and risks of using the material. It ensures that the necessary methods to control the risks are set up before the material is purchased. The items approved are given an organization approval code which is cited each time a purchase request is made.

The members of the control group vary according to the needs and the situation. The group customarily includes people with the following responsibilities:

Purchasing—identify sources of supply, obtain material safety data sheets, administer the approval code.

Safety—evaluate the risks to people and property under organizational conditions; advise on exposure control, and audit controls for the material.

Chemical and trade names
Chemical ingredients and exposure limits
Physical data such as vapor pressure and solubility
Fire and explosion hazard data
Reactivity with other chemicals and stability data
Health hazard information
Spill containment and clean-up
Disposal procedures for waste
Handling and storage requirements
Protective equipment and facilities required
Emergency treatment for exposure
Medical screening and special precautions

Figure 12-4. Information contained in a material safety data sheet.

Engineering—research the use of the process.
Training—advise on the education and training of employees.
Personnel—coordinate preplacement and health maintenance medical examinations when appropriate.
Medical—assess needs for emergency care facilities and equipment; advise medical facilities on the exposure when employees are referred for treatment.

Chemicals Inventory

Another important part of loss exposure identification is the inventory of chemicals. Purchasing controls should include a current inventory of all chemicals used or stored within the organization. This inventory can be used to identify the materials under different trade names which are redundant, increasing the exposure; materials which are outdated and possibly unstable; materials which are seldom or never used; and materials which could cause reactivity problems.

Limiting Exposure

When it is necessary to use a hazardous material, there are several ways to minimize the risk of accidents. These are:

1. *Minimum quantity.* Buy the smallest practical quantity. The amount depends on the rate of use, availability of storage, size of containers for customary packaging, and quantity pricing.
2. *Working supply.* Keep at the work site no more than one day's working supply or the smallest suitable container.
3. *Segregated storage.* Store hazardous materials away from operating areas. Separate buildings, fire barriers, or similar devices to block other hazard exposures are preferred. Separate materials which react with each other.
4. *Materials movement.* Select routes for movement of materials from receiving to storage, to use, to disposal so that the exposure of people and property to the risk is minimized.
5. *Emergency controls.* Provide systems and equipment to combat emergencies at appropriate locations where the material is stored, handled, used, and disposed of. These systems should include fire control, spill control, and health hazard control.

Communicating Hazards

Employees who might be exposed to hazardous materials must have full knowledge of the hazards. This is managed in several ways:

1. *Material safety data sheets.* Employees are to be educated on how to read safety data. A data sheet for each material purchased is to be obtained and made available for reference in the work area.
2. *Placards.* Standard symbols to indicate the hazard are to be put up in each area of exposure and on each container and vessel.
3. *Labels.* Standard labels giving the name of the substance, hazards, emergency treatment, and manufacturer are to be affixed to every container holding the material for storage, processing, or movement.
4. *Procedures.* Information on hazards and control measures, such as use of personal protective equipment or splash barriers, is to be included in work procedures and manuals.
5. *Signs.* Signs reminding people of the hazards and the control measures are to be placed at the point of exposure. These reinforce the basic education on the hazard and the job procedure.

GENERAL MATERIALS

Many materials not classified as hazardous substances can still cause safety problems. The material may be structurally weaker than expected, such as wood planking with knots or chains with metal defects or of inadequate size. The material may be changed by certain conditions, such as a normally skid-resistant floor tile that becomes slippery when wet or oily, or a cloth that builds up static electricity in dry air. The material may have unforeseen hazards, such as plate glass which becomes jagged and sharp when broken.

Other materials may create hazards when packed for transportation. Items like paper, which is easily carried in reams, is heavy when packed in cartons. Cartons or even single items easily handled by the manufacturer can lead to back strains or other injuries when handled manually by the user.

In managing purchasing controls for general materials, several factors need to be considered before preparing and issuing purchase orders. Some of these are:

1. Will the material be used differently than is customary? Does this difference create a hazard?
2. Will environmental conditions affect the material and create a hazard?
3. Will abnormal conditions affect the material and create a hazard? What is the likelihood of the abnormal condition?
4. How will the material be handled? Must it be lifted or carried manually? Will it be stored in cramped spaces or on high or low shelves?
5. Does the material vary significantly in content? Is there likely to be a variation at some future time? Will the variation create any problem?

Many organizations develop standard specifications for general-use materials. It is highly recommended that standard specifications be prepared and made easily accessible to the requester and the purchasing department. The specifications should include all the consideration indicated above which can affect the operation or the facility.

TOOLS AND EQUIPMENT

Most of the legislated safety standards pertain to tools and equipment. Hazards result from power sources and points of operation. Often the purchasing of equipment is separate from the purchasing of maintenance and operating materials. Thus, these can easily escape the purchasing controls procedures.

Some of the ways hazards occur with tools and equipment are as follows:

1. Variations exist between equipment purchased at different times or from different manufacturers. For example, a worker uses a machine with a hazard exposure not present on the machine that he or she was trained on.
2. A machine in an out-of-the-way location is overlooked when safety modifications are made to others of that type.
3. A safeguard devised to satisfy a new standard makes certain work or maintenance inefficient or more hazardous.
4. Spare parts for older machines become unavailable or costly, and the machine's condition gradually deteriorates.
5. Spare parts in stock for older machines are used on newer models, where they cancel new safety features.
6. Changes in processes cause abnormal wear, which is not perceived and not taken into account in preventive maintenance and replacement schedules.
7. Machines of similar types but made by different manufacturers have varying controls, which confuse operators and maintenance personnel and result in errors and accidents.

Equipment Purchase Program

To be most effective in providing safe tools and equipment, an organization should use a systems approach:

Life cycle. When choosing equipment, consider not only the operation but also the installation, testing, maintenance, potential modifications, and disposal.

Interface. Examine the interfaces with materials-handling, servicing, and other process equipment in the environment and facility where the equipment is to be used.

Training. Examine the requirements for training of operating and maintenance personnel, as well as supervisors. Look for places where people

PURCHASING CONTROLS 327

might have to learn a different method to operate the new equipment safely.

Inspections. Determine the need for periodic inspections of critical parts and safety features such as guards, controls, and energy sources.

Housekeeping. Examine systems for removal of wastes, scrap, and environmental hazards such as dust or noise.

A common purchasing plan to promote safety with new equipment purchases is:

1. Search standards, codes, and regulations to ensure compliance and to realize the benefits from industry accident experiences.
2. Examine the process's history to identify common and unique accident experiences. Remedial actions taken on previous accidents can identify critical specifications for new equipment.
3. Search other sources of help to supplement the basic research. Ask vendors about safety problems and specifications that other customers have proposed. Ask to visit workplaces where the equipment being considered is in operation. Observe the operation for safety as well as production qualities.
4. Hold a preliminary prepurchase conference. Get inputs from maintenance, operations, and safety on desirable features of the new equipment.
5. Develop requirements lists from the analyses and studies. Determine what trade-offs must be made and how these will affect the safety of use.
6. Develop flow diagrams to see how the process will work with the new tools and equipment.
7. Establish priorities and timetables for purchasing, installing, testing, and operational testing of the new tools and equipment.

An abbreviated form of this purchasing procedure can be used for smaller tools. It is important, however, to include the search of standards, the examination of the accident history, the conference on requirements, and the safety reviews and operational tests.

SUMMARY

Purchasing controls provide an opportunity for cost-effective safety management, as well as greater effectiveness in accident prevention. Through purchasing controls, hazards can be kept out of the workplace, and the risks accepted can be better assessed and understood. A purchasing controls program requires a policy that all purchases be made through cental buying authorities; that safety and health criteria for desired items be researched and specified to vendors; that there be safety reviews and approvals of purchase requisitions; and that prospective vendors be screened to find the most suitable one(s). Some of the most critical safety problems occur with hazardous substances and equipment. Purchasing controls are important for these exposures, but they can also be used for greater safety and cost effectiveness in buying general materials and ordinary tools.

BIBLIOGRAPHY

Aljian, George W., *Purchasing Handbook,* New York, McGraw-Hill, n.d.

Ammer, Dean D., *Materials Management,* Homewood, IL, Irwin, n.d.

Bird, Frank E., and Robert G. Lotus, *Loss Control Management,* Loganville, GA: Institute Press, 1976.

Heinritz, Stuart F., and Paul V. Farrell, *Purchasing Principles and Applications,* 7th edition, Englewood Cliffs, NJ, Prentice-Hall, 1986.

Mackie, Jack B., and Raymond L. Kuhlman, *Safety and Health in Purchasing,* Loganville, GA, Institute Press, 1981.

Chapter 13

Materials Handling

INTRODUCTION

No matter what business an organization is in, it has certain operations in common with all organizations. One of these, found in every business from the large industrial complex to the small branch office, is materials handling. The particular materials handling operations may be large and complex, such as those found in a heavy manufacturing plant, or simple, such as the ones found in the stockroom of a three-person office. Yet all these operations have certain problems and hazards in common, and they are much more far-reaching in their safety and health implications than we may realize.

Materials handling is carried out between every department, division, and plant of a company. Every worker in the industry is involved, whether it is his or her only duty or just a part of the regular work. The worker may do it either by hand or with the help of mechanized equipment.

The terms "material transport" or "material movement" are often used as well. In some expert circles, the term "materials handling" applies only to material which is handled manually. In this chapter, however, we use the term "materials handling" in the broader sense to include all handling of materials, either by hand or by mechanical means.

The basic philosophy of materials handling safety and health is deceptively simple. The best solution to materials handling is to eliminate it altogether. In other words, the less materials handling has to be done, the safer the operation will be. In initial design, or in redesign of existing facilities, the object is to eliminate as much unnecessary handling of materials as possible. This can be done by careful facility layout or equipment design. For example, the receiving area for certain raw materials may be located directly adjacent to the machinery that uses those materials. Where materials handling cannot be eliminated altogether, the next best solution is to use mechanical equipment in place of manual handling.

MATERIALS HANDLING, SAFETY, AND EFFICIENCY

Not long ago, unguarded machines and unprotected equipment were the chief causes of industrial accidents. While these are still major considerations, we now see the problem somewhat differently. For example, we consider that unsafe practices, as distinct from unsafe equipment, are involved in nearly 90 percent of accidents. This high percentage includes unsafe practices in handling materials.

Materials handling, by all estimates, accounts (at least in part) for some 25 to 30 percent of accidents. This is not surprising when we consider just how extensive materials handling operations are. The sheer amount of materials handled is enormous. An average industry handles some 30 to 50 tons of material for each ton of production. From the manager's standpoint, these figures and percentages alone make this an area where major waste, loss of resources, and loss of productivity occur due to accidents. By reducing or eliminating accident-causing hazards and practices in this area, management can effect major improvements in productivity and profitability.

A key to reducing materials handling accidents is to make the process as safe and efficient as possible. One of the most successful ways of preventing these accidents is through improved methods of operation, with increased emphasis on safety-conscious supervisory methods in materials handling. Management should consider the following questions:

Can the job be engineered to eliminate manual handling?
Can the material be conveyed or moved mechanically?
How can the material being handled cause injury or damage?
Can employees use handling aids (such as properly sized cartons, lift trucks, or hooks) that would make the job safer?
Will protective clothing or other personal protective equipment help prevent injuries?

THE COST OF MATERIALS HANDLING ACCIDENTS

We know that handling materials with the hands alone is the slowest and least efficient method. More important, it is often the most hazardous method as well. As in other areas, industry has been turning increasingly to automatic or mechanized handling systems. This is not a new development, but it is being employed with increasing frequency. Although mechanized methods create a new set of hazards, the new result is fewer injuries.

A great variety of hand-operated accessories have been developed, including such simple things as special hand tools, bars, racks, dollies, and wheelbarrows. All of these call for careful storage, repair, and use. A piece of equipment or an accessory should not be put into use until its safe operation is fully understood by all users. Above all, both the supervisor and the worker should clearly understand the hazards involved in both manual and mechanized methods.

A common injury in materials handling involves the human back. Injuries often result from lifting too heavy a load or from using an unsafe procedure. The problems of lifting loads have existed ever since civilization began; they still account for around 20 percent of human injuries.

Most back injuries are not serious in themselves, keeping the injured employee away from work for an average of 3 weeks. However, the sheer number of back injuries make it a factor that must be acted upon. In the 1970's, back injuries cost U.S. industry an estimated $1 billion a year! In 1989 the cost was $3.5 billion per year.

The cost also goes up considerably when combined with other types of materials handling injuries, such as overexertion or bodily reaction to injuries. It has been estimated that such injuries now average 140 disability days per injury, at a cost of $5000 each. We have every reason to suspect that these figures are too low. If you consider the indirect costs of accidents, a conservative estimate would place the cost at about four times the $5000 figure, or a total of $20,000 (including direct costs) per injury!

To see what this really means to a business, let's take a simple example. Suppose that, in your operation, the total direct and indirect costs of an accident are about $10,000. If you are making a 10 percent profit on every dollar of business transacted, it will be necessary to do $100,000 worth of additional business to cover the cost of a single accident. If the figures from your local chain grocery store can be believed, their profit margin is only one cent on a dollar's worth of business. A $10,000 accident on their loading dock, then, would require $1 million worth of additional business to compensate for the costs of that one accident alone! And keep in mind that these figures are thought to be conservative.

Another factor for the manager to consider is that some aspects of the OSH Act apply directly to materials handling practices, so that this is not an entirely voluntary process on management's part. OSHA contains specific regulations regarding the use of industrial trucks, cranes, derricks, hoists, and other mechanical equipment. Management is also expected to provide safe equipment with which to handle materials. The senior manager can expect the foremen and supervisors to know all that is required for OSHA compliance in connection with materials handling. That implies that management must provide foremen and supervisors the opportunity to acquire the necessary knowledge and put it to work.

ERGONOMIC OPTIMIZATION

Since the body, particularly the back, is weaker than mechanical equipment, its proper use should be an important part of accident prevention programs. The prevention of back injuries, for instance, has traditionally been attempted by (1) careful selection of workers, (2) good training in safe lifting practices, and (3) designing the job to fit the worker (ergonomics). Several studies indicate that these common techniques are not very effective in controlling back injuries. Ergonomic redesign of tasks to reduce manual handling exposure is a partial approach to the effective control of low back injuries. Unfortunately, physical differences make it impractical to establish safe lifting limits applicable to all workers. One commonly used approach is strength testing, in which a person's physical capabilities are matched to a job's physical demands.

Ergonomics is much more complex than that. It provides important input into the design of safe materials handling systems, especially by providing the design interface between worker, machine, equipment, and environment. Ergonomics interacts with both traditional engineering and management. Its object is to provide a proper approach to materials handling by matching the technical and human systems.

Basically, three things are considered in the "ergonomic optimization" of materials flow:

1. Designing the material flow system to fit the people involved.
2. Improving existing equipment and the selection of new equipment.
3. Improving existing facilities.

The first management task in ensuring ergonomic optimization is to secure the proper human engineering expertise for design and redesign of materials handling systems, and to apply ergonomic expertise to personnel selection and training. In 1975, Armine highlighted a few techniques to help achieve efficient and safe materials handling. These are:

1. Optimize materials transport methods.
2. Optimize material flow.
3. Use ergonomic rules in manual handling.
4. Eliminate the need to lift or lower manually.
5. Eliminate the need to push, pull, or carry.
6. Eliminate the need for manual handling.

The human engineering expert and the materials handling consultant will have several approaches for each of the listed items. For example, consider item 4. If complete elimination is not feasible, the human engineering expert would then attempt to reduce the amount of energy or force required by:

Reducing the size or weight of the load.
Using a dolly, truck, or conveyor.
Using wheels, casters, or ramps.
Applying air-bearing devices.
Improving the maintenance of equipment and floors.

This list is by no means comprehensive; it merely suggests that there are numerous alternatives available to the expert. This approach can be just as easily applied to any of the six areas mentioned above.

MATERIALS HANDLING FUNCTION

While the senior manager cannot hope to become an expert on the finer points of ergonomic optimization, he or she should try to achieve a comprehensive overview of materials handling functions in order to determine where human engineering expertise may be needed. Such an overview is presented by Kroemer 1983. He breaks materials handling down into seven separate functions or phases.

1. *Transport of raw material.* Any material delivered to a plant requires people and equipment to move it. Even when this material is delivered by an outside carrier, there is an important interaction with the manufacturing process.
2. *Receiving materials.* Receiving materials from the transportation mode and bringing them to the workplace also involves handling. The materials must also be inspected for quantity, quality, weight, and other specifications, which requires additional time and handling.
3. *Materials flow.* The flow of materials to or between work stations creates problems of worker time, cost, and accident exposure. The materials can be moved many times, often as single elements, combined subsystems, or systems. Storage of the materials between operations is critical, as it often involves potentially hazardous situations.
4. *Workplace handling.* Workplace handling is a costly operation with built-in hazards resulting from the movement between fixtures and jigs, into and out of machines, and handling of finished products and scrap.
5. *Packaging materials.* Handling of packaging materials and waste stemming from the manufacturing and assembly processes presents many additional problems. Materials must be moved, stored, disposed of, processed, sorted for reuse, recycled, or discarded. These are most frequently manual operations.
6. *Warehousing.* The warehousing function involves order selection, assembly, packing, and movement out of the warehouse.
7. *Distribution.* This is the last function of the cycle and can involve the transportation of materials by road, rail, water, or air. Fortunately, many methods have been developed to reduce manual handling during this phase.

This is just one possible breakdown of the materials handling function, but it can be quite useful to the manager. In particular, it lends itself to ready assignment of accident prevention accountability by operational function.

It should be obvious that all these phases require a safe workplace. This workplace control is management's particular responsibility. Organization and administration of a safety program are, after all, based on management's efforts to provide a safe workplace. Proper materials handling equipment and techniques have an enormous effect on workplace safety. For this reason, before management can make many improvements in the workplace, it should conduct a thorough survey of both the plant and the workplace injury and accident experience. Since familiarity with the plant may make it difficult to recognize hazards, outside assistance for the survey may be practical. The time and cost involved in such a survey will pay off, not only in an improved and safer working environment but in increased overall efficiency, as well.

SOME PRINCIPLES OF MATERIALS HANDLING

Even when the probability of a materials handling accident is very low, continual manual handling will eventually lead to such an accident. The message is clear:

1. Do not have people handle materials unnecessarily.
2. If materials handling cannot be avoided, reduce the personnel involvement and let machines do the work. Use automated machinery that does not require people, drivers or helpers as far as possible.
3. If the system cannot be automated, then use machines and other equipment that will take over the heavy, difficult, and dangerous tasks of lifting, pulling, pushing, raising, lowering, and holding of material. The operator should control only the movement of the machine itself.

Several principles of materials handling can be listed. Examples of their application are shown in parentheses.

1. *Planning.* Careful planning should be done to obtain efficient and safe movement of materials (graphic and tabular aids).
2. *Systems.* Work for a coordinated system of operations from receiving through processing to distribution.
3. *Materials flow.* Plan all sequences and equipment to optimize the flow (process charts and diagrams).
4. *Simplification.* Reduce materials flow by combining movements and equipment (using die-cast parts instead of making several pieces and then welding them together).
5. *Gravity.* Use gravity, whenever possible, to move materials (inclined roller conveyors and gravity feed bins).
6. *Space utilization.* Use building cubes (high racks with narrow aisles, stacking trucks, overhead conveyors), taking care not to create more hazardous conditions than are being eliminated.
7. *Unit size.* Increase either the quantity or the size of the load so that equipment must be used, or reduce the size and weight so that one

person can handle it (pallet-sized loads, lightweight materials, or reduced bulk).
8. *Automation.* Use automation whenever feasible (automated stackers).
9. *Mechanization.* Whenever possible, operations should be mechanized (powered conveyors).
10. *Equipment selection.* Consider all aspects of the material flow in selecting equipment (fork trucks with different forks).
11. *Standardization.* Use standardized methods and standard types and sizes of equipment (standardized pallets, cartons, and containers).
12. *Adaptability.* Utilize methods and equipment that can perform a variety of functions, rather than specialized equipment (variable-speed conveyors and multipurpose forklifts).
13. *Dead weight.* Reduce the ratio of dead weight to load carried (low-density materials).
14. *Utilization.* Plan for optimum use of equipment and personnel (radio control of drivers).
15. *Maintenance.* Plan for scheduled and preventive maintenance (periodic functional checks of trucks).
16. *Obsolescence.* Replace old equipment and methods with new ones when practicable (keep up-to-date on methods and equipment).
17. *Control.* Seek improved control of production, inventory, and order handling (computerize flow control).
18. *Capacity.* Use equipment to help raise production capacity (replace manual operations with mechanized ones).
19. *Performance.* Determine effectiveness in terms of cost per unit (computerized quality/quantity control).
20. *Ergonomics.* Apply ergonomic principles to best fit and use human capabilities.

Each of these principles can introduce changes into the system, which must then be evaluated for safety and health. Although such changes generally are made in the name of more efficient production and handling, all should increase safety and health as well, particularly if they are applied early in the design phase.

EQUIPMENT

There is an endless variety of materials handling devices, most of which have specific requirements for safe operation and design. Nearly all require specialized training, sometimes even extended courses and certification. We will discuss only a few representative types of equipment, including cranes, industrial trucks, ropes, chains, slings, conveyors, railroad cars, and trucks. These brief overviews are intended only to alert the senior manager to the extent of the safety considerations involved in materials handling equipment.

Cranes

Cranes are so important that only persons thoroughly familiar with their construction and operation should be involved in their use. In the United States, cranes are generally constructed to meet the strength and service standards of the American National Standards Institute (ANSI). Operating rules have been developed from several sources. We recommend those of the Crane Manufacturers Association of America.

Crane operators require special training. A thorough training program should include:

1. A review of crane accidents due to overloading, failure to use safety devices, abnormal conditions, faulty slings, signals, faulty erections, and dismantling.
2. Design procedures, including stability, safety factors, height, wind and crane types, along with their advantages and disadvantages; rail-mounted cranes; safe load indicators; and cutouts.
3. Erection, testing, and dismantling.
4. Steel wire and fiber ropes, including their proper selection and use; slings and their correct use; and maintenance of ropes.
5. Chains, hooks, and lifting gear, including construction, storage, and inspection; materials and their recognition; eyebolt positioning and special lifting gear.
6. Maintenance, inspection of, and access to cranes; periodic inspection and lubrication; tire and track care; brakes, clutches, hooks, and sheaves.
7. Safe operation of cranes, including the controls, use of load indicators, overhead obstructions, traveling and positioning, raising and lowering loads, and tandem lifting.
8. All of the applicable regulations, with emphasis on operator duties, accident and dangerous condition reporting, penalties, and highway obligations.
9. Methods of slinging.

All of this information should be formally presented, with practical demonstrations and applications at each step. The duties of the craneman and the chainman (who fastens the load to the hook and chain) are complex, extensive, and important. Training should include at least an annual review of crane operations, with refresher training for the craneman and chainman.

Even operations that do not involve the use of large cranes frequently make use of relatively simple floor cranes or of portable, hand, or "multipurpose" cranes, as they are sometimes called. These cranes are moved by hand and are not to be confused with the larger, motorized "compact" cranes. Most of them are in the 100- to 6000-pound range. We mention them

MATERIALS HANDLING   337

particularly because they are so often involved in accidents. Most of these accidents are of two kinds: (1) tipping over when a load is extended on the boom or (2) tipping over when the front wheels are obstructed and a load is on the boom. Figures 13-1 and 13-2 illustrate what happens in these instances. Restricting load movement by means of a tether is one sound approach to avoiding such accidents. Impressing crane operators with the need to be alert to these two conditions is also essential.

Certain safety rules should be observed for all cranes, as well as for hoisting equipment. While these rules are very general, they will give the manager some idea of the main safety considerations involved. Keep the following in mind:

1. The rated load must be legibly marked on each side of the equipment, and employees should be aware of the load weight.
2. The equipment must have a self-setting brake applied to the motor shaft or some part of the gear train.
3. For powered cranes and hoists, holding brakes must be applied automatically when the power is off.
4. Hooks, chains, and all functional operating mechanisms must be inspected daily for any indication of damage, wear, and tear. Monthly records should also be maintained.

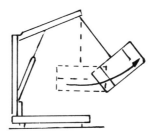

Figure 13-1. Extended load on a boom could cause tipping.

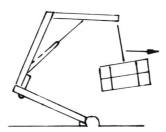

Figure 13-2. Obstruction of front wheels can also cause tipping.

338 SAFETY AND HEALTH MANAGEMENT

5. Loads should never be carried above the heads of people.
6. The operator must test the brakes each time a near-capacity load is handled. This is easily done by raising the load a few inches and then applying the brakes.
7. The hoist rope or chain must be free of kinks or twists and should not be wrapped around the load.

Industrial Trucks

Powered industrial trucks may be classified by power source, operator position, and means used to engage a load. Power sources include electric motors energized by batteries, gasoline, L-P gas, or diesel fuel, or a combination of these. The power source used depends on the work environment. For example, it may be too hazardous to use some power trucks where dangerous concentrations of flammable vapors, gases, or dust may exist. Conventional trucks give off carbon monoxide, making them unacceptable in poorly ventilated areas. Diesel trucks, which are widely used, give off a foul odor that workers believe is dangerous to their health. Provisions for safe operation, maintenance, and design of powered industrial trucks are found in ANSI Safety Standard for Powered Industrial Trucks, B56.1. Three categories of powered industrial trucks are:

1. Those controlled by a riding operator and commonly referred to as a "forklift," since they lift loads by means of a fork-like mechanism that is often higher than the operator. Also in this category is the straddle truck, which straddles the load, and the order-picker truck, which lifts an operator to a desired height to reach storage bins.
2. The motorized hand truck is controlled by a walking operator.
3. Becoming common is the electronically controlled vehicle which travels, without an operator, along prearranged routes and is guided by frequency sensors, a light beam, or induction tape in the floor.

The first type, forklifts, receive the most attention and are discussed briefly here. Here is a selection of forklift mishaps that illustrate common problems:

1. An untrained forklift operator drove his loaded forklift forward, down an incline, at night. He raised the load fully and turned, thus overturning and fatally injuring himself.
2. An untrained, unsupervised worker was using a stacker-type forklift in a warehouse. He collided with a double tier of paper reels, which fell on a machine operator, killing her as she sat at her machine.
3. A worker who was permitted to drive a forklift truck even though he had received no training reached through the mast to operate a release

lever. He accidentally operated the mast tilt control, and his head was crushed between the mast and the canopy frame.
4. A warehouseman was directing the driver of a sideloader forklift truck in placing bundles of steel tubing on a stack. The bundles slid off the forks and killed him.
5. A forklift without a load was carrying a passenger, although not equipped for it. The track was traveling with the forks raised, obstructing forward vision, when it hit and killed another worker.
6. A 17-year-old employee in his first week of work was being shown how to drive a forklift. He had received no formal training and had not been given a proficiency test. He made a tight turn near a truck being loaded and crushed the truck driver against his own truck, killing him.

A few general rules follow that may apply to all industrial truck operations.

1. High-lift rider trucks must be fitted with an overhead guard (see Figure 13-3).
2. Developing and using effective training methods to instruct operators in the safe operation of powered industrial trucks is a must.
3. When a powered truck is left unattended (that is, when the operator is more than 25 feet from the truck), the forklifts must be fully lowered, the control level positioned in neutral, the power shut off, and the brakes set. The wheels must be blocked if the truck is parked on an incline (see Figure 13-4).

Figure 13-3. The load on a high-lift rider truck can injure an operator unless there is an overhead guard.

340 SAFETY AND HEALTH MANAGEMENT

Figure 13-4. Truck wheels must be blocked to prevent rolling if the truck is parked on an incline.

Figure 13-5. Do not travel forward when the load obstructs the view.

4. Industrial trucks must be examined daily for any conditions adversely affecting their safety. This inspection should be made before they are placed in service for the day. If a truck is used around the clock, it should be inspected after each shift.
5. If the load being carried obstructs the forward view, the operator is required to travel with the load trailing (see Figure 13-5). If the powered truck is controlled from the ground or floor by a handle, the operator must be careful not to place himself in the path of the rearward-moving truck.

We cannot stress too strongly that operators should be physically qualified and thoroughly trained in the operation of the equipment. Such training must include the following principles of safety. Qualifying training and certification training are readily available in most communities. Only qualified personnel should be allowed to operate the trucks, and riders should never be permitted.

The U.S. Army found that six cause factors were involved in over one-half of its forklift mishaps:

1. Lack of attention by operators.
2. Operators and maintenance personnel working on running equipment.
3. Workers placing themselves between fixed objects and operating equipment.
4. Operators concentrating on the load and failing to use ground guides in congested areas.
5. Slips and falls in getting on and off the equipment.
6. Passengers falling off forklifts and workers standing on elevated tines or using them for ladders.

Good training programs and careful operation can prevent these and most other types of forklift mishaps. For expanded treatment on preventing forklift mishaps, the reader is directed to the *Accident Prevention Manual for Industrial Operations: Engineering and Technology,* 9th edition, cited at the end of this chapter.

Hoisting Apparatus

A crane, by definition, is a machine for lifting and lowering a load and moving it horizontally, with the hoisting mechanism an integral part of the machine. A hoist, then, may be part of a crane, exerting a force for lifting or lowering. Some hoists are designed for small loads; others can lift hundreds of tons. The hoists most used in industry include air, electric, and hand-powered types, ranging in capacity from 0.25 ton to 2 tons.

Much of what has been said about cranes applies to hoists. Detailed inspections are needed on a regular basis, with special attention to load hooks, brakes, and limit switches. Moreover, hoists should never carry people, unless in combination with approved safety devices that meet all relevant codes.

Conveyors

There is sometimes confusion about what constitutes a conveyor. Definitions are found in ANSI Standard B.20.1, "Safety Standards for Conveyors and Related Equipment," also known as the "conveyor code." Typically, a conveyor is a horizontal, inclined, or vertical device to move or transport bulk materials, packages, or objects, and having points of loading or discharge. The definition specifically excludes industrial trucks, tractors, trailers, tiering machines, moving walks or stairways, cableways, pneumatic devices, and other transportation devices.

Conveyor systems fall into two general categories: unit load-handling and bulk load-handling systems. For in-plant operations, we are mostly concerned with unit load-handling conveyors because of their wide use in manufacturing and distribution. They often differ greatly in design, as well as in the safety requirements for each system. In fact, the designs of conveyor systems differ so widely that it is wise to design safety programs for the specific types of conveyor in use. Unit load-handling conveyors fall into two basic categories: gravity and power.

Gravity conveyors, which are simpler and safer, "mechanize" an operation that does not justify more costly powered equipment. Most common are the gravity wheel and gravity roller conveyors.

Power conveyors include belt-in-roller, live roller, padded chain-driven roller, slider bed belt, and automatic pressure. A variety of directional change units and sorting systems are used. A sorting system consists of a sequence of "diverters" and a control system which activates them. Diverters transfer materials from one conveyor line to an intersecting line. There are many types of diverters.

Safety in conveyor systems starts with systems that meet the ANSI B.20.1 standard. It has good guidelines for safe design and for operator and maintenance team safety. Meshing the conveyor safety program with the overall safety program can be a challenge that involves the vendor, purchaser, and user. While a discussion of the details of conveyor safety is beyond the scope of this book and a matter for experts, the twelve safety rules in the standard are worth noting:

1. Keep clothing, fingers, hair, and other body parts away from conveyors.
2. Know the location and function of all start/stop switches.
3. Keep all start/stop devices free of obstructions for easy access.
4. Operate the conveyor with trained personnel only.
5. Make sure that all personnel are clear of obstructions.
6. Keep the area around the conveyor clear of obstructions.
7. Service conveyors with only authorized maintenance crews.
8. Report all unsafe practices around conveyors.
9. Don't service conveyors until the motor disconnent is locked out.
10. Don't load a conveyor beyond its design limits.
11. Don't climb, sit, step, or ride on the conveyor at any time.
12. Don't remove or alter conveyor guards or safety devices.

Ropes, Chains, and Slings

There are many special precautions to be taken in the use of ropes, chains, and slings. A safety professional or consultant will be familiar with their use and should be called on to furnish specific information and input.

Safety factors involved in the use of ropes are affected by several things.

One is the material of which the rope is made. Synthetic fibers are more popular than fiber ropes for several reasons. In recent years, manufacturers have found out more about the properties of synthetic fibers and have matched them to market needs. Splices can be readily made of synthetics, and they are much stronger than natural fibers. Nylon, for example, is two and a half times stronger than manila rope and has four times its working elasticity. It is well suited to shock loading and is highly resistant to mildew. It performs well when wet and does not lose strength when either wet or frozen.

Today wire rope is also widely used instead of fiber rope. Its physical properties make it preferable because:

1. It has greater working strength and durability under severe working conditions.
2. Its physical characteristics do not change in varying environments, such as very hot, cold, damp or dry conditions.
3. It has controlled and predictable strength characteristics.

Wire rope does have certain drawbacks, however. It requires frequent, thorough inspection, as it does deteriorate for a number of reasons. Deterioration is brought on by wear, tear, corrosion, kinks, fatigue, drying out of lubricants, and overloading. The last factor especially refers to dynamic overloading from rapid stops, overwinding, and mechanical abuse brought on by careless handling.

Chains have a fairly standard construction, alloy steel being used because of its high resistance to abrasion and its immunity to failure. Most of the causes of chain failure can be detected before failure actually occurs if good inspection procedures are followed.

Chain slings are usually constructed of synthetic web or metal mesh. Synthetic web is generally used where the load must be protected (such as with easily scratched metal rollers). Slings are rated, just as chains and ropes are. Normally, the sling should be able to handle five times the rated load of the rope or chain. The *Wire Rope Users Manual*, published by the American Iron and Steel Institute in Washington, D.C., provides most of the information needed in selecting and using wire ropes, chains, and chain slings.

Railroad Cars and Trucks

Not long ago, heavy industry was possible only where adequate railroad systems existed for moving raw materials and finished goods. Even today, many operations have their own spur lines or sidings and routinely receive large rail shipments, sometimes several carloads at a time. Handling railroad cars and the freight they carry poses particular and unique materials handling problems.

Switch engineer crews generally take care of spotting the railroad cars. However, it is sometimes necessary for company employees to move a railroad car as well. This is a dangerous operation and should be done only under the supervision of an experienced foreman. The following factors must be considered:

1. The facilities or equipment available to move the car.
2. The distance the car must be moved.
3. The number of cars to be moved.
4. Whether or not the track is level or on a grade.

After weighing these factors, the supervisor can decide whether to move the car by hand or whether to take the material to or from the car without relocating the car itself. If the car is to be moved without power equipment, a car mover can be used. These devices allow one to apply leverage under a car wheel; their safe use requires sound guidance and knowledge. It may take two car movers working with opposite wheels in some cases.

Several general rules should be observed in dealing with cars:

1. Never use an inch or crow bar to move the cars.
2. Never try to stop a car with a piece of lumber or metal on the track.
3. Never allow a car to move toward the main line of the railroad if the slope of the track is in that direction.
4. Never use a truck or other road vehicle to push or pull a car.
5. Never leave a car mover in place under a wheel. Under pressure, it may fly out when a train moves the car.

Far more than the railroad car, the commercial truck is a universal feature of modern business operations. In general, its safe maintenance and use falls into the domain of traffic or vehicular safety, a field requiring expert knowledge in itself. There are, however, several considerations we must mention specifically for materials handling crews working with truck trailers during loading and unloading operations.

Most accidents during loading and unloading operations involve hand or industrial trucks going through gaps between the trailer and the loading dock. The trailer should always be guided into place and the truck parked in gear, with the brakes on. In the event that the truck moves away and leaves the trailer, the trailer should be both braked and well choked, usually under the rear wheels. Do not consider moving lift trucks into a trailer or truck until bridge plates are adequately secured and locked in place, and until the operator inspects them to ensure security and adequate bridging of any gaps, noting especially that they are properly overlapped.

MATERIALS STORAGE

Materials storage is a job in itself, requiring proper training and experience for safe performance. It may be helpful for the manager to be able to recognize certain conditions essential to safe and efficient materials storage.

First, both temporary and permanent storage areas should be kept neat and orderly. If the business is large enough to have a warehouse supervisor, this should be his task. The warehouse supervisor is also the one to direct raw material storage, with particular care for those materials which will be kept on hand for longer periods.

Storage problems are by no means confined to the warehouse, however. The production supervisor stores limited amounts for short periods close to the operations area. In an organization with a high volume of incoming and outgoing materials, there may be time constraints on how long the production supervisor's materials can be kept in storage. It is not unusual for deadlines to be set, requiring production to pick up items within a day or two of their receipt. Seldom will more than a week's delay be accepted.

Regardless of who stores material or where, care must be taken not to obstruct fire alarm boxes, exits, sprinkler system controls, first aid equipment, fire extinguishers, walkways, or aisles. Standard clearances are set for most of these areas by law.

The use of storage bins and racks can reduce hazards, as well as taking up less space. Material that is stored on racks, pallets, or skids can be moved easily from station to station with less material handling and, thus, fewer potential accidents or injuries. Different types of storage require special precautions, which the appropriate supervisor should recognize and understand. He should be able to handle special problems associated with the aisles in storage areas, the height and construction of bins, and the types of pallets, barrels, and kegs used. He may need to know special handling methods for rolls of paper and uncrated stock such as lumber, piping, bar metal stock, sheet metal, and so on. There are special procedures for handling most of these types of material.

Well-planned materials storage arrangements can reduce the handling necessary to bring materials into production, as well as to remove finished products from production to shipping. And the less handling that is required, the safer and more injury free the operation will be.

MATERIALS FLOW

One useful technique for eliminating materials handling hazards is to analyse material flow. There are several methods available for examining these hazards in detail, including the job safety analysis, as discussed earlier in this book. Most of the techniques of interest to us are "before-the-fact" techniques.

That is, they are applied before an accident happens. The "after-the-fact" techniques involved in accident investigation can also be useful, though obviously they are not preferable. Neither set of techniques should be neglected. Before-the-fact techniques may be better in that they allow management to take action before someone is injured or valuable property is damaged. Management is not, in effect, waiting for proof that a hazardous situation exists, but is taking steps to ensure that the hazardous condition never develops. Few things reflect more poorly on management's efficiency, credibility, and effectiveness than accidents stemming from known and existing causes. If accidents involving materials flow do happen, sound investigative techniques (as discussed in detail in Chapter 10) will be of great assistance.

Preventing the accident is more desirable than investigating it afterward, however. Attention to certain details will lead the way. In-plant routes should be laid out and organized to avoid blind corners and to minimize turns. Planning for one-way traffic and carefully positioned, designated crossing routes can remove many of the hazards. The width of routes should be clearly posted, as should hazard warning signs.

Better materials handling can also be achieved through:

Reduction of transportation distances.
Simplifying the handling operations.
Using modern material-moving equipment.
Making the material flow independent of external elements such as machining and assembly time.

The big challenge in materials handling, as pointed out earlier, is to reduce handling to a minimum. The next challenge is to mechanize material handling as much as possible, getting rid of the hazardous, strenuous part of the job. Start in the planning stages to:

1. Design facilities requiring the minimum movement of material.
2. Automate materials movement that cannot be eliminated.
3. Mechanize the process so that people control, but do not handle, materials.

Where material is already being handled by people, consider how machines can handle the flow or maintain a smooth flow, or how the flow can be redesigned to reduce manual movement.

Even under the best conditions, it is difficult to describe and analyze a complex materials handling process, since it may interact with numerous equally complex manufacturing and storage processes. Several techniques have been developed which can aid in this description and analysis process. Most techniques involve three steps:

1. Break down the process into separate functions, from receiving to distribution.
2. Diagram and tabulate activities to determine the elements of the activities involved.
3. Take preventive and corrective action.

Step 2 is carried out using graphic and tabular control aids. While these usually require some expert knowledge and skill to perform and apply, the basics are familiar to most managers. Usually the process begins with a flow diagram, a two-dimensional picture or sketch of the activities involved, indicating the sequence of events. Naturally, the more complicated the process, the more involved and specific the diagram must become. The flow chart is a tabular listing of the activities, including time flow and information on impinging activities. At this point, it is important to realize that materials flow is a specialized task and that there are many experts in the field. The manager should make use of expert knowledge and training wherever feasible or necessary.

From the author's viewpoint, this is an ideal time to bring in expert assistance. It would be hard to find a better time to increase efficiency, maximize profits, and control hazards. Dollars spent on consultation here will yield better and more profitable results than in many other areas.

The materials handling expert will seldom be a safety and health expert, so all planning should be carried out in conjunction with the safety specialist, with an eye to designing hazards out of the operation. It may also be appropriate to bring plant time-and-motion personnel into the review process. Several responsible organizations have combined their principles to give us this listing, and it can be soundly applied in every department and operation.

HAZARDOUS MATERIALS

Special attention must be given to materials handling operations that involve any type of hazardous material. Perhaps we are fortunate in having many guidelines already established in this area, as such mandated guidelines provide at least a starting place and a set of minimum standards and practices. Keep in mind, though, that these are merely minimum guidelines. An efficient, well-planned materials handling safety program will not stop at mere minimum compliance. The Merritt Company's *OSHA Reference Manual* provides checklists for these minimum standards, and the manager should take the time to become familiar with them before planning or reviewing the company's own hazardous materials handling operations.

348 SAFETY AND HEALTH MANAGEMENT

The supervisor should also be familiar with all current regulations and practices for the hazardous materials his or her department handles. The supervisors of shipping and receiving will naturally get the lion's share of the responsibility, since they must be familiar with the regulatory details of floors, ramps, aisles, lighting, stock selection, pallets, dockboards, bridge plates, special shipping and receiving equipment and tools, steel strapping, sacking, and so on—in addition to the specific regulations governing the shipping and receiving of hazardous materials. Obviously, these supervisors will need the cooperation and assistance of a number of staff and management people for full compliance.

Special problems exist with hazardous materials, some of which are unique. With hazardous liquids, for example, we may be concerned with pipelines, portable containers, and tank cars. In the case of hazardous solids, there may be special bins and portable containers. Explosive storage requires specialized knowledge, including the application of special regulations. The handling of oxygen, nitrogen, argon, helium, and hydrogen is common but does require special knowledge of their properties and the rules of safe handling. Other more exotic materials are becoming more and more common as well. Storage and handling of cryogenic liquids, for example, is a growing problem.

One of the keys to handling cryogenic liquids is actually a matter of simple economics: the cost of delivering and storing liquids is often less than that of delivering and storing the same substance in compressed gas form. Nitrogen, to take just one example, occupies *700 times* as much space at atmospheric pressure as it does in liquid form. The reduction in the cost for containers, demurrage, shipping, and storage is obviously enormous. However, handling liquefied gases that are stored and used at very low temperatures requires special knowledge and precautions. The plant engineer and his or her assistants must know the specific properties of each liquid and its compatibility with other materials. There are many special precautions to be observed, in addition to the ordinary safety rules. General safety practices for handling cryogenic liquids include the following:

1. Always handle cryogenic liquids carefully, since they can damage the skin or eyes.
2. Stay clear of boiling or splashing liquids and their vapors.
3. Never allow any unprotected part of the body to touch uninsulated pipes or vessels containing cryogenic liquids.

There may also be special precautions to take, depending on the liquid involved. As one example, corrosive materials require preventive measures that include safe handling, spillage prevention, and the use of protective equipment and clothing, providing for emergency showers, eyebaths, and respiratory equipment. Preventive measures should also consider substituting less toxic or corrosive materials where possible, isolating the hazardous process by the use of enclosures, providing adequate exhaust ventilation, and so on.

The National Safety Council has developed a checklist of precautions for use when handling hazardous liquids. It is particularly useful for the senior manager, since it does not require specialized knowledge of the material, storage, or handling procedures. The following precautions are recommended:

1. Protective equipment should be worn.
2. Floors should be clean and not allowed to become slippery.
3. Drains, syphons, ejectors, and other emptying devices should be emptied completely before removal from carboys. (Carboys are small containers with special trucks which are stored on specially designed racks. They require constant and thorough inspection to ensure against leakage.)
4. When diluting acids, always add the acid to the water (not the water to the acid) slowly and stir constantly with a glass implement.
5. In case of an accident, give first aid, flush with water, and call a doctor.
6. Do not force hazardous liquids from containers with compressed air.
7. Do not wash or clean a container before returning it unless this is specifically required by its label.
8. Do not store hazardous chemicals in glass or other containers near steam or heat pipes or where strong sunlight will strike them.
9. Do not stir acids with metal implements.
10. Do not store hazardous materials in poorly lit areas where the wrong chemical might be selected.
11. Do not pour corrosives by hand from a carboy.
12. Do not move carboys that are not securely stoppered and wired.
13. Do not stack drums of hazardous liquids. Place them in special racks suitable for each type of drum and segregated by contents. If liquids are flammable, they should be grounded when stored and bonded when being transferred to smaller containers.
14. Storage areas should be well ventilated, and spills should be cleaned up immediately.
15. Ensure that flood showers and eyewash fountains are immediately available where caustics or acids are stored.

This list is by no means complete; it covers only the most general rules. However, the manager can readily use it to check the general safety and health precautions observed in handling and storing hazardous materials.

CONCLUSION

Nearly every operation involves materials handling of some sort. This chapter has dealt with the subject from the standpoint of the manufacturing process, but the same principles apply to any business. A materials handling accident stemming from careless lifting of a carton of typing paper can cost just as much in lost time and compensation as one stemming from operating an overhead crane; storing a case of rubber cement thinner involves the same basic safety practices as storing a drum of corrosive acid.

Materials handling is a separate specialty, but in a small company the senior manager or plant engineer may need to double as the expert. Even where specialists are employed or consulted, the manager will want to be sure that the safe practices recommended and established are followed in day-to-day operations.

Because materials handling is involved in so many workplace accidents and injuries, it is a fruitful field for preventive planning and action, as well as a prime area for increasing efficiency and lowering operating costs. Safety and health programs are not complete unless they give careful attention to all the hazards and activities involved in materials handling operations.

REFERENCES AND BIBLIOGRAPHY

Accident Prevention Manual for Industrial Operations, Engineering and Technology, 9th edition, Chicago, National Safety Council, 1988.

American Iron and Steel Institute, Committee of Wire Rope Producers, 1000 16th Street, N.W., Washington, D.C. 20036. *Wire Rope Users Manual,* 1979.

Anton, Thomas J., *Occupational Safety and Health Management,* New York, McGraw-Hill, 1979.

Bunk, Kenneth H., "The Crane Paradox," *Professional Safety,* July 1978, pp. 18, 28.

Grimaldi, John V., and Rollin H. Simonds, *Safety Management,* 4th edition, Homewood, IL, Irwin, 1984.

Handley, William, *Industrial Safety Handbook,* 2nd edition, London, McGraw-Hill, 1977.

Heinrich, H. W., Dan Petersen, and Nestor Roos, *Industrial Accident Prevention,* 5th edition, New York, McGraw-Hill, 1980.

Kroemer, K., *Material Handling, Loss Control Through Ergonomics,* Chicago, Alliance of American Insurers, 1983.

Odom, S.S.G., and C. Eddy, "Safety Profile—Forklifts," *Countermeasures,* 1987, pp. 1-2.

OSHA Reference Manual, Santa Monica, CA, The Merritt Company, quarterly updates, 1990.

"Powered Industrial Trucks," *Accident Prevention Manual for Industrial Operations: Engineering and Technology,* 9th edition, Chicago, National Safety Council, 1988, pp. 155-176.

Supervisors' Safety Manual, 5th edition, Chicago, National Safety Council, 1978.

Tuinstra, Bernard J., "Effective Conveyor Safety Programs Must Be Tailored to Specific Design," *Occupational Health and Safety,* October 1989, pp. 104-106.

"What's the Big Deal on Forklifts?" *Safety Management Newsletter,* Santa Monica, CA, The Merritt Company, April 1987, pp. 4-5.

Chapter 14

The Record-Keeping Function

Somewhat vital, often necessary, usually costly, and frequently overwhelming, the record-keeping function cannot be avoided. In this chapter, we will not always be able to draw the line between what is required by regulation and what is simply required for efficiency. Literally hundreds of records are required for a variety of reasons and purposes—far too many to be discussed in detail. This chapter will thus be confined primarily to the record-keeping requirements for safety and health. Further, we will deal with those records most likely to be involved in the operation of small- to medium-sized businesses. But bear in mind that the same principles and practices apply equally to all organizations, whatever their size.

The chapter has two main sections. The first deals with record keeping under the OSHA Act. The second covers other workplace reporting concerned with safety, health, and the workplace environment. The requirements, particularly those related to the environment, are not always precise and are subject to rapidly changing court decisions. Some are dealt with in other chapters.

WHY RECORD KEEPING?

Safety- and health-related records are vital for several reasons. First, they are kept because they are required. Second, and too often overlooked, is the fact that the survival of the business depends on records. Without records, management simply doesn't know how the business is going. Without records on accidents and injuries, they may be unable to remedy hazards and build a sound prevention program. Moreover, without the legally required records on safety and health, they may be subject to civil penalties. For example, if management fails to keep proper records in connection with products, a court may consider this as evidence of laxness and make a product liability award far beyond the company's ability to handle. Failure to maintain certain employee health records, likewise, may result in the company's being held liable for a former employee's health problems, even decades after the alleged exposure took place.

Unfortunately, wisdom in safety and health record keeping is too often a matter of hindsight. Anticipating what records will be needed, as well as knowing what record keeping is required, must be decided by the company's management. The excuse of not recognizing a needed record-keeping function is equally unacceptable in a court of law or a stockholders' meeting. It is sometimes impossible to draw a line between what is good to have versus what is absolutely necessary in record keeping.

RECORD KEEPING UNDER OSHA

In general, OSHA requires simply that companies record information on illnesses and injuries that pertain to the workplace. Basically, it is necessary to record all information that involves a fatality, lost workdays, certain other injuries that do not involve lost workdays, and injuries that require medical or first aid attention. This is not to say that information on non-work-related injuries and illnesses need not be reported. In the past, injuries sustained on the plant parking lot or while the worker was out to lunch or on a company errand have also been considered work related by the courts. If there is any question about whether an accident is or is not work related, a report should be made and covering documents should be retained until there is no further possibility of action.

Handling the OSHA record-keeping function is hardly trivial, since the act states that making a false statement can be punished by a fine of as much as $10,000 and/or 6 months in prison.

Types and Forms of Records

Two main forms are used for OSHA record keeping. One form, OSHA No. 200, serves two purposes:

1. The Log of Occupational Injuries and Illnesses records the occurrence, extent, and outcome of cases during the year.
2. The Summary of Occupational Injuries and Illnesses summarizes the log at the end of the calendar year to meet the employer's posting needs. This "Annual Summary" is the section to the right of the vertical dotted line on OSHA No. 200 (Figure 14-1).

The other form, the Supplemental Record of Occupational Injuries and Illnesses, OSHA No. 101 (Figure 14-2), provides more data on each case in the log.

Forms are available from the nearest OSHA or Bureau of Labor Statistics office or from state OSHA agencies. Where state law does not require any specific format, the forms can be of the company's own design if they include the required data. Figures 14-1 and 14-2 show OSHA No. 200 and No. 101 and how to complete them.

The OSHA standards also require accurate and current records on:

Occupational injuries and illnesses
Fire protection
Materials handling and storage
Machinery and machine guarding
Welding, cutting, and brazing equipment
Training

Each of these has its own type of backup records, all discussed in more detail in the OSHA standards themselves. State plans must meet or exceed the federal requirements.

Some employers are exempt from OSHA record keeping, except for reporting fatalities and multiple injuries. If you have no more than ten employees at any time during the calendar year, the federal record-keeping requirements do not apply. However, some states require that records be kept on a smaller number. Employers in low-hazard industries in retail trade, finance, insurance, real estate, services, and certain Standard Industrial Codes (SICS) are also exempt. Don't forget that being exempt from federal regulations may not exempt the company from state requirements, which may be stricter.

The Log and Summary of Occupational Injuries and Illnesses (No. 200) (Figure 14-3) is a combination form, a cumulative calendar year record, that is completed and posted by February of the next year. It must stay posted until March 1.

The Supplementary Record (No. 101) provides more data. This completed form must be in the workplace within 6 workdays after the injury or illness. Entries are later transferred to the Log and Annual Summary from this form.

Accident investigation records are also required, consisting of an oral or written report to the nearest or regional OSHA office whenever there has been a fatality or if five or more employees have had to be hospitalized.

Record keeping is an important part of any radiation protection program; ionizing and nonionizing radiation are covered by separate OSHA standards. A summary of the radiation-related records includes the following:

1. A daily record of the contamination levels at the entrance of each radiation-restricted area.
2. A record of daily accumulated exposures on pocket devices.
3. A record of weekly or semimonthly accumulated exposures on film badges.
4. A record of nonroutine exposures monitored by special devices.

Bureau of Labor Statistics
Log and Summary of Occupational
Injuries and Illnesses

INJURY / ILLNESS LOG

Page 1 (a) of 1

NOTE: This form is required by Public Law 91-596 and must be kept in the establishment for 5 years. Failure to maintain and post can result in the issuance of citations and assessment of penalties. (See posting requirements)

RECORDABLE CASES: You are required to record information about every occupational **death**; every nonfatal occupational **illness**; and those nonfatal occupational **injuries** which involve one or more of the following: loss of consciousness, restriction of work or motion, transfer to another job, or medical treatment (other than first aid). (See definitions.)

A CASE OR FILE NUMBER	B DATE OF INJURY OR ONSET OF ILLNESS	C EMPLOYEE'S NAME	D OCCUPATION	E DEPARTMENT	F DESCRIPTION OF INJURY OR ILLNESS
Enter a nonduplicating number which will facilitate comparisons with supplementary records.	Enter Mo./day.	Enter first name or initial, middle initial, last name.	Enter regular job title, not activity employee was performing when injured or at onset of illness. In the absence of a formal title, enter a brief description of the employee's duties.	Enter department in which the employee is regularly employed or a description of normal workplace to which employee is assigned, even though temporarily working in another department at the time of injury or illness.	Enter a brief description of the injury or illness and indicate the part or parts of body affected. Typical entries for this column might be: Amputation of 1st joint right forefinger; Strain of lower back; Contact dermatitis on both hands; Electrocution — body.
					PREVIOUS PAGE TOTALS ⟶
1 01	1/3	MACON A. FOLGER	LABORER	So. Pump Sta.	LACERATION OF HANDS
2 02	1/14	HAROLD T. PLUMBER	WAREHOUSEMAN	Gen. Warehouse	CONTUSION, TOP L. HAND
3 03	4/19	NANCY A. WYATT	ACCOUNTANT	GEN. OFFICE	SPRAINED RT. ANKLE
4 04	5/13	DOLORIS GARCIA	DISHWASHER	CAFETERIA	CONTIN. DERMATITIS
5 05	7/19	JOHN J. STONE, JR.	ELECTRICIAN	MAINTENANCE	ELECTROCUTION - BODY
6 06	8/22	JAMES C. MAYOR	RECEIVING CLERK	SHIPPING	BURSITIS - LT. SHOULDER
7 07	10/10	ANNE N. LORETO	SUPERVISOR	DISPATCHING	FOOD POISONING
8 08	12/1	JOSE DEL CASTILO	MECHANIC	TRANSPORTATION	SUNSTROKE
9 09	12/20	EDDY C. JOHNSON	INSPECTOR	QUAL. CONTROL	FRACTURE, LFT. COLLAR
10					
11					
12					
13					
14					
15					
16					
17					
18					
19			.		
20					
					TOTALS (See instructions) ⟶

OSHA No. 200

Figure 14-1. OSHA No. 200 (Log and Summary of Occupational Injuries and Illnesses).

U.S. Department of Labor

INJURY / ILLNESS SUMMARY

For Calendar Year 19 _88_   Page _1_ (b) of _1_

Form Approved O.M.B. No. 44R 1453

Company Name: O.K. JONES OIL CO. — STARS & BARS MFG. CO.
Establishment Name: SAME
Establishment Address: 142 SMITH ST., SOMEWHERE, TEXAS 77773

	EXTENT OF AND OUTCOME OF INJURY						TYPE, EXTENT OF, AND OUTCOME OF ILLNESS												
	FATALITIES	NONFATAL INJURIES					TYPE OF ILLNESS							FATALITIES	NONFATAL ILLNESSES				
	INJURY RELATED	INJURIES WITH LOST WORKDAYS				INJURIES WITHOUT LOST WORKDAYS	7 CHECK Only one column for each illness							ILLNESS RELATED	ILLNESSES WITH LOST WORKDAYS				ILLNESSES WITHOUT LOST WORKDAYS
	1	2	3	4	5	6	a	b	c	d	e	f		8	9	10	11	12	13
	Enter DATE of death Mo./day/yr.	CHECK away from work	CHECK restricted	Enter number of DAYS away from work	Enter number of DAYS of restricted work activity	CHECK if no entry in cols 2 or 5	Occupational skin diseases or disorders	Dust diseases of the lungs	Respiratory conditions due to toxic agents	Poisoning (systemic effects of toxic materials)	Disorders due to physical agents	Disorders associated with repeat trauma	All other occupational illnesses	Enter DATE of death Mo./day/yr.	CHECK away from work	CHECK restricted	Enter number of DAYS away from work	Enter number of DAYS of restricted work activity	CHECK if no entry in columns 8 or 9
1		✓	✓	1	2														
2		✓			3														
3						✓													
4							✓								✓	✓	1	8	
5	7/19/89																		
6											✓				✓			4	
7												✓							✓
8									✓						✓			1	
9		✓	✓	½															
10–20																			
Totals	1	3	2	1½	5	1	1	–	–	–	1	1	1	—	3	1	1	13	

Certification of Annual Summary Totals By _FRED BAKER_   Title _SAFETY DIRECTOR_   Date _1/12/89_

OSHA No. 200   POST ONLY THIS PORTION OF THE LAST PAGE NO LATER THAN FEBRUARY 1.

Figure 14-1. (continued)

Bureau of Labor Statistics
Supplementary Record of
Occupational Injuries and Illnesses

U.S. Department of Labor

This form is required by Public Law 91-598 and must be kept in the establishment for 5 years. Failure to maintain can result in the issuance of citations and assessment of penalties.	Case or File No. 02	Form Approved O.M.B. No. 1220-0029

Employer

1. Name   O.K. JONES OIL CO. — STARS & BARS MFG. Co.
2. Mail address (No. and street, city or town, State, and zip code)   142 SMITH STREET, SOMEWHERE, TEXAS 77773
3. Location, if different from mail address

Injured or Ill Employee

4. Name (First, middle, and last)   HAROLD TOM PLUMBER   Social Security No. 1 2 3 4 5 6 7 8 9
5. Home address (No. and street, city or town, State, and zip code)   789 SPRUCE STREET, ELSEWHERE, TEXAS 77773
6. Age   34   7. Sex: (Check one)   Male ☒   Female ☐
8. Occupation (Enter regular job title, not the specific activity he was performing at time of injury.)   WAREHOUSEMAN — PIPE FITTER
9. Department (Enter name of department or division in which the injured person is regularly employed, even though he may have been temporarily working in another department at the time of injury.)   GENERAL WAREHOUSE

The Accident or Exposure to Occupational Illness

If accident or exposure occurred on employer's premises, give address of plant or establishment in which it occurred. Do not indicate department or division within the plant or establishment. If accident occurred outside employer's premises at an identifiable address, give that address. If it occurred on a public highway or at any other place which cannot be identified by number and street, please provide place references locating the place of injury as accurately as possible.

10. Place of accident or exposure (No. and street, city or town, State, and zip code)   142 SMITH STREET, SOMEWHERE, TEXAS 77773
11. Was place of accident or exposure on employer's premises?   Yes ☒   No ☐
12. What was the employee doing when injured? (Be specific. If he was using tools or equipment or handling material, name them and tell what he was doing with them.)   LIFTING PIPE

13. How did the accident occur? (Describe fully the events which resulted in the injury or occupational illness. Tell what happened and how it happened. Name any objects or substances involved and tell how they were involved. Give full details on all factors which led or contributed to the accident. Use separate sheet for additional space.)   EMPLOYEE LOST GRIP ON PIPE AND DROPPED 2' LENGTH ON PIPE ON HAND

Occupational Injury or Occupational Illness

14. Describe the injury or illness in detail and indicate the part of body affected. (E.g., amputation of right index finger at second joint; fracture of ribs; lead poisoning; dermatitis of left hand, etc.)   CONTUSION OF TOP OF LEFT HAND
15. Name the object or substance which directly injured the employee. (For example, the machine or thing against or which struck him; the vapor or poison he inhaled or swallowed; the chemical or radiation which irritated his skin; or in cases of strains, hernias, etc., the thing he was lifting, pulling, etc.)   LENGTH OF PIPE

16. Date of injury or initial diagnosis of occupational illness   1—14—88   17. Did employee die? (Check one)   Yes ☐   No ☒

Other

18. Name and address of physician   DR. A.L. BETTER — 128 GRANDVIEW, SOMEWHERE, TEXAS 77773
19. If hospitalized, name and address of hospital

Date of report 1—14—88	Prepared by FRED BAXTER	Official position SAFETY DIRECTOR

OSHA No. 101 (Feb. 1981)

Figure 14-2. OSHA No. 101 (Supplemental Record of Occupational Injuries and Illnesses).

Figure 14-3. Annual Summary of Occupational Injuries and Illnesses.

358   SAFETY AND HEALTH MANAGEMENT

5. A record of exposures for employees working under special work permits.
6. A record of visitors' exposure records.

Discretion is allowed in the design of training records, the formats of accident reports, and the records of plant safety committees. In the last case, for example, the various safety committees may include in their minutes such matters as safety projects, safety promotional activities, and any items that require management action. While specific details may be required for the records, they also allow unusual freedom for the company's safety consultant or manager to provide direction.

Definitions

Certain terms recur frequently in OSHA standards and directions, particularly those that pertain to record keeping. For clarity, here are some brief definitions of these terms:

Occupational injury is any injury, such as a cut, fracture, sprain, or amputation, which results from a work accident or from an exposure involving a single incident in the work environment. (Note: Conditions resulting from animal bites, including insect or snake bites, or from a one-time exposure to chemicals, are considered injuries.)

Occupational illness of an employee is any abnormal condition or disorder (other than one resulting from an occupational injury) caused by exposure to environmental factors associated with employment. It includes acute and chronic illnesses or diseases which may be caused by inhalation, ingestion, or direct contact.

Medical treatment includes treatment of injuries by a physician, registered professional personnel, or laypersons (i.e., nonmedical personnel). It does not include first aid treatment (one-time treatment and later observation of minor scratches, cuts, burns, splinters, and so forth, which do not ordinarily require medical care), even though provided by a physician or registered professional personnel.

Establishment is a single physical location where business is conducted or where service or industrial operations are performed (e.g., a factory, mill, store, hotel, movie theater, restaurant, farm, ranch, bank, warehouse, or office). Where distinctly separate activities are performed at a single physical location (such as construction activities taking place at the same physical location as a lumber yard), each activity shall be treated as a separate establishment.

Work environment consists of the employer's premises and other places where employees are present as a condition of their employment. It includes not only physical locations but also the equipment or materials used by the employee during the course of work.

Injury and Accident Statistics

Although OSHA-associated statistics on illnesses and injuries have been kept since the inception of OSHA, the available information greatly understates the magnitude of safety and health problems. Prior to OSHA, no systematic attempt had been made to collect data on the incidence of occupational illness and disease; even the data on accidents was grossly inadequate. Although the OSHA Act brought new record-keeping requirements, there is every reason to believe that we still have not collected sufficient information to identify problems properly or take effective corrective actions.

One of the main reasons for OSHA record keeping is to determine the magnitude and nature of safety and health problems. However, official statistics are just the tip of the iceberg. While this lack of information is a national problem, its greatest impact is on smaller businesses, which lack guidance for safety and health efforts.

Businesses are understandably reluctant to accept more reporting requirements, since they question the cost effectiveness of such measures. Cost/benefit analysis on new reporting requirements has already resulted in a lowering of those requirements deemed to be a burden to the smaller business. A ready, practical solution is not likely for the near future. Some organizations have found that proper record keeping enables them to pinpoint safety and health hazards or problems, thus allowing them to improve future performance or reduce costs. For these businesses, the cost/benefit question has already been answered. But for many organizations, it has yet to be seen or demonstrated.

Quick Reference

The variety of records, forms, logs, and reports may seem almost too complex and extensive to grasp. How is the manager to know if all the proper forms and records are being kept? As a convenient way of answering this question, Figure 14-4 presents a quick reference checklist developed by the Merritt Company for its *OSHA Reference Manual.* The chart lists the types of records, logs, and reports that should be kept, as well as the necessary training and instruction required. The form is self-explanatory; however, the reader should note that the second column, "subpart," refers to the section of the *Reference Manual* in which the subject in question is found.

For example, the checklist shows what is required in the case of respiratory protection/respirators. The subject is covered in subpart I of the *Reference Manual,* so more detailed information should be sought there. Further

QUICK REFERENCE – RECORDS – TRAINING CHECKLIST

Description of Activity, Material or Operation	Subpart	OSHA Standards Number	Records · Logs · Reports on Activities & Material											Training & Instr		
			Drawings/Specifications at Job Site	Periodic Maintenance, Inspections, Tests	Inspection Tag, Dating	Inspector's Report	Current Material Inventory	Environmental/Personal Monitoring	Establishment of Regulated Area	Exposure(s) Records and/or Reports	Operations/Incidence Reports to OSHA Area Director	Pre-Assignment Med. Exam	Period Med. Exams	Warning Signs/Labels & Signals	Training Materials & Instructions Job Hazards & Safe Work Practices	Operating Procedures/Emergency Plans/Maintenance Manuals
1,2-Dibromo-3-Chloropropane	Z	1910.1044					X	X	X	X	X	X	X	X	X	X
2-Acetylaminofluorene	Z	1910.1014					X		X	X	X	X	X	X	X	X
3,3'Dichlorobenzidine and its salts	Z	1910.1007					X		X	X	X	X	X	X	X	X
4–Aminodiphenyl	Z	1910.1011					X		X	X	X	X	X	X	X	X
4-Dimethylaminoazobenzene	Z	1910.1015					X		X	X	X	X	X	X	X	X
4,4' Methylene bis (2-chloroaniline)	Z	1910.1005 (deleted)	–	–	–	–	–	–	–	–	–	–	–	–	–	–
4–Nitrobiphenyl	Z	1910.1003					X		X	X	X	X	X	X	X	X
Accident Prevention Signs and Tags	J	1910.145													X	
Acrylonitrile	Z	1910.1045					X	X	X	X	X	X	X	X	X	X
alpha-Naphthylamine	Z	1910.1004					X		X	X	X	X	X	X	X	X
Anhydrous Ammonia Handling	H	1910.111												X	X	
Asbestos	Z	1910.1001						X		X			X		X	
Benzidine	Z	1910.1010					X		X	X	X	X	X	X	X	X
beta-Naphthylamine	Z	1910.1009					X		X	X	X	X	X	X	X	X
beta-Propiolactone	Z	1910.1013					X		X	X	X	X	X	X	X	X
bis-Chloromethyl Ether	Z	1910.1008					X		X	X	X	X	X	X	X	X
Carbon Dioxide Fire Extinguishing Systems	L	1910.161		X		X					,				X	
Chain Saw Operations	R	1910.266		X											X	
Coke Oven Emissions	Z	1910.1029		X		X		X	X	X	X	X	X	X	X	X
Cotton Dust	Z	1910.1043						X		X		X	X	X	X	X
Cranes, Overhead and Gantry	N	1910.179		X											X	
Crawler Locomotives and Truck Cranes	N	1910.180		X											X	
Derricks	N	1910.181		X												
Diving, Commercial	T	1910.401	X	X	X	X				X		X	X	X	X	X
Employee Emergency Plans	E	1910.38														X
Ethyleneimine	Z	1910.1012					X		X	X	X	X	X	X	X	X
Explosives and Blasting Agents/ Handling/Facilities, etc.	H	1910.109		X			X							X	X	
Fire Brigades	L	1910.156											X		X	X
Fire Extinguishers, Portable	L	1910.157		X	X										X	
Fire Extinguishers, Fixed Dry Chemical System	L	1910.157		X		X									X	

(continued on next page)

Figure 14-4. Quick reference checklist (reprinted from the *OSHA Reference Manual*).

QUICK REFERENCE – RECORDS – TRAINING CHECKLIST

Description of Activity, Material or Operation	Subpart	OSHA Standards Number	Records - Logs - Reports on Activities & Material										Training & Instr		
			Drawings/Specifications at Job Site	Periodic Maintenance, Inspections, Tests	Inspection Tag, Dating	Inspector's Report	Current Material Inventory	Environmental/Personal Monitoring	Establishment of Regulated Area	Exposure(s) Records and/or Reports	Operations/Incidence Reports to OSHA Area Director	Pre-Assignment Med. Exam	Period Med. Exams	Warning Signs/Labels & Signals	Training Materials & Instructions Job Hazards & Safe Work Practices	Operating Procedures/Emergency Plans/Maintenance Manuals
Flammable and Combustible Liquids, Storage	H	1910.106		X			X								X	
Forging Machines	O	1910.218		X											X	
Helicopters	N	1910.183												X	X	X
Inorganic Arsenic	Z	1910.1018					X	X	X	X	X	X	X	X	X	X
Ionizing Radiation	G	1910.96						X						X	X	X
Laundry Machine Operations	R	1910.264		X										X	X	X
Lead	Z	1910.1025					X		X		X	X	X	X	X
Manlifts	F	1910.68		X	X										X	X
Mechanical Power Presses	O	1910.217		X							X				X	X
Medical Services and First Aid	K	1910.151													X	
Methyl Chloromethyl Ether	Z	1910.1006					X		X	X	X	X	X	X	X	X
N-Nitrosodimethylamine	Z	1910.1016					X		X	X	X	X	X	X	X	X
Non-Ionization Radiation	G	1910.97												X	X	
Open Surface Tanks	G	1910.94													X	
Powered Industrial Trucks	N	1910.178		X											X	
Powered Platforms	F	1910.66		X											X	
Pulpwood Logging, Operations, Facilities	R	1910.266		X											X	
Respiratory Protection/Respirators	I	1910.134		X	X									X	X	X
Scaffolding, Tube and Coupler	D	1910.28(c)(4)	X													
Slings	N	1910.184		X		X									X	X
Tanks, Flammable and Combustible Liquid Storage	H	1910.106		X			X								X	
Telecommunication	R	1910.268		X				X						X	X	X
Vinyl Chloride	Z	1910.1017					X		X	X	X	X	X	X	X	X
Welding, Cutting and Brazing	Q	1910.252		X											X	

Figure 14-4. (continued)

reading and the actual language of the regulations can be found in Part 1910.134 of the OSHA Act. The chart further shows that records are required on periodic maintenance, inspection, tests, inspection tags, and dating. It also notes that training is required in at least three areas.

Location, Retention, and Access to Records

All employers must maintain records at each establishment or workplace. (Where different activities are performed at the same physical location, each activity must be treated as a separate establishment, as noted previously in the example of construction activities carried out in a lumber yard.)

Transportation or construction firms may operate at widely dispersed locations. The place to which workers report each day is considered a separate establishment. Separate records of illnesses or injuries must be maintained at each establishment. If the employees do not regularly report to the same place, a central location from which they are supervised will suffice as the establishment.

Forms required to be posted must be posted even if there have been no injuries or illnesses. In that case, the forms will simply show zeroes under each heading. In the case of Annual Summary totals, the person responsible for the summary will verify it by signing it at the bottom, thus indicating that the totals are true and complete.

All records must remain in the establishment for 5 years after the year to which they relate. If a business changes hands, the new owner must preserve the records for that required period. Exposure records and data analysis, however, must be kept for 30 years, and medical records must be maintained for the duration of employment plus 30 years. Certain states or individual agencies (such as Workers' Compensation Boards) may have even more stringent requirements.

Authorized federal or state agencies have the right to inspect or copy OSHA records at any reasonable time. Exposure and medical records may be seen and copied by responsible representatives of OSHA or the National Institute for Occupational Safety and Health (NIOSH).

CONSUMER PRODUCT SAFETY

The Consumer Product Safety Act, which established the Consumer Product Safety Commission (CPSC), combined the functions of several other agencies and acts, namely:

>   The Hazardous Substances Act
>   The Toy and Children Protection Act
>   The Flammable Fabrics Act

The Refrigerator Safety Act
The Poison Prevention Packing Act

Many managers do not consider industrial products as consumer products. However, two out of three workplace injuries involve industrial products. Workers hurt by faulty industrial products are sometimes judged to be consumers and are able to sue the original manufacturer. In addition, many manufacturers produce consumer products. Thus, the company has a stake in many consumer products. Our main concern here is with record keeping under the Act. Note that CPSC representatives may:

1. Enter
    a. Any factory, warehouse, or place where consumer products are manufactured or held, in connection with distribution in commerce, or
    b. Any conveyance being used to transport consumer products in connection with distribution in commerce.
2. Inspect such transports or those areas of any factory, warehouse, or workplace where such products are made, held, or transported and which may relate to the safety of such products.

The act goes on to state that anyone who is a manufacturer, private labeler, or distributor of a consumer product shall establish and maintain records, make reports, and provide data that the CPSC requires. The CPSC can inspect such records, reports, and data.

We still have not clearly defined what a "consumer product" is, but we have excluded some things. These include articles not produced for consumer use, tobacco products, motor vehicles, poisons, aircraft engines, boats, drugs, cosmetics, and food. Other laws cover these items.

When the company works with consumer products, management must find what records are required and made available. Unfortunately, it is not clear which records are required and which ones should be available. This is a good time to note that a main reason for consumer product record keeping is the possibility of liability.

Manufacturers may ask, "Why do we have so much record keeping connected with product liability?" The following list shows some sources of the records required by law or in the interests of liability, sound management, decision making, or production monitoring:

Materials control
Process engineering
Reliability

Package engineering
Repair
Advertising and sales
Quality assurance
Marketing
Customer feedback

Through record keeping, producers or sellers have a defense against some of the claims brought against them. Here are some reasons for successful suits:

Faulty design
Faulty manufacture
Faulty components or materials
Faulty inspections and specifications
Incomplete instructions
Lack of warning about misuse
Inadequate warnings about hazards from use
Unclear instructions due to translation or poor copy

Since the CPSC can send top management to jail or recall a company's products, and since a liability suit can put a company out of business, a positive attitude toward record keeping is important. See Chapters 20 and 21 for more information.

MEDICAL RECORDS

The question of medical records not only pertains to OSHA but has general application as well. Under OSHA, employers are required to keep accurate records and make reports of work-related deaths, injuries, and illnesses. The same records can also aid management by providing data for use in job placement, establishing health and physical standards for certain jobs, documenting workers' compensation cases, providing data for epidemiological study, and in general aiding management in the task of improving company safety and health practices.

Data collected from an employee from any medical interview, or on any first aid visit, helps to establish a medical profile of the person. Such record keeping often uncovers chronic or recurrent conditions and allows the company to take effective action before a disability or injury claim results. In addition, many occupations require physical examinations before placement, on an annual basis, and on termination. These apply to persons working with asbestos, lead, cotton dust, and other substances, particularly those that are hazardous, toxic, or radioactive. The medical records on these individuals must be maintained with precise accuracy to observe the legal requirements.

ENVIRONMENTAL PROTECTION AGENCY (EPA)

The rapidly changing requirements of the EPA and associated legislation make it necessary to say much more about this matter. Chapter 18 has three parts dealing with hazardous waste management:

1. Hazardous waste management
2. Responding to Chronic and Acute Episodes
3. SARA—Superfund Amendments and Reauthorization Act (see "Hazard Communication" immediately following)

Several federal agencies work in these areas, and there is heavy involvement at the state, county, and community levels. The several layers of government agencies makes this one of the heavier concentrations of record-keeping and reporting requirements. It is a burden especially for the small user, distributor, transporter, handler, storer, and seller of hazardous materials. The country's mood regarding environmental protection and health hazards makes this a sensitive target area for public scrutiny. Recent hazardous materials catastrophes such as those in Bhopal, India, and Mexico City, and the oil spills such as those in the English Channel, Alaska, and the Antarctic, have raised public awareness and often created ugly moods regarding all parties—government and private. When radioactive matter is present, near mass hysteria results.

Unfortunately, government agencies trying to cope with hundreds of federal laws are not perceived as successful in protecting the public interest. This invariably results in a still greater record-keeping burden for business and industry.

CHEMICAL HAZARD COMMUNICATION

The Hazard Communication Standard (Hazcom) has broad recording and reporting requirements. It applies to all employers except federal, state, and local governments. It gives employees working with hazardous chemicals the right to know about such chemicals that they might encounter in the workplace. To accomplish this, manufacturers must make available to all employees material safety data sheets, which identify hazardous chemicals and state their hazards. Employers must also educate workers about potential workplace hazards and ensure that hazardous substances contain warning labels. They must also develop warning labels and material safety data sheets that include data about the physical and chemical characteristics of the substance, known health effects, exposure limits, and more.

Some confusion exists on the differences between Hazcom and community right-to-know, which is an adjunct to Hazcom. Community right-to-know legislation ensures that residents of communities near a chemical plant know about the potential hazards in case of a leak and develop response plans to meet chemical emergencies. It involves a host of local reporting requirements and nearly all businesses. Although handled through state and local governments, the reporting demands are federally mandated.

DEPARTMENT OF TRANSPORTATION (DOT)

The Secretary of Transportation is responsible for regulating interstate commerce. As part of its responsibility, DOT has issued advisory standards which, while not mandatory, are useful to companies engaged in transport. One part in particular, "Cargo Security Advisory Standards," deals with seal accountability, high-value commodity storage, and internal accounting procedures. These standards suggest that certain records are directly connected with safety as well as security.

For the fleet safety director, there are three major types of records to maintain:

1. Driver records
2. Accident records
3. Maintenance and compliance records

In general, driver records should include the following:

1. Application for employment
2. Medical examiner's certificate
3. Responses from agencies and past employers on driver's record
4. Certificate of driver's road test and a copy of the classified license
5. Certificate of the written test
6. Certification of traffic violations, submitted annually
7. Annual record review copy
8. Safety and service awards
9. Accident record
10. Commendation letters
11. Violation of rules and regulations
12. Disciplinary actions
13. Training and retraining records
14. Road patrol reports

Accident information records are developed to help determine the causes of accidents, as well as their solutions. They frequently serve as evidence in defending suits or supporting damage claims. Accident reports must be filed with federal, state, and local agencies (depending on the situation) and with the workers' compensation agency when injuries are involved. In the last case, an accident report must also be filed, according to OSHA requirements.

Maintenance records are also a vital part of fleet safety records. The American Trucking Association has developed a system suitable for large or small fleets, for either manual or computerized application. The manager is urged to become familiar with these guidelines.

FIRE REPORTS

Certain fires are reportable under OSHA, though such reports are not sufficient for most safety program purposes. A good fire report can be the basis for fire prevention action, determining financial losses, meeting insurance qualifications, and rebuilding or restoring a facility after a fire. Unfortunately, there is no single standard for fire reporting, though Standard 901 of the National Fire Protection Association is a step in the right direction. This format will almost certainly be accepted throughout the country, which certainly argues in favor of making it the company's standard.

Also important are the written reports of fire inspections. These should always be discussed and reviewed with management to arrange corrective action. The caliber of these reports has a direct influence on fire prevention activities.

NUCLEAR REGULATORY COMMISSION (NRC)

The NRC is responsible for regulating the conduct of licensees who handle special nuclear material in the course of their activities. Most licensees are electrical companies operating nuclear generating plants, but the category also includes nuclear manufacturing, reprocessing, and storage facilities, as well as laboratories and other research facilities. The basic source of regulatory material is found in CFR Parts 50 and 73, copies of which can be secured from the NRC in Washington, D.C. The NRC has also issued regulatory guides which are advisory or nonbinding, though they are relied on extensively.

RELIEF COMING?

Even this very brief view of record-keeping requirements makes one thing obvious: record keeping is a major task involving considerable expertise, not to mention enormous amounts of time, paperwork, and money. Over the past few years, the burden has become increasingly heavy. Some relief from federal record-keeping regulations may be expected because of the Small Business Regulatory Flexibility Act of 1980. Among other things, this act requires each federal agency to analyze the impact of its rules and regulation on small, privately owned businesses. One portion of the act calls for careful consideration of the reporting burden imposed by any new regulation and a review of existing rules to see if the reporting load may be reduced. It is a long-term process, but it does offer some hope, at least for the small business. Yet even so, the medium-sized or large operation, or the public corporation, is guaranteed no such relief. For most managers, the record-keeping task remains one of the most time-consuming and fast-growing areas of administrative work.

The answer to the question "Is relief coming?" is "No." Relief is not in sight either for the public seeking ever tighter control over business and industry performance or for business and industry hoping for, and seeking, relief from reporting and record keeping.

CONCLUSION

This chapter has presented a very brief summary of the record-keeping requirements connected with safety and health, particularly as they affect the average manager. It is not as in-depth reference but a general guideline that defines the scope of the field. Hopefully, this brief review underlines the importance of a thorough documentation and record-keeping program, above and beyond the minimum compliance requirements. Considerations of product litigation, quality control, preventive action and decision making all require records and reports far beyond the scope of those called for by OSHA or even the CPSC. The conscientious manager will quickly understand that the cost of this record keeping may be more than justified by the money it could save the company in the long run, should an official citation or liability suit arise.

Fortunately, the bleak outlook conveyed by this chapter is balanced by a wave of computerized approaches to handle most record-keeping requirements efficiently. The programs to do this are mostly good, available, and reasonably priced. There is some discussion on this subject in Chapter 4, Part 2. While such programs are not the best solution (easing record keeping would be better), they appear to be the best, perhaps the only, approach to record keeping.

BIBLIOGRAPHY

Accident Prevention Manual for Industrial Operations, 9th edition, Chicago, National Safety Council, 1988.
Anton, Thomas J., *Occupational Safety and Health Management,* New York, McGraw-Hill, 1979.
Ashford, Nicholas A., *Crisis in the Workplace,* Cambridge, MA, MIT Press, 1977.
Gray, Irwin, *Product Liability,* New York, American Management Association, 1975.
Grimaldi, John V., and Rollin H. Simonds, *Safety Management,* 4th edition, Homewood, IL, Irwin, 1984.
Leary, Margaret M., "Systems Response to Hazcom Relief Pressures for Companies," *Occupational Health and Safety,* September 1989, pp. 15, 19.
Lenz, Matthew, Jr., *Risk Management Manual,* Santa Monica, CA, Merritt Company, current.
OSHA Reference Manual, Santa Monica, CA, Merritt Company, 1987-89.
Petersen, Dan., *Techniques of Safety Management,* 2nd edition, New York, McGraw-Hill, 1978.

Pope, William C., "Organizing Your Safety Department," Alexandria, VA, Safety Management Information Systems, 1979.

Recordkeeping Guidelines for Occupational Injuries and Illnesses, Washington, DC, U.S. Department of Labor, 1986.

Rosenfield, Harry N., "Wire from Washington," *National Safety News,* November 1980, p. 66.

"Training Requirements of the Occupational Safety and Health Standards," Washington, DC, Department of Labor, n.d.

Whitman, Lawrence E., *Fire Prevention,* Chicago, Nelson-Hall, 1979.

Wynholds, Hans, and Lewis Bass, *Product Liability,* Princeton, NJ, ECON, Inc., 1977.

Chapter 15 / Part 1

Medical Services

ELEMENTS OF AN INDUSTRIAL MEDICAL SERVICE

The American Medical Association defines an occupational health program as one usually provided by management to deal constructively with the health of employees in relation to their work. Since 1911, every state has enacted workers' compensation laws requiring employers to compensate employees for occupational disability and to provide medical care for work-induced injuries. Most of these laws also require employers to provide medical care for employees with occupational diseases.

These laws have given employers an obligation, as well as an incentive, to maintain safer and healthier working environments. Since World War II, the exploding and increasingly complex technology, with its new, potentially hazardous physical, chemical, and biological agents, has served as an additional stimulus to the development of occupational health and safety programs. The earlier concepts of curative medicine have been broadened to include prevention and maintenance in health as well.

MEDICAL STAFF AND FACILITIES

To fulfill its mission, the medical service requires full management support and access to considerable company resources. Ideally, the director of the service should be a member of the president's staff. In practice, in all but the largest companies, the medical service is located in the industrial relations branch; in smaller companies, in the personnel department. If warranted by the size and type of the operation, a full-time physician, trained in occupational medicine, should be employed and should be designated as the medical director. At the very least, companies engaged in factory-type operations employing 300 to 500 workers should utilize the services of a full-time, industrially trained registered nurse. A centrally located first aid station staffed by a nurse will generally service 4 to 7 percent of the work force during a single work shift, almost equally divided between work- and non-work-related complaints. In addition, the nurse will be the chief health counselor, management health advisor, and industrial hygienist. Further, in the absence of a safety engineer, the nurse will probably assist in the overall safety effort as well. Small companies that cannot justify or support in-plant medical facilities can contract with local medical clinics, physicians, or even hospitals for various medical services. Regardless of size, the arrangement should ensure that a doctor, or at least a nurse, visits the plant periodically to provide expert guidance to management.

Most manufacturing plants employ a full-time physician for every 3000 to 5000 employees. In some foreign countries, the law states that companies must employ a physician for every 1000 employees. In the United States, there are currently over 500 board-certified occupational physicians and a total of 10,000 full- or part-time physicians working in industry. Ideally, the medical director or one staff physician should be board certified; the others should be encouraged to obtain certification. Management should provide an incentive to obtain the necessary training and certification. Today the qualified practitioner of occupational medicine must have training and knowledge in epidemiology, toxicology, counseling, and health education, as well as rehabilitation, compensation, and preventive medicine.

Industrial Hygienists

Thousands of synthetic chemicals are currently being used in industry, with new ones being added almost daily. Many of these organic chemicals have unknown toxic hazards. Physical hazards, such as ionizing and nonionizing radiation, noise, vibration, temperature extremes, and biological agents, are being encountered in the workplace with increasing frequency. Federal and state laws requiring that exposure to these chemical, physical, and biological agents be controlled and monitored make it mandatory that companies either employ full-time industrial hygienists or have access to specialists in industrial hygiene and toxicology. These specialists are in very short supply, though a number of colleges and universities now offer short-term training or degrees in this area. If the company employs an industrial hygienist, he or she should be part of the medical service department and should work closely with the medical director. In some companies, the hygienist is located in the safety organization. If so, liaison and communication with the medical staff are essential.

Nurses

There are approximately 40,000 nurses employed in industry. Relatively few have had formal training in occupational medicine or nursing, however. More and more short courses in various aspects of occupational medicine and nursing are now being offered by universities, hospitals, and professional societies. Management should encourage company nurses to avail themselves of these opportunities, especially nurses who work in single-nurse plants or who lack the opportunity for the programmed, on-the-job training provided by larger companies.

The industrial nurse is usually the first and frequently the only medical person to interact with the sick or injured worker. All too often, the medical outcome of the case is determined by this initial contact. A competent, tactful, and pleasant nurse with a genuine concern for the employee will elicit a grateful response from the worker and minimize subsequent legal and union difficulties in many cases.

Administrative and Secretarial Help

The need for administrative and secretarial help will vary with the size of the medical program. Larger medical departments should have an administrator who handles personnel, budgets, purchasing, facilities planning, and staff duties. A nursing supervisor is indicated for departments with six to ten nurses. An insurance coordinator may be desirable if the department is involved in processing workers' compensation or insurance claims. A medical counselor or rehabilitation counselor may be assigned to the department if the company operates an alcohol and/or drug rehabilitation service or another medical counseling service. Finally, one or more nurses may be designated as a visiting nurse, with responsibility for notifying relatives when employees become seriously ill or injured during work, are sent to hospitals, or die during the work shift. These nurses may make occasional house or hospital calls on seriously ill or injured employees. Depending on the need and on state laws, some companies also employ medical, laboratory, or x-ray technicians or audiometrists. More recently, nurse practitioners, physician assistants, or other paramedical personnel have been utilized, especially for health examinations and initial treatment of injuries and illnesses.

Adequate clerical staffing is essential, particularly to meet governmental regulations in maintaining medical records and reports. These same records help to protect the company should litigation ensue. Computerization of records, wherever possible, will help reduce the clerical load, conserve space, and control costs. However, the medical organization should be provided with whatever clerical backup is still needed.

All companies should maintain a roster of local medical specialists to be used for referral. Few companies can justify the full-time employment of specialists, especially those whose services are in greatest demand—orthopedists, dermatologists, and ophthalmologists. Others, required less frequently, are neurosurgeons, general surgeons, otolaryngologists, chest physicians, psychiatrists, and cardiologists. When warranted by the volume, arrangements can be made for specialists to visit the plant at periodic intervals. Consultants should be invited to tour the plant and to inspect the factory operations before they undertake any work for the company.

BASIC FUNCTIONS
Medical Examinations

To ensure safety and health on the job, it is essential that applicants for employment be placed in positions compatible with their capacities and state of general health. This will naturally involve a medical assessment, which is usually accomplished by a preplacement physical examination. The scope of the examination will be determined by the medical director or advisor, and will vary according to the type of operation, plant environment, governmental requirements, cost restraints, available medical facilities, and personnel. The examination should assess total capabilities and/or limitations, as well as suitability for the specific job. In addition, data needed to establish baseline health parameters (breathing capacity, blood chemistries, and so on) should be obtained prior to possible exposure to hazardous agents.

If the examination cannot be performed in the plant, applicants may be referred to local physicians. Mobile examination services which travel to smaller plants and perform screening examinations on site might also be considered. In addition to the physical examination, it is essential that a good medical history and an occupational history be obtained. It is estimated that about 70 percent of the information derived from the examination is obtained from the history. This is particularly true of conditions which affect consciousness and cerebration, such as epilepsy, mental disorders, the use of drugs, and certain congenital disorders.

Medical standards for placement have traditionally been promulgated by the medical director, although they frequently reflect management's philosophy and the company's safety history. Unrealistic and overly stringent standards of fitness for employment defeat the purpose of health examinations and prevent maximum utilization of the available work force. Antidiscrimination laws barring the arbitrary rejection of employment applicants on the basis of sex, age, or physical handicap have made many standards obsolete and dictate a more individual health evaluation for each job. This does not remove the need for or desirability of selective placement based on health, nor does it imply mandatory employment for the handicapped and aged. Certain jobs require strength, stamina, manipulative skills, and normal sensory perception. Those lacking these particular attributes are not qualified to perform the job, regardless of their state of health, sex, or age.

Periodic hazard exposure examinations are mandated by both federal and state laws. These examinations, which vary from a simple procedure such as an audiometric hearing test to more complex testing for exposure to radiation and chemical carcinogens, are performed to determine the effects of such exposure on the worker's health and to detect early evidence of any disease. The testing is also used to learn whether workers are utilizing proper protective measures or to determine the effectiveness of the personal protective equipment in use in the company. In addition, serial health examinations provide continuous medical surveillance of the employee's health status, revealing the effects of the total environment on his or her health. Logic dictates that the employee be informed immediately of any abnormal findings, especially if they relate to workplace exposure. This should be done in writing, with a copy kept in the medical files. If the abnormality is related to work, it is vital that the company initiate treatment at once and take prompt steps to avoid further exposure. If it is not related to work, the significance of the findings must be emphasized and proper recommendations made for future care. At this time, confusion exists concerning the employee's right of access to medical records, as well as about how long the company should retain such records. It is hoped that these issues will be clarified in the near future.

Diagnosis and Treatment

In the past, safety and health emphasis in industry was on the avoidance and treatment of traumatically induced injuries. This emphasis is changing due to government pressures and to the states' increasing acceptance of work-related diseases under workers' compensation laws. Cardiovascular disorders, malignancies, lung diseases formerly associated with smoking, mental disorders, digestive system problems, arthritis, and even alcoholism are being related to work stress. Complicating the medical picture are such factors as aggravation of existing conditions or the effects of repeated trauma.

Since most diseases become clinically manifest only when the individual's immune system fails, even minimal exposure to a secondary stressor may result in the aggravation of a preexisting condition or in the appearance of a new disorder usually resistant to the secondary stressor phase. Periodic health examinations may detect underlying, potentially correctible diseases which may respond to early treatment. In addition, an examination that brings an employee in contact with a doctor and emphasizes health allows the doctor to educate and motivate the examinee to use good health and safety practices both on and off the job.

Diagnosis and treatment of work-related injuries and illnesses is an essential element of the medical program. In fact, it is the basis for the original establishment of most occupational health progams. The company's responsibility in providing this treatment is dictated by law. However, some state laws allow employees to seek medical care from a physician of their own choice, provided he or she meets certain conditions. Without relieving the company of its financial, legal, and moral responsibilities, an ever-increasing number of states are shifting this privilege to the employee. However, many private practice physicians are unfamiliar with the work environment, so treatment by company-employed or -designated specialists is preferable.

Whether employees accept the proffered care depends upon their confidence in the medical department and on management's attitude toward the injured worker.

The great majority of industrial injuries are minor and can be treated by a nurse. However, injuries that are potentially serious, such as back, knee, and eye injuries, should be examined by a doctor. Tendon or major nerve injuries, bone fractures, head injuries, or major lacerations requiring sutures, as well as similar injuries, should be referred to a doctor. Depending on the availability of diagnostic equipment and the skills of the staff doctors, treatment may be limited or extensive. It is generally wiser to refer severely injured workers or those requiring extensive diagnostic studies, hospitalization, and/or major surgery to specialists for the necessary care and treatment. Under no circumstances should management pressure doctors to perform more than they feel that they can do safely and competently.

Most employees claiming an occupational disease should be referred to a doctor for a definitive diagnosis and possible treatment. The majority of these cases will be skin disorders, although illnesses induced by physical factors such as noise, temperature extremes, and radiation are fairly common, as are skeletal and neuromuscular problems resulting from repeated trauma. Diseases of the lungs induced by toxic agents and whole-body poisoning are next in frequency. Claims for heart and blood vessel diseases, as well as emotional problems, are increasing. These are usually classified under the "continuing trauma" provision instituted in a number of states. As the statute of limitations does not apply until the person becomes aware of the causal relationship between the illness and the work environment, legal action may not be filed until long after the employee has been removed from contact with the offending agent. In some cases, the problem does not surface until years after the worker has left the company.

Monitoring

The industrial hygienist has the responsibility for work site inspection and monitoring to detect and appraise health hazards. Through periodic inspections of the work environment, knowledge of the processes used, and an understanding of current regulations, the hygienist is ideally qualified to make appropriate recommendations for corrective measures.

The medical department also has a key role in monitoring, particularly monitoring and controlling absenteeism due to illness. Employees who request medical leave after reporting to work may be referred to medical personnel for evaluation and possible treatment. Guidelines authorizing absences should be established and applied in a consistent manner. Employees returning from leave may also be required to get clearance from medical personnel before beginning to work again. In larger companies, it may not be feasible to require all returning employees to clear through the medical department. However, those who have been on extended leave; others who return with evidence of restricted activity (crutch, cast, or eye patch, for example); and those who have been hospitalized, have had surgery, or have been treated for infectious disorders should be reevaluated by the medical department before being cleared to resume work. Physicians' requests for employee leaves, if they include a diagnosis, should always be kept in the medical department to ensure confidentiality. Medical personnel should also evaluate all requests for leave extensions. Statistical data concerning disease categories should be maintained for trend evaluation, handicap data, and future health education and prevention activities.

The importance of maintaining complete, legible medical records cannot be overemphasized. Not only is it required by law, but it is also essential in defending claims. The records should be repositories of factual information gleaned from the employee's earliest examination or treatment to the time of termination of employment. Medical personnel should avoid entering frivolous and nonrelevant data and opinions which reflect the emotions of the examiner. Medical personnel should respect the confidentiality of the records and avoid releasing any information without the written consent of the employee, except as required by law.

ANCILLARY FUNCTIONS

The great majority of companies with in-plant medical facilities offer free treatment for non-work-related illnesses. Although not required by law and rarely incorporated in union agreements, it is the usual policy to offer limited services. The extent of such services will vary with management policy; available facilities and skilled medical help; cost; location of the plant; and accessibility of community medical resources. Procedures for handling major medical emergencies should always be established. Since the majority of these involve heart attacks or strokes, provisions for oxygen and rapid transport to a nearby hospital should be established. In-plant facilities may include an emergency cardiac care center containing sophisticated recording and resuscitation equipment. While such equipment is not required by law, it should be properly maintained and inspected and should be operated only by trained personnel. At the very least, the nurse or designated first aid person should be well trained and certified by the American Heart Association or the Red Cross in cardiopulmonary resuscitation techniques, as well as in other rudimentary first aid procedures. This latter requirement is usually met by training fire fighters in first aid. Ambulances should be available for rapid transport to nearby hospitals. With the growth of paramedical services, skilled help and transport are readily accessible in many areas. If not, arrangements should be made with local private ambulance companies to provide transport. Whenever possible, the employee's personal physician should be contacted and advised about the emergency.

Most companies provide treatment for whatever illnesses arise during the work shift. These are generally minor problems such as colds, headaches, gastrointestinal problems, muscle cramps, and tension. Treatment is simple, inexpensive, and effective. The employee is happier, is better able to concentrate on the work, is less apt to have an accident, and is likely to finish the work shift. Such treatment is generally provided on a one-time basis, and companies do not attempt to treat chronic diseases or disorders. The latter should be dealt with by the employee's own physician, and employees should be encouraged to seek professional help for the diagnosis and/or treatment of such persistent or chronic complaints. In any event, the legal department should be consulted concerning the company's potential liability in providing various treatments and services.

Occasionally, companies will offer more definitive treatment, particularly when plants are located in isolated areas or when local medical services are not easily attainable. These treatments might include administration of drugs, desensitization shots for allergies, physiotherapy, skin medication, and others. Management should be aware of the full scope of this practice and should thoroughly review the proposed program before granting approval.

Inoculations are frequently offered to employees without charge and are administered on company premises during working hours. People will accept inoculation most readily when it is made easily available. It is to the company's advantage to have as many employees as possible immunized against tetanus and to receive influenza vaccinations. Without such vaccinations, companies have experienced absenteeism rates of 15 to 30 percent during an epidemic.

Key Personnel Services

Periodic health examinations are also performed on select groups of employees in many companies. The so-called executive or management health examination is designed to assess the health status of key management personnel who must maintain reasonably good physical and mental health at all times and whose absence or replacement would be costly for the company. Eligibility for such an examination, where it will be performed, its scope and frequency, and the handling of the results vary with company philosophy and policy, as well as with its ability to sponsor the exam. The examination should be performed without charge, and the examinee must always be informed of the results and properly counseled. Examinations performed in-house have the advantages of cost containment, minimal time involvement, and (if performed by a regular staff physician) knowledge of the examinee's regular work environment and job demands. In addition, the physician is available to the examinee for consultation and follow-up where needed or desired.

Some companies prefer to have these examinations performed off site, especially when they involve top executives. These exams may be conducted at major medical clinics throughout the country and allow the executive to enjoy a few days away from work. They may take several days, involving extensive diagnostic procedures and many special tests. The cost of such examinations is naturally higher than that of in-house exams. In some cases, the executive is allowed to select a physician, including a personal physician.

There is no consensus on how to handle the results of such examinations. Since much depends on the history provided by the subject, a voluntary program with an assurance of absolute confidentiality will achieve the best results. A copy of the examination findings should be forwarded to the employee's personal physician. If approved by the employee, a copy should also be sent to the company physician. Management frequently argues that since the company sponsors the examination, it should be informed of the findings. Most industrial physicians, however, feel that the value of the examination will be lessened if the examinee knows in advance that the results will be released to nonmedical sources or that the information may influence his or her job assignmnent or promotion.

Other Examinations

Occasionally, a company will sponsor periodic examinations for all employees, hoping for early detection of diseases and/or greater health awareness among employees. Employees and unions sometimes question the motive for these exams, especially if they are performed in-house or if the results are released to the company. Unions have frequently negotiated increases in medical benefits in their contracts to compensate for such examinations. This gives the employee a choice of doctor, as well as control over the release of information from the examination.

Other examinations may also be performed, depending on the company's needs and philosophy. A multinational organization, for example, will generally require thorough physical examination before an employee begins an overseas assignment. This examination should consider the duration of the travel or assignment, the geographical area involved, its health and sanitary conditions, culture shock, and work demands, as well as specific required immunities or health measures. Reexamination may also be performed on return, depending on the duration of the stay, the health and sanitary conditions to which the employee has been exposed, and the employee's personal medical history.

Special examinations may also be needed if the company operates its own aircraft, as the Federal Aviation Administration (FAA) requires certifying examinations for flight crews. However, these are minimal examinations, omitting many essential diagnostic procedures. Pilots who fly nonpaying passengers need only an annual Class 2 commercial examination. Since neither an electrocardiogram, chest x-ray, or blood count is required, the company may wish to perform its own examination, including the above procedures. The fact that a pilot possesses an FAA medical certificate does not rule out disqualification by the company physician.

Termination examinations are routinely performed by some companies. These have the advantage of detecting and recording physical attributes of employees when they leave the company, which can be essential in defending legal suits filed after termination. Since the statute of limitations does not start running until the subject becomes aware of the work-related disorder, it is conceivable that years may pass before the former employee files a claim. It becomes the company's responsibility to disprove the claim—something that is very difficult to do in the absence of written evidence to the contrary. Large companies with heavy employee turnover may find the cost prohibitive, however. In addition, the vagaries of workers' compensation laws being what they are, termination exams may represent an exercise in futility.

Certain occupations may also require a certifying examination to ensure minimal proficiency levels of performance. Generally, these are occupations involving visual acuity and depth perception, coordination, and motor skills. The company should set realistic certifying standards, relevant to the task and defensible if challenged. These examinations need not be repeated annually, but they should be performed at reasonable intervals—for example, to measure changes in physical parameters that may result from aging or from illnesses or injuries that occurred since the last examination.

Supervisors should be encouraged to refer the employee for medical evaluation whenever they suspect the existence of a health problem which may make it unsafe for the worker to continue on the job or may represent a hazard to others. This should be done selectively, however. Medical referrals should not be used to solve the supervisor's administrative problems. Difficult employees are not necessarily sick, nor are elderly employees who simply cannot keep up the work pace. For this reason, employees should know why they are being referred to the medical department, and every attempt should be made to help the employee resolve the problem before referral. It may be a simple matter of the employee's cutting down on an overly ambitious schedule of after-work activities that produce chronic exhaustion; assigning the worker to a less stressful project; or merely encouraging the worker to change a negative attitude to a more positive one. In any event, information concerning the employee and the apparent work problem should be forwarded to the medical department before the referral. This will permit the medical department to research the employee's medical records. All contacts with the employee's personal physician should be made by the company's medical personnel.

SPECIAL PROGRAMS

Emotional Problems

Emotional problems are among the most prevalent disorders affecting employees. These may be manifested by a variety of symptoms and behavioral patterns that are readily recognizable to the trained eye. The problems may result from unresolved stress originating in the employee's domestic life, personal health problems or worries, or job-related pressures.

An employee, after all, spends a relatively small portion of his or her time on the job. The rest of the time, the employee is subject to all the other stresses of living. However, he or she brings to the workplace a mental attitude and outlook formed by the interaction of all these environmental stresses. An employee who is normally easygoing may blow up unexpectedly simply because he was up all night with a sick child or had an argument with his wife over breakfast. Most emotional problems are minor ones of this sort, with superficial, short-lived causes. They can usually be cleared up by giving the worker a chance to talk to a sympathetic listener. However, many supervisors are under pressure themselves, and either cannot or will not take the time to counsel employees.

Acute or severe stress, maintained over a long period of time, can trigger physiological changes in the body. These changes themselves add to the employee's anxiety. Eventually, this cumulative stress takes its toll in sickness, accidents, absenteeism, termination, and/or death. The industrial nurse or doctor seldom sees employees until late in the evolution of this disease process. Frequently, they are too busy to take the necessary time to get to the root of the employee's problem. They treat the symptoms, not the underlying cause.

Larger companies can afford the services of a health counselor, preferably a psychologist or medical social worker trained in counseling. Some registered nurses are obtaining additional training in psychology and, in smaller companies, are doubling as the counselor and doing excellent work. Since the supervisor or manager is rarely trained in this field, the counselor can be a valuable resource, training and educating supervisors to recognize emotional problems or assist them individually to solve their problem cases. In some companies, the counselor also serves as the alcohol rehabilitation counselor, and may provide invaluable assistance to recovered alcoholics in resolving some of the emotional problems associated with drinking, which affect their work performance as well as their private lives.

Alcoholism

Alcoholism is one of the major health and safety problems in industry today. It is conservatively estimated that 6 to 9 percent of the average work force are alcoholics. "Alcoholics" are defined here as persons whose drinking interferes with major phases of their lives. Alcoholics have a higher incidence of illness, more on- and off-the-job accidents, make more mistakes in their work, and have a higher incidence of absenteeism. The net result interferes with plant operations, increases the accident rate, and drains health and accident insurance. Since alcoholism is a progressive disease, it may require 5 to 15 years of increasingly heavy drinking before it reaches the problem stage and comes to the company's attention. Thus, the typical alcoholic is a senior employee, usually with a good work record, a respected family member, and a long-time friend of many other workers. Because of this, the supervisor may hesitate to identify the alcoholic at an early stage, when he or she is most amenable to recovery. This, along with the average alcoholic's unwillingness to admit to the problem and the ability to manipulate friends, delays the identification and makes treatment more difficult.

In recent years, industry has begun to recognize more clearly the problems precipitated by the alcoholic employee, and has been increasingly pressured for action by both the government and the community. Rather than simply fire potentially valuable workers, companies have begun to take positive steps to help them. This must begin with management's recognition of alcoholism as a disease similar to any other. In many cases, granting full sick benefits to alcoholics is essential in encouraging them to seek help. Establishing a company rehabilitation program for alcoholic employees is the next logical step. This usually involves hiring an alcoholism or employee counselor (frequently a recovered alcoholic with experience in Alcoholics Anonymous) and assigning this person to the medical department. In some cases, the counselor may be assigned to the personnel department instead. Union cooperation is essential and is generally forthcoming. Organized labor, too, is beginning to recognize alcoholism as a major problem and is frequently helpful in getting the alcoholic into a rehabilitation program. Some unions even sponsor their own education, counseling, and rehabilitation programs. In any event, whatever facilities or support are available, the company must set definite guidelines for the education of plant personnel, referral and evaluation procedures, and final disposition of cases.

All plans have the following procedures in common:

1. Education of line supervisors to assist in early identification of possible alcoholics.
2. Referral to the counselor for evaluation and disposition.
3. Treatment, including utilization of community resources for definitive care.
4. Follow-up interview and evaluation after the employee returns to work.

Community resources are generally available and include physicians with special experience with alcoholics, special hospitals or special facilities within the community hospital, Alcoholics Anonymous, the Veterans Administration, and government health agencies, to name just a few. Utilizing these procedures and resources, company programs have achieved a success rate of 60 to 80 percent. This represents not just a tremendous benefit in human terms, but also savings of thousands of dollars in health insurance costs, on-the-job accident and error costs, lost work time, and reduced rework of jobs, plus the retention of skilled, potentially valuable men and women.

Since alcoholism affects employees at all levels, it does not touch the shop worker or manual laborer alone. It can also affect professional personnel, including members of top management. These men and women, in fact, may represent even more valuable resources to the company, and their inability to function at full capacity may be far more costly, particularly where they occupy decision-making positions.

It must be emphasized that *alcoholism is a disease that cannot be cured.* It can merely be arrested or controlled by the alcoholic's complete abstention from alcohol. Constant attention is needed to prevent relapses, and the alcoholic may need follow-up support and help in staying sober. Such support and help are generally readily available. Many recovered alcoholics volunteer their services in helping other alcoholics, either through Alcoholics Anonymous or in other programs. A number of companies go so far as to provide facilities for group meetings, as well as counseling for the families of recovering alcoholics. Companies too small to employ their own counselors may contract with private counselors or seek help from the National Council on Alcoholism or from similar community resources. Many companies rely on a combination of these—company counselors and programs, community facilities, local meetings of Alcoholics Anonymous, and whatever other resources are available in the area.

Drug Abuse

In recent years, drug abuse has become a source of increasing concern throughout society. Industry is clearly affected by the problem and its enormous costs. The drug problem in industry, however, is not as clearly defined as alcoholism, nor is management as willing to accept it in the same manner. To begin with, most drugs are illegal, and their use in the workplace violates both security regulations and civil laws.

Attempts to screen applicants in order to eliminate those with drug problems have raised a number of difficulties: charges of selective discrimination unless tests are performed on all applicants, handling of positive results, inability to test for certain drugs, the lack of a cheap, simple test for others, and the cost and time involved in processing applications for new employees. In addition, recent antidiscrimination laws concerning previous drug abusers make it necessary to evaluate every former drug abuser who seeks employment on an individual basis. Former drug abuse alone cannot be used as a blanket disqualification for employment under the law.

The real problem begins, however, when the company discovers that some employees do, in fact, have problems of drug abuse. These problems drain the company in exactly the same way that alcoholism does: in higher accident rates, increased illness and absenteeism, and costly workplace errors. In addition, it may sharply increase employee thefts, especially if the worker needs money to support an expensive addiction. Like the alcoholic, the employee with a drug problem is a potentially valuable asset to the company, and it is in the company's best interest to rehabilitate rather than terminate this person.

Some companies with alcoholic rehabilitation programs have expanded them to incorporate drug rehabilitation as well, and community resources have become more widely available in recent years. The following section on testing includes alcohol as a drug and applies to alcohol abuse.

Testing for Employee Drug Abuse. The annual cost to industry of drug abuse is estimated at $16.4 billion annually. Drug abuse is not a victimless crime. Victims include the employer, the drug-using employee, his or her family, and other employees. The drug abuser endangers his or her own health and safety, affects his or her family, and reduces the company's profits. Worse yet, a drug-using employee may turn to drug sales or workplace theft to support a drug habit. Clearly, the victims of drug abuse are many.

Symptoms of Drug Use. What are the symptoms to look for? On a companywide basis, there will be security problems (increased thefts), safety problems (more accidents), quality control and production problems, and medical problems (increased illnesses). Supervisors should expect higher absenteeism and accident rates.

Drug use in the workplace is often noted first by security personnel when new cases of employee theft are found. Next, the medical department may be alerted to the problem, followed by the personnel department and supervisors. Quite often, management is the last to discover a workplace drug problem. On an individual basis, detection is more difficult. There is no strict category that describes the average drug user. This person may be young or old, blue collar or white collar, male or female, and of any ethnic background. The symptoms of drug use vary according to the substance used.

The drug used most often at this time is marijuana. Marijuana can impair hand-eye coordination, reflex actions, sense of time, thinking, and memory. Because of its prevalence, marijuana is sometimes seen as a socially accepted drug. However, management must make it clear that the use of any drug is unacceptable and is grounds for disciplinary action or dismissal.

The use of barbiturates and hallucinogens is relatively rare in the workplace compared with marijuana and cocaine. Amphetamines are used mostly by truckers and students, who need to stay awake for long periods.

Employees using cocaine or amphetamines are usually anxious and excited. They experience decreased appetite, rapid breathing, dilated pupils, lack of fatigue, and increased alertness. These conditions, while perhaps briefly boosting productivity, may result in injury (the person thinks he can lift more than he really can) or illness (the person goes without food for too long). Also, when the drug wears off, the person feels extreme fatigue, production deteriorates, and there is a greater risk of accidents. Thus, supervisors should look for significant changes in productivity.

A cocaine "high" usually lasts for only 45 minutes to 1 hour, but repeated use can sustain it. Supervisors should be alert for long, frequent trips to the restroom or parking lot. The average cocaine user spends $200 per day on the drug. To support such a habit, the user must steal from the workplace or deal to fellow employees. Supervisors should notice an increased incidence of theft.

Employees using drugs may become psychotic, paranoid, and dangerous. They may react violently. The supervisor should look for bizarre, erratic behavior. Massive doses of drugs may cause convulsions, coma, and even death. Depression and mood swings are two of the symptoms of long-term drug use, as are weight loss and greater susceptibility to infection and illness.

Drug Testing and Accuracy

One possible answer to the growing drug problem in industry is the testing of employees suspected of being drug users. Drug testing is not considered the answer to the country's drug problem, but it may be used to minimize drug-related losses to industry.

Drug tests are highly accurate. Some tests can give immediate results with 95-99 percent accuracy. Immediate results are especially helpful to employers. Formerly, disciplinary action had to be considered while waiting for test results. After an initial positive result, if the employee still denies using drugs, the specimen may be sent to a laboratory for chromatography/spectrometry tests to confirm the positive results.

The blood alcohol concentration (BAC) scale has been developed to indicate alcohol intoxication. Although no urine test can prove that an individual is intoxicated with drugs at a given time, there are definite parameters. For example, if a test for barbiturates is positive and the subject admits to having taken a sleeping pill the night before, the lab tests can determine whether the level of drug found is consistent with therapeutic or abusive levels.

The tests for marijuana will be positive only if the person used the drug (with 95-99 percent accuracy). Therefore, passive inhalation or merely being in a room where others are smoking marijuana is no defense for a positive result.

Company Policy. Preemployment drug testing is legal as long as all employees are tested equally. The trend is toward increased testing at the preemployment level. Management should first have the prospective employee sign a statement denying drug use. Then, if the test is positive, the employer has both the test results and the falsified statement as evidence of the company's absolute right to refuse employment.

Although routine testing of job applicants is legal, probable cause is required to test current workers. The tests cannot be given on an individual or haphazard basis; otherwise, the employer risks charges of discrimination and unfair labor practices.

It is important that a company have a written policy regarding employee drug use and communicate that policy to employees. Not until then may drug testing be used legally in the workplace. The company policy must outline the circumstances under which employee drug testing may be performed and what the consequences of a positive result are (dismissal, suspension, mandatory rehabilitation, etc.). The policy must be uniform for the entire company, not just for certain departments or employment levels. A policy that provides for testing of production personnel but not office personnel invites trouble. Once a companywide policy has been effectively communicated to employees, it may be enforced.

The person responsible for enforcing the testing program must be well trained in handling the drug-using employee. This training must be provided for supervisors as well. No one should ever make accusations which may later prove false. For example, if a supervisor notices a line employee whose slurred speech, lack of coordination, and glassy, reddened eyes suggest substance abuse, the supervisor must not confront the employee with the statement "You are obviously drunk. I am taking you to be tested for blood alcohol content." First of all, the employee may be using barbituates, which cause the same symptoms as alcohol intoxication. In this example, since an accusation of drunkenness was made, only alcohol testing may be performed. Furthermore, if the employee is not drunk, the company may be sued for libel.

Instead, the supervisor should approach the employee and say, "I do not think you are performing your job correctly or safely. Is there any reason you are acting in this manner? If not, I am taking you to the company doctor in order to determine the cause." The supervisor should keep the employee calm and, with the possible exception of one witness, keep the interview confidential. The interview must be limited to advising the employee that the supervisor is adhering to company rules, and is proceeding in a uniform and correct manner.

The company policy may require that employees with a positive test result be immediately fired. However, the company may lose valuable employees in this manner. Therefore, a possible alternative is to have a one-chance policy. Such a policy mandates rehabilitation the first time an employee shows a positive test result. A second positive result, however, results in discharge.

Legal Considerations. The lower courts have consistently held that demanding a urine specimen from an employee is legal as long as the company policy is uniform and fair, and there was a reasonable, possible cause for testing that employee. It is important to document as much as possible in case of a lawsuit. Witnesses from both labor and management should be present when the employee is tested. The poor job performance should be documented as well as possible. A refusal to take the drug test should be witnessed and documented. Additionally, the procedure must be kept confidential.

The company risks libel charges if an employee is acused of substance abuse based on anything but job performance. If a suspected drug user is caught stealing, disciplinary action should be based on the theft. When a degeneration of job performance is noted, it should be documented, if possible, and then the company should act with regard to drug use.

Worse than not knowing about a drug problem is knowing about it but not acting on it. The company is responsible for its employees' actions. Under no circumstances should an employee suspected of being under the influence of drugs or alcohol be allowed to operate machinery, including an automobile. If an employee tests positive and company policy is to send a drunk or drug-using employee home, someone else should drive that employee home. Otherwise, the company's potential liability may be astronomical.

Conclusion. The benefits of a drug-testing program can be great. The U.S. Postal Service reported savings of over $135 million on the $15 million investment in employee drug testing. Firestone claims to have saved $1.7 million annually in reduced accidents and absenteeism alone. Conrail has estimated its savings at $947,000 with a cost of $247,000, a return of nearly $3.50 on each $1.00 invested.

It is important to consider whether your company has an employee drug problem. Examine the accident, productivity, and illness rates. Are some employees exhibiting the symptoms of short-term or long-term drug use?

The answer to this problem involves developing a company drug use policy and program. Some key points to remember are these:

> Seek advise from legal, medical, and security professionals in order to design the policy best suited to your workplace.
> Develop a uniform, fair company policy regarding employee drug use.
> Inform employees of your policy.

Train supervisors in the correct manner of handling employee drug use, including the issue of probable cause.
Arrange for medically approved testing.
Keep the drug screening procedure confidential.
Document as much as possible, including declining job performance, testing, refusals, and so on.
Arrange for professional rehabilitation for drug- or alcohol-using employees.

The manager who is interested in learning more about the scope of both alcohol and drug problems in industry, and effective ways in which the company can deal with them, will naturally want to seek more detailed information. The Merritt Company's Protection of Assets Manual devoted four full chapters to the problems and suggested program, all from a management and policy-making standpoint, while Merritt's Risk Management Manual presents both an overview of the problem and actual sample company policies and programs that have proven effective in use.

Prevention and Research Programs

The public in general is beginning to demand more emphasis on preventive medicine, and the medical profession is responding with efforts both to preventing disease and to keep people healthy. Industry has much to gain by encouraging these efforts. The company is in an ideal position to assist, since it has a "captive audience" of relatively healthy adults in an ideal age range. Preventive medicine starts with health education, promoted through company publications, company-sponsored displays, lectures and films, periodic examinations, and health counseling services. Liberalization of health insurance programs to pay reasonable costs for "well-person examinations" is also a great aid. Providing company facilities for employee physical conditioning programs (such as gyms, tracks, athletic fields, and swimming pools) and sponsoring company athletic teams can pay off handsomely. Many companies find that the costs of these facilities are compensated for by the fact that employees show improved health and morale, both of which mean less absenteeism and higher production.

Companies can also encourage and conduct medical research in association with a responsible university- or hospital-affiliated project or group. Indeed, much health research has been conducted with industry cooperation. The information gained about the physical cost of performing various job acitivities has been useful both to companies and to doctors—for example, in evaluating the ability of heart patients to return to their jobs.

Industry is also an ideal setting in which to perform practical health research. Employees are continuously being exposed to various chemical and physical agents, including noise, vibrations, accelerations, thermal variations, and ionizing and nonionizing radiation. Laws already require health surveillance of employees and their work sites. Research programs can readily be incorporated into the surveillance program, including both short-term and long-term studies on the effects of such exposure, and the relative effectiveness of protective devices and techniques. With many employees remaining at their jobs for 20 years or more, epidemiological studies can reveal the cumulative effects of environmental stressors. It behooves companies to keep accurate records on all employees who die or suffer serious illness during their employment in order to determine trends in illnesses and possible environmental interaction. Such records can be a rich source of research material.

Medical personnel may also have capabilities that will be of value to other employees. Engineering and design personnel may need human factors information when designing company products and equipment. Inputs from knowledgeable and interested medical personnel may be invaluable in improving new-product designs or in improving production efficiency. An understanding of ergonomics may reduce employees' physical complaints, improve morale, and prevent injuries and illnesses.

PHYSICAL FACILITIES

Medical facilities may consist of a single first aid station or of many well-equipped nursing stations and clinics staffed by doctors. Very small plants may have only first aid cabinets or designated stations with one or more first aid-trained employees in charge. Most companies have at least a centrally located nursing station equipped with a treatment counter, bed, oxygen, heat lamp, basic examining tools, and simple medications. Unless closely supervised by a licensed doctor, the nurse should not give injections, apply sutures, remove embedded foreign bodies from the cornea, or even administer prescription drugs. In all cases, the nurse should be given written operating procedures and should be supervised by a consulting physician. She should have ready access to the doctor to discuss questionable cases, and should clearly understand her limitations as defined by the state's nursing laws, as well as by company policy. She should also have access to a variety of local consultants to whom she may refer cases. She will probably need to work closely with the personnel or employee benefits department, which interface with the workers' compensation carrier. In many cases, she will fill out forms and reports required later by both the company and the insurance carrier.

Additional facilities may be needed if a part- or full-time physician is to be employed. The medical facility may be called a "clinic," "dispensary," or "hospital," though local or state regulations should be checked before so naming it. There may be legal requirements for a clinic that would not apply to a dispensary. In addition to an examining room and an office for the doctor, a small operating room and x-ray facilities may be incorporated, as well as audiometric testing equipment, vision testing equipment, a spirometer for measuring breathing capacity, and a small laboratory. More extensive physiotherapy equipment may be acquired if needed. Two or more nurses should be employed, as well as a medical secretary. The doctor will be responsible for procurement of prescription drugs and narcotics. Many states require certification of x-ray, laboratory, and audiometric technicians, and these personnel may be difficult to locate. Since these people may be needed only sporadically, it may be wise to employ a registered nurse who is also certified in one or more of these skills.

Many larger industrial medical facilities include emergency cardiac care rooms. These centers contain sophisticated heart monitoring and therapeutic equipment, including defibrillators, pacemakers, drugs, piped-in oxygen, and the capability to transmit electrocardiograms by telephone. Since the great majority of serious medical emergencies occurring during work are heart related, every effort should be made to prevent heart deaths. Unfortunately, relatively few deaths have been prevented by industrial medical teams. Except for immediate institution of manual cardiopulmonary resuscitation, treatment of the fibrillating heart requires a team effort by highly skilled professionals backed by sophisticated medical technology. This is best provided in a hospital setting by teams of people trained and experienced in resuscitation techniques. The necessary equipment is used infrequently in industry. Thus it is rarely maintained and inspected as required, and there is generally a shortage of experienced and knowledgeable personnel to operate it. All medical personnel, however, should be certified in manual resuscitation techniques. Promptly calling the fire department, local paramedics, or the emergency services of a nearby hospital will probably save more lives.

Large companies with scattered facilities usually require several nursing stations, which are located in the more heavily populated buildings or areas. These, in turn, may feed into a central clinic or into several smaller clinics. In addition, there may be a medical warehouse or supply facility supervised by the medical administrator or a pharmacist. The latter is responsible for purchasing drugs, pharmaceuticals, and surgical supplies. Capital equipment, however, may still go through the purchasing department's ordinary channels. For budgetary purposes, the cost of drugs and supplies may be anticipated by calculating the per capita cost for the previous year, adding an inflation factor, and multiplying by the anticipated employee count for the coming year.

COMPANY-COMMUNITY INTERACTION

Because it controls policy and funding, management determines both the type and the effectiveness of the medical services in the long run. Management should therefore recognize the potentials of its medical service and use this valuable resource to its fullest. At the same time, it should fully support its efforts. Medical professionals should have access to top management in order to facilitate decisions and offer the best possible care. It is essential that management recognize that the medical professional is somewhat different from other company employees. The medical professional's first obligation is not to the company but to the patient being treated. This person is more closely bound by the ethics of his or her profession than by company policy, and in conflicts between the two, the responsible medical professional must decide in favor of medical ethics. Realistically, management must recognize this fact in framing company policy and must fully support the professional integrity of medical personnel.

Generally, the line supervisor is the person most intimately involved with the ill or injured employee. He or she is usually the first one on the scene of an injury and the one who sends the worker to the medical facility. Cooperation between the supervisor and the medical department is absolutely necessary. Management should make it clear that the supervisor must carry out the recommendations of the medical department, and should not allow the supervisor to override medical decisions. During their training programs, supervisors should be acquainted with the medical personnel, and with the resources available to them and how best to use them.

It is essential that the medical and safety departments cooperate fully with each other. The safety engineer is generally concerned with the physical hazards of the job or process. He or she may also be responsible for safety training, accident investigation, and preventive activities. Frequently, the safety engineer serves as the company's representative to state and federal safety agencies, and is responsible for maintaining the required safety records. The chief safety engineer and the medical director should clearly understand the scope and extent of each other's responsibilities and should have a close working relationship. Ideally, if not located in the same department, they should at least report to the same line of authority.

The medical department's interaction with the community is also essential. The plant physician is the company's representative to the medical profession. He or she should be a member of the local, regional, and national medical associations. In addition, the physician should be on a local hospital staff (even if he or she does not admit patients), and should attend staff meetings to meet local practitioners and update personal knowledge. It is necessary to be on good terms with the local health department personnel, for their services may later be invaluable. Likewise, the physician should attempt to get an appointment at one or more local medical schools for the same reasons. In addition, it is important to be aware of the activities of the local community volunteer groups such as the Red Cross, Heart Association, National Counsel on Alcoholism, Mental Health Society, and others. Finally, the physician should establish a panel of consultants representing the various medical disciplines, drawing on the most competent and ethical doctors in the area and cultivating both their professional cooperation and their friendship. Occasional visits of these doctors to the plant or to the executive dining room, and presentation to top executives, are most helpful. In short, the company should choose as its medical director someone who has not only the proper medical qualification but also an ability to interface with the community's medical resources and establishment. Ideally, this person both represents the company and helps establish strong ties to the community's health and medical resources.

MEDICAL CASE MANAGEMENT

A health care costs review usually means reviewing total admissions and services costs to obtain the big picture. The case management approach focuses on the smaller groups of patients who account for an undue large share of health care costs, according to Dr. Phillip Polakoff (1989).

In one company, two researchers were tasked to find out how much the company was spending to take care of high-cost patients. For the base year 1986-1987, the results were revealing:

- Only 1 percent of the employees generated 34 percent of the claims costs. The average claim was $23,000, and none was less than $10,000.
- About one-half of the high-cost patients were hospitalized more than once during the year.
- Only about one-fifth of the patients received home health care, even though many could have been managed at home.
- The high-cost patients saw an average of six physicians, raising the question of who was coordinating their care.
- Many of the patients had multiple hospitalizations during the review period.

The company decided to expand its medical case management approach and developed a plan to do it. Now there are many questions to answer and management decisions to make.

- What is the purpose of the program? Is it to save money, enhance care, or something else?
- What types of patients will be targeted?
- What are the up-front costs, especially if outside vendors (costing from $50 up to several hundred dollars per hour) are involved?

Savings range from a proven $0.22 up to a claimed $5 for each $1 spent. There are also many questions about vendor quality control. In-house disadvantages include liability exposure in unfavorable outcomes and a perception that care is offered on behalf of the company, not the patient. This forces most companies to use an outside vendor for medical case management.

The case management approach has four phases:

1. Identification and referral of patients who can benefit from the service. These patients are selected as a result of their large annual claims, use of several physicians, multiple hospital admissions, and other high-cost factors.
2. Screening and assessment. The vendor gets enough information to decide if case management is needed. If so, a treatment plan is drawn up and sent to the company, sometimes with a cost savings estimate.
3. Active case management. The designated case manager and physician work closely to build the right plan and change it to reflects the patient's progress.
4. Closure. When the case manager and the company's health advisor decide that the patient no longer needs case management, all parties are notified of the forthcomig closure and the case is closed.

Medical case management seems a simple and logical process. However, the patient too often bounces around in the company system, getting services as needed or requested and sent to various physicians and hospitals as he or she seeks to get well or prolong the lack of attention. Cost containment and active management of the patient is more fortuitous than planned. While savings are not proven to be great, there are tangible savings and many intangible ones.

WELLNESS

Employee well programs (EWPs) have sprung up across the country. You've read about them in safety publications and heard about them on the evening news, but what exactly is an EWP? The concept of wellness varies from person to person. While we all think that we know what it is, interpretations vary. Wellness is not merely an absence of disease, but the active pursuit of good health through improvement in lifestyle. From a business standpoint, if people are well, they are on the job and productive.

Cynthia Davidson, former editor of the Merritt Company's *OSHA Reference Manual,* provides a quick test to see how much you really know of employee well programs.

1. EWPs are only for large companies with many employees.

False. Because large companies usually have the resources, they have pioneered the development of EWPs. However, many smaller companies have since followed the lead and developed successful EWPs suited to their smaller size, budget, and employee population. EWPs can range from a health literature stand in a central workplace location to comprehensive, multifaceted programs.

2. EWPs are expensive to develop and maintain.

False. While beginning an EWP does require an initial investment, EWPs can be scaled to meet any company's size and budget.

3. There are few indications that EWPs are cost effective.

False. Most EWPs result in clearly reduced health costs, and where they don't, they at least reduce the amount of yearly increase. Other substantial savings are realized through reduced absenteeism, reduced turnover, and increased productivity. Of course, EWPs are primarily designed to reduce health care costs over the long-term.

Consider, for a moment, worker safety. The emphasis is on accident prevention (before-the-fact emphasis), certainly not on paying for accidents as they occur! Accident prevention is accomplished through safety programs, worker awareness, education (training), and safety regulations. Now apply the same logic to employee health. Each year, companies spend billions of dollars curing employee illnesses after they occur. Wouldn't it make more sense to emphasize employee illness prevention (before the fact) by emphasizing employee wellness? Health care costs could be significantly reduced through health programs, worker education, and healthier worker lifestyles.

In 1950, health care expenditures represented 4 percent of the gross national product. This has risen to 10.8 percent and increases each year. An estimated 50 percent of the increase in hospital costs is lifestyle associated — the way we eat, live, work, and play today. We have accepted stress, and there is a price to pay for living that way. Health promotion is one of several efforts being made to reduce these costs.

Medical Costs Not Expected to Be Reduced

Billions of dollars are spent annually in medical costs, with no dramatic relief in sight. More than 6 percent of all American workers are engaged in health-related occupations. We are currently spending an average of $1200 per year on health care for every man, woman, and child in this country, and that cost is projected to rise to over $4000 by the year 2000. In fact, at the current rate at which health care costs are increasing, we will reach that figure much sooner. Of the $350 billion spent annually on health care, more than one-fourth is subsidized by employers.

Current theory underlying the wellness movement states that we are now dying from very different diseases than affected people at the turn of the century, when infectious diseases were the primary causes of death. We have largely replaced those early conditions (e.g., tuberculosis, pneumonia, diarrhea) with what are called today "diseases of choice." Major killers today are heart disease, cancer, stroke, and accidents. Factors contributing to these diseases are, for the most part, those over which individuals have some control.

The investment in traditional medical care (emphasis on treatment) is less effective with these new diseases, and we have reached a point of diminishing returns. The response has been an interest in wellness and health promotion, on the assumption that it is cheaper to prevent illness and disease than to treat them. Treatment of illness rather than prevention is but one variable contributing to escalating health care costs.

New Programs Are Effective

A flurry of programs have been developed recently in response to the dramatic annual increase in health care outlays, such as claims control and coordination of benefits. One program to reduce costs highlights consumer incentives that encourage employees to reduce the cost of their health care. These include incentives to stay healthy and to be judicious about the choice of health care services. Bonuses for health lifestyles, deductibles for service usage, and selection of health maintenance organizations are examples. Programs to reduce employee selection of services have limited savings, since they frequently cost less than $1000. The more expensive procedures (costing over $1000) comprise 78 percent of the claims.

Health promotions, as a cost containment effort, fall in the long-range category, requiring 5, 10, or more years before significant, tangible savings will be recognized. However, the potential savings from short or middle-range efforts are even greater.

The advantage of strategies directed to lifestyle improvement, rather than to more efficient health care systems, is shown by the relative impact of the four major factors contributing to premature death. These four factors are heredity, health care, environment, and lifestyle. The effects of lifestyle far exceed those of the other three factors when considering the causes of death. Thus, strategies that concentrate solely on rearrangements of traditional health care delivery programs can only have limited, short to medium-range impacts on organizational costs.

Until recently, there was little documented evidence to show that wellness programs produced the expected cost savings. This is not surprising, given that the primary benefit of lifestyle change is the reduced risk of degenerative diseases that occur over long periods of time, sometimes 20 to 30 years. Health promotion requires long-term, gradual changes, making cost/benefit analysis difficult.

Case History of a Working Wellness Program

The Control Data Corporation has a sophisticated program called "Staywell," which is self-insured and self-administered, so the company is in a good position to research its costs and savings. There are 22,000 employees in fourteen cities who participate. Because it is a long-term program, cost results are not available; however, there is already firm evidence that employees are changing their lifestyles as a result of it. Through computers, sophisticated programs have been developed to track the costs and savings, and preliminary results have encouraged Control Data to carry on with the program. Part of the program consists of identifying high-health-risk employees, since the greatest savings will come from programs tailored to that group.

As data supporting the investment in wellness continue to accumulate, organizations will begin to see the cultural and behavioral changes within their organizations as a high priority.

Benefits of EWPs

Many companies have discovered that there are numerous benefits of EWPs. As discussed previously, some cost savings are realized almost immediately, especially in smoking cessation and weight loss programs. Smokers have more illnesses, more absenteeism, and greater rates of hospitalization than nonsmokers. Overweight people are more prone to illness as well, so weight control programs can quickly reduce employee health and absenteeism rates.

More and more executives are considering the wellness of their employees as an important factor in the success of their company. They realize that, for the most part, the corporation and the employees have the same vested interest in health improvement.

Fitness programs result in cardiovascular wellness, reducing the risks of heart disease and heart attacks among employees. Reducing hypertension through fitness and stress reduction programs also reduces the incidence of heart attacks. Physically fit people are more resistant to illnesses and disease, and are less likely to be involved in injuries involving strains and sprains. Healthy employees are more productive, have a better attitude toward work, and work more safely. They view the EWP as an important fringe benefit and as evidence that the company is concerned and people oriented.

Starting a Wellness Program

Here is a checklist to help the company get moving on a wellness program:

1. Make health literature available to employees.
2. Conduct health screening.

3. Encourage or sponsor employee participation in community health and fitness activities, such as "Y" programs, aerobic classes, and running events.
4. Replace candy and soda vending machines with fruit and juice machines.
5. Use company publications to promote a healthy lifestyle.
6. Make sure that employees are involved from the start.
7. Conduct a survey of the activities employees are most interested in and tailor the initial program to meet the needs and requests of the majority.
8. Form a planning committee and set realistic goals.
9. Promote the program and encourage employee participation.
10. It is often a good idea to allow employees' families to participate, especially when lifestyle changes are being taught (smoking cessation, nutrition, weight control).

A small company interested in starting a wellness program must have management's commitment. A long-term plan is critical. Employees must know that the program will continue to be offered so as to be motivated to join. The following ideas may be helpful in getting started.

While exercising equipment, lockers, and showers are highly desirable, if the company does not have the resources, that is not the end of the wellness program. Aerobics classes can be conducted in the cafeteria by pushing tables out of the way, providing exercise mats, and hiring a qualified aerobics instructor. Smoking cessation classes and nutritional/weight counseling may also be conducted in the cafeteria or other meeting rooms. While encouraging off-premises community activities may be a good way to begin a wellness program, it should be noted that employee participation is much higher when programs are conducted on site. (Note: If your company is planning to move to a new location, this is an ideal opportunity to design fitness facilities into the new building.)

With a well-established, smooth-running program enjoying employee participation, the company may be able to justify some of the following additions: showers and locker rooms, a fitness center with stationary bicycles and weight-resistance equipment, an outdoor fitness track, yoga and meditation classes, strength testing, healthy back programs, and seminars on stress reduction, hypertension, so on. The wellness library can be expanded to include films and videotapes. More health screening—such as blood pressure, cholesterol, and cancer—can be offered. Classes in first aid and cardiopulmonary resuscitation can be provided. By this time, the company may be ready to install a gymnasium or outdoor tennis and basketball courts.

A full-time exercise instructor may be necessary at this point, since qualified supervision is a must. A qualified instructor can also design an individual fitness and nutritional program for each employee and can help employees track their progress.

When the initial EWP is in place, it is time for an evaluation. What is the extent of participation—10, 20, or 50 percent? Could participation be increased by a change in program scheduling? How else can the existing program be improved? Are there employee needs that cannot be met by the current program? Perhaps some previous employee suggestions were not implemented. Conduct another survey to assess fitness needs and interests. Now may be the time to expand to meet those needs.

Will It Work?

The New York Telephone Company evaluated the results of its nine health promotion and disease prevention projects. It found a reduction in the risk factors of smoking, cholesterol, and hypertension and an increase in fitness. After subtracting the cost of the programs, the company showed an annual gain of $2.7 million.

The only approach that has worked to help reduce workplace injuries and their associated costs is a comprehensive, long-term program addressing the full scope of activities. Programs such as rehabilitation alone offer little chance of seriously changing the picture.

A solution to the problem may be to take the savings from short-range cost-cutting measures and invest them in health promotion and wellness programs. The organization forgoes some immediate gains to establish long-term, permanent, broader-based successes. Failure to go beyond the short-term measures will leave a company on a gradient of escalating costs.

MANAGEMENT'S ROLE

The company's medical service is a tremendous resource whose full potential is seldom used by management. Since management controls both policy and budget, it defines the scope and role of the service. The manager must understand that the medical service can contribute greatly to a healthy and safe working environment. If intelligently organized and operated, it can increase the efficiency of performance, improve employee morale, reduce lost work time, and help cost containment. It may even improve the company's products through human factors inputs, and greatly enhance the corporate image in the community and marketplace. Management's understanding and support are essential if these results are to be achieved. Careful review and (if necessary) revision of company policies and practices can come only from top management, along with direction for establishing special programs and services. The senior manager must recognize the medical service as a potentially valuable resource and give serious thought to its fullest and most profitable use.

BIBLIOGRAPHY

Davidson, Cynthia, personal conversations and written material during conferences in Los Angeles, June-July, 1985.

────── "Employee Drug Testing," *Safety Management Newsletter,* Santa Monica, CA, Merritt Company, 1986.

Essential of Medical-Nursing Services in Industry, Chicago, Council on Industrial Health, American Medical Association, n.d.

Guiding Principles and Procedures for Industrial Nurses, Chicago, Council on Industrial Health, American Medical Association., n.d.

Guiding Principles of Occupational Medicine, Chicago, Council on Industrial Health, American Medical Association, n.d.

Johnstone, R. T., and S. E. Miller, *Occupational Diseases and Industrial Medicine,* Philadelphia, W. B. Saunders, 1960.

Opatz, Joseph P., "Wellness Is a Cost Containment Strategy," *National Safety News,* June 1985, pp. 66-70.

OSHA Reference Manual. Santa Monica, CA, Merritt Company, June 1985.

Patty, F. A., ed., *Industrial Hygiene and Toxicology,* Vol. 1, 3rd edition, New York, Wiley, 1979.

Polakoff, Phillip D., "Medical Case Management Provides a Closer Look at Healthcare Costs," *Occupational Health and Safety,* September 1989, pp. 43, 70.

Strachta, Bruce J., "Teleconference Focuses on Cost Containment," *National Safety News,* July 1985, pp. 57-58.

Chapter 15 / Part 2

Stress in the Workplace

INTRODUCTION

"Stress" is best defined as the reaction of the body to any demand made on it. The demand (also known as a "stressor") may be pleasant or unpleasant and may evoke conscious or unconscious reactions in the body. In most cases the reaction is mild and transient, with no adverse or lasting effects on the person's health or ability to perform. In certain people, however, stresses that are too intense or prolonged may cause anxiety, discomfort, disturbed body function, and, finally, actual disease or damage.

In general, the body's adaptive mechanisms are flexible enough to meet and resist the stresses imposed by the environment. Indeed, most of us are even stimulated by the challenge of meeting each day's many needs and changes. For the most part, we may even be unaware of our responses to stress. Automatic changes continuously occur in the nervous system and glands, adjusting the body functions to meet these demands. Rarely do these adjustments disturb the normal stability and balance of bodily systems.

This is not always the case, however; there are times when stress does interfere with normal functioning. This often occurs when the stress represents a threat of some kind. It may be the threat of losing control in handling a situation, or a feeling of guilt, or the actual danger of physical harm. All of these factors may exist in both work and nonwork environments. In either setting, only a small number of people will show signs of stress-induced illness or disease. Usually, several factors must be present to reduce the ability to cope with any given situation. It is essential that the manager understand these factors, particularly their interactions, and learn to recognize the early signs of stress or maladaptation. Only then can effective preventive and modification techniques be incorporated into the overall safety and health program.

BODY RESPONSE

To function normally, the body must operate with very narrow biological and chemical limits. The normal operation of the cardiovascular, respiratory, gastrointestinal, nervous, and urinary systems depends upon automatic adjustments in the glandular and nervous systems in response to stimuli that arise within the body itself. These changes may affect us in any of several ways:

Psychologically or emotionally
Functionally
Biochemically
Pathologically

Psychologically, when we feel excessive stress, we may develop depression, difficulty in concentrating, loss of appetite, insomnia, fatigue, and withdrawal. Functionally, we may experience headaches, muscle aches, nausea, constipation, diarrhea, irregular heartbeat, and frequent urination. Biochemically, there is an increase in the output of the adrenals, or stress glands. This is accompanied by an increase in blood fat levels, as well as by changes in blood sugar and mineral levels.

Chronic tension may result in the increased use or even dependence upon tobacco, alcohol, and other drugs. This type of abuse depletes the normal body constituents and interferes with its normal defense mechanisms. Eventually this can lead to the destruction of body tissues, resulting in disease. Such pathological disorders as migraine, coronary heart disease, high blood pressure, bronchial asthma, peptic ulcers, colitis, neurosis, some skin diseases, and possibly arthritis are frequently associated with chronic stress.

PREDISPOSING FACTORS

Obviously, some people are better at handling stress than others. Individuals vary widely in their ability to tolerate stress, either on a short-term or a prolonged basis. While there is no handy formula for predicting the exact degree to which a prospective worker will be able to handle stress, we do know that numerous factors predispose some people to react more strongly than others. The major predisposing factors are heredity, environment, and the general state of health or illness.

The exact influence of heredity in determining stress tolerance is somewhat uncertain. However, we do know that congenital defects in the body systems involved in the adaptive process can reduce stress tolerance. Defects of the brain, autonomic nervous system, and glandular system are the most critical, though an abnormality in any major body organ may be significant. Even where no physical abnormality exists, the psychological environment of the growing child can have an enormous influence on his or her later ability to handle stressful situations. Psychological factors include the relationship with parents and siblings, the role in the family unit, interactions with peers, the influence of teachers or social institutions, and the opportunities provided for maturation and growth. Physical illnesses during the growth period (especially those involving the adaptive body systems) may greatly reduce an individual's later ability to combat threatening stressors.

WORK ENVIRONMENT
Job Demands

There is a certain amount of stress involved in every work situation: the need to perform, to meet deadlines, to cope with normal changes or unexpected events. The universality of stress may obscure its importance as a factor in safe, efficient operations, and the importance of identifying and mitigating it as much as possible.

Work stress may be measured by the worker's reaction to any situation that is consciously or unconsciously perceived as threatening. Such threatening situations are almost invariably associated with change. For example, the new employee wonders whether he will succeed on the job. He may be concerned about the effects of the work environment on him. Will he fit in socially with his fellow workers? Will the job require certain new skills or physical abilities applied in a unique environment? There may be a strong demand for concentration, attention, sensory alertness, perception, and precision. Unless the worker is already knowledgeable and experienced in the job, the fear of failure, loss of self-esteem, and possible job loss can pose a real threat. In the absence of strong support from the manager and fellow workers, this threat may easily become exaggerated.

The nature and amount of work itself may also create stress. Understimulation and overload on the job may be equally stressful. Some workers relish deadlines as challenges and perform best under pressure. For others, the very thought of having to meet a deadline creates enormous anxiety. Some workers view menial tasks or monotonous assembly line work as meaningless and unfulfilling. They develop feelings of frustration and hostility toward themselves (for their inability to control their lives) and toward the company (for its apparent lack of interest). Other workers perform the same repetitious jobs without experiencing any apparent conflict. They find security in knowing exactly how to perform their work without having to make difficult choices, decisions or readjustments. Most workers probably fall somewhere between these two extremes, able to tolerate a certain amount of dull, routine work while enjoying some challenges or changes in routine.

The effects of the physical demands of a job are often difficult to assess, since much depends on the physical condition of the workers. Work requiring unusual postures, heavy or repeated lifting, repetitious assembly line tasks, or frequent travel, as well as work considered dangerous, all pose greater threats to the average worker than safer, more routine or sedentary work. Hence, these jobs may be perceived as more stressful. Unless the worker can choose his or her hours, mismatching may also occur with respect to personal desires, home or school demands, and body clock. A small number of people are "night workers," at their best on late shifts; others are more alert early in the day. It is well known that frequent shift changes or rapid travel through many time zones can disrupt the biological rhythms of the body. Young people are able to adapt more rapidly to shift changes without bodily injury or significant changes in work performance. This is not necessarily true for older workers, for those with marginal health, or for those taking certain types of medication. Even when there is no physiological problem, personal or domestic reasons may create difficulties when the worker must deal with late shifts or sudden or repeated shift changes.

Environmental Stress

Many studies have been performed on the effects of certain environmental stressors. We have known for years that noise above 85 to 90 dB can cause permanent damage to hearing unless earplugs are worn. It also interferes with voice communication, which can cause great annoyance, especially to those exposed for the first time. Most workers can quickly make the psychological adaptation, but a very small group cannot adapt and must be removed from further noise exposure. This group may experience irritability, insomnia, headaches, and tension, which disappear once the worker has been removed from the noise. More recently, it has been alleged that excessive noise exposure can also cause vascular problems, including high blood pressure. Most investigators believe that noise may be a nonspecific stressor to those who are not accustomed to it. They state that an initial increase in blood pressure may occur but will subside in time.

Moderate background noise is not necessarily a stressor, nor is it counterproductive. In some occupations, large numbers of employees work in close proximity. Where production and safety are not sound dependent, production may actually increase. It has also been noted that workers doing repetitive small-assembly bench work may perform better and with less fatigue if allowed to use radios with individual earphones.

Ultrasonic acoustic energy, generated by many machines currently used in industry, does not seem to constitute a hazard to workers. However, vibrations below 20 Hz are very annoying, as they may cause resonance in various body organs and interfere with muscular action, vision, speed, and coordination. Muscular aches and fatigue may persist after exposure. However, with the exception of a few occupations and exposures in which workers have sustained permanent vascular and neuromuscular injuries, vibrations should have no permanent effects.

Lighting may also be a significant stress factor. Illumination of the proper intensity and direction, without glare, is essential to avoid eye strain. Both the muscles and the lens of the eye are involved in making the necessary adjustments for different light levels. Under adverse conditions, they may have to work continuously, resulting in eyestrain. Headaches, aching eyeballs, fatigue, and irritability are the common complaints associated with eyestrain. Generally, eyestrain does not mean permanent eye damage, but it can seriously impair worker comfort and productivity.

Proper temperature is also essential to controlling stress. With the exception of noise, there are more worker complaints about workplace temperature than about any other factor. The first hot days are difficult to tolerate, especially if the humidity is also high. This environment may pose health and comfort problems. Depending on the level of physical work performed, temperatures in the 68° to 75°F range, with relative humidity between 25 and 70 percent, are well tolerated. One or two weeks may be necessary to acclimatize workers to greater heat, after which they feel less discomfort. In most factory and office environments, air flow and temperature are primarily comfort and performance considerations. However, they can also be stress considerations to some workers.

Potential hazards abound in most factory operations. Chemicals, radioactive substances, electrical and electronic equipment, and new materials and processes all represent potential physical and health hazards to workers. The recent OSHA emphasis on work-induced diseases, especially cancer, has made workers more aware of workplace hazards and may add to their anxiety on the job. To a certain extent, this anxiety can be allayed by presenting adequate information concerning the exact nature and extent of the hazards, the measures in use to prevent harmful exposure, and the availability of immediate assistance in the event of a harmful exposure.

Morale

Properly motivated and contented workers are most productive and have fewer complaints. Any alteration in the work environment can precipitate a stressful response, whether the change is a promotion or demotion, possible layoff, transfer, loss of friends or co-workers, a new supervisor or manager, a change in work activity or process, net shift, impending company difficulties, or even rumors about such changes. Anticipating such changes, preparing for them, and providing both co-worker and managerial support for the worker may help minimize the impact of the change.

Stress can also be reduced by specific changes in the physical appearance of the workplace. The presence of outside windows and the use of soft colors, comfortable furniture, and carpeting can all reduce stress. Other comfort and convenience factors such as nearby parking, elevators, ample washroom facilities, clean and efficiently run eating facilities, and company-sponsored recreation facilities may not have a direct bearing on productivity. However, they all tend to encourage and support workers and to instill in them a sense of well-being, which reduces overall stress.

With the current emphasis on providing speedy, vital medical treatment for life-endangering illnesses or injuries, workers may be greatly concerned about the availability of emergency medical services. Companies located near community-operated paramedical services or hospitals are assured of a quick response which may help reduce employee stress or anxiety. The presence of a company doctor, nurse, or first aid technician may have a similarly reassuring effect. This can be especially important in industries or operations with a significant risk of injury.

CONTRIBUTING FACTORS

Personal Health Habits

The ability to tolerate stress depends on both the intensity of the immediate stimulus and the susceptibility of the body at that time. So far, we have spoken of some of the stress factors that may be found in the workplace. However, most of us spend only a small portion of our time at work. In addition, we live in a 24-hour environment, where many other factors produce stress. Some of these physical and emotional stresses may be carried over into the workplace.

Some of these events and circumstances are beyond our control; others are self-induced and can be modified. Our resilience and state of well-being at any moment can be greatly influenced by what we eat, how much we drink or smoke, whether we take any drugs, how much sleep or rest we have had, what exercise we take, and what pleasurable pursuits we enjoy.

Much has been written about nutrition and its role in the causation of heart and blood vessel disease, high blood pressure, and diabetes. The role of high blood fat levels and obesity in causing these diseases is still under investigation, but there is little doubt that they do constitute risk factors. Anxiety, fear, and increased tension appear to increase both circulating blood fats and, initially, blood sugar. In recent times, much emphasis has been placed on the condition known as "hypoglycemia," or low blood sugar, which may cause light-headedness, weakness, or even unconsciousness. Both diet and workplace stress can contribute to a hypoglycemic reaction. The stress of stimulation can cause the system to produce too much insulin, which, in turn, causes the blood sugar level to drop too low. This sudden drop in blood sugar may cause the worker to become faint, panicky, or unconscious. Fasting, missing meals, or eating refined sugars which leave the blood very rapidly may lay the groundwork for a hypoglycemic reaction when paired with a sudden stressful stimulus to the body.

Consumption of alcohol, particularly if done regularly or often, can also have a radical effect on the ability to deal with stress both on and off the job. Alcohol should be thought of as a depressant drug with the same addictive potential as any other dangerous drug. It is a water-soluble drug which diffuses into all fluid compartments of the body and can interfere with the normal oxygenation of body cells. This produces its most sought-after effect—relaxation, often accompanied by a release of inhibitions or anxieties. It thus interferes with judgment, self-evaluation, and reaction time, any one of which could be crucial in the work environment.

There is no evidence that small quantities of alcohol, even if consumed daily, are detrimental to health or longevity in a healthy person. However, some individuals can become addicted to alcohol, developing increased tolerance and consuming progressively greater amounts of it. Along with this, the alcoholic experiences toxic symptoms of withdrawal (such as tremors), and will drink even more to avoid these symptoms. If alcoholism continues unchecked, it eventually leads to early death, whether due to accidents, infectious disease, or liver or heart problems.

Since alcohol tends to dull reality and release inhibitions, it is frequently used as an escape from stress or problems. Increased consumption of alcohol is often noted among people under stress. It represents an immature coping mechanism in persons with low self-esteem. However, abstinence from alcohol as such does not automatically imply that the recovered alcoholic will develop better or more mature means of coping with stress.

Drugs are used for many of the same reasons as alcohol. Other than simple analgesics and cold preparations, the drugs most frequently consumed by workers are tranquilizers, sedatives, and antidepressants. All act on the nervous system, either depressing electrical activity or blocking transmission of certain impulses. In addition to their sedative effect, many of them are addictive and may have unpleasant side effects, including withdrawal symptoms. The habitual use of these drugs indicate an underlying emotional problem, which in itself tends to increase the person's sensitivity to other stressors. Increased dependency on these drugs is an index of emotional turmoil and a fear of losing control without them.

Increased smoking is another sign of an inadequate response to stress. The role of smoking in the development of respiratory cancer has been well publicized, especially among workers exposed to various chemicals. Less well publicized has been the increase in coronary heart disease among smokers. Despite this, many workers continue to smoke, indicating a disregard for their own health or an addiction to nicotine.

It is impossible to feel well when the body is not functioning at maximum efficiency. Maximum efficiency, including increased tolerance for fatigue, rests on an appropriate program of sleep, rest, relaxation, and exercise. Sleep is an essential physiological process during which waste products are removed from the body and all organs are rested and reenergized. The average person requires 7 to 9 hours of sleep, with slightly more for young persons and less for the elderly. Both the quality and the quantity of sleep are important. Sleep cannot be stored up, and it may take 2 or 3 nights of normal sleep to make up for one night's lack of sleep. Insomnia is a common complaint of persons under tension, indicating an inability to relax at bedtime. The insomnia, in turn, contributes to both mental and physical fatigue, lowering resistance to other stimuli. This sets up a vicious cycle of reduced tolerance to stressors, higher tension, and increased sleep difficulty, which further reduces tolerance to stress and continues the cycle.

Lack of adequate rest and the inability to relax during and after work contribute to fatigue. Fatigue may be either mental or physical, slowing both mental processes and physical responses. In addition, fatigue is characterized by irritability, anxiety, poor concentration, and withdrawal. It renders the worker more susceptible to stimuli that he or she would normally be able to ignore or cope with.

Lack of adequate exercise also decreases the body's efficiency, hastening the onset of fatigue. Many factory jobs involve simple, repetitive body movements but little whole-body activity. Boredom and stagnation due to lack of activity are common contributors to fatigue. For this reason, many workers find that a short, brisk walk during break periods is more invigorating than a snack or a smoke.

Domestic and Social Stresses

Domestic and economic problems may engender stresses that are equal to or greater than those associated with the workplace. The death of a close family member, a divorce or separation, a serious family illness or injury, sexual problems, and the assumption of a mortgage all represent significant changes. The inability to cope with them results in uncertainty, severe anxiety, and reduced resistance to additional stress.

We are also influenced by world, national, and local events, often to an extent we do not recognize. Situations which threaten our security, whether threats of war, neighborhood crime, pollution, inflation, lack of transportation, housing inadequacies, or energy shortages, can affect us emotionally and physically. When these insecurities are severe, and we feel unable to control them or our environment, we develop a feeling of helplessness. If severe or persistent, this feeling can ultimately degrade our coping ability and render us susceptible to other stresses.

Most of the factors and conditions discussed so far are not work related, and managers may have little knowledge of or control over them. They cannot, after all, see to it that every worker gets enough sleep, cuts down on smoking or drinking, or solves his or her domestic problems. Managers should, however, realize that these factors can create serious problems in the workplace and may explain why a worker's previously satisfactory performance suddenly deteriorates. Lecturing the worker may simply increase the stress and further erode the performance. By contrast, an effort to pinpoint the difficulty and take preventive or ameliorative action may improve it.

PREVENTIVE MEASURES
Proper Placement and Motivation

The new employee brings certain talents and capabilities to the job, as well as certain hopes, aspirations, desires, needs, and fears. Generally, during the interview, the employer will try to match the worker to a job that is compatible with his or her skills, training, experience, and aims. The worker may then be required to undergo a physical examination to determine his or her state of health and establish the physical limitations. The extent of the preemployment physical examinations vary with the company, but they are usually quite superficial. Rarely are they designed to detect emotional problems. In addition, much of what is gleaned from the examination is information volunteered by the subject; it is unreasonable to expect job applicants to volunteer information which they feel may jeopardize their chance of employment. Hence, very important facts may be suppressed or even falsified, only to surface later on the job.

With increasing government emphasis on health and safety in the workplace and the steady liberalization of workers' compensation laws, the employer has an urgent need not only to conduct thorough preemployment examinations but also to expand their scope. Matching a worker's physical abilities to a job is much easier and often less crucial then emotional matching. The latter problem, in fact, cannot be resolved using most present methods of job matching. A skilled employment interviewer is often better at assessing the applicant's personality, dependency needs, desires, and hopes than are most health care personnel. A private meeting, during which the interviewer expresses a genuine interest in the applicant, can do more to allay fear and anxiety than other, more sophisticated testing.

Many relatively simple things can be done to reduce or eliminate workplace stress. Many of these, you will notice, are simply sound management practices. Assuming that the worker is placed in the proper job, it is still important to provide proper motivation. Workers must feel that they are being supported by both their co-workers and their supervisor. Job requirements and objectives must be clearly detailed, yet workers must be given some personal control over their activities. If possible, they should be allowed to make decisions relevant to their work. Above all, they must be made to feel that they are important to the effort, that they are making a significant contribution to achieving certain goals. Meaningful rewards should be given frequently. (We must emphasize that rewards which enhance the worker's self-esteem are often more reinforcing and motivating than mere cash awards, though those too have their place.) Major changes in assignments, procedures, personnel, or benefits should be announced well in advance to allow adequate preparation or readjustment; few things produce greater insecurity and stress than the unexpected. Workers with greater potential should be encouraged to develop their skills and given more responsibility where possible. Promotions should never be arbitrary, especially if they involve giving one worker responsibility over others.

Early Detection of Stress

Managers are not doctors and cannot be expected to act as such. They should, however, learn to detect the early signs of stress in subordinates and know the actions they can reasonably take to alleviate it. Most work-related stress reactions, if detected early enough, can be resolved at the manager-worker level before they create serious problems. Frequently, this simply means listening to a worker's complaints or worries sympathetically. On other occasions, relatively minor adjustments in work detail or work site can resolve a worker's frustration. Unless information is volunteered by the worker, however, the manager should make no attempt to delve into the details of the worker's personal life. When the primary stresses are related to nonwork conditions, the manager can take very little direct action to alleviate them. He should, however, know what company or community resources are available to the worker, and should offer both the information and encouragement to seek this help.

Early stress reactions are most readily noticed in changes in the employee's work performance, behavior, or appearance. Unusual delays in completing work, increased frequency of errors, increased accidents or sick time, lower standards of performance, and general apathy on the job may be the first signs of stress. Similarly, the unexplained appearance of poor personal hygiene, tremors, sudden weight change, argumentativeness or hostility, chronic confusion or hesitation, suspiciousness, and increased smoking may signal personal problems that need attention.

The manager should not take on the responsibilities of a qualified medical diagnostician or skilled psychological counselor, which could have serious repercussions for both the manager and the company. If personnel counseling services are available in or through the company, the manager should take advantage of this resource. Suspected medical problems should be referred to the company nurse or doctor. Most corporations have clear policy guidelines and legal constraints on what the manager can and cannot do in these cases. Other company personnel may be available to help in the areas of economics, career planning, and additional training if those seem indicated. A list of community agencies in the area should be maintained—if not by the manager, then by the personnel counseling or medical staff. Ideally, the company should establish a preventive medical program including the following:

1. Education in recognizing and avoiding stress.
2. Use of tension-reducing techniques, such as exercise, progressive muscle relaxation, biofeedback, and meditation.
3. Screening programs to detect high blood pressure and diabetes.

4. Alcohol and smoking avoidance programs.
5. Periodic examinations of key personnel.
6. In-plant counseling services.

Keep in mind that stress is as disruptive to the worker as it is to the workplace. Most people are happy to cooperate when offered some means of reducing or coping with uncomfortable stress.

MANAGEMENT STRESS

Causative Factors

Managers are a select group of employees who, by demonstrated ability and/or training, have been assigned to positions of leadership and responsibility. Their jobs invariably involve a certain unavoidable amount of stress, though most managers have already demonstrated the ability to cope with this stress by the time they reach a position of responsibility. They are generally in decision-making jobs, which give them a significant degree of control over their activities. While management positions, therefore, are highly satisfying, they may also involve great tension. Stress reactions of varying degrees do occur among even the most competent and well-organized managers. Most of these reactions are directly related to the following stressors in the work environment.

One of the primary stressors is work overload, which may be both quantitative and qualitative. Quantitative overload occurs when the manager feels the pressure of having too much work to do, of being unable to stay current or to achieve control of the job in the time available. Qualitative overload is the feeling that there is never enough time to do the job well, or as well as the manager wishes. Both types of overload occur frequently and are associated with stress reactions.

Another prime stressor is role conflict. This arises when others, whose jobs are functionally interdependent with the manager's, make incompatible demands on the manager. Consider, for example, the manager who is being pressed by higher management to get production up in her department, while being pressured by others to reduce the accident rate or level of waste. Or consider the manager who must simultaneously upgrade his subordinate's earning power and control the department's payroll. The tension involved in these role conflicts can increase even more when the gap between the demands made and the manager's ability to deliver becomes too great.

Job ambiguity is another stressor. This can be defined as the discrepancy between the amount of information a person has and the amount he or she needs to perform the job. A poorly defined job description, being assigned to manage a department whose functions are entirely unknown, or having to step in and take over a job with inadequate briefing can all produce feelings of dissatisfaction, anxiety, and depression.

Being responsible for people, rather than things, evokes greater feelings of tension and anxiety. Things, after all, are more predictable and controllable than people. Decisions affecting the future of subordinates are especially stressful to those with a strong conscience and a high fear of failure. (Ironically, conscientious people, who tend to be selected for the most responsible positions, are the very persons who have a higher incidence of coronary heart disease, diabetes, high blood pressure, peptic ulcers, and other stress-related ills.)

Age is another stress factor that is too often overlooked. The most important changes in job status usually occur between the ages of 35 and 45. The individual's sense of insecurity and fear of failure may increase at that time, when he or she may also be facing the personal problems of midlife transition. At this time, the need for self-esteem, autonomy, and self-actualization generally has a higher priority than money. The conflict between the manager's personal needs at this time and the company's needs may lead to increasing stress reactions, particularly when the work environment does not offer the expected gratification or the needed assurance.

Recognition of Stress

Stress reactions usually do not appear overnight. There are ample warning signs that the individual is experiencing increased stress and is having difficulty handling it. Increased attention to work, long work hours, excessive travel time, interrupted home hours and vacations, and physical fatigue (often coinciding with changes inherent in the aging process) are all stress signs. None of them are conducive to maintaining a stable personal or home life, which may further increase the stress. Unless supported by an empathetic mate, children and friends, the manager's tensions may be aggravated by a fear of failure as a spouse and parent, adding to the feeling of inadequacy. In a vicious circle, the manager may try to escape from family tensions by increasing his or her work load, which further increases the problem.

Identifying stress in the executive may take longer than with the lower-level worker, but the symptoms are similar. Disruption in family life, economic and health problems, and changes in personality or personal habits soon become evident. A physician familiar with the company's operations, its executive personnel, its policies, and its current problems can frequently detect evidence of stress, either through a physical examination or a careful interview. Having regularly scheduled key-personnel examinations as part of company policy can greatly aid in detecting excessive stress before it becomes damaging to the executive.

We should mention that managerial stress is not simply a personal matter. When a high-ranking executive is under unmanageable stress, his or her tensions can have a detrimental effect on the performance of all subordinates, even whole departments or divisions. An inexplicable increase in stress-related disorders such as headaches, peptic ulcers, and hypertension, all occurring in a single department, may well alert the company physician to tensions induced by a distraught department or division head. Bringing this to senior management's attention promptly can save the company millions of dollars.

Stress Prevention

Stress prevention in managers requires that the company understand their dependency needs, desires and fears, as well as their psychological and physiological limitations. The need for good communication, responsible and clear delegation of authority, and sensitivity to work overload is clear. The company should also consider the stress to the manager's family when sudden or drastic changes in the manager's duties or location are contemplated. Excessive pressure or problems in the family can seriously affect the manager's performance on the job. The company should provide full support, using all of its resources to ensure maximum job satisfaction and minimum extraneous conflict. Managers, after all, are among the company's most valuable resources. The company can certainly justify the considerable expenditure in time and funds required to keep these key people functioning at their best.

Managers should be encouraged to have periodic health evaluations, performed by a physician who understands workplace stress in general and (ideally) the particular situation in which the manager works. The examination should be paid for by the company, and the executive should have the absolute assurance that the results will be released only to the company's physician and the executive's own doctor. These periodic examinations may show early changes or trends that can be reversed before they become damaging to the manager's health or performance.

Personal counseling services can also play a vital role in an executive stress management program. Counseling services should be available for discussing personal as well as job-related problems. No stigma should ever be attached to these visits. In fact, top management should actively encourage managers to make use of them. Some companies have found that group sessions for managers and industrial psychologists, seminars in handling stress, role playing, and relaxation techniques are equally valuable. Wherever possible, the company should make every effort to deemphasize the "crisis" aspects of these counseling services and activities in favor of "personal enhancement" or "self-actualization."

Managers should be encouraged to maintain maximum physical health as well. Exercise facilities, including jogging tracks, handball courts, swimming pools, and gyms, can pay handsome benefits in executive health and satisfaction. Relaxation techniques, including whole-body exercise, progressive muscular relaxation techniques, biofeedback training, and meditation, are being taught and used in many companies, with good results. Some of these techniques can be practiced several times a day without disrupting daily work activities, effectively reducing stress in both managers and their subordinates.

CONCLUSION

The complete avoidance of stress is neither possible nor desirable; a certain amount of stress is inherent in any business or work activity, whether making a decision involving the corporation's direction or simply making one's quota in the machine shop. Such stress is natural and normal. However, when allowed to escalate, stress can produce substantial problems for the individual and the company, ranging from errors of judgment to increased sick time, higher accident rates on the job, and the costly waste of valuable human resources. A safety and health management program that deals only with physical factors, ignoring the crucial emotional and psychological aspects of safety and health associated with workplace stress, is only half a program. It may, in fact, be ignoring safety and health problems that will be far more costly in the long run, eroding efficiency and profits without a visible accident or injury to bring it to management's attention.

BIBLIOGRAPHY

Caplan, R., et al., *Job Demands and Worker Health,* No. 75-160, Cincinnati, National Institute for Occupational Safety and Health, 1975.

Friedman, M., and R. H. Rosenman, *Type A Behavior and Your Heart,* Greenwich, CT: Fawcett Publications, 1975.

Jacobson, E., *Tension Control for Businessmen,* New York, McGraw-Hill, 1963.

Margolis, B., et al., "Job Stress: An Unlisted Occupational Hazard," *Journal of Occupational Medicine,* 1974.

Poulton, E. C., *Environment and Human Efficiency,* Springfield, IL, Charles C. Thomas, 1970.

Rahe, Richard H., *Life Crisis and Health Change,* Report No. 67-4, San Diego, CA, Naval Medical Neuropsychiatric Research Unit, 1967.

Selye, Hans, *Stress Without Distress,* New York, Lippincott, 1974.

Chapter 15 / Part 3

Off-The-Job Safety Program Management

One element, off-the-job safety, is often missing from an organization's safety and health program. The reasons for this omission are logical, but at the same time somewhat erroneous. Many managers feel that preventing accidents on the job should have first priority, and others believe that they cannot impose on their employees' freedom when they are not in the workplace. Some also fail to realize what an impact off-the-job accidents have on the organization's productivity.

JUSTIFYING OFF-THE-JOB SAFETY PROGRAMS

The time and effort devoted to off-the-job safety are resources well spent. Studies by leading safety organizations, such as the National Safety Council, the International Loss Control Institute, and the DuPont Company, have shown that off-the-job injury experiences are four to twenty-six times those at work. These accidents are also costly in terms of medical treatment, insurance, and absenteeism. In addition, these studies revealed that employees appreciated honest efforts on behalf of their personal safety. Another interesting fact is that after developing comprehensive off-the-job safety programs, many organizations found their job accident rates decreasing as well. The activities inherent in an off-the-job safety program create a higher capability for working safely on the job too.

A MATTER OF ENVIRONMENT

Managers often wonder why employees have such high accident rates off the job. It seems, on the face of things, that when there is a good safety and health program in the workplace, employees should carry an attitude of safety with them as they walk out the door at the end of their work shifts. They usually do, but attitude is just an adhesive that holds the program together. The essential "materials"—knowledge and safe conditions—must exit before a program can be considered effective.

When a comprehensive program exists in the workplace, it represents a complete environment, and the "materials" mentioned are the program activities detailed in the chapters of this book. They include, among others:

> Senior management's leadership of the program: setting policy, defining activities, and setting standards.
> Managers, trained in safety fundamentals, setting an example of safe behavior and directing program activities.
> Engineers designing work areas and processes to comply with safety codes and standards based on years of accident research.
> Purchasing staff and materials control groups researching safety and health factors in order to buy the safest possible equipment and materials.
> Incorporation of manufacturers' operating instructions, along with safety analyses, in written procedures for jobs to be done.
> Preventive maintenance properly performed to keep equipment and facilities in safe operating condition.
> Regular workplace inspections to detect unsafe conditions and practices.
> People oriented to hazards in the workplace and trained in safe work methods.
> Reference libraries maintained, periodicals screened, and seminars attended to keep safety knowledge up-to-date.

OFF-THE-JOB COMPARISON

Now let's look at the environment off the job. The structured environment doesn't exist there. The employee must be prepared to be the executive, middle manager, supervisor, safety coordinator, engineer, purchasing manager, human relations director, trainer, and researcher all at the same time. These are the capabilities that the off-the-job safety program must create. Putting up a few posters, having a couple of holiday safety briefings, and giving out a few pamphlets on home and recreational hazards won't do the job. They won't even come close.

LOSS EXPOSURE IDENTIFICATION

An effective safety and health program starts with an inventory (or determination) of loss exposures. To do otherwise is to manage by blind guessing. This includes surveying off-the-job accident experiences and what they cost. The cost summary should include medical treatment and insurance payments to employees and their families, days lost from work, and the cost of compensatory temporary or overtime wages. This is a preliminary exposure survey. Studies show that there are about sixty near-accidents for every injury, so many exposures do not show up in the cost summaries. It would be a very rare workplace that had employees report near-accidents off the job. Therefore supplemental methods must be employed to get an accurate picture of the potential exposure.

As you know, the problems we must solve are today's; so, the bulk of the exposure survey involves determining the potential accidents that employees and their families could be involved in. To do that, it is necessary to learn what activities people pursue off the job and what hazards those activities involve.

Data on past losses can readily be obtained from several sources within (or serving) the organization. These include:

The risk management or insurance department or the benefits administrator. Company offices may have information on payments made for nonwork and dependent injuries; if not, they can get this data from the insurer.

Personnel or payroll. Absences are recorded on timekeeping forms. These forms are (or can be) coded to show when absence is due to an off-the-job accident.

Supervisors. Most organizations require an absence to be reported and explained to the supervisor. The line managers may, or can, keep data on absences caused by off-the-job or family accidents.

Medical staff. The reasons for treatment given by occupational health personnel are usually entered in treatment records.

A commonly used worksheet format for collecting and analyzing data on off-the-job injuries is shown in Figure 15-1. This worksheet divides the cases into three groups: transportation accidents, home accidents, and public accidents. Within each group, data is broken down by common modes. Smaller organizations may combine several of these modes to simplify the worksheet and reveal more meaningful trends.

There are several ways information can be collected on the current loss exposures of employees and their dependents. These include family activity surveys, employee interviews, and supervisors' knowledge of employees' activities. The survey form shown in Figure 15-2 can be used as a data worksheet for any of these methods. The methods differ only slightly in the way the data is obtained.

The Family Activity Survey is performed by giving a copy of the form to each employee. It is completed and returned to the manager, the safety office, or the project team organizing the program. Handout surveys normally have a very poor response, so the purpose of the survey must be explained when the forms are distributed. This can be done at a safety meeting or at a similar gathering of the employees. Salaried people need to be included in the survey; they are often overlooked. A convenient return method can increase the rate of response. Collection boxes, return-addressed forms, or mail envelopes can be provided if personal collection by the supervisor is not desirable.

The employee interview is a more reliable way to get the data on off-the-

(OTJ) INJURY REPORT FORM

Off the Job Safety Program Management

MONTH	YEAR
DIV.	LOCATION

Total No. of Off-Job Inj. This Month: _____
Year To Date: _____
Total No. of Fatalities This Month: _____
Year To Date: _____
Total Calendar Days Lost This Month: _____
Year To Date: _____

Total Medical & Ins Cost for OTJ Inj.: _____
Year To Date: _____
Total Benefits Paid on Death/Disability: _____
Year To Date: _____
Total Wages Paid on Days Lost: _____
Year To Date: _____

	1	2	3	4	5	6
	Current Month			Days Lost Old Cases	Total Days Lost This Mo.	Total Cost
Transportation	No. Cases	No. Fatalities	Days Lost			
Auto, Truck						
a) Collision						
b) Repair work						
c) Pedestrian						
d) Miscellaneous						
Bus, Streetcar						
Bicycle						
Motorcycle, Moped						
Railroad						
Aircraft						
Watercraft						
Other						
TOTAL						
Home						
Fall, Slip						
Fire, Explosion						
Firearm						
Animals, Insects						
Exposure: Heat, Cold						
Lifting						
Pushing, Pulling						
Struck By Object						
Struck Against Object						
Toxic Material						
Electricity						
Machinery, Tools						
Sharp Objects						
Other						
TOTAL						

Col. 1 **Number Cases.** Include all off-the-job injuries occurring this month when one or more working days are lost.
Col. 2 **Number Fatalities.** Include all fatalities occurring this month.
Col. 3 **Days Lost.** Total calendar days lost on injuries occurring this month.
Col. 4 **Days Old Cases.** Records **additional days** lost this month due to injuries previously reported.
Col. 5 Total calendar days lost this month (Cols. 3 + 4).
Col. 6 Total wages & salaries paid for days lost.

Figure 15-1. Injury report form.

OFF-THE-JOB SAFETY PROGRAM MANAGEMENT 417

(OTJ) INJURY REPORT FORM (Cont)

MONTH _____   YEAR _____

DIV. _____   LOCATION _____

Total No. of Off-Job Inj. This Month: _____
Year To Date: _____
Total No. of Fatalities This Month: _____
Year To Date: _____
Total Calendar Days Lost This Month: _____
Year To Date: _____

Total Medical & Ins. Cost for OTJ Inj.: _____
Year To Date: _____
Total Benefits Paid on Death/Disability: _____
Year To Date: _____
Total Wages Paid on Days Lost: _____
Year To Date: _____

Public	1	2	3	4	5	6
	Current Month			Days Lost Old Cases	Total Days Lost This Mo.	Total Cost
	No. Cases	No. Fatalities	Days Lost			
Fall, Slip						
Fire, Explosion						
Firearm						
Animals, Insects						
Exposure: Heat, Cold						
Lifting						
Pushing, Pulling						
Struck By Object						
Struck Against Object						
Toxic Material						
Fight, Assault						
Sports						
Recreation						
Other						
TOTAL						

Col. 1. **Number Cases.** Include all off-the-job injuries occurring this month when one or more working days are lost.
Col. 2. **Number Fatalities.** Include all fatalities occurring this month.
Col. 3. **Days Lost.** Total calendar days lost on injuries occurring this month.
Col. 4. **Days Old Cases.** Records **additional days** lost this month due to injuries previously reported.
Col. 5. Total calendar days lost this month (Cols. 3 + 4).
Col. 6. Total wages & salaries paid for days lost.

Figure 15-1. (continued)

job activities, but it is time-consuming. First, people must be instructed on how to conduct the interview. They must talk to employees individually, explain the purpose, ask the questions on the form, note the answers given, and then ask follow-up questions for more details. The interview is usually a seven-step process. The steps are:

1. Explain the purpose: tell the employees that the company is interested in doing more to help keep them safe off the job. State that in order to provide useful information, the company needs to know a little about the person's activities and interests.

FAMILY SAFETY SURVEY

The company is interested in improving its program to help you keep safe off the job. You can participate in this program as much or as little as you wish. In order to provide practical help and information, we need to know a few things about your interests and activities. You do not need to put your name or department on this survey.

Please check the appropriate answers and return this form promptly. Mark all boxes that apply to you and your family living with you. Thank you.

1. What type of residence do you live in?
   - ☐ HOUSE
   - ☐ ONE-STORY
   - ☐ DUPLEX
   - ☐ TWO-STORY
   - ☐ OTHER _____
   - ☐ APARTMENT
   - ☐ MULTIPLE-STORY
   - ☐ CONDOMINIUM
   - ☐ TRAILER

2. What type of home care/maintenance do you do yourself?
   - ☐ GENERAL CLEANING
   - ☐ WOOD REPAIR
   - ☐ ELECTRICAL REPAIR
   - ☐ PAINTING-INSIDE
   - ☐ MASONRY REPAIR
   - ☐ PLUMBING REPAIR
   - ☐ OTHER (DESCRIBE) _____
   - ☐ PAINTING-OUTSIDE
   - ☐ ROOF REPAIR
   - ☐ APPLIANCE REPAIR
   - ☐ GARDENING
   - ☐ INSECT SPRAYING
   - ☐ LAWN MOWING

3. What appliances and tools do you use?
   - ☐ OVEN/RANGE
   - ☐ MIXER
   - ☐ WASHER
   - ☐ HAIR DRYER
   - ☐ TABLE SAW
   - ☐ MICROWAVE OVEN
   - ☐ GRIDDLE
   - ☐ DRYER
   - ☐ POWER DRILL
   - ☐ LATHER
   - ☐ OTHER (LIST) _____
   - ☐ FOOD PROCESSOR
   - ☐ ELECTRIC KNIFE
   - ☐ VACUUM
   - ☐ POWER SAW
   - ☐ WELDER
   - ☐ TOASTER
   - ☐ OUTDOOR GRILL
   - ☐ POLISHER
   - ☐ ROOTER
   - ☐ CHAIN SAW

4. What type of heating and cooling do you use?
   - ☐ FORCED AIR
   - ☐ SOLAR
   - ☐ RADIANT
   - ☐ FIREPLACE
   - ☐ FRANKLIN STOVE
   - ☐ ELECTRIC

5. Do you have any arts or crafts hobbies?
   - ☐ CERAMICS
   - ☐ WOODWORKING
   - ☐ OTHER (LIST) _____
   - ☐ PHOTO DARKROOM

Figure 15-2. Family safety survey.

2. Ask the general questions: find out what home and recreational activities the person engages in.
3. Ask about specific items: types of residence, appliances, hobbies, and so on; find out what risks are involved in each. Learn what equipment, materials, and safeguards are used.
4. Ask about the hazards: check how well the person recognizes the dangers in the activity. Be objective, not judgmental.

FAMILY SAFETY SURVEY (Cont)

6. Do you participate in any sports?

    - ☐ ARCHERY
    - ☐ FOOTBALL
    - ☐ HANDBALL
    - ☐ SKIING
    - ☐ FISHING
    - ☐ FLYING
    - ☐ BOWLING
    - ☐ SOCCER
    - ☐ SWIMMING
    - ☐ SNOWMOBILING
    - ☐ HUNTING
    - ☐ CURLING
    - ☐ TENNIS
    - ☐ OTHER (LIST) _____

    - ☐ SCUBA DIVING
    - ☐ CANOEING
    - ☐ CAMPING
    - ☐ GOLF
    - ☐ HIKING
    - ☐ RUNNING
    - ☐ BOATING
    - ☐ SKY DIVING
    - ☐ BASEBALL
    - ☐ RACKETBALL
    - ☐ WEIGHT LIFTING
    - ☐ SAILING
    - ☐ HANG GLIDING

7. How often do you participate in sports?

    - ☐ DAILY
    - ☐ ALMOST DAILY
    - ☐ WEEKLY
    - ☐ MONTHLY
    - ☐ OCCASIONALLY

8. Do you regularly take vacation trips by personal auto or camper, or do you regularly make pleasure drives?

    - ☐ NO
    - ☐ YES, ABOUT 100 MI EACH WAY
    - ☐ OVER 1000 MI
    - ☐ OVER 200 MI
    - ☐ OVER 500 MI

9. Have you or your family had an accident, within the past three years, in which someone was injured or property damaged?

    YES: ☐ NO
    - ☐ HOME
    - ☐ RECREATION
    - ☐ TRANSPORTATION
    - ☐ SPORTS
    - ☐ HOBBY
    - ☐ OTHER _____

    PLEASE DESCRIBE BRIEFLY. ADD PAGES IF NEEDED.

    _____
    _____
    _____
    _____
    _____
    _____

10. Have you had (in the past three years) any near-accidents off-the-job which could have been serious?

    YES: ☐ NO
    - ☐ HOME
    - ☐ RECREATION
    - ☐ TRANSPORTATION
    - ☐ SPORTS
    - ☐ HOBBY
    - ☐ OTHER _____

    IF YES, PLEASE DESCRIBE BRIEFLY. ADD PAGES IF NEEDED.

    _____
    _____
    _____
    _____
    _____

Figure 15-2. (continued)

5. Make notes and summaries; use the form to produce accurate records. Show the person that you care about what he or she is saying by making note of it.
6. Give some feedback: summarize key comments to make sure that you understand them. Give your appraisal of the risk—high, medium, or low—and see if the person agrees.
7. Leave the door open: ask the person to tell you later if he or she thinks of any additional exposures or near-accidents.

Supervisors' knowledge of their people is another source of information. Supervisors can be asked to fill out a form on each of their employees. Management's interest in the workers is a major source of motivation. If supervisors already take a personal interest in their employees, they will know a lot about their activities. If not, this provides an opportunity for them to become personally involved. Supervisors should be instructed to mark down only what they are sure of. Erroneous data on employees' activities could orient the off-the-job safety program in a wrong direction.

Once the exposures have been inventoried and management knows what activities are most likely to cause off-the-job accidents, they are ready to start an off-the-job safety program.

Major Categories of Loss Exposure

Within the major categories of off-the-job accidents—home, transportation, and public—there are several key common exposures. To help start the exposure inventory, these are as follows:

Home Safety
Slips and falls
    Footwear and clothing
    Floors, stairs, and other walking surfaces
    Bathtubs and wet enclosures
    Ice, mud, and other weather hazards
    Running and rushing
    Stools, ladders, and climbing devices
    Illumination
    Open and weak floor surface covering
    Protective railings and edge warnings
Lifting and carrying
    Back strain—manual lifting
    Hernia prevention—weight carrying
    Posture for back care
    Mechanical lifting and transporting devices

OFF-THE-JOB SAFETY PROGRAM MANAGEMENT 421

Chemicals
   Household cleaning materials
   Pesticides
   Airborne contaminants from toxic materials
   Corrosives
   Prescription medications and other drugs
   Dermatitis and poisonous plants
   Children's ingestion of chemicals
   First aid for chemical exposures
Electrical Safety
   Color coding for electrical connections
   Basic current flow and conductors
   Grounding
   Defective electrical equipment
   High-voltage lines and equipment
   Fire safety
   Flammable substances
   Combustible materials
   Heat sources
   Chemical reactions
   Home fire detection and evacuation
   Home maintenance
   Lawn mowers, edgers, and trimmers
   Hand tools
   Power tools
   Ladders
   Repair of buildings
   Repair of appliances

Transportation Safety
Private motor vehicle operation
   Defensive driving
   Driving emergencies
   Severe-weather driving
   Vehicle maintenance
   Influence of drugs and alcohol
Aircraft operations
   Rules of the air
   Emergency actions
   Weather factors
Boating
   Right of way on the water
   Severe weather
   Mechanical breakdowns

Public Safety
Recreation
  Children's play
  Competitive sports
  Individual sports
  Outdoor recreation
  Hobbies and crafts
  Recreational vehicles
Other employment
  Volunteer work
  Club and association activities
  Self-employment
  Outside employment
Spectator recreation
Travel on public conveyances
Civil disturbances

Planning Your Program

Once the general exposures are known, one can begin to plan the best way to control risks. Some risks are adequately controlled with knowledge of safe practices, so the program calls for education. Others need skill, and the program calls for ways to get supervised practice, such as through a recreation club or sports team. Sometimes the hazard can be controlled with personal protective equipment, and the program requires sources of equipment, proper fitting, and instruction on equipment use. At times, other safeguards need to be employed, such as electrical insulation or machine guards, and the program must provide guidelines for purchasing these items.

Some hazards may require the employee or a family member to know a special first aid technique. Others may require family rules to limit the use of certain equipment or materials to adult or older teen-age family members. Various other risk management techniques will be used. The controls for risks off the job vary and call for different types of measures, just like those in the workplace. Program planning should include an analysis of the forms of control needed by different employees.

Starting the Plan

Planning the program consists, first of, creating a general outline on how to transmit the appropriate controls to employees and their families. Possible program activities to consider are as follows:

    Selling personal protective equipment for off-the-job use in the company's safety supply store. Smaller companies may be able to arrange discount prices with local merchants. In either case, the plan could include suggesting standards for such equipment and letting employees know where it can be obtained.

Supporting or encouraging hobby and recreational clubs. Providing space for meetings, aiding in organizing the clubs, and helping to obtain qualified instructors or speakers are invaluable aids.

Conducting management training in off-the-job safety for first-line supervisors so that they can serve as resource people for employers.

Training employees in how to organize a personal safety program.

Providing reference materials, such as safe practices data sheets on common home and recreational activities.

Providing safety talk materials and information reinforcing pamphlets for use by supervisors.

Organizing special safety promotion activities, like theme campaigns, family safety fairs, or first aid training.

Guidelines for Effective Programs

Because each employee's personal life is different, the safety program must be tailored to the individual if it is to be truly effective. Family size, ages of family members, type of home and furnishings, domestic activities, recreational interests, and personal travel all differ. Thus, every employee's safety program will be different. However, there are some universal guidelines for effective off-the-job safety programs conducted by organizations. These are:

1. Overall direction should be given by an experienced professional. Line supervisors trained in the fundamentals of off-the-job safety can give employees advice on their problems. The safety coordinator is an additional resource when more help is needed.
2. Management's commitment must be visible and consistent at all levels. Senior, middle, and first-line managers must understand and believe in the off-the-job safety program. They must budget the time and other resources needed to perform the vital program activities.
3. A well-organized motivates individual participation and produces the best results. The major activities need to be planned and budgeted so that the necessary personnel and resources will be available.
4. Employee participation in activities should be spontaneous and voluntary. Each program activity should include an invitation for employees to participate in planning, preparing, presenting, and following through. In the beginning, selected employees may have to be asked to help, but a quality program motivates continuing participation.
5. The program needs a broad loss-control approach; it should go beyond personal injury prevention. To protect all of the employees' interests, it should include personal property damage, fires and explosions, personal and property security, and personal liability. It could even be

expanded to include personal insurance management, helping employees determine insurance needs, deductible limits, and so on.
6. The program should be family oriented. The family is the key group for sharing ideas and changing behavior. Spouses influence each other's actions. Parents influence children and vice versa. Studies show consistently better results when the message gets into the home and when activities include families.

The Team Approach

Teamwork often makes the difference between success and failure. An off-the-job safety program provides opportunities for forming and using several types of teams made up of managers and employees. These teams organize and conduct the overall program, develop special promotion activities, solve specific problems, and present recommendations to management. The types of teams used include steering committees, promotion teams, and safety circles.

The Steering Committee

Any effective off-the-job safety program starts with upper management involvement. The steering committee provides management control of the program. It:

Identifies the work to be done in the program.
Sets standards for managers' actions. In doing so, it authorizes the time, people, space, and money needed for the program.
Measures performance to the standards by those who have responsibility for the program.
Evaluates the program's progress toward goals and objectives.
Commends the people who have contributed to the program and corrects the behavior of those who have not met its standards.

The steering committee meets on policy and decisions. It should be chaired by the senior manager. It is usually made up of key functional managers, such as those of operations, maintenance, and personnel, plus employee representatives. The committee's activities should be restricted to the off-the-job safety program; it must be a dynamic group with a well-thought-out agenda.

The steering committee's initial job is to develop effective project teams, even though later, its role changes to that of a review group. It is important that team members be well versed in both safety fundamentals and the concepts of off-the-job safety programs. Therefore, training of the team members is the first order of business. As members are replaced, their votes should be withheld until they have completed qualification training. Once competent, the committee directs the making of the preliminary loss exposure inventory. It then sets the program's objectives and standards, and begins to recruit people for project teams and safety circles.

After the program is organized, the committee concentrates on review and compliance. It hears recommendations presented by project teams and safety circles. It approves goals and budgets to meet program objectives. It coordinates program activities with other functions of the safety and health program. Finally, it audits managers' compliance with program standards. All these functions require high levels of authority to commit people and resources. Consequently it is vital that senior managers not delegate committee responsibilities to subordinates.

Project Teams

Off-the-job safety breaks down into several major aspects, such as highway auto and motorcycle safety, fire prevention and protection, boating, home maintenance, various hobbies, team and individual sports, electrical, aircraft, gliding, diving, and so on. In addition, there are also off-the-job safety problems that occur in various modes. Slips and falls are examples. A comprehensive safety program is needed for each topic that is appropriate to employees' exposures.

One effective way of developing comprehensive topic safety promotions is the use of project teams. Superficial programs have little, if any, effect. Comprehensive promotion programs require a good deal of work—too much for one or two people. The team approach is useful and gets managers and employees involved in the program. In most cases, the core of the team should consist of middle managers drawn from the operations, maintenance, engineering, purchasing, personnel, and training departments. Equally important is having employees from functional areas. Each of these people can add specific knowledge of both problems and methods. Combined, their talents are most effective.

A team should be organized for each problem. Many of the exposures are seasonal, and the rest can be scheduled according to the degree of risk. Six topics a year is reasonable for most organizations. Team memberships should also be rotated to give all an opportunity to participate, to bring fresh ideas to the program, and to allow rest periods between projects for the team leaders. A person may be a team member on one project, however, and a team leader on another one 4 or 6 months later.

How a Campaign Begins

A typical promotional campaign might include:

1. Make a survey to verify the loss exposures and potential risks in order to confirm that these were properly assessed in the original inventory.

2. Write motivational messages for the senior manager and department heads to announce the theme and stimulate employees' interest.
3. Develop guidelines for employees to use in their personal safety programs. These can include safety features to look for when buying tools or materials, self-inspections for compliance with safety standards, and safe practices to learn or to teach to family members.
4. Make up employee and family contests that can be used to stimulate education regarding, and performance of, safety activities.
5. Develop or obtain educational material to improve people's knowledge related to the theme and to their personal hazard exposures.
6. Develop guides and visual aids for supervisors to use in safety talks with their employee groups.
7. Develop lesson plans for special educational programs to teach people how to identify related hazards or to teach standard safe practices.
8. Make up displays, posters, and signs to reinforce the information presented in the planned educational activities.
9. Outline key points for briefings to senior management on the content and conduct of the theme campaign.

Sample Chart

The agenda for a typical theme campaign includes the appointment of team members, training of the team, preliminary discussion of the topic, survey of loss exposures by reviewing the inventory and talking with employees, outline of the theme campaign activities, assignment of responsibilities, project research, development of alternatives, briefings to management, preparation of final materials, conduct of the campaign, and critique of program results. The critique should include individual appraisals and commendations for work to and above standards.

Three essentials of the successful project theme campaign are a definite beginning, adequate preparation time, and a definite end. The campaign should have a kickoff date at which it is announced to employees through various media, such as newsletters and posters. The scope of the campaign and the activities planned should be introduced and preliminary materials distributed. Depending on how much time team members can devote to the campaign during their work, the preparation must start 2 to 3 months ahead of the kickoff date. Finally, the campaign needs to build a climax and have a distinctive closing date. It should not be allowed to die of old age and lack of interest. Employees will continue the activity in their personal safety programs, and supervisors will continue to review and reinforce the key points, but the campaign as such should end and all the special notices, posters, and displays should come down. If appropriate, the theme can be repeated at a later time, with a slightly different approach and renewed emphasis.

Safety Circles

Continuing off-the-job safety problems are handled by safety circles. The circle is a small group of people with a common interest or problem. This interest may be a specific sport or hobby, a domestic activity, a community activity, or a technique such as first aid. The circle is not structured with management. People belong because of their common interest. Although there is a leader to move the discussions along, there is no authority figure — hence the term "circle." All members are equal. Circles may be organized within departments or across departmental lines.

Common Denominators

Certain factors are common to all successful circles. They include the following:

Management and employee commitment and support
Clearly defined goals, roles, procedures, and relationships for all participants
A people-building philosophy
A climate of trust, respect, and concern
Adequate training of members
Recognition of achievement

The factors which usually doom safety circles to failure are:

Predetermined objectives by the organization or leader
Inadequate training
Lack of leadership in the facilitator
Lack of management support
Overemphasis on quick, short-term results
Supervisors' perception of circles as a threat to their authority
Lack of recognition and encouragement

A circle facilitator is designated to lead the program. This person is the liaison between the circles and the steering committee. He or she has several vital functions:

Organizing the circles; recruiting leaders and members
Training circle leaders to conduct the activities of their groups
Acting as a resource person to the circles; prompting initial actions, advising on sources for researching problems, assessing the strengths and weaknesses of circles, and facilitating members' development

The circle leader is the facilitator. He or she keeps the circle moving toward its objectives, prompting participation from the members. The leader also helps people develop their problem-solving abilities. This includes:

Helping to recruit and train circle members
Scheduling circle meetings
Prompting the identification of critical problems
Ensuring that meetings are oriented to problem solving
Encouraging individual involvement
Keeping minutes and records or having them kept
Coordinating expert assistance
Scheduling briefings on problem studies and recommendations

Organizing safety circles must not be a hurried process. Attempts to put a large number of circles into operation at the same time will probably guarantee failure. One or two circles should be started and their success used as the catalyst for others. A few circles will let the facilitator divide his or her time in order to give each circle meaningful help in organizing. With too many circles to handle, the facilitator tends to give them "laundry lists" of things to do between meetings with them. The members sit and wait for the facilitator, doing trivial things in the meantime, and the facilitator rushes from one circle to another, able to give only superficial help.

A suggested course of action should include:

1. Picking critical loss exposures. The steering committee selects a few exposures that have the highest potential for off-the-job accidents.
2. Selecting circle leaders who are best qualified or most experienced in the activity associated with the critical exposures. These people should also have the personality characteristics to be facilitators rather than directors.
3. Training circle leaders in basic problem-solving techniques, group dynamics, personal and group communications techniques, accident causes and controls, and safety circle procedure.
4. Recruiting circle members by identifying employees who are interested in the topic, are good team workers, and have a good reputation for job safety.
5. Training team members. This is a learn-by-doing approach. After an initial orientation in the circle procedure, the leader prompts the selection of a problem to work on. The first problems should be simple ones so that the circle can concentrate more on learning the method. Keep teams small enough to be manageable; seven to ten members is ideal.

6. Conducting the circle activity. Have members rotate in leading a brainstorming activity. Prompt them to participate by asking open questions to those who hold back. Ask members to volunteer for researching various topics, for bringing in resource people for questioning, and for analyzing the data gathered.
7. Giving briefings on problems. After a problem has been researched and recommendations have been decided by the circle, have members take turns preparing and giving briefings to appropriate upper-level managers.
8. Keeping progress data. Periodically, report on each circle's activities to the steering committee. Present the problems identified and the extent of research conducted.

The three types of teams provide for the involvement of people at all levels within the organization, as well as for meaningful study and analysis of off-the-job safety problems. The product of the work of these teams is then presented to employees. Through direct learning as a circle member or through various educational media, the employee has the opportunity to enhance his or her personal and family safety programs.

Developing Personal Safety Programs

Off-the-job accidents occur outside the environment of the organization and its management channels. The company's approach, therefore, needs to be one of getting employees interested in starting personal safety programs and then providing the tools they need. These tools include education in safety and health fundamentals, information on how to organize personal safety programs, assistance with the more difficult problems, and resources for ideas and materials.

Everything that exists is built piece by piece; nothing appears as a whole. The same is true of a safety program, whether corporate or personal. It is built piece by piece. The more pains taken with each part, the better the quality of the final product. The parts of a personal safety program are called "elements." Each employee builds the parts according to his or her motivation and desire for effective personal safety and budget for time and materials. People can choose whatever elements of their personal programs they wish.

The common elements used are:

1. *Leadership.* Whether a family or a group such as a sports team, every program needs leadership. One person must inspire and challenge the others to act safely. The leader sets a safety policy and encourages the setting of standards for safe conditions and practices. The leader prompts the others and ensures that they are adequately equipped and educated. The leader also sets the example.

2. *Training and education:* Lack of knowledge or skill is a basic cause of many accidents. Training activities ensure that everyone acquires the knowledge needed for safety in all activities. Sometimes this occurs by formally teaching safe practices; at others times, it happens through self-study.
3. *Engineering controls:* There are safety standards for all types of facilities, whether homes, vehicles, or recreation areas. These standards were bought with someone's blood; they are the result of accident investigations. Someone must research and verify the safety of facilities.
4. *Purchasing controls:* Many tools and materials used in homes and in hobbies or other forms of recreation have hazardous properties. People must become acquainted with them, buy the things that are least hazardous, and tell family members or teammates about the hazards that exist.
5. *Personal protective equipment:* Some hazards can be controlled through proper personal protective equipment. This can range from clothing to protect against poisonous plants to face shields or goggles to protect against chemicals or flying objects. Everyone needs to study the specific hazards and know about suitable protection.
6. *Emergency preparedness:* Natural disasters and technological accidents can affect personal safety; the effects vary from one situation to the next. People need to consider potential disasters, make emergency plans, and hold emergency drills for families or other participating groups.
7. *Care of the injured:* First aid can prevent complications of injuries. Suitable first aid kits need to be obtained and people trained in first aid techniques.
8. *Inspections:* Periodic examinations of facilities, equipment, materials, and practices need to be made to ensure that they continue to meet safety standards.
9. *Group meetings:* People need to be reminded about key aspects of safety and the control of accidents. Family, team, or other group meetings need to be held frequently to discuss safety in various off-the-job activities.

The Company Program

Like personal programs, the organization's off-the-job safety program is made up of elements and, in the same way, an effective company programs is built piece by piece. The core program elements and the resulting activities are the same for most organizations:

1. *Leadership and administration.* At the outset, there must be a policy statement explaining and emphasizing the organization's interest in

the employees' safety off the job. It should include senior managers' visible support of the program, all managers' participation in program activities, program standards for activities, regular discussions of off-the-job safety at management meetings, a program reference manual, annual program objectives, the delegation of program coordination responsibilities, and a library of program reference materials.
2. *Management training.* The effectiveness of a company program depends greatly on how well senior, middle, and first-line managers understand off-the-job safety and their responsibilities in the program. This understanding is achieved through orientation and formal training.
3. *Employee and family education.* Ignorance of the consequences never motivates anyone. People need a good deal of knowledge to desire a personal safety program and to understand safe practices. Obtaining this knowledge requires systematic education and reinforcement.
4. *Group meetings.* Conditions change constantly. Updating and reinforcement of knowledge is important. The most economical way to do this is through group meetings which include common off-the-job safe practices.
5. *Safety awareness.* Most personal safety programs have a lot of common ground. Much education on safe practices, introduction to safety equipment, distribution of educational materials, and sharing of workable ideas can be done through special safety promotions in the form of family safety fairs, displays, educational programs, community service activities, and media such as company newsletters.
6. *Engineering and purchasing controls.* Company technical staff can provide valuable assistance to employees by aiding in the research of safety standards, by making special safety data available in an understandable form, and by identifying sources of personal protective and other safety equipment. Helping employees buy the safest tools and equipment, and teaching them to use directions and label information, promotes more economical as well as safer buying.
7. *Safety analysis and safe practices.* The most effective time to analyze the causes of accidents is before they occur. This also provides early recognition of accident potential so that employees can be made aware of the hazards and possible controls.
8. *Personal protective equipment.* This is one element where the job safety program has a direct carryover effect. If employees are adequately educated in selecting protective equipment, including its fitting, use, and care, they can apply this knowledge to their off-the-job exposures. However, they do need information on where equipment can be bought for home and recreational use.
9. *Inspections.* This element picks up unsafe practices and conditions not foreseen in safety analyses but before losses occur. Involving employees in job safety inspection programs can introduce them to

the value of these techniques. It can also train them to use inspection checklists. The company can distribute checklists appropriate to home, vehicle, and recreational hazards. A comprehensive program should include general, preuse, critical part, and preventive maintenance inspections.
10. *Emergency preparedness.* Through its emergency plans, a company may provide some protection for the community. It can also be a source of emergency assistance in various disasters. The instructions in emergency actions, and appropriate drills, can include emphasis on the importance of a home emergency plan and of making temporary emergency plans when traveling.
11. *Emergency care of the injured and ill.* Studies by the St. Johns Ambulance service in Ontario show that emergency medical training by itself prompts greater awareness of safety. Company programs to provide first aid and cardiopulmonary resuscitation training for employees increase the quality of the job safety program. They also make people more effective in their personal safety programs.
12. *Incident investigation and analysis.* Comprehensive reporting, investigation, and analysis of off-the-job and family injuries can help an organization determine where employees need help with their personal safety programs. One organization found that merely having employees record off-the-job accidents for their own use promptly led to a 60 percent decrease in such accidents. Awareness of the impact on themselves and on their jobs prompts employees to be more conscientious in their personal safety program.

Employee Off-the-Job Safety Training

People need to know safe practices before they can be expected to take actions that will avoid accidents. Off the job, this requires either the motivation for self-study and safety analyses or comprehensive education in safe practices. One way to educate employees would be to tell them about all of life's safe practices, but that method has two flaws. First, there are literally millions of accident possibilities. To cover each of them even once would be almost impossible. Second, the method assumes perfect memory and instant recall of the appropriate information, and no one's memory is that good.

A better way might be to educate people on how to recognize hazards and how to develop effective control measures. The ability to find most of their own problems and to determine workable controls is both more satisfying and more effective. Of course, the amount of time any organization can give to such education will never make any employee achieve a risk-free life. Even the most experienced and well-educated supervisors or safety professionals won't find every hazard, but they will discover the most critical safety problems. The reliability of the trained employee will be higher and his or her lifetime accident experience lower.

The training needed by a family or team safety leader can be divided into three parts. The first part is *basic safety education*: how accidents occur, what causes them, how to recognize hazards, how to control the severity and frequency of accidents, and how to make safety part of everyday life. The second part is *special exposure control*. This includes fire safety, slips and falls, chemical hazards, electrical safety, recreational safety, and defensive driving. The third part is *follow-up training*. This progresses from the basic course to give competence in more specific areas, such as vehicle operation, recreation, or particular hobbies.

A good foundation for personal safety leadership can be provided in one day, although, from an educational perspective, it would be better to give the course in two half-day segments. There is a lot of material to cover, and most employees are not used to sitting in a classroom. A half-day at a time is less stressful both physically and mentally.

Here is a suggested outline for an 8-hour course:

Introduction to and scope of off-the-job accidents
How to identify and control risks
Methods of accident prevention
Applied safety
   Slips and falls
   Overstress and back injuries
   Fire safety
   Electrical safety
   Chemical and health hazards
Safety inspections and analyses
Organizing a personal safety program

Once employees have a true understanding of safety fundamentals, they can start to develop their personal safety programs. As a result, they often find that they need and want more information. The basic course makes them aware of how little they really know about safety; that will motivate many of them to try to learn more. The opportunities for continuing education can be made available through the safety circles. There employees can grow and learn, as well as work on solving specific problems.

BIBLIOGRAPHY

Accident Prevention Manual for Industrial Operations Administration and Programs, 8th edition, Chicago, National Safety Council, 1981, chapter 3.

Home Guide to Fire Safety and Security, 4th edition, Norwood, MA, Factory Mutual Engineering Corporation, 1983.

Johnson, William G., "The Family Approach," in *Directions in Safety,* ed. Ted Ferry and D. A. Weaver, Springfield, IL: Charles C. Thomas, 1976.

Kuhlman, Raymond L., *Effective Off-the-Job Safety,* Loganville, GA, Institute Publishing Division, 1986.

Kuhlman, Raymond L., "The Need for Off-the-Job Safety Programs," *Professional Safety,* March 1986, pp.

"Learn and Live," Pittsburgh, Medical and Health Resources Division, Gulf Oil Corporation, 1980.

"Off-the-Job Safety Training," Wilmington, DE DuPont Company, 1982.

Safety in Your Lifestyle, Toronto, Industrial Accident Prevention Association of Ontario, 1982.

The Off-the-Job Safety Program Manual, Chicago, National Safety Council, 1984.

Chapter 16 / Part 1

Industrial Hygiene

INTRODUCTION

The field of industrial hygiene, like that of safety, is continually changing. It has evolved from one of investigation and response to incidents to a discipline dedicated to preventive program management. For the corporate manager or safety director, this is a welcome development. A 1983 guest editorial in the *Journal of the American Industrial Hygiene Association* illustrated the changes that have taken place in the past few years and set the stage for the role of the industrial hygienist in the future. The editorial, directed to industrial hygienists, stated:

> Over the years, we have tended to measure our progress by the growth of membership in AIHA and by the increased utilization of industrial hygienists in industry, organized labor, and government. However, the real measure of success is how well our profession protects the health of the worker. This effort requires influencing the appropriate levels of management who control the allocation of resources. While professionals who entered the field 25 years ago did not expect to have direct access to upper management, industrial hygienists today routinely advise key management personnel on issues related to employee health.

Generally, the professional training and experience of most industrial hygienists has not prepared them to participate effectively as members of the management team. As a result, hygienists are competing for limited resources and do not fully understand the techniques required for integrating their goals with those of the organization. Hygienists must acquire knowledge in areas outside of industrial hygiene and develop the skills necessary to becoming effective members of the management team.

INDUSTRIAL HYGIENE AND CORPORATE MANAGEMENT

The following are examples of techniques which industrial hygienists might consider in order to interact more effectively with managerial personnel:

1. Learn management jargon. The ability to communicate with senior management in their own language is essential.

2. Learn to identify management's needs and sensitivities in order to bring about needed action most effectively.
3. Improve knowledge and skills in such areas as human relations, management theory, and group dynamics.
4. Become familiar with financial analysis techniques such as cash flow and discounting of money, payback periods, and return-on-investment criteria.
5. Know your organization. What are its objectives? How are resources allocated? How are decisions made? What indicators are used to evaluate managers?
6. Participate as needed in areas such as marketing, sales, engineering, finance, and long-range planning. Besides expanding the hygienist's background, such involvement will help key members of the management team to see the importance of industrial hygiene and the positive impact it can have on productivity and employee relations.

A good understanding of the needs and conditions within the organization, and of effective motivational and management techniques, will create a better forum for the development of a positive, proactive industrial hygiene progress and less reliance on "scare tactics" to give impetus to effective industrial hygiene programs.

Unfortunately, whenever management fails to encourage an expanded role for the industrial hygienist, this helps to maintain the common misperception of hygienists as sample collectors. With that limited perception, a valuable management resource can easily be overlooked. Management should consider the contributions of the industrial hygienist along with those of other safety professionals when allocating resources to protect the worker.

To effect these changes, the hygienists themselves must assume greater responsibility. The closing comments of the editorial writer mentioned earlier emphasized that "the opportunity to become members of the management team is a reality. The challenge is ours."

The Industrial Hygienist and the Safety Professional: A Historical Perspective

In order to place this discipline in perspective, one must look at the evolution of the industrial hygiene field and that of the safety specialist. In the United States, industrial hygiene grew out of the medical profession and the industrial hygienist traditionally reported to the physician. Most schools offering degree programs in industrial hygiene combined their training with that of the physician. Thus a strong bond has long existed in their professional relationships. The field of the safety specialist, on the other hand, had its roots in an entirely different area. The safety specialist usually came up through the ranks of industry and emerged through on-the-job training and practical experience. Thus, the two fields were originally quite separate and distinct, with very little integration.

Fortunately, this situation has changed. Today, both the industrial hygienist and the safety specialist are recognized, well-trained professionals, and both roles have emerged to the point where a better description of both might be "occupational health and safety professional." Both fields, however, still maintain their respective areas of specialization.

As the editorial cited pointed out, the field of industrial hygiene is no longer dominated by the medical profession. Industrial hygienists, as well as safety professionals, play an important role in protecting the health and safety of the worker. The *Safety Management Newsletter* of August 1983 emphasized the importance of interaction between safety and hygiene. It told of an article, appropriately titled "The Marriage That Must Work," that stated, in summing up:

> The real direction of the safety and health interface for the 80s is a search for the most effective system of managing industrial hygiene or occupational health and safety. Senior managers can do much to assure that the two disciplines work together, rather than grow further apart, while seeking to establish their own professionalism. Your most valuable safety and health professional is one who knows both the safety and health sides of the house and can manage a combined effort.
>
> Safety and health are both known as staff functions, but a better approach is to consider the safety and health managers as staff resources, with responsibility for safety and health of workers and the workplace being a line function. To emphasize the point, safety and health of workers and the workplace is a line responsibility of managers and supervisors. They carry out the functions with the help of expert staff or consultants. . . .
>
> It quickly becomes apparent that safety and health are so closely involved that they continually overlap. When they are separated, it is often to protect someone's turf, or because of a misunderstanding of the relationships. Part of this misunderstanding comes from a lack of appropriate background on the part of safety practitioners, and part comes from a tendency of health associated professionals to consider themselves somehow related to medical practice. Without arguing the merits of either view, it is suggested instead that the most effective management of safety and health problems is a coordinated approach.

RESPONSIBILITIES OF HEALTH AND SAFETY PROFESSIONALS

In reviewing the general approach to industrial hygiene, one must keep in mind that the conservation of human resources is a continuing process that involves all levels of management and every employee. The ultimate goal of the occupational health and safety professional is appropriately stated in the preamble to the Occupational and Safety Health Act of 1970: "to assure safe and healthful working conditions" for "every working man and woman in the Nation."

Management's Responsibility

In this context, it is management's responsibility to establish overall policies and guidelines to meet this goal, and it is the responsibility of the safety and health manager to see that these goals are met.

A safety and health department plays an integral part in a business's overall health and safety program. It is responsible for (1) conducting an effective safety program by coordinating educational, engineering, and enforcement activities according to guidelines established by management; (2) providing educational materials; (3) assisting supervisors in teaching safety and health rules and procedures; (4) conducting surveys of potentially hazardous areas to ensure that proper practices and procedures are followed; and (5) recommending changes to keep pace with technological advances.

Supervisor's Responsibility

Similarly, it is each supervisor's responsibility to maintain safe working conditions within his or her department or group and to directly implement the safety program. Program objectives should include (1) maintaining a work environment that ensures the maximum safety for employees; (2) instructing each new employee in the safe performance of his or her job; (3) ensuring that meticulous housekeeping practices are developed and employed at all times; (4) furnishing all employees with proper personal protective equipment and enforcing its use; and (5) informing the safety department of any operation or condition which appears to present a hazard to employees.

Employee Responsibility

Next, we come to the employee or staff member. Every individual is responsible for contributing to the success of an environmental health and safety program. Each person's responsibility includes (1) observing all safety rules and making maximum use of prescribed personal protective equipment and clothing; (2) following the practices and procedures established to conserve health and safety; (3) notifying the supervisor immediately when certain conditions or practices are discovered which may cause personal injury or property damage; and (4) developing and practicing good habits of personal hygiene and housekeeping.

One of the first considerations in a properly oriented occupational health and safety program is creating an awareness of health and safety in each employee through an effective educational and training effort.

STRESS IN THE WORKPLACE

Industrial hygiene focuses on the recognition, evaluation, and control of those environmental stressors—chemical, physical, or biological, and ergonomic—that can cause sickness, impaired health, or significant discomfort to employees or residents of the community.

Stress is caused by the external environmental demands placed upon an individual. If external stressors are increased beyond a worker's tolerance, his or her performance and productivity will drop. Usually, subjective feelings of discomfort will precede actual changes in performance.

Some stressors interact to produce an additive effect greater than that of one stressor alone. For example, studies have shown that neither vibration nor low illumination alone has much effect upon reading ability; however, when these two stressors are combined, reading performance deteriorates.

In other cases, combined stressors counteract each other to produce a smaller effect. For example, studies have shown that lack of sleep combined with a high noise level is compensatory; the effect of noise, in this case, is to increase the individual's level of arousal.

Effect on Productivity

One of the objectives of the industrial hygienist is to determine (1) the level of environmental demands in the workplace and (2) workers' limitations and tolerance for stress. In general, longtime work under extremely low-stress conditions produces lack of alertness and work under high-stress conditions produces fatigue. Both factors play a leading role in causing accidents. The goal then is to determine for each worker an optimal stress level, which will result in increased productivity and performance and in decreased accidents.

In carrying out routine accident prevention activities, the safety professional is frequently required to determine the allowable degree of health hazard in a given industrial operation. In emergency situations and in the absence of an industrial hygienist, it becomes the safety professional's duty to see that proper action is taken to evaluate and contol these hazards.

The variety of substances and processes that present occupational health hazards is steadily increasing. Recently developed raw materials and methods of manufacture, and new combinations of their older counterparts, create new environmental stressors. Improved techniques for the prevention and control of existing hazards and stressors are also being developed.

In some instances, a survey of raw materials, by-products produced intentionally and unintentionally, sources and methods of dispersing airborne contaminants, and exposures to physical agents will indicate the effectiveness of existing control measures. Some complex problems will require a complete, detailed study by a qualified industrial hygiene engineer.

ENVIRONMENTAL FACTORS AND WORK PROCESSES

In order to recognize environmental factors and stressors which influence health, the industrial hygienist must be familiar with work operations and processes. The most important categories of stressors are:

1. Chemical stressors—liquid, dust, fume, mist, vapor, and gas.
2. Physical stressors—electromagnetic and ionizing radiation, noise, vibration, and extremes of temperature and pressure.
3. Biological stressors—insects, mites, molds, yeasts, fungi, bacteria, and viruses.
4. Ergonomic stressors—body position in relation to the task, monotony, boredom, repetitive motion, worry, work pressure, and fatigue.

The industrial hygienist recognizes that such stressors may immediately endanger life and health, accelerate the aging process, or cause significant discomfort and inefficiency.

THE HYGIENIST AS EVALUATOR AND PROTECTOR

Evaluation of the magnitude of environmental factors and stressors arising in or from the workplace is done by the industrial hygienist, aided by quantitative measurement of chemical, physical, biological, or ergonomic stressors. These measurements, when combined with training and experience, help the industrial hygienist to give an expert opinion on the general healthfulness of the environment, either for short periods or for a lifetime of exposure.

Corrective procedures, when necessary to protect health, are based on the industrial hygienist's quantitative data, experience, and knowledge. He or she can prescribe a control method, such as isolation of a work process, substitution of a less harmful material, or another measure.

Functions of the Industrial Hygienist

The functions of an industrial hygienist are to:

1. Direct the industrial hygiene program.
2. Examine the work and external environments by:
   a. Studying work operations and processes to get full details of the nature of the work, materials and equipment used, products and by-products, number of employees, and hours of work.
   b. Measuring the magnitude of the exposure or nuisance to workers and the public. In doing so, it is necessary to:
      (1) Select or devise methods and instruments suitable for such measurements.

(2) Obtain such measurements, either personally or through others under direct supervision.
(3) Study and test materials associated with the work operation.
  c. Studying and testing biological materials, such as blood and urine, by chemical and physical means when such examinations will aid in determining the extent of exposure.
3. Interpret the results of the examination of the work and external environments in terms of the ability to impair health, nature of the impairment, workers' efficiency, and community nuisance and/or damage, and present specific conclusions to appropriate parties such as management and health officials.
4. Determine the need for or effectiveness of control measures and, when necessary, recommend procedures which will be suitable and effective for both the work and external environments.
5. Prepare rules, regulations, standards, and procedures for the healthy conduct of work and the prevention of nuisances in the community.
6. Present expert testimony before courts of law, hearing boards, workers' compensation commissions, regulatory agencies, and legally appointed investigative bodies covering all matters pertaining to industrial hygiene.
7. Prepare appropriate text for labels and precautionary information for materials and products used by workers and the public.
8. Conduct programs for the education of workers and the public in the prevention of occupational disease and community nuisances.
9. Direct epidemiologic studies among workers and companies to discover the presence of occupational disease, and to establish or improve threshold limit values or standards as guides for the maintenance of health and efficiency.
10. Conduct research to advance knowledge of the effects of occupation upon health, and to improve the means of preventing occupational health impairment, community air pollution, noise, nuisances, and related problems.

INDUSTRIAL HYGIENE: SOME HISTORICAL NOTES

Evolution of the Field of Industrial Hygiene and Related Legislation

As early as 300 B.C., Hippocrates recognized certain dangerous practices in specific trades. He was also responsible for making the first record of lead poisoning and for changing medicine from a superstition to a scientific practice. About 500 years later, Pliny the Elder recorded the dangers of working with zinc and sulfur, and even recommended that workers in dust-producing trades wear animal bladders over their faces as respirators to prevent the inhalation of dust. In the second century A.D., Galen developed several theories on anatomy and pathology. He was the first to record the dangers of acid mists to copper miners but gave no solutions, writing simply that certain occupations seemed to be a menace to health.

The Middle Ages

During the Middle Ages, little was done to improve work standards. The only advance made during this time was the provision of assistance to sick guild members and their families. The universities of the twelfth and thirteenth centuries underwrote considerable experimentation. However, the study of occupational hazards was ignored. There were few advances in the field of safety and industrial hygiene until 1473, when Ulrich Ellenbog published a pamphlet on occupational diseases, including his occupational hygiene instructions. During the fifteenth and sixteenth centuries, the hazards of mining were widely written about. Two important authors during this period were Phillipus Paracelsus and Georgias Agricola. Paracelsus wrote a treatise on occupational diseases in mining, describing disturbances of the lungs, stomach, and intestines that resulted from the digging and washing of gold, silver, salt, alum, sulfur, lead, copper, zinc, iron, and mercury. He theorized that these miners' diseases were caused by vapors and that contact with these metals should be avoided, as these diseases were incurable. Paracelsus was also opposed to certain theories and practices then in use, and recommended specific remedies instead of bleeding. Agricola wrote *De Re Metallica,* which told of health dangers to miners. He described the miners' diseases, as did Paracelsus, attributing them to improper ventilation, and went as far as recommending the use of protective masks.

About 100 years later, Bernardo Ramazzini wrote the first complete book dealing with occupational diseases, entitled *De Morbis Artfactum Diatriba.* Ramazzini believed in the welfare of the worker. He visited workshops and factories in all trades, noting the diseases common to each trade, and observed the workers' habits. His book discussed several diseases resulting from the inhalation of dust and fumes, and recommended reforms to alleviate many hazards, but he was ignored for more than a century. Ramazzini was the first to advise physicians to ask a sick man, "Of what trade are you?"

Beginnings of Labor Legislation — Labor Customs

Labor legislation has played an important part in the progress and development of industrial hygiene and safety. Many of England's working customs date back to the fourteenth century. The Statute of Labourers of 1351 required that every able-bodied man or woman who did not own land or was already engaged in other work to accept any job offered. This law was in effect until the turn of the nineteenth century. The laws and customs of

England were inherited by its colonies and were practiced even after they were removed from the books.

One objective of charities in the seventeenth and eighteenth centuries was to found houses of industry in which small children, some less than 5 years old, were trained for apprenticeships with employers. The apprentice system gave rise to the labor legislation which swept through the nineteenth century. One result of this legislation was recognition of the need to control work hazards; this became a basic principle of organized safety programs. A movement to determine the causes of work-related injuries, and then to eliminate them, had finally begun.

Powered Machinery and Child Labor

The rapid development of power-driven machinery and its effects on manufacturing led to the increased use of children in English textile factories. In 1784, an outbreak of fever in the mills of Manchester led to the first governmental safety action. By 1795 Manchester had formed a city board of health, which recommended legislation to reduce work hours and to alleviate improper working conditions in the factories. The first legislative step toward the prevention of injury and the protection of laborers in English factories occurred in 1802 with the passage of the Health and Moral Apprentices Act. This act was designed to end the abuses of the apprentice system, which then involved a large number of pauper children who worked long hours under the worst possible conditions. Inspectors were appointed to enforce the act and to "direct the adoption of such sanitary regulations as they might think proper." The conditions of child workers in England were thus a major factor in prompting safety legislation.

Women in the Early Work Force

Possibly due to the regulation of child labor, a work force of women emerged in the textile mills in the nineteenth century. In 1844, Parliament passed legislation reducing the maximum work time for women to 12 hours a day. And for the first time ever, this legislation contained specific provisions for worker health and safety. Compensation for preventable injuries caused by unguarded mining machinery was also initiated with the passage of the Mines Act of 1842. This act prohibited women, girls, and boys under 10 years of age from working underground; it also called for mine inspections. By 1843, the first mine inspectors had made their report; 2 years later, however, women were still working underground. It became obvious that mere legislation was not the answer; a method of enforcement was necessary. The amount of safety and labor legislation continued to rise; however, so did the accident rate.

Mine Safety Inspections — Start of a Movement

Because of the increasing accident rate in the mining industry, the British government organized a mine safety inspection program in 1850. By 1855, a new mine safety act listed specific areas of safety regulation for the safety inspectors to investigate: ventilation, guarding of unused shafts, proper means of signaling, correct gauges and valves for steam boilers, and indicators and brakes for powered lifting equipment. After several disastrous mining accidents and explosions, the mine act was extended to the Mines Act of 1860. However, investigations revealed evidence of incompetent management, as well as a total disregard of the existing safety laws. Because of these investigations, a new regulation was passed, the Coal Mines Act of 1872. That same year, health and safety laws for the metal mines were codified in England. This was the beginning of a complete industrial code regulating specific dangers to health, life, and limb. This 1872 act extended the safety regulations of previous acts, and included requirements for the employment of competent, certified managers and an increased number of inspections. Other new regulations included the mandatory use of safety lamps and requirements for securing roofs and sides. This act also stated that willful neglect of safety provisions by employers, as well as by miners, would be punishable by imprisonment at hard labor.

Workshop Safety in the 1800s

During the early nineteenth century, factory and workshop legislation continued to develop. In 1833, Parliament passed the Factory Act, establishing mandatory factory inspections and limiting workers' hours. Lord Ashley's "Great Factory Act" of 1844 required the fencing of gears and shafts. This was the first law requiring the guarding of hazardous machinery. Several additional trades were brought under control with the Factory Act of 1864. This act was the first to respond to the results of experiments of medical and sanitary observers by requiring ventilation for the removal of gases, dusts, and fumes originating from machinery. The Workshop Regulation Act of 1867 made the factory acts applicable to all places in Great Britain which made or finished articles or parts for sale.

Work Hours Regulation and Reform

France and Germany also passed laws during the nineteenth century to regulate labor hours and to protect workers in some of the most dangerous trades. In the United States, factory inspections were initiated in Massachusetts by 1867. Two years later, Massachusetts established the first state bureau of labor statistics, which was charged with researching the causes of accidents. As a result of the bureau's research on deaths and injuries from industrial accidents, Massachusetts passed the first state law requiring dangerous moving machinery to be guarded. Research also revealed the part that fatigue plays in causing accidents. This aided in the fight to shorten the work day to 10 hours.

Introduction of Workers' Compensation

The great legislative reform which initiated the organized safety movement in the industrialized world would not have progressed very far without the passage of workers' compensation laws. In 1885, Bismark of Germany, led the way. Although his law covered only sickness, it was the first compulsory compensation act for workers. Earlier, Great Britain had required compensation for deaths or injuries from boiler explosions, but this legislation was designed to protect the passengers of steamboats, not the workers. Workers' compensation laws, more than anything else, made employers appreciate the unrecoverable losses suffered when the services of their finest workers were lost in accidents on the job. Great Britain, Austria, Hungary, France, Italy, and Russia soon followed Germany's example in requiring some form of workers' compensation.

It is often said that the passage of workers' compensation laws in the United States encouraged early efforts in accident prevention. The first compensation law was passed around the beginning of the twentieth century. Obviously, these laws awakened employers to the fact that accidents were costly and so should be prevented, if only to reduce losses. But as pointed out earlier, the idea that little or no safety programs existed before 1900 is false. With the passage of compensation laws and the high price of medical care, most companies had developed well-organized safety programs prior to the passage of the compensation laws. According to David Stewart Beyer's *Industrial Accident Prevention,* as cited in Russell DeReamer's *Modern Safety and Health Technology,* even before 1914, machine guards and safety devices were marketed by safety-product manufacturers. Such products as hard hats, safety goggles, sweep guards, circular saw guards, and safety cans for flammable liquids were also available.

Employee Benefit Association

In addition to developing formal safety programs, several companies had begun paying compensation to employees disabled in work accidents. In 1908, the International Harvester Company initiated such a program, called the Employees Benefit Association. Membership in the association was voluntary; however, a high percentage of the employees participated. This program received mixed reactions from those outside the company. It was viewed as either visionary and liberal or as a revolutionary attempt to head off all legislation designed to benefit the worker.

Common Law Doctrine and Worker Injury Suits

It must be remembered that, at this time, most suits brought by injured employees were based on common law doctrines. These doctrines divided industrial accidents into two groups: those in which no one was at fault and those in which either the injured employee, a fellow employee, or the employer was identifiably at fault. Also, the employer was permitted to use these common law defenses in refuting liability:

1. *Assumption of risk:* the employee knew about the risk at the time of employment and accepted it.
2. *Contributory negligence:* the employee contributed in some way to the accident, relieving the employer from liability.
3. *Fellow servant rule:* the employer could not be held liable if one or more fellow employees was the cause of the accident.

From this, it is easy to see that a court would rarely hold the employer liable, and the possibility of an employee's gaining any form of compensation from the employer was very small. Although court decisions involving common law defenses were proper in principle, they were lacking in social justice.

Workers' Compensation in the United States

Recognition of these inequalities led to the passage of workers' compensation legislation. The first compensation law in the United States to be declared constitutional was passed by the federal government in 1908. Two early state compensation laws—Maryland's in 1902 and New York's in 1910— were declared unconstitutional because of conflicts with the "due process of law" provisions of the Fourteenth Amendment. In 1911, Wisconsin and Washington passed the first state compensation laws that were declared constitutional. Soon afterward, several states followed; by 1928, forty-three states plus Hawaii, Puerto Rico, and Alaska had passed workers' compensation laws.

The workers' compensation laws eliminated any consideration of fault. Under these laws, the employer is held liable if an injury arises "out of and in the course of" employment. This last statement is open to interpretation. Due to liberal court interpretations of that statement, plus rising medical costs and increases in workers' compensation benefits, insurance costs for workers' compensation coverage has risen dramatically. The total cost of workers' compensation insurance premiums paid by U.S. companies rose from $732,000,000 in 1948 to more than $14.6 billion in 1981. These rising costs have had two main effects. Management is becoming more concerned with the effectiveness of safety programs, and safety personnel have been given greater responsibility for proper accident investigation and reporting.

Safety Organizations — Their Beginnings

In 1912, 200 men interested in accident prevention met in Milwaukee. This meeting, titled the First Cooperative Safety Congress, was sponsored by the Association of Iron and Steel Electrical Engineers. The following year, a committee was formed and the National Council of Industrial Safety was organized. The goals of the council were broadened over the years and the name was changed to the National Safety Council (NSC). Today, the NSC is the leading safety organization in the United States. Its interests include traffic, home, farm, school, and industrial safety.

American Society of Safety Engineers (ASSE). In 1911, paralleling these other developments, three insurance company representatives met in New York City to form the United Association of Casualty Inspectors. This was the beginning of the organization that today is known as the American Society of Safety Engineers (ASSE). The ASSE has become an important influence on the careers of safety engineers and has advanced fundamental safety knowledge.

American National Standards Institute (ANSI). The American National Standards Institute (ANSI) was founded in 1918 and provides research for the development and publication of standards, including safety and health standards. ANSI's health standards are advisory; only after adoption by a federal, state, or local health agency do they become mandatory.

Professional Hygiene Associations. In 1938, the National Conference of Governmental Industrial Hygienists began. Its purpose was to provide a forum for the discussion of difficulties and experiences of industrial hygienists. The American Industrial Hygiene Association, also organized in 1938, was established to encourage the exchange of information on different sciences, such as chemistry and toxicology, and their application to industrial hygiene. This organization is the nucleus of industrial hygiene development and is responsible for the recognition of industrial hygiene as a science.

These organizations are mentioned to convey some idea of how professional groups develop in response to a need. There are many other safety organizations and professional societies in the United States, and throughout the world, that provide a forum for the exchange of ideas, through which professionals are able to learn from one another. In Chapter 5, these and a number of other organizations are discussed.

The United States — A Leader in Health and Safety Legislation

Recent legislation has made the United States one of the leaders in industrial hygiene and safety. The three major pieces of legislation to date are the Metal and Nonmetallic Mine Safety Act of 1966, the Federal Coal Mine

Health and Safety Act of 1969, and the Occupational Safety and Health Act (OSHA) of 1970.

The Metal and Nonmetallic Mine Safety Act details mine health and safety standards. This act created the Federal Metal and Nonmetallic Safety Board of Review and provides for mandatory reporting of all mine accidents, injuries, and occupational diseases.

The Federal Coal Mine Health and Safety Act is an attempt to establish the highest possible standards of health protection for the coal miner. This act gives the federal government the power to withdraw miners from any mine it considers unsafe. The act also establishes requirements intended to ensure working conditions free of dust and fumes while the miners are underground. Among these are standards regarding dust control, use of respiratory equipment, proper ventilation, and medical examinations for the miners.

OSHA became law on December 29, 1970, and took effect April 28, 1971. OSHA has been both criticized and praised. When President Nixon signed the bill, he declared that it was one of the most important acts ever passed by Congress. The then Secretary of Labor ranked it with the National Labor Relations Act of 1935, the Social Security Act of 1935, and the Fair Labor Standards Act of 1938 for protection of the rights of the worker. On the other hand, President Ford described it as an example of a well-intentioned regulation with an expensive price tag which has not yet solved our problems. The act has also been called a national example of overregulation.

Late in 1977, the then Secretary of Labor, F. Ray Marshall, and the Assistant Secretary of Labor for Occupational Safety and Health, Dr. Eula Bingham, declared a redirection of the Carter administration's priorities, concentrating resources on the most serious health and safety programs. The administration's new goals included directing 95 percent of the inspections to industries believed to have poor safety records and hazardous operations. Also, educational and consultative services were expanded to help small business organizations. Finally, an attempt was made to eliminate unnecessary safety regulations and to simplify others.

Health Standards Development

Enforceable federal standards have been developed by the ANSI the National Fire Protection Association, and the National Institute for Occupational Safety and Health (NIOSH). NIOSH was established within the Department of Human and Health Services under the provisions of the OSHA Act. NIOSH is located in Atlanta, Georgia, at the Center for Disease Control. This is the principal federal agency engaged in research, education, and training related to occupational safety and health.

The primary functions of NIOSH are to (1) develop and establish recommended occupational safety and health standards; (2) conduct research experiments and demonstrations related to occupational safety and health; and (3) conduct educational programs to provide an adequate supply of qualified personnel to carry out the OSHA Act. NIOSH also provides on-site evaluations of potentially toxic substances, technical information concerning health and/or safety conditions, technical assistance for controlling on-the-job injuries, technical assistance in the areas of engineering and industrial hygiene, and assistance in solving occupational medical and nursing problems at the request of the employer.

Legal Challenges to Health Standards

Three recent court decisions have changed the focus from safety to health issues. These decisions are based on OSHA's cotton dust standard, benzene standard, and lead standard. In the cotton dust decision, the United States Supreme Court upheld OSHA's cotton dust standard and declared that the agency is not required to conduct a cost/benefit analysis in promulgating health standards. The majority opinion of the Court noted that Congress defined the basic relationship between costs and benefits by placing the benefit of the worker's health above all other considerations, including costs. In the benzene standard decision, the Court declared that OSHA must provide a more substantial basis of evidence for the determination of future regulations. This decision marks the first time that courts have turned away from reliance on OSHA findings. Justice Thaddeus Stevens noted that the basic problem with the benzene standard was that OSHA had based its determination on assumptions. The U.S. Supreme Court of Appeals ordered OSHA to reassess the feasibility of the standard. So, from these decisions, cost/benefit analysis is not required in the promulgation of a standard; however, the standard's feasibility must be considered.

CONCLUSION

Safety and industrial hygiene principles have come a long way, especially since the mid-nineteenth century. However, each year, accidents still cause many fatalities and disabling injuries to millions of people. The NSC's statistics for 1982 showed 93,000 deaths and over 9 million accidents. The overall cost was $88.4 billion. Of these, 11,200 fatalities were work related, down about 900 from the previous year, and there were 1,900,000 disabling injuries. As many as 70,000 of the work-related injuries resulted in permanent impairment. The cost of the 1982 work-related incidents came to $31.4 billion. The NSC translates this figure into an annual cost of $320 per worker in production and services.

Obviously, the present concepts are not working completely. Professionals in the safety and health field need to increase their emphasis in two areas: first, in the area of safety awareness on the part of every employee and, second, in a continuing effort to monitor and evaluate the health and safety standards of the general population. Also, newly developed and existing threshold limit standards must be reasonable, feasible, and, above all, medically documented and administered by properly trained and dedicated industrial hygienists.

BIBLIOGRAPHY

Accident Preventional Manual for Industrial Operations, Engineering and Technology, 9th edition, Chicago, National Safety Council, 1988.

Aubeie, James P., "The Marriage That Must Work," *Navy Lifeline,* March-April 1982, pp. 6-7.

DeReamer, Russell, *Modern Safety and Health Technology,* New York, Wiley, 1980.

Grimaldi, John V., and Rollin H. Simonds, *Safety Management,* 4th edition, Homewood, IL, Irwin, 1985.

Insurance Facts, Best's Aggregate and Averages, Rensselaer, NY, Insurance Information Institute.

McCall, Brenda, *Safety First—At Last,* New York: Vantage Press, 1975.

Occupational Safety and Health Reporter, Vol. 10, No. 25, Washington, DC, Bureau of National Affairs, November 20, 1980.

Occupational Safety and Health Reporter, Vol. 11, No. 4, Washington, DC, Bureau of National Affairs, June 25, 1981.

Occupational Safety and Health Reporter, Vol. 11, No. 5, Washington, DC, Bureau of National Affairs, July 2, 1981.

Patty, Frank A., ed., *Industrial Hygiene and Toxicology,* 2nd revised edition, New York, Interscience Publishers, 1963.

Selders, Thomas, *Journal of the American Industrial Hygiene Association,* February 1983.

United States Department of Health, Education and Welfare, Government Printing Office, 1973, *The Industrial Environment—Its Evaluation and Control,* Washington, DC.

Chapter 16 / Part 2

Radiation Safety

INTRODUCTION

The use of radioactive material and ionizing radiation has become widespread throughout industry and commerce. At the same time, the misunderstanding and fear associated with "radiation" have grown. Now more than ever, management must have a basic understanding of the problems to be solved when working with radioactive material and other sources of ionizing radiation. This chapter briefly covers the structure of a basic radiation safety program in handling either radioactive sources or radiation-producing (X-ray) machines. Since some of the words and concepts used throughout this chapter may be unfamiliar to many readers, a glossary of terms has been included at the end. Hopefully, we have included all of the terms that may raise questions in the reader's mind.

The safety procedures and legal requirements for using radioactive material and ionizing radiation have generally evolved along with the growing use of such agents. The radiation safety field in the United States dates back at least to the days of the Manhattan Project during World War II, and the first major compendium of federal law on the subject was developed in 1954. The regulations and safety procedures created in parallel with the development of the radiation industry have produced an excellent safety record throughout the more than 45 years of work with ionizing radiation.

The scientists and other technical professionals who study the properties and effects of ionizing radiation, and who develop methods and procedures for the safe handling of radioactive materials and/or ionizing radiation, are called "health physicists." These scientists have been studying the short- and long-term effects of radiation for many years, and the properties and hazards of radiation in the workplace are at least as well known as those of any other physical or chemical agent. Further, the procedures (and often the regulations) have been continuously updated as more knowledge has been acquired.

Elements of a Radiation Safety Program

A generic radiation safety program can be broken down into four basic elements or components:

1. *An administrative program* to ensure that the use of radioactive material and/or radiation is in compliance with all applicable state and federal laws and regulations.

2. *An operational program* to ensure the safety and health of employees and members of the public who may potentially be exposed to the radioactive material and/or radiation used in the facility.
3. *A record-keeping program* to show compliance with all laws and regulations; to document employee exposures; to document releases to the environment; and to generate a file of evidence for potential defense of liability suits.
4. *An education and training program* to teach employees how to handle radioactive material and/or radiation safely and, in some cases, to inform the public of what the company is doing and the fact that it is operating safely.

The remaining sections of this chapter will outline in greater detail the necessary components of each of these program elements. Obviously, the detailed design of a comprehensive radiation safety program is beyond the scope of this book. In many cases, it requires the services of a professional consultant.

ADMINISTRATIVE PROGRAM

The group responsible for a company's radiation safety program will in most cases be directed by the radiation safety officer (RSO), whose name appears on the license to obtain and use radioactive materials, and who is technically and often administratively responsible for the program. This person will range from a part-time technical consultant (for a small program) to a full-time, technically trained professional. It is important that the RSO not be a part of, or report directly to, the user group within the company. That is, the RSO must be independent of the people who actually use the radioactive material and/or ionizing radiation. This point cannot be emphasized too strongly. In many cases, the RSO has reported to one of the users, and this arrangement has never worked.

In most cases, in addition to the RSO, a radiation safety committee (RSC) must also be established; however, very simple uses of one or two radiation sources will usually not require such a committee. The RSC should have a minimum of three members, excluding the RSO. The members should be chosen from fairly high levels of management and may include some of the direct users of the sources of ionizing radiation. The RSC reviews plans and procedures proposed by the RSO; approves uses of radiation sources within the organization; and provides management support to the RSO in dealing with users.

The basic law governing the use of radioactive material is found in 10 CFR 20, which is administered by the Nuclear Regulatory Commission (NRC). Federal law allows the NRC to enter into agreements with individual states to enable a state to set up and administer its own radiation control program.

Currently, some twenty-six "agreement states" have entered into such an arrangement. Note that the NRC is authorized to control only artificially made radioactive material. All regulations governing the use of radiation-producing machines or naturally radioactive materials (e.g., thorium, radium) are promulgated by the state, regardless of whether or not it is an agreement state.

In order to obtain (purchase) and use radioactive material, an organization must have a license issued by either the NRC or the agreement state. If this license is from the NRC, it will specify compliance with 10 CFR 20; if it is issued by the state, it will specify compliance with that state's appropriate laws. In all cases it will be equal to, or more restrictive then, federal regulations. The license will also include specific requirements dealing with the unique conditions for each particular applicant. The RSO will be named on the license, and he or she will be responsible for making sure that the organization meets all of the requirements established by that license.

The quantity, variety, and complexity of use of radioactive material will determine whether a consulting, part-time, or full-time RSO is required; by the same token, the size and qualifications of the support staff will be determined by the requirements of the particular license. Note, however, that the license never specifies staffing requirements. The conditions of the license must be met, but the licensing agency leaves the method of compliance to the discretion of management.

The NRC and/or agreement states perform periodic inspections and audits (usually annually or biannually) of the licensee's premises and programs. Any violations of the general or specific license conditions are citable. Such citations may include a monetary fine if the violation is considered flagrant or is likely to cause a serious safety hazard. If a large number of very serious violations (e.g., endangering the public near the facility) are found, the license may be suspended or revoked.

The acquisition of radiation-producing machines (e.g., X-ray machines) generally does not require a license. However, particular states may require the registration of such machines, the licensing of machine operators, and annual or biannual inspection programs. Generally, if an organization uses only radiation-producing machines, it will not be required to have a formally designated RSO or a highly organized and complex radiation safety program. Radiation exposure monitoring of workers and periodic calibration and/or surveys of machines will, however, be necessary.

OPERATIONAL PROGRAM

A properly designed radiation safety program ensures that appropriate controls and procedures are in place so that radiation workers will not be exposed to harmful amounts of ionizing radiation from external or internal sources. In addition, the program must ensure that radioactive material is not routinely released to the environment (except for certain "allowed" situations) and that members of the general public are not exposed to sources of ionizing radiation from either machines or radioactive material.

In 1956, the National Commission on Radiological Protection (NCRP) recommended permissible doses of radiation for both radiation workers and members of the general public. They are summarized in the following table.

Occupational Dose Limits

	NO. OF REMS PER CALENDAR QUARTER	NO. OF REMS PER YEAR
Whole Body	1.25	5
Hands and forearms, feet and arms	18.75	75
Skin of whole body	7.50	30

These doses are legally permissible and, based on the best scientific evidence, will produce no observable effects over a worker's 50-year working life. Nonetheless the basic philosophy of any radiation safety program is to keep all doses "as low as reasonably achievable" (ALARA). In recent years, the NRC has required all radiation safety programs licensed by either the NRC or an agreement state to contain an approved ALARA program. Note, however, that ALARA is not a number, and what may be reasonably achievable for one licensee may not be reasonable for another.

External Exposure Control

Whether the source of ionizing radiation is a machine, a large sealed source (e.g., from industrial radiography), or an unsealed source of radioactive material (e.g., those used in research), the initial consideration in designing a radiation safety program is to minimize the probability of an employee's being exposed to harmful doses of radiation through routine work or by accident.

Planned exposures, for either repetitive, routine, or occasional maintenance operations, may be controlled by using one or more of the following three procedures:

Time: Limit the time that a worker is exposed to the source of radioactivity. This technique is probably most useful for nonroutine operations where the work steps can be carefully preplanned and timed and/or where several workers do various parts of the job in order to keep all personnel doses in the permissible range. In addition, when working with radioactive sources (particularly unsealed sources), it is good practice to run through the procedure at least once, without the presence of radioactive material, to reduce workers' exposure time and increase their knowledge of how to do the job.

Distance: Increasing the distance from a radioactive source reduces the exposure rate by the square of the ratio of the two distances (i.e., doubling the distance reduces the exposure rate by a factor of 4, tripling the distance reduces the exposure rate by a factor of 9, etc.). This rapid decrease in dose rate with distance is the reason why remote-handling tools (even relatively short tongs) are so effective in reducing the dose delivered to workers.

Shielding: Any material interspersed between the radiation source and a person will absorb part of the radiation and reduce the dose to that person. For most types of ionizing radiation used in industry (except neutrons), the denser the material, the better the shield. Radiation shields may consist of fixed lead or concrete walls, or of lead-filled steel casks for holding a source, or of portable shields that can be positioned as needed.

Accidental exposures to radioactive sources are generally prevented by access controls of some type. Simple locks with controlled access to the key(s), or electrical and/or mechanical interlocks, are often used to restrict access to high-radiation areas. For example, radiation detectors interlocked with access doors are frequently used to prevent accidental entry into X-ray machine rooms, accelerator vaults, or radioisotope irradiation facilities. High-radiation alarms are also used to warn personnel of a dangerous condition.

Internal Exposure Control

One of the more important radiation safety problems involves work using unsealed (i.e., open) sources of radioactive material that can penetrate the worker's body. This is a serious situation, since once radioactive material is deposited in the human body, it cannot easily be removed or "flushed out" by medical treatment. The material stays in the body and continues to irradiate the tissues until it either decays or is removed by normal metabolic processes.

Foreign materials can enter the body by inhalation, ingestion, absorption (through the skin), or through a wound (e.g., a puncture). Therefore, when working with an unsealed source of radioactive material, it is necessary to evaluate the potential for each of these possible pathways into the body and take appropriate steps to minimize that potential. Detailed planning for such eventualities may require the services of a professional health physicist. It is useful to consider the concept of "contamination control" at this point.

"Radioactive contamination" refers to radioactive material located in some undesired or unexpected place. Such contamination can generally be easily cleaned up if it is known to exist. The problem arises when radioactive material is unknowingly transferred from its correct location to some other location where it is not supposed to be. In this case, the potential for ingestion or inhalation of the material, either by a worker or by a member of the public, rises sharply. Hence, very careful contamination control measures are needed to ensure that the unsealed radioactive material is not transferred to an unexpected location.

Environmental Discharges

In general, a radiation safety program governing the use of unsealed radioactive material should try to prevent the discharge of such material to the environment (sewer, air, ground, etc.). Even though the regulations do allow the discharge of radioactive material into the atmosphere and/or sewer lines under carefully controlled and monitored conditions, it is only prudent to avoid such methods. This is particularly true in today's political climate, where any discharge of a pollutant to the atmosphere, even if legal, may not be worth the adverse public reaction.

Monitoring

The program areas discussed so far require extensive and detailed monitoring to document that exposures are within permissible limits and meet whatever ALARA guidelines the company (licensee) has adopted.

External dose monitoring is done by using an integrating dosimeter, such as a film badge or thermoluminescent dosimeter. These dosimeters are changed, usually monthly or quarterly, depending on the perceived potential for receiving doses that approach the permissible limit. Personnel radiation dosimeters are available from certified commercial vendors. The dosimeters are all about the same, but the type of record reporting varies greatly—from typewritten quarterly reports to in-house video display terminals connected to the vendor's mainframe computer.

It is important for management to ascertain that all employees assigned a personnel dosimeter wear it at all times while on the job and take good care of the device. The records generated by the personnel dosimetry system are primary in the defense of any liability claim alleging radiation injury. If such records are not available or are incomplete, defense of a liability claim becomes much more difficult, partly because lack of a record reflects poorly on management's commitment to safety.

Internal dosimetry involves much more difficult and complex measurements than external dosimetry. Various bioassay techniques (urinalysis, thyroid count, whole body count, etc.) are used, as appropriate, to obtain data on the amount of radioactive material deposited in a person's body. The description of such a program is beyond the scope of this chapter, except to note that complete, accurate records are again of paramount importance.

Finally, if there is any potential for the release of radioactive material to the environment, it is essential to keep records documenting measurements that substantiate what was or was not released.

In all monitoring programs, results showing a zero reading are just as important as positive results and should be recorded.

Pregnancy and Radiation

A special problem occurs if there are potentially pregnant women among the employees exposed to ionizing radiation. It must be emphasized that the problem does not affect all fertile women, but only those who have decided to bear a child. The concern arises because the human fetus is most sensitive to radiation during the first trimester of growth, particularly during the first 4 to 6 weeks of life—precisely the time when the woman may not know that she is pregnant. The generally accepted safe dose to the fetus is a total of 0.5 rem (similar to the annual permissible dose to a member of the general population). Therefore, if a pregnant or potentially pregnant woman limits her monthly dose to 0.05 rem, her fetus can receive no more than 0.450 rem (0.05 rem $\times$ 9 months).

To achieve this dose limitation, a woman may want to restrict her exposure to ionizing radiation or transfer to another job while she is pregnant or may become pregnant. It is the responsibility of the employer to ensure that all women working with ionizing radiation sources are given this information. It is the woman's responsibility to make the decision about her job exposure after she has been properly informed.

Radioactive Waste Disposal

At this time, the legal disposal of low-level radioactive waste is, a politically volatile issue. It is not technically difficult to dispose of such waste safely; the problem is almost entirely one of public perception and politics. Currently, there are three disposal sites in the United States (Richland, Washington; Barnwell, South Carolina; and Beatty, Nevada). All low-level radioactive waste must go to one of these three sites; it must contain no free liquid; it must be packaged in strong, watertight containers (usually 55-gallon drums); and each container must be appropriately labeled and described on the shipping manifest.

The cost of low-level waste disposal is high (approximately $35 per cubic foot currently) and will undoubtedly increase rapidly in the next 5 to 6 years (before new sites come on line). Therefore, any firm contemplating the generation of such waste is advised to carefully consider methods to minimize waste generation and to reduce the volume of waste after it is generated (e.g., by compaction, incineration, or radioactive decay).

Here again, the maintenance of accurate and complete records is very important. Improperly labeled or manifested shipments can be grounds for the refusal of a site to accept further shipments.

RECORD-KEEPING PROGRAM

Any use of sources of ionizing radiation (machines or radioactive material) requires that many detailed records be generated and preserved for long periods of time. For example, each required record of a dose received by individuals, and of medical examinations and bioassays, must be preserved indefinitely (or until the licensing agency authorizes disposal). These record requirements arise from at least two sources: (1) licensing agency regulations and (2) as a defense against liability suits of employees or the public.

The records required by regulations generally concern activities that generate real or potential radiation exposures to individuals. The following six categories are examples of the types of records generally required by law:

1. Radiation exposure of all employees for whom personnel monitoring is required. In this case, personnel monitoring includes external monitors such as film badges and bioassay tests (urinalysis, thyroid count, etc.) used to assess internal exposure.
2. Measurements of radioactive material in effluent discharges to the atmosphere or to water (sewer system). Such discharges generate potential exposure to the public; hence, records are required.
3. All disposal of radioactive waste.
4. Each receipt, transfer, and disposal of a source of radiation (including machines).
5. Calibration records of all instruments used to measure or survey radiation fields or amounts of radioactive material (e.g., in effluent discharges).
6. A running inventory of radioactive sources on hand, along with leak test records for any sealed sources of radioactive material.

This list is not intended to be all-inclusive, but it does provide a sense of the types of records required. The following two record categories are examples of data which should be kept to defend against potential liability suits:

1. Records detailing the type of training each employee has received; the dates; the name of the instructor; and so on.
2. Work area survey records with measurements of area radiation fields, use of protective equipment (gloves, anticontamination clothing, fume hoods, etc.), and results of contamination surveys.

In view of the increasingly litigious nature of society, record keeping has become an essential part of any safety program. The various regulatory agencies (e.g., NRC, EPA, OSHA) are assessing large monetary fines at a rapidly increasing rate for lack of proper records. It is almost a truism that "any action or measurement not properly recorded has not been done."

EDUCATION AND TRAINING PROGRAMS

In the preceding sections of this chapter, the basic components of a radiation safety program were briefly described. However, unless the employees and often the neighboring public are adequately informed of what is being done to ensure their safety, much of what is done will be relatively ineffective.

Liability lawsuits today often start on the basis of someone's perception of what has or has not been done to ensure that people are protected from the harmful effects of some material, agent, or product. In order to reduce the potential for such suits, and to put the risks of ionizing radiation exposure in proper perspective relative to other everyday risks, all employees (actual users and nonradiation workers) should be given some education about the properties of ionizing radiation.

Licensing agencies require all users to undergo training at a level appropriate to their use of or exposure to ionizing radiation. Such training may be as simple as a 1- or 2-hour orientation lecture or as extensive as a 2- or 3-day course. It is also a good idea to require a brief quiz at the end of any training session, and it is always necessary to keep, as a permanent part of each employee's personnel record, information pertaining to the date, length, and subject matter of each training session.

Minimum information to be given to employees who work near sources of ionizing radiation (but do not actually use such sources) should include:

1. Properties of radiation.
2. Methods of controlling exposure (i.e., time, distance, and shielding).
3. Permissible doses.
4. Dose rate conditions likely to be found at the particular facility.
5. Biological effects of radiation.
6. Relative risks of radiation exposure and relation to natural background radiation.

The people who actually work with radiation sources must receive all of the information listed above, as well as the following:

1. Principles of operations and use of survey equipment.
2. Use of personnel dosimeters.
3. Control of contamination from unsealed sources (if appropriate to the specific facility).
4. Field use of radiation sources (if the business involves field radiographic operations).
5. Review of all applicable laws and regulations pertaining to the facility's license to use ionizing radiation.

Users should also receive refresher training on a periodic basis. Annual retraining is usually considered adequate.

It may be necessary to prepare an environmental impact statement (EIS) if the facility in question will discharge any radioactivity to the environment (air or water); this may involve public hearings. The process gives management an opportunity to educate the public about the programs and procedures used to ensure public safety in the vicinity of the facility. Even if public

460 SAFETY AND HEALTH MANAGEMENT

hearings are not required, a public information program may be desirable to dispel fears which generally arise from lack of information and/or misunderstandings about the hazards of radiation sources.

In summarizing this section, it is appropriate to reiterate the essential need for education and training of radiation workers and fellow employees. While such training costs money in terms of time away from productive work, there is no question that it is necessary even if the law did not require it. Training provides workers with the information they need to assess the real risks of their job intelligently and gives them a basis for making informed and safe decisions.

CONCLUSION

This chapter has presented a brief overview of the basic components of any safety program designed for persons who handle radiation sources. More detailed information can be obtained from the NRC offices in Washington, DC 20555 or from individual state health departments. In practice, every radiation safety program is designed to meet the particular needs of the organization. Except for the simplest cases, this requires the services of radiation safety experts. Often such people can be located through the local chapter of the Health Physics Society or sometimes through the offices of county or state health departments (radiological safety branches).

GLOSSARY OF TERMS

Agreement State: A state which has entered into an agreement or contract with the federal government to administer and control the use of sources of radioactive material within that state.

Anti-C Clothing: Protective clothing designed to prevent the body from coming into direct contact with radioactive material (gloves, coveralls, lab coats, shoe covers, head covers, etc.).

Becquerel: See *Curie.*

Bioassay: A measure of radioactive doses resulting from internal contamination.

Contamination: Radioactive material located where it is not wanted (particularly on the outside of containers, on work surfaces, floors, walls, tools, or human skin).

Curie: The historical unit of radioactivity. One curie is the quantity of radioactive material that produces 3.7×10 disintegrations per second. Note: the new unit of radioactivity is the becquerel, One becquerel is equal to one disintegration per second.

Dose: A quantity of radiation energy absorbed per unit mass. The unit of biological dose is the rem.

Dose Rate: The intensity of a radiation field in terms of the rate at which energy is delivered to an absorbing body. The unit of biological dose rate is rems per hour or mrems per hour (see *Millirem*).

Dosimeter: An instrument which measures the accumulated energy to which one may be exposed (i.e., noise, radiation, etc.).

Erg: The unit of work needed to impart to a mass of 1 gram an acceleration of 1 centimeter per second.

Film Badge: A personnel monitoring device consisting of a small packet of photographic film held in a holder suitable for wearing on the body. The film records exposure to radiation and hence measures the dose to the body.

Fume Hood: A cabinet open on one side through which air is drawn and exhausted to the outside, to provide an environment for capturing and diluting small amounts of radioactive material released into the air during certain manipulations of such material.

Ionizing Radiation: Radiant energy that is capable of removing an electron from an atom. Ionizing radiation includes X-rays, gamma rays, alpha and beta particles, and certain other nuclear particles, but does not include radio waves, light, infrared, or ultraviolet light.

Millirem (Mrem): A subunit of dose equal to 1/1000th of 1 rem.

Natural Background: Radiation which originates in space (called "cosmic radiation"), plus the radiation that comes from natural radioactive materials such as thorium and uranium. The earth has always had this naturally occurring radiation background.

Occupational Dose: A dose of radiation received by any individual in the course of employment, education, training, or other activities which involve exposure to radiation (excluding medical or dental diagnostic or therapeutic exposures).

Personnel Monitoring Equipment: Device(s) worn or carried by an individual to measure the radiation dose received (e.g., film badge or thermoluminescent dosimeter).

Rad: A dose corresponding to the absorption of 100 ergs of energy per gram of material for any radiation.

Radiation (as used in this chapter): Shorthand for "ionizing radiation."

Radiation Machine: Any device capable of producing ionizing radiation.

Radioactive Material: Any material which spontaneously emits ionizing radiation.

Rem: The unit or dose of any radiation to body tissue in terms of its estimated biological effects relative to a dose of 1 rad of X-rays or gamma rays. (This accounts for the fact that certain types of radiation can produce biological damage more effectively than other types.)

Sealed Source: Any radioactive material permanently encapsulated.

Survey: An evaluation of the radiation hazards incident to the production, use, release, disposal, or presence of sources of radiation.

Thermoluminescent Dosimeter: A personnel monitoring device consisting of a solid-state crystal held in a holder suitable for wearing. The crystal is sensitive to radiation and hence measures the dose to the body.

User: Any person who is licensed to possess and/or use radioactive material or other registered sources of radiation.

Chapter 17 / Part 1

Emergency and Disaster Planning

Emergencies, by their very nature, are the unexpected, unpredictable events that disrupt the best plans and operations. The senior manager, however well organized and farsighted, can do nothing to prevent a freak event, a natural disaster, or any of the dozens of emergency situations that are beyond human control. No one, especially an executive who is used to directing and controlling activities, likes to think that he or she is helpless to prevent something, especially when that event can be catastrophic. A single disaster, after all, can destroy an entire organization, cost millions of dollars in damage and human misery, and take years to recover from.

If we discount Noah's flood, the worst disaster on record seems to have been a flood along China's Hwang Ho River in 1887, in which over 900,000 lives were lost. As the worst disaster on record in the United States, the Galveston, Texas, tidal wave of 1900 claimed some 6000 lives. These are extreme examples, of course; few disasters or emergencies are this costly in terms of life. But a single earthquake can put an entire factory out of operation, and even a small flood can cost a company millions of dollars in damage.

Obviously, no one can prevent a flood, an earthquake, or a windstorm. However, there are effective countermeasures and preventive plans that can keep such events from doing unlimited damage or prevent a bad situation from being worse. Here the manager does have some degree of control. This chapter focuses on such control measures—the preparedness and planning that can be activated immediately in an emergency and lessen the impact on the organization.

Many emergency preplanning measures can keep an event from becoming a disaster in the first place. For example, immediate evacuation of an area filling with toxic vapors may keep workers from being killed or injured. The quick, efficient action of an in-house fire-fighting team may keep a fire from getting out of hand and igniting explosive or toxic chemicals. Such is the nature and intent of emergency and disaster planning.

A MATTER OF SURVIVAL

Preventing disasters is obviously a prime concern, but it is only half of the picture. Of nearly equal importance to the manager is making certain that the firm's business can be conducted, resumed, or directed into new channels after an emergency or disaster. The precautions that ensure survival must be brought together in an emergency/disaster plan.

Naturally, if such a plan attempts to deal with all possible emergencies and their ramifications, it will be incredibly complex. Just how complex it can become is shown in Figure 17-1. This figure lists four areas that must be considered in any thorough emergency/disaster plan:

1. Types of emergencies or disasters.
2. Necessary steps in developing a plan.
3. Elements contained in a complete plan.
4. Functionaries, agencies, and individuals involved.

TYPE OF EMERGENCY	PLANNING STEPS	PLAN ELEMENTS	RESPONSIBLE PARTIES
Fire	Assess Probability	Policy	Executive Board
Explosion	Assess Hazards	Purpose	Chief Officer
Flood	Designate Coordinator	Authority	Coordinator
Hurricane	Develop Plan	Control Center	Function Committee
Tornado	Approve Plan	Charts	Unions
Earthquake	Train Personnel	Organization	Neighbors
Civil Strife	Test Plan	Fire Equipment	Fire Agencies
Sabotage	Revise Plan	Functional	Security
Strike	Coordinate	Emergency Desc.	Safety
Accident		Maps	Medical
Spills		Plan	City
Radiation		Emergency Equipment	County
Weather		Medical Equipment	State
LPG		Shelters	Federal
Energy		Evacuation	Red Cross
Kidnapping		Communications	Mutual Aid
		Shutdown	Utility Companies
		Visitor Control	
		Security/Guards	
		Repair/Restore	
		Transportation	
		Training	
		Personnel Assignment	
		Employee Welfare	
		Lighting	
		Alarms	
		Protection Equipment	
		Coordination	
		Vital Records	

Figure 17-1. Emergency/disaster plan considerations.

All four of these elements must be thoroughly considered, fully incorporated in the plan, and coordinated with each other. Obviously, this is a large task for management to undertake.

There is a bright spot or two, however. Not every organization needs to plan for every conceivable emergency. For example:

1. In a very small plant with few resources, it may be impractical to plan for every emergency.
2. A company located on high ground or in a desert area may not have to consider the possibility of floods.
3. In some parts of the country, the possibility of a hurricane or blizzard is so remote that it need not be planned for.
4. In some cases, the location of a facility may be such that neighbors, government facilities, or a parent corporation may share the expense of emergency preparation—a concern it might have to face if isolated or alone.
5. For most emergencies, model plans and precedents already exist (almost every organization has a fire plan, for example). These may only need to be adapted to a particular situation or incorporated in a larger plan, greatly easing the planning task.

On the other hand, we don't want to paint an overly optimistic picture. A small emergency, or a small omission in a disaster plan, can rapidly lead to other, more catastrophic events. One of the major aims of a disaster plan is to contain the effects of an event. This is generally termed the "ameliorative function" of a disaster plan.

TYPES OF EMERGENCIES AND DISASTERS

Ordinary, routine activities may have the potential to cause great losses. However, it is the unexpected events, or the ones that are beyond our normal capacity to handle, that interest us here. A "disaster" has been defined as a great, sudden misfortune resulting in a loss of life, serious injury, or property damage. Strictly speaking, such a misfortune can occur even on an individual level. Having your house burn to the ground is a personal emergency or disaster. In current usage, however, the term generally refers to a sudden occurrence that kills or injures a relatively large number of people. One prominent listing of disasters includes only those occurrences that involve the loss of twenty-five or more lives. The NSC's *Accident Facts* compiles only those disasters that involve 100 or more fatalities. The issue is further confused by the use of the term "catastrophe." The Metropolitan Life Insurance Company says that a catastrophe is an accident that claims five or more lives.

466 SAFETY AND HEALTH MANAGEMENT

Looking at these various definitions, it may seem that one person's misfortune is another's disaster—or catastrophe. Whatever the definition, however, planning is needed to ensure survival and to provide amelioration of the consequences. Figure 17-2 shows the types of emergencies that might be considered and the major steps involved in developing a plan to meet them. A discussion of these various emergencies, and the steps involved in dealing with them, form the bulk of this chapter. Neither listing is all-inclusive, but it provides a simple, proven approach to the general problem. We will begin with a look at some of the emergencies that an organization may face.

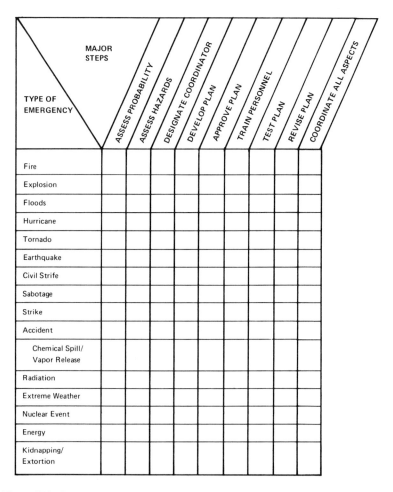

Figure 17-2. Types of emergencies and major steps in developing a plan to meet them.

Fire

Fire is probably the most common emergency or disaster situation, but that does not make it the most harmless. The manager cannot assume that local fire services will be able to assist with every fire emergency. For example, in a widespread disaster, fire organizations may not be able to respond because of disrupted traffic, lack of equipment, or commitments with a higher priority. Alternatively, a responding unit may need the assistance of a plant fire brigade, or the plant fire brigade may be the only resource available at all.

Having automatic fire protection systems, such as sprinklers or fire doors, does not remove the need for a plant fire brigade. The automatic devices may fail or may not protect the particular area involved, or a fire may simply be beyond their capacity to control. Every organization should have a fire brigade to deal with such emergencies. The size of the brigade, the kind of training and equipment it needs, and its organization will vary from one company to another. In small companies, the senior manager may handle the details of the fire effort personally. More often, someone with the responsibility reports directly to him. If a single person evaluates and manages all risks, that person should be considered the plant fire chief. At larger plants, fire protection may be so complex that it is regarded as a separate function or department. This is particularly true for operations that involve an unusual risk of fire or explosion (petrochemical plants, those that use highly inflammable or explosive mixtures, and so on).

When a fire starts, time is of the essence. Plans should provide for quick action by trained in-house groups, with immediate and automatic notification of larger resource units. The point is that small fires should be kept small and handled as soon as possible. The first few minutes are especially critical. No matter how small the fire is, warning and evacuation should begin immediately. Waiting until the fire is out of hand could be disastrous. Quick in-house action is needed to keep small fires from becoming major disasters or catastrophes.

Since a fire can start in almost any location, the fire plan should be thoroughly tested for all areas. In the larger company, tests can be arranged by the fire chief. In the smaller company, someone who has had volunteer fire company experience or other training should be in charge of the testing. The smaller the plant, the more likely it is that a senior executive will be in charge of the fire program. In plants that do not have a fire chief or manager, the services of an insurance company or consultant should be used for planning and testing.

The makeup of an emergency fire organization will depend on the particular situation: the size and fire hazards of the property, the type of fire protective equipment on hand, and the availability of public fire protection.

In a typical, automatically protected plant, each operating shift should have:

1. A person in charge.
2. A team or fire squad for each operating department.
3. A sprinkler control person.

Other assignments should be made depending on the size, type, and location of the plant.

Explosion

Explosions are usually accompanied by other emergencies. In some cases, the explosion is the result of a fire or other unexpected event. In others, the explosion itself causes a fire and other attendant disasters. The onset is sudden, and there is seldom any advance warning. Therefore, the need for fire-fighting and other emergency action is urgent, requiring a faster reaction than normal. The suddenness of an explosion should never be forgotten in any aspect of disaster planning.

Flood

For planning purposes, floods should be divided into two categories: flash and normal. Like explosions, flash floods are sudden and seldom give advance warning; like explosions, they require a faster than usual action. Normal floods generally build more slowly, threatening for days before they actually occur. There is usually enough time to take protective measures.

Where floods are frequent, permanent protective measures are often possible. For example, an operation located in a probable flood area might be protected by dikes. Floods, too, can be accompanied by other disasters, including windstorm and fire. A complete flood plan should include measures for dealing with these attendant disasters as well. As an example, the fire portion of the flood plan should include:

Moving vehicles and tank cars containing hazardous materials to high ground, particularly if the materials interact with water.
Keeping automatic sprinklers in working condition, since fire may be a side effect and the premises may have been vacated due to the flood threat.
Shutting off gas supplies at the main control valve, since flood waters can break distribution piping.
Filling vertical tanks of flammables with water to prevent their floating away (if there is no adverse reaction with water) or otherwise anchoring them securely.
Closing discharge valves of flammable containers.
Extinguishing ignition sources such as boiler fires.

Hurricane

Hurricanes are generally confined to certain geographical regions, and those regions are usually alerted to a hurricane's possible occurrence well in advance. Well-developed meterological tracking and warning systems go into action as much as several days in advance, and so should the organization's emergency plan.

When the danger reaches a certain point, the operation can close down in an orderly manner and precautions can be taken to minimize the expected wind and water damage. Some measures cannot be taken on short notice. For example, battening down openings, putting shutters in place, or anchoring tall chimneys and signs may take a lot of time and should be undertaken as soon as possible.

Hurricane emergency plans should also take into account the fact that state and local emergency services may be overtaxed and unable to respond to a call. The organization should be prepared to handle attendant damage or emergencies with its own resources.

Tornado

Most parts of the country also have tornado tracking systems that allow at least some time to prepare. While the specific path of a tornado is highly erratic and unpredictable, tornado warnings are generally broadcast to the entire area of possible threat. This means that there is usually time to make preparations before a tornado strikes. Even if a tornado does not touch down in a given area, there may be strong winds that cause a broad path of damage, interruption of electrical and other power sources, unavailability of overtaxed local emergency services, or other attendant emergencies.

Tornado plans should include:

Procedures for getting personnel to a safe location immediately if advance warning is given.
Procedures for personnel to take for their own protection if only a few seconds' warning is given.
Trained emergency crews who can handle downed lines, broken pipes or mains, and can safely work without injury to themselves in hazardous conditions. Plans should be made for food and rest for these crews, who may need to work for long hours under stressful conditions.

Earthquake

Earthquakes happen so quickly and so unpredictably that there is seldom time for ameliorative action, unless it is has been done well ahead of time. In an industrial area, earthquakes can release combustibles, break fire cutoff devices, and ignite fires or explosions. Unless sprinkler systems have flexible couplings, they may break away and become useless. The fire hazards of an earthquake are often aggravated, since water supplies (particularly municipal ones) are frequently casualties of the quake.

Fire is only one of the principal types of damage attendant on an earthquake. The collapse of walls, buildings, and other structures is also a major hazard. Fortunately, we now know more about construction techniques, which enables us to design buildings better able to withstand earthquake shocks. This type of construction is only a recent standard in earthquake country, however. Thus, many people (particularly those occupying older, prestandard buildings) are still faced with a survival problem.

Employees should clearly understand that the safest place during an earthquake is *within the work area under preselected cover.* Every employee should know exactly where he or she will seek immediate cover in case of a quake. Reinforced door frames or sturdy metal work tables may give protection against falling objects.

In planning for an earthquake, survival is the main consideration. After a severe earthquake, most personnel are anxious to go to the aid of their families. However, after a general disaster, roads, bridges, communications, and public transportation systems are usually out of order. Employees may be exposed to additional hazards if they attempt to leave the work area, and the organization should try to protect them as much as possible. Employees running to get to parked cars may be exposed to such hazards as downed power lines and falling debris. If a stampede or traffic tie-up occurs, still others may be injured. Where it is impossible for employees to leave the plant area due to transportation and communication failures, their welfare becomes the organization's concern. Food and shelter, as well as first aid treatment, may have to be provided during the continuing state of emergency.

Civil Strife

No area of the world is immune to riots and civil disturbances. In recent years, they have been included in emergency and disaster planning. This type of emergency/disaster, after all, can be as costly and damaging as any other disturbance and should receive full consideration in planning.

Planning must take into account the general social setting and the probability of false reports and rumors. Personnel should be urged to avoid areas where disturbances are taking place, leaving them to the police and other agencies to handle. Wherever possible, the organization should seek to isolate its employees during such a situation.

If a plant or its premises are invaded by outsiders, the protection of grounds, facilities, and personnel is important. However, a major question for any company is "How much right do you have to protect your personnel and property?" Consideration must also be given to protecting invaders against the use of excessive or unnecessary force, which could have disastrous repercussions later—repercussions often more costly than the original incident. The company's attorney should be involved in the planning phase for this emergency, as should the company's security organization.

Sabotage and Terrorism

The public perceives terrorism and sabotage as much the same thing. Sabotage is more intent on destroying resources, while terrorism aims to intimidate, demoralize, and subjugate. Both may achieve the same destructive results by similar techniques.

For planning, the rational defense against persons bent on sabotage or terrorism is to keep them off the premises in the first place. Since a saboteur or terrorist may belong to the organization, internal security measures are needed to reduce the risk. Where the action involves danger to life or property, such as a bomb threat, emergency plans must be able to go into immediate action. Generally, these plans use detection rather than disposal of an explosive device. The plans definitely should provide for orderly evacuation. Removing or disarming of explosive devices should be left to the government agencies that are trained and equipped to handle the operation safely. In training personnel to handle these situations, they must understand not only what to do but also what not to do.

Terrorism and sabotage, at least in an overt form, are more likely to be encountered in offshore operations, but the threat of activities within the United States must not be ignored. Sabotage by disgruntled company employees is a consistent, if not life-threatening, activity and requires constant attention.

Strikes

Strikes may or may not have emergency or disaster aspects; they may or may not involve sabotage, threats to property or personnel, and disruption of operations. Even when a strike is carried out in a peaceful, orderly fashion, however, the company should be ready to implement emergency measures at a moment's notice. Should some other emergency occur while a strike is in progress, whether a fire or an earthquake, advance preparation is doubly necessary.

Accident

Preplanning is a major feature of effective accident prevention and investigation. Much of the organization's safety and health effort will naturally be concentrated on accident prevention measures as a matter of course, and investigative procedures will already be established. These will also play a part in emergency and disaster situations, as well as in more routine or ordinary events. However, there are some special factors to take into account when an accident falls into the disaster or catastrophe category.

First, one of the urgent problems is that rumors explode in chain reaction fashion when the facts are not immediately known, especially if the accident involved many people or the loss of life. The neighborhood may be as much affected as company personnel. For example, an accident at a plant that uses hazardous chemicals may be blown out of proportion in the neighborhood, causing panic. In the larger community, people may believe that a company product was at fault, and consumers may react strongly. Reduced sales and canceled orders may seem to be a slow reaction, but they can be as costly as an in-plant explosion. Quick, effective communications can minimize both in-plant and community reactions. A public information program should be an integral part of the emergency plan, and responsible management should incorporate an employee communications plan as well.

Chemical Spill/Vapor Release

It is estimated that over 250,000 hazardous materials shipments move through the United States each day. This amount is expected to double within the next decade. In addition, we can only guess at the millions of in-plant handlings of hazardous materials each day. Reported accidental releases of hazardous materials numbered about 16,000 in 1979, but most of them were in the public eye. We cannot estimate how many more of these accidental releases were never reported and never came to the public's attention. Only a few of the 1800 regulated commodities were responsible for accidents that involved casualties. Those most often mentioned were (1) flammable liquids, (2) pressurized gases, and (3) corrosive liquids. These involved nearly 80 percent of the fatalities.

While liquefied petroleum gas (LPG) disasters are rare (at least where large amounts are involved), those events have been so devastating that every effort has been made to keep them from happening. They account for about 21 percent of the fatalities in this classification—a disproportionately high percentage. Hundreds of fatalities, complete devastation of property, and rapid spread to neighboring facilities are common in LPG incidents.

When a chemical is released or spilled, time is of the essence. Immediate information on the chemical involved is essential so that appropriate actions can be taken. There is now a 24-hour toll-free telephone emergency information service available through the Chemical Transportation Center (CHEMTREC) in Washington, D.C. They also publish emergency guides, a sample page of which is shown as Figure 17-3. The telephone service and the emergency guide have one object: to give quick information on first aid, fire control, spill or leakage cleanup, and other pertinent facts.

Further information on the toxicity of chemicals can be obtained from a

Hydrogen Chloride

(Nonflammable Gas, Corrosive)

Potential Hazards

Fire: —Cannot catch fire.

Explosion: —Container may explode due to heat of fire.

Health: —Vapors extremely irritating. Contact may cause burns to skin and eyes. *If inhaled, may be harmful.*
—Runoff may pollute water supply.

Immediate Action

—Get helper and notify local authorities.
—If possible, wear self-contained breathing apparatus and full protective clothing.
—Keep upwind and estimate *Immediate Danger Area.*
—Evacuate according to *Evacuation Table.*

Immediate Follow-up Action

Fire: —**Small Fire:** Dry chemical or CO₂.
—**Large Fire:** Water spray or fog.
—Move containers from fire area if without risk.
—Cool containers with water from *maximum distance* until well after fire is out.
—Stay away from ends of tanks.

Spill or Leak: —Do not touch spilled liquid.
—Stop leak if without risk.
—Use water spray to reduce vapors.
—**Large Spills:** Dike for later disposal.
—**Small Spills:** Flush area with water.
—Isolate area until gas has dispersed.

First Aid: —Remove victim to fresh air. Call for emergency medical care.
—If victim is not breathing, give artificial respiration.
—If breathing is difficult, give oxygen.
—If victim contacted material, immediately flush skin or eyes with running water *for at least 15 minutes.*
—Remove contaminated clothes.
—Keep victim warm and quiet.

For Assistance Call Chemtrec toll free (800) 424-9300

In the District of Columbia, the Virgin Islands, Guam, Samoa, Puerto Rico and Alaska, call (202) 483-7616

Additional Follow-up Action

—For more detailed assistance in controlling the hazard, call Chemtrec (Chemical Transportation Emergency Center) toll free (800) 424-9300. You will be asked for the following information:
- Your location and phone number.
- Location of the accident.
- Name of product and shipper, if known.
- The color and number on any labels on the carrier or cargo.
- Weather conditions.
- Type of environment (populated, rural, business, etc.).
- Availability of water supply.

—Adjust evacuation area according to wind changes and observed effect on population.

Water Pollution Control

—Hydrogen Chloride is water soluble and can kill fish. Prevent runoff from fire control or dilution water from entering streams or drinking water supply. Dike for later disposal. Notify Coast Guard or Environmental Protection Agency of the situation through Chemtrec or your local authorities.

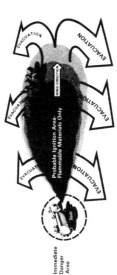

Evacuation Table — Based on Prevailing Wind of 6-12 mph.

Approximate Size of Spill	Distance to Evacuate From Immediate Danger Area	For Maximum Safety, Downwind Evacuation Area Should Be
200 square feet	125 yards (150 paces)	1/2 mile long, 1/2 mile wide
400 square feet	180 yards (216 paces)	1 mile long, 1/2 mile wide
600 square feet	225 yards (270 paces)	1 mile long, 1 mile wide
800 square feet	260 yards (312 paces)	1 1/2 miles long, 1 mile wide

In the event of an explosion, the minimum safe distance from flying fragments is 2,000 feet in all directions.

Figure 17-3. Sample page from the *Hazardous Materials Emergency Action Guide.*

hotline run by the National Library of Medicine at Rockville, Maryland. It is known as TOXLINE, for the Toxicology Information Conversation Network.

Of further use to management, the toxicology committee of the American Industrial Hygiene Association has published a series of emergency exposure limits (EELs), defining how large a single brief exposure to airborne contaminants can be tolerated without causing permanent toxic effects. These limits do not attempt to replace accepted safe practices, but they can be used in emergency planning, to help determine what type of emergency action to take, and to assist in the proper selection of personal protective equipment.

Radiation

Emergency planning cannot be too strongly stressed in the case of ionizing radiation. When radioactive materials are being handled, typical emergencies involve spills, contamination, leakage from containers, or a fire involving radioactive materials or the areas in which they are used or stored. Emergency procedures for such incidents should be developed even where only small amounts of radioactive material are handled. Immediately washing the injury and the implement that caused it with a great deal of water is important. This, of course, calls for strategically placed showers and eyewashes. For leaks or spills, the main concern is to minimize exposure, confining it to the persons first involved. These persons should be monitored for contamination before being allowed to mingle with others or go into other parts of the plant. Except for the most minor incidents, the affected areas should immediately be cleared of people.

Extreme Weather

By "extreme weather," we mean sustained conditions that cause normal operations to fail. This can call for the protection of personnel, customers, passers-by, and even nearby residents. If a sudden blizzard closes highways near a plant, the company may be faced with buses and carloads of stranded travelers, as well as company personnel. A severe ice storm may interrupt all power in an area, bringing operations to a standstill. A flash flood can wipe out a critical access bridge, stranding employees and others at the plant. In some areas of the country, dust or sandstorms can create similar emergency situations, making travel from the plant impossible.

Planning cannot cover all weather emergencies, but the most probable ones can be considered. Moreover, if a flexible attitude is maintained, plans for one weather emergency can readily be adapted to another (e.g., an unexpected dust storm can be dealt with using plans originally developed for blizzard conditions).

Nuclear Events

Nuclear events for this discussion may be either accidental events such as Chernobyl or Three-Mile Island, or an intentional use of a nuclear device such as a terrorist group might employ.

Remote as it may seem in peacetime, the possibility of an unexpected nuclear event does exist, and the responsible manager will ensure that the organization has included it in disaster planning. In an armed conflict, any place is subject to nuclear attack and its consequent additional disasters. For those near nuclear generators or other large-scale radioactive sources a well-organized plan is essential for survival as well as for safe and efficient recovery. While the results of a nuclear event can reach remote areas, we are most concerned with population and industrial centers, which are more likely to be adversely affected on a large scale.

Planning for a possible nuclear event requires expert knowledge, both of the nature of the immediate threat and the most effective survival techniques. Some governmental agencies specialize in providing information and advice on the subject, and every state (as well as some cities and counties) can provide organizational direction. It is important for the corporate plan to fit into governmental or municipal plans for mutual support.

Two points must be made in connection with nuclear events.

1. Should a plant survive such an event, the first 24 hours are the most dangerous. This is the critical period of radioactive fallout. Any emergency plan should place heavy emphasis on getting through this period.
2. Many of the emergency plans developed for other disasters can be adapted to nuclear emergencies. Equipment and training for other emergencies are applicable as well. The need for first aid, shelter, emergency communication facilities, and so on applies equally to any disaster.

The following checklist can be used on both an individual and a corporate level. Survival rests on the following:

1. Know the local emergency plans.
2. Understand the hazards of a nuclear attack or event.
3. Know the warning signals for such an event.
4. Know the location of fallout shelters.
5. Know what type of protection is needed.
6. Know how to improvise protection if a fallout shelter is not available.
7. Have emergency supplies ready, and know where they are located.
8. Conserve emergency supplies as much as possible.
9. Maintain sanitation and sanitary practices.

10. Reduce fire hazards.
11. Know the basics of emergency medical or first aid care.
12. Follow official instructions.

Energy Emergencies

Energy emergencies or shortages are becoming increasingly a part of the average person's awareness, if not of his or her actual experience. If anything, they are even more critical for industry than they are for the individual.

Energy emergencies include sudden fuel shortages caused by embargos and slowly progressing shortages caused by rising fuel prices and dwindling reserves. There may also be periodic imbalances in supplies at particular times or seasons, such as a higher demand for heating fuels during severe winter weather. Geographic imbalances can also exist. A gasoline shortage, for example, may be less critical in a community where most workers live within walking distance of the plant than it is, say, in some western states, where workers may have to travel 50 or more miles to work each day. A shortage of one type of fuel (such as natural gas) can cause subsequent shortages of substitute fuels.

The problem of supply is usually handled at the community level, and the company may have little control over it. Such shortages may still impose a special burden on executives, however; securing supplies and solving problems caused by the shortage can take time away from other pressing duties.

There are four major categories of problems caused by energy emergencies:

1. Quickly developing safety and health problems (stoppage of iron lungs, elevators and life support systems).
2. Heating problems.
3. Electric power problems.
4. Natural gas problems.

Any of these may involve key personnel, who may have to concentrate their energies on solving plant or community problems, lessening their availability for other duties.

Government agencies can provide guidance in handling energy emergencies. In fact, in some energy emergency situations, the company may be required to work closely with local and state agencies or conform to emergency regulations. It is important to keep in mind that an energy emergency will rarely affect just a single plant or company, and that the need to cooperate with larger plans may supersede parts of the corporation's formal plan.

Kidnapping and Extortion

Ordinarily, this section of emergency planning is kept separate from the regular emergency plan. First, only a few key personnel are generally expected to be involved. Second, making the plan known acknowledges the possibility of threat and lessens the plan's effectiveness. The specific features of the plan should be confidential. The plan, after all, may contain proprietary information that should not be general knowledge.

When a call that threatens any kind of damage is received, a plan of action must go into effect immediately. If the threat involves a hidden device, a search should be made at once, preferably by someone familiar with the area. The most rapid and efficient search could be made by supervisory personnel and employees who work in the area. On receiving an alert, all supervisors should initiate a search of all parts of their assigned areas. Most employees will cooperate for their own protection. Areas that are not normally staffed should be searched by trained teams. The search should be guided by search plan folders prepared in advance and handed out to the team leader. It should be clearly understood that the objective is to search and find only. Under no circumstances should a searcher touch, move, or try to remove a suspicious object or container.

The decision to evacuate a facility is difficult, but evacuation is preferable to exposing workers to injury or death. Experts have developed profiles of letter writers and callers; this expert knowledge can be of help in determining how serious a threat may be and whether evacuation is desirable. In all cases, it is better to err on the side of caution and to evacuate a threatened facility than it is to assume that the threat is a hoax and risk having employees killed or injured.

Kidnapping, particularly for ransom, has been more and more frequently reported in the news. Keep in mind that many kidnappings never reach the public eye and that the threat of executive kidnap for monetary or nonpolitical motives is a rapidly growing problem to which no corporation or manager is immune. There are special techniques for handling all phases of a kidnapping or extortion threat, ranging from advance planning, to training courses on how executives can minimize the danger of kidnapping, to negotiation guidelines, and even providing for postkidnapping psychological adjustment. Few organizations have such expert knowledge in-house, but specialists are readily available for consultation. This is an area where expert advice is essential in setting up an effective plan of action.

CIVIL PREPAREDNESS: A NEW PERSPECTIVE

Traditionally, disasters and emergencies have been a case of "every man for himself" and, far too often, every company for itself. While it is still vital for the organization to have plans, personnel, and equipment for dealing with its own emergencies, there are other resources on which it may be able to call.

More and more, governments are willing and able to assist in emergency and disaster planning, particularly for emergencies that can affect an entire city, state, or area. Too many people still think of civil preparedness (or "civil defense," as it is sometimes called) solely in terms of the bomb drills of the

1950s. If the executive has not reviewed current policies and resources for civil preparedness, he or she may be ignoring a vital source of help when an emergency strikes. A new perspective may be needed.

Civil preparedness is not a separate function from the normal responsibilities of government. On the contrary, civil preparedness operations occur whenever a local government responds to any extraordinary emergency, such as flood, major fire, tornado, or earthquake. It also, but only incidentally, involves the most massive emergency: a nuclear attack.

In any extraordinary emergency, the local government forms the nucleus around which doctors, hospital staffs, the news media, industry, volunteers, and other groups organize. Extraordinary emergencies require coordinated operations, including all civil and governmental functions, to minimize danger and damage. The official ordinarily in charge remains in charge during the state of emergency. (This principle offers a good example to industry when establishing emergency plans: as much as possible, maintain the ordinary chain of command and authority.)

If a state of emergency affects a large enough area (such as, a nation in time of war), the surrounding communities are likely to have their own problems and should not be counted on to provide support or aid. Industrial firms must therefore rely more heavily on their own resources in what is loosely called "industrial civil preparedness." This requires the company to fend for itself in terms of medical assistance, shelter, fire fighting, security, communications, and so on. Company personnel can be trained for industrial civil defense through the Federal Emergency Management Administration, which has offices in major population centers.

The object of industrial civil defense is to ensure that industrial and commercial enterprises make appropriate preparations for protecting life and property during a major emergency, whether it is a war or a widespread weather disaster.

Government and local civil defense organizations may be able to supply training materials, model plans, consulting services, and information to the company emergency/disaster plan coordinator. In addition, the local civil defense organization can help the company to:

1. Ensure its own survival and that of its work force.
2. Minimize danger to its facilities and operating capabilities.
3. Actively interface with the community or area program.
4. Maintain vital functions during an emergency.
5. Accelerate recovery after a disaster.

Interestingly, all of the organization's safety and health plans can serve as a base for an emergency/disaster plan. For example, the company plan that

EMERGENCY AND DISASTER PLANNING

covers the handling of flammable liquids can be extended to cover an emergency involving flammable liquids as well. An effective safety and health plan makes the establishment of an emergency/disaster plan immeasurably easier. The executive, on careful review, may find that the company's safety and health program already provides for many of the trained people, much of the equipment, and many of the resources that will be needed in an emergency.

PLANNING STEPS

Essentially, the following nine steps are involved in emergency and disaster planning:

1. Assess the probability of an event.
2. Assess the hazards involved.
3. Designate an emergency/disaster coordinator.
4. Develop an emergency/disaster plan.
5. Obtain top management's approval of the plan.
6. Train the necessary personnel.
7. Test the plan.
8. Revise the plan as needed.
9. Coordinate all aspects of the plan.

These suggested steps are only one way to approach the development of an emergency/disaster plan. It is essentially a systems approach to ensure that all aspects are considered. Figure 17-2 shows the relationship of these steps to various emergencies in a checklist format. A similar relationship exists between these steps, the plan elements, and the plan functionaries. These last two will be discussed in Chapter 17, Part 2.

Turning to these nine steps, let us take a brief look at what each of them involves from the manager's point of view.

Assess the Probability

The first step is to decide how likely it is that the company will be faced with a given emergency or disaster. What is the probability of the event? We might use a simple ranking procedure, dividing all possible events into categories such as (1) most likely, (2) quite possible, (3) unusual but possible, (4) remotely possible, (5) has never happened but is conceivable, and (6) practically impossible. These possibilities can be weighed both subjectively and objectively for decision-making purposes. Normally, the cost and consequences of the event, should it happen, would also be considered. This merges into the step of hazard assessment to a certain extent.

The objective at this stage is to answer the following questions:

1. What events should the company prepare for?
2. What commitment of resources should the company make?
3. What company activities will be excluded from the plan?
4. What level of activity will plan implementation require?

Assess the Hazards

Assessing the hazards merges with the probability assessment step to a certain extent. This step must consider the costs of various hazards, such as the likely cost of catastrophe and the likelihood of multiple fatalities or even a great number of minor injuries. The assessment must also take into account all company resources likely to be involved, including facilities and market share.

To help the reader distinguish between these two steps, let's consider a simple example: the probability of an earthquake. In step one, the planner would consider how likely it is that an earthquake would occur. For instance, the manager of a manufacturing operation in Southern California might rightly conclude that an earthquake will be very likely. In step two, he would look at his operations and assess the specific hazards and their likely impact. He might include, for instance, the fact that the plant stores and uses large quantities of highly flammable material which could burst into flames following an earthquake; the fact that the plant was constructed prior to the development of earthquake construction codes; the knowledge that the plant is the sole source of raw materials used by four other company plants; and the fact that it is located far away from the nearest medical facility. Taken together, these facts would lead him to the conclusion that an earthquake is likely to happen and will potentially be very costly to the company in terms of property damage, injuries or fatalities, and interference with company operations elsewhere. He would therefore assign a very high priority to developing a thorough and effective earthquake plan.

Designate a Coordinator

This person, preferably an official, should be high enough in the organization to have the clout not only to see the plan through development and testing but through actual implementation as well. His or her duties will be discussed in greater detail in Chapter 17, Part 2, in the section "Plan Functionaries."

Develop a Plan

In developing the plan, it is vital to consider everything that could possibly affect it. The plan might be a single document or a series of manuals, each covering a different emergency or function, with suitable appendices. It might be a combination of both. That is, a master emergency plan covering all functions and emergencies could be developed. Individual sections could then be broken out and a manual created by emergency type, department, or function.

The best advice is to keep the plan as simple as possible. Unfortunately, even "a simple plan must be somewhat detailed if most emergencies are to be considered. Regardless of its size and depth, the plan must be easy to follow and understand. Even the best plan will not work if people do not understand it. Try to keep in mind who the ultimate user of the document will be, and write on a level, and in language, that he or she will be able to grasp.

Approve the Plan

An emergency/disaster plan ultimately involves company survival. Therefore it must have the approval of top management, including the board of directors and the chief executive officer. It is equally important that specific portions of the plan be approved at the operating level where the activity will take place. This should not be a mere formality indicating that the person involved has received and read the plan. It is a vital step in gaining feedback and verifying that the plan has been thoroughly read and understood.

Train the Personnel

If an emergency/disaster plan is to work successfully, all participants must be trained in the responsibilities and duties required by the plan. Training objectives may range from knowing the evacuation route from a certain section of the plant, to knowing the location of all emergency equipment, to taking charge of the welfare of all personnel for an extended period of time. For some, it will involve drills and wallet-sized instruction cards. For others, it may mean a series of training courses away from home and continuous coordination with outside agencies. The training is discussed in greater detail in Chapter 17, Part 2.

Test the Plan

Waiting for an actual emergency to find out if the plan works is too risky, too costly, and too late. Testing the plan is the only sensible way to find out if it will work. First, testing can point out flaws or gaps that can quickly be corrected. Second, it can serve as a training and refresher device.

Testing exercises can be designed to try out selected elements of the plan (such as the fire or evacuation plan) or to test the entire disaster organization. Tests of individual elements should be carried out first, and deficiencies corrected, before the entire plan is tested. In no case should the test endanger personnel or resources in any way. This is an important point, since an unpracticed or hurried response to tests is a fertile ground for accidents.

Tests should be discussed and reviewed with the safety manager before they are carried out.

Revise the Plan

Any test of a plan should be critiqued immediately afterward. In this way, corrective measures can be implemented before the problems are forgotten. The critique process should consider the comments of everyone involved in the test, since even minor failures can jeopardize successful implementation of the entire plan.

This critique process should not be reserved for test runs. Any actual activation of the plan should also be thoroughly critiqued to see if things went as planned in actual operation. It is particularly important to uncover problems that involve management deficiencies or that require management action to correct. Generally, three deficiencies are uncovered in both tests and actual emergency plan activation:

1. There are no clear guidelines on when to stop activities and get help.
2. There has been no practice in actual emergency shutdowns.
3. Management has not been alert to changes in personnel, processes, or facilities that require adjustments in the plan.

The feedback and critique processes, as well as the testing itself, allow management to make the necessary revisions to the plan. It must further be emphasized that once revisions are made, the plan should be tested again, both to make sure that the revised plan now functions properly and that no new problems have been created by the changes.

This last point is essential. Change is a constant in most operations. Trained personnel leave or are transferred; departments are moved from one area of the plant to another; partitions, exits, and first aid cabinets are moved to accommodate new machinery or equipment. Management must revise the plan accordingly each time such a change is made and then retest the whole. To a certain extent, this overlaps with the next step: coordinating all aspects of the plan.

Coordinate All Aspects

Every time a change is made in a plan, the change must be coordinated with all involved parties and proper adjustments made in the plan itself. Imagine the consequences of not having, for instance, the new telephone number of the medical response agency printed on all wallet cards or the confusion that might result if a new type of alarm was being used in case of a flash flood. The few extra minutes needed to find the right number or ask the right question could cost lives and valuable property. Moreover, any changes in operations, facilities, personnel, or equipment that may affect the plan should be reviewed and the plan changed accordingly.

As you can see, the emergency/disaster plan requires continual review and updating; possibly new tests and training from time to time; and printing of new evacuation or access maps, equipment lists, wallet cards, and so on. So many small but critical details are involved that it is essential to have one person in a position to coordinate and check them all.

DEALING WITH THE MEDIA IN EMERGENCIES

Whether it is an emergency situation or a normal encounter, confronting the media may sometimes feel as though the sky is falling. This is particularly true if the issue involves an accident and the environment. However, if a good relationship has already been developed, confronting reporters in an emergency may be much easier.

Reporters don't intend to humiliate their subjects. However, they want news that will stimulate readers, and it is often bad news that makes the headlines. Hence, if the company wants its side of the story handled accurately, cooperation is necessary. Making the effort to work with the media will create a better base for fair treatment, particularly after a mishap.

It is important to show cooperation before an emergency occurs. Show the media that the company is a solid source of information by giving timely and honest reports. This may result in exposing embarrassing facts on company mistakes, but it can be counteracted by emphasizing what the company is doing right. Make it a policy to send the media fact sheets on the company and the steps the company is taking to ensure a safe, healthy environment. The company's chances of receiving fair coverage will also increase if it has an emergency action plan dedicated to giving the media timely and proper information. Here are a few suggestions:

- If the press wants information, talk to them or provide another company representative. Refusing to meet the press will only result in bad relations.
- Be careful not to mention what the company doesn't want on the record, and don't attempt to remain "anonymous." A story is more credible when the source is named.
- The main task of the media is to inform the public. Working with the press is a quick, efficient way to inform the public on future company plans (e.g., a new process that will reduce odors or traffic noises). Such plans should reach the public before the issue becomes a matter of concern to the community. The aim is to educate the media before they fill in knowledge gaps with nonfactual information.
- Radio and television interviews call for special techniques. The best

answers are short ones. Be sure that the answer closely fits the original question, which might be cut in the broadcast.

Often the press will change topics and ask a question on a different subject that the company is not prepared to discuss. Don't be forced into an answer. Merely say that you are not prepared to cover the new subject.

Reporters and editors are always short of time and space, so expect that your comments will be cut. However, reporters and editors have an obligation to be fair, and if you feel that you have been treated unfairly, let them know.

Always remember that being prepared for emergencies builds the best rapport, ensuring fair treatment when you need it.

CONCLUSION

While effective emergency/disaster planning may involve a large commitment of time and resources, the manager should not attempt to cut corners or stint in the effort. The consequences of even a small emergency, if the company is unprepared to meet it, can be costly at best and catastrophic at worst. More than any other area of safety and health management, emergency/disaster planning involves critical needs and situations. It requires more than ordinary management acumen, since, in the last analysis, it is a matter of survival.

BIBLIOGRAPHY

Reid, J. L., *What to Do When the Sky Starts Falling: Emergency Planning for the Construction Industry,* Northglenn, CO, Secant Associates, 1987.

Chapter 17 / Part 2

Emergency and Disaster Planning

In Part 1 of this chapter, we focused on the types of emergencies and disasters an organization may face and the steps involved in developing and preparing an adequate plan to meet them. Several key questions have been left unanswered, however. Among these are "Exactly what should the plan contain?" and "Who will perform the actions it outlines?" These questions touch on many elements, ranging from necessary equipment and personnel to complex interfaces and backup systems. This part of Chapter 17 attempts to describe in greater detail the two major areas needed to complete our overview of emergency and disaster planning:

1. Plan elements.
2. Plan functionaries.

PLAN ELEMENTS

Clear, thorough planning for emergencies and disasters is possible only if the plans are in writing. First, the written plan can delegate specific responsibilities, so that all persons know exactly what is expected of them. Second, a plan cannot be well tested in advance if it is not precisely detailed in writing. The elements of a thorough plan and the responsible functionaries are shown in Figure 17-4. We will look first at the elements of the plan.

Policy

This is a general statement by the functional head of the corporation establishing the corporation's attitude toward emergency planning. It should be brief and general enough to give leeway for various operations but specific enough to outline general responsibilities for the development and functioning of the plan. It may actually provide for several plans to meet corporate needs. It should ensure some flexibility, since it is impossible to foresee every contingency and since the plan must be capable of quick modification if needed. It must, however, provide for someone to be in charge at all times under all conditions. For this reason, it may be wise to specify an order of succession. Some companies naturally consider this order very sensitive and decide that is should not be part of the policy statement. However, it is still necessary to inform the persons directly involved

486 SAFETY AND HEALTH MANAGEMENT

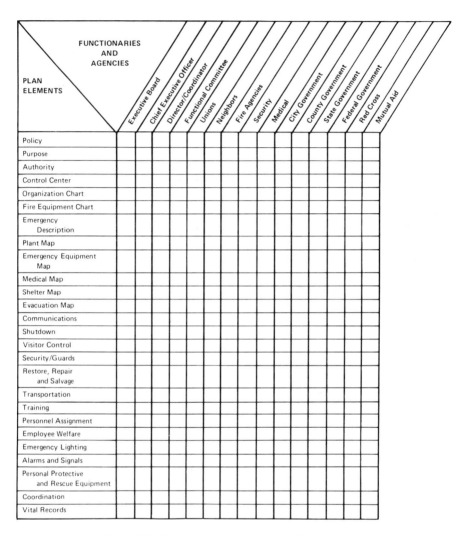

Figure 17-4. Emergency plan elements and functionaries.

in the succession. The main point is: keep the policy statement as lean as possible.

Purpose

No single emergency plan will be able to do all things for all organizations. However, it is vital to spell out the purpose of a particular plan so that the wrong direction is not taken.

Nothing disorders a normal situation more quickly than not knowing who is in charge or who is running the organization. This situation is made far worse by an emergency.

Because a disaster or emergency can occur at any time of the day or night, during the week, or on weekends, or on holidays, a plan should provide for someone to be in charge at all times. Top management is not always available, particularly on short notice. Overall responsibility in an emergency should be a function of the executive who is usually in charge. Naturally, a coordinator/director will be involved in developing the plan and perhaps implementing it, because he or she will be the person most familiar with it. However, someone at the top must still be in overall charge. A management succession list will ensure that someone is always available, regardless of the time or circumstance.

Some organizations take action to ensure that the board of directors can still function in a large-scale disaster, even when it is not possible to get a quorum. The company bylaws should be written to allow for this contingency, possibly providing a means to return the board to full strength.

Authority

Effective plans do not just happen. They are shaped, built, and implemented by people who are capable of doing the complex, long-range planning required. The functional committee (see Figure 7-4) should see to it that someone is put in charge of both building the plan and implementing it in any emergency. Appointing a director/coordinator to perform these functions is an early step. As mentioned earlier, the director/coordinator should be a member of top management who can speak and act with authority on a wide range of matters. He or she should be a cool, quick-thinking person, robust enough to stand the pressures and long hours of an emergency. It is equally imperative to name an alternative director/coordinator who can step in during emergency situations. The alternative should be kept fully informed of plan details and all aspects of managing an emergency.

Due to the wide range of responsibilities involved, the director/coordinator may require special training. This person will not have to handle all the details, but he or she must be familiar with them in order to see that they are properly carried out by others.

Control Center

Having a centrally located control center from which to direct emergency operations is vital. Its location should be carefully chosen, keeping several things in mind. First, it should be located in a place that is likely to survive most emergencies. Second, it should be large enough to accommodate the necessary people. Third, it should have communications and access to both the inside and outside of the plant. Some corporations have established special installations that meet all these criteria, since it is unusual to have a ready-made location that has all these characteristics.

Keep in mind that it is always possible that a disaster will render a control center unusable or inaccessible. For this reason, it is important to designate an alternative headquarters or control center. The disaster plan should list the alternative location, as well as spelling out who should report to it and under what conditions.

Fire Equipment Chart

This chart should identify all the types of fire protection available and their capabilities. To be useful, this chart should probably include several pieces of information about the equipment. It should state which equipment is portable, mobile, or fixed; what types of fires each may be used on; and where the equipment is normally located. Such equipment should also be identified in the emergency/disaster plan.

Emergency Description

This document clarifies what constitutes an emergency, and describes the various types of emergencies. It should, for example, define what is meant by a radiation emergency or spell out the difference between a "flash flood" and a "normal flood" emergency.

Plant Map

In a very simple plant layout it may not be necessary to have a separate map of the entire installation, since other maps will be included (as described in the following paragraphs). Such a plant map, however, has many uses, particularly for breaking the plant into areas of responsibility for emergency handling purposes. The overall map should be supplemented by more detailed maps of each designated area. This would be useful, for example, in a bomb search emergency. The overall map would divide the entire installation into areas of responsibility, while the detailed maps would show every nook and cranny of the area to be searched.

Emergency Equipment Map

While much of the emergency medical equipment will be kept at or near the medical facility, some may be dispersed for better access and handling of emergencies. A map showing the location of all such equipment will facilitate both emergency actions and regular inspections and maintenance checks. The map can also designate temporary hospital facilities for victims who cannot be moved or taken to a local hospital immediately. (Normal medical facilities in a plant do not provide such accommodations, so space should be provided during the plan development phase.)

Management should consult the industrial doctor, nurse, and safety professional in developing any plan involving injured persons. These professionals can be of assistance in:

1. Selecting and training auxiliary medical personnel for nursing and first aid duties.
2. Transporting and caring for the injured, including transporting the seriously injured to hospitals or other medical facilities.
3. Coordinating medical plans with security, personnel, safety, police, road patrols, fire departments, and community groups.
4. Making advance preparations to obtain additional medical supplies from local medical supply houses or from other facilities in the area.

Shelter Map

While shelter areas are most commonly associated with enemy attack, they may also be necessary for such disasters as floods, hurricanes, and tornadoes. There is some psychological value in having shelters in places that are normally used for other purposes, such as a well-designed and protected cafeteria. However, it may be necessary to designate several different shelter areas, since the nature of the emergency itself may render one location more vulnerable than another. A well-designed and protected basement, suitable for protection from a nuclear attack, may be the first place to be flooded. Likewise, a glass-walled cafeteria located on high ground may be ideal as a shelter during a flood but particularly vulnerable during a windstorm or nuclear attack. In choosing the shelter, both location and construction should be carefully considered. Height above flood lines, presence or absence of breakable glass, and flexibility in withstanding wind or earthquake shock should all be considered. If an enemy attack is considered, blast and radiation protection are especially critical.

Evacuation Maps

In an emergency, it is essential to guide employees and visitors to safety, direct employees away from hazardous areas, and avert panic. In particular, every effort should be made to get and keep people away from the scene of the disaster or emergency, wherever possible. To achieve these aims, evacuation plans and evacuation maps are essential elements of any emergency disaster plan.

An evacuation plan covering most emergencies can be shown most clearly via a map. Here again, an overall facility map supplemented by area maps will probably be most useful. Small area maps can simplify the evacuation and eliminate the possibility of error.

First, when developing exit and evacuation plans, the National Fire Protection Association's standards and local codes should be consulted. Each building should have enough exits to permit the prompt escape of all personnel in an emergency. If any exit may be blocked due to fire or debris, an alternative exit should be designated clearly on the map. Exits should be kept unobstructed and accessible at all times. Exit doors should swing in the direction of travel, particularly if a large number of persons will evacuate via that exit. Moreover, exits should be well marked in highly visible lettering that is clearly legible.

Every office or section should have designated monitors who will ensure that persons are guided to safety. These monitors should be clearly designated in the emergency plan. They should participate in all drills and tests so that they will be prepared when the real emergency occurs. However, while the monitor may guide an evacuation, it is up to the person in charge to authorize the evacuation, as well as the return to the building.

Communications

The communications portion of the plan must consider two basic needs: internal communications and external communications.

Internally, it is necessary to make certain that the plan has been distributed to all those who will need it. Portions of the plan should be sent to those who do not need the complete plan. Distributing it to senior management alone does not ensure that the working level will get the word. The cost of distributing the plan to line supervisors is amply justified by the knowledge that those who need the information will have it. Saving a few hundred dollars by not distributing the plan to supervisors may cost millions of dollars in a real emergency. Prompt, effective action in an emergency depends on a knowledge of what to do, and that requires detailed information in advance.

The control of personnel during an emergency is always a difficult problem. A system of badges to identify those who are allowed in an area will ease the control and communications problems. Such badges can also identify emergency medical or rescue personnel or the functionaries to whom these personnel should report. In addition to badges, emergency personnel should be issued wallet-sized checklists or instruction cards, which can carry reminders of emergency duties, key locations or phone numbers, or the names of the persons in charge.

A standard alarm signal should identify the emergency. All employees should be familiar with it in advance. Proper handling of emergency calls is another key part of the communications plan. The switchboard operator should have a list of key personnel to be notified in an emergency, as well as any necessary emergency phone numbers (such as fire and police stations, ambulance services, and so on). Emergency plans should not, however, depend solely on switchboard communications, since the regular phone system may be knocked out by the emergency. Portable communications equipment should definitely be included in the emergency supplies.

External communication plans must also include public information and public relations. At least five groups must be kept informed:

1. Employees and their families.
2. Customers.
3. Suppliers.
4. The general public.
5. Stockholders.

Following a disaster, employees and their families will want to know several things: who is hurt and how badly; whether the plant will be closed, either temporarily or permanently; how long it will take to rebuild or repair the plant; whether the employee will have a job when it reopens; what the employee should do in the meantime; what the employee should do about hospitalization; and so on. Customers will want to know about deliveries they were expecting, probable delays involved, and whether they should seek other sources of supply in the meantime. Likewise, suppliers will want to know about existing and future needs. Should certain services or supplies be discontinued for a period, or should they simply continue at another location? Arrangements may be necessary to restore and repair utility services, as well as those provided by private firms.

In addition to all of this are the needs of the news media, which have a legitimate interest in the emergency or disaster. If they are not fully informed of the situation, they will draw their own conclusions. Having partial or distorted reports published or broadcast can have far worse results than a complete and honest account of the event.

Shutdown

An orderly and controlled shutdown may be necessary for the protection of company resources, the lives of employees, and the welfare of neighbors. This is particularly true in certain industries, such as engineering and research laboratories, storage facilities, and high-technology operations. Explosives, fumes, high-voltage lines, or other hazards may constitute an imminent danger following a disaster or emergency. Failure to execute an orderly shutdown could result in the destruction of the entire facility—or, worse, cause widespread injury and death among workers and those in the neighborhood.

An adverse condition may develop with or without attendant disasters. Management should be able to identify such a condition as soon as it becomes abnormal and dangerous, and before it has had time to cause injuries or damage. With suitable emergency plans, such injuries and damage can be avoided. This interim period is the most critical time for emergency action. There are three ways the situation can be handled:

1. Continue normal operations while taking steps to remove or correct the danger.
2. Shut down the entire operation until the dangerous situation can be corrected.
3. Shut down only malfunctioning equipment. This can be readily done if:
   a. The equipment is not essential to the overall operation.
   b. There is redundancy in equipment.
   c. The equipment has finished its function and is not required at present.
   d. A temporary substitute or backup is available.

Naturally, the most desirable approach is the first: continuing normal operations while steps are taken to remedy the abnormal condition. However, this is not always possible or safe, so an emergency plan must contain provisions for partial or complete shutdown as well.

Keep in mind that some equipment may take several hours to shut down fully and that a flood, fire, explosion, or earthquake will probably not give that much warning.

In many cases, the shutdown crew may be the last to leave the plant. For that reason, it is vital to provide maximum protection for these people. Crews should be kept as small as possible and provided with personal protective equipment, as well as the equipment needed for actual shutdown. Careful training and good simulation drills are essential. These drills greatly speed the actual operation and ensure that the crews will not take any unnecessary risks.

Generally, the responsibility for executing an orderly and efficient shutdown rests with one person. This is usually someone from the plant engineering office who is familiar with the facilities to be shut down. Sometimes referred to as the "utilities officer," this person should help develop designs and checklists spelling out the steps to be taken for orderly shutdown. A simple matrix can be used to show the shutdown task, the time required for it, the name of the person or persons assigned to the task, and the date or time assigned and checked. The person in charge of the shutdown will also:

1. Identify all processes that would prove dangerous in an emergency and determine what should be done about each.
2. Make a list of lines and control devices and what is to be done about each in any emergency.

3. Know who to contact in the locality regarding protective action in case of failures involving lines, valves, and switches.
4. Identify the employees best qualified to evaluate and report conditions following a disaster and those best able to take corrective actions.

These steps are all fairly time-consuming. However, keep in mind that emergency plans can be developed from existing information and that the person in charge of the shutdown may already have such information available, ready to be put into a formal emergency plan format.

Visitor Control

"Visitor control," closely related to the evacuation process, refers to getting visitors to a safe location, preferably off the premises. Visitors' safety may be more difficult to ensure, since visitors probably know nothing of the corporation's emergency plans and do not know the evacuation routes or the location of the shelters. It is the duty of the monitor to take care of the visitors in his or her area. Equally important, there should be a firm policy and procedure to keep other visitors from entering a facility or area after an emergency situation has developed.

Security Guards

If the plant is large enough to have a full-time security service or personnel on the premises, they can play a major part in any emergency plan. Integrating them into the plan is the responsibility of the head of security operations. Special training may be needed by security personnel, both for safeguarding the facilities and for protecting other personnel during an emergency.

In general, contract security personnel (such as watchmen who are employed by an outside agency and assigned to the contracting organization) will be less useful than company-employed security people. For one thing, they have only a minimal stake in the company's survival. For another, it is impossible to know how thoroughly they have been trained, screened, or selected by the agency employing them.

Restoration, Repair, and Salvage

Getting things back to normal, or at least into shape to resume operations, calls for special skills and knowledge. This, too, should be provided for in the emergency plan.

Before restoration, repair, or salvage can be undertaken, the plant engineering crew should survey the building and grounds thoroughly, paying particular attention to the structural integrity of buildings and installations. Typically, they should look for unsupported walls, precarious equipment, live or exposed wiring, escaping gases or liquids, dangerous chemicals, and damaged containers of volatile materials. Any part of the shutdown process that was not completed before the emergency should be taken care of at this time. Only after a thorough investigation should the utilities or processes that were shut down be turned on again.

The survey also provides a basis for the repair of facilities and equipment. It can determine what can and cannot be repaired and how much effort will be required to do so. Whatever cannot be economically restored to working order should at least be salvaged. That is, it should be retrieved and disposed of as profitably as possible.

In the abnormal or disrupted conditions following an emergency, the repair, restoration, and salvage operations can be very hazardous. They should be performed only by personnel who are properly qualified, trained, and equipped to do so, and they should be adequately supervised throughout the operations. Only these qualified work crews and supervisors should be allowed in the area; all other persons should be kept out until work is completed and has been inspected by engineering and safety personnel. Like the shutdown crews, repair, restoration, and salvage crews should be kept as small as possible.

Transportation

Like communications, transportation is essential in handling an emergency promptly and efficiently. It serves such vital roles as getting people out of endangered areas, moving necessary emergency equipment to the right locations, and transporting injured or disabled persons to treatment facilities.

All emergency transportation responsibilities should be assigned to one person for action and coordination. In addition, all of the organization's available transportation should be inventoried and its use in an emergency detailed. Advance arrangements should also be made for securing additional vehicles from leasing companies or garages, or from employees where needed.

Getting employees to and from work after a disaster may present substantial problems. Some of these can be met by car pools, leased buses, or the use of other company vehicles. Drivers assigned to vehicle duty should be given appropriate training in advance. Advance planning should also include provisions for emergency fuel supplies and sources. A thorough emergency transportation plan should include maps, diagrams, and instructions for drivers, including the locations of emergency facilities, routes to nearby hospitals, locations of fuel supplies, and so on.

A special high-priority aspect of emergency transportation is the transport of victims to hospitals or treatment centers. Besides an inventory of area ambulance facilities, company vehicles should be inventoried for the task and designated for that purpose in an emergency.

Training

Faced with disaster, especially with fear for their lives, most people can be counted on to respond energetically, participating actively in rescue and recovery operations. How well they do this, however, depends on preplanning, particularly advance training for emergencies and disasters.

Personnel assigned to emergency duties must know what to do and how to do it. Only training provides this information. Emergency plans should involve training small groups of people to handle specific tasks in case of an emergency or disaster. It should provide for checking their performance periodically, during drills or simulated emergency conditions. A small, well-trained group can serve as a nucleus for expanded activities, both in training programs and in actual emergencies.

In keeping with the times, the role of women should not be overlooked. They can be trained for any emergency task that is within their physical capabilities and are capable of quick, efficient responses in emergencies. While women may not be able to handle large pieces of equipment, they are certainly capable of handling security duties, transportation activities, monitoring duties, shutdown procedures, training duties, first aid stations, communications centers—in short, most of the vital functions involved in an emergency.

Provisions should be made for realistic training, or at least the realities of an emergency should be stressed. However, tests should not be so realistic as to endanger anyone, nor should the plan be tested so often that it becomes routine. How often should a plan be tested? Perhaps once every 6 months, to make certain that personnel don't foget the procedures, or when there have been many changes in personnel or in the plan itself since the last test.

We have mentioned the need to issue wallet-sized cards outlining the essential duties and actions to perform in an emergency. These should be reviewed periodically to check for deficiencies in the plan or to serve as a checklist at emergency time. Larger versions of the checklist can be used as posters and placed where they can be easily consulted. Cards and posters, however, are not enough. Personnel will have to be motivated, questioned, and tested, not once but repeatedly.

Personnel Assignment

The fact that people behave in an amazingly responsible manner during emergencies is good reason to assign an emergency task to many of them. Moreover, an employee who has a certain task to perform will be less likely to succumb to panic or confusion. Be certain that all employees with an assigned emergency task are included in tests and critiques of the plan. Their involvement will be a key to the amelioration of the emergency. The main point here is that every person who has a specific task to perform during an emergency should know and understand it well in advance of the need. If the person has no specific assignment, he or she should still know exactly what to do: proceed to the nearest exit or shelter, report to a particular person, or simply wait for instructions.

Employee Welfare

While is it desirable to have employees involved in the disaster plan, keep in mind that their welfare is the responsibility of management. They must be protected from additional hazards, and are not likely to understand those hazards unless management provides that information.

Employees should know what to do and not do after a disaster. They should be warned and periodically reminded of the hazards that might be encountered, such as live wires, broken gas mains, or collapsing buildings. They should be given specific instructions, and reminders should be posted in prominent locations.

Following a widespread disaster, employees may be in the care of management for an extended period. Food, sleep, shelter, dry clothing, and medical care may be needed. Cold weather may be a particular problem if sources of heat are damaged or unavailable.

Most employees are anxious to leave company shelters: they want to find out how their families are or how their homes survived the disaster. Communications and transportation may be their first concern, even before food, shelter, or first aid. The company should be aware of this and should attempt to provide safe means of transport as soon as it is possible to do so.

Emergency Lighting

An emergency of any size is apt to involve the loss of electrical power. This, in turn, means that evacuation, shutdown, and rescue must be performed in darkness unless emergency lighting is provided. The company should at least provide sources of light that come on automatically when the power fails, if only to allow for safe evacuation. These would include portable power generators, battery-operated lanterns and flashlights, and so on. The location of such equipment would be shown on the emergency equipment map, and all personnel who might have occasion to use it should be fully instructed in the operation of the various lighting devices.

Alarms and Signals

Warning and alarm systems must be included in the emergency plan. Special attention should be given to the warning time likely to be available. Except for flash floods and earthquakes, there will usually be some warning time, allowing employees to get to shelters or emergency stations. Above all, the warning time should be used to carry out the most critical parts of the emergency plan.

Existing signal systems should be used to give the warning: the plant whistle, intercom system, fire alarms, or the like. There should also be some means of sounding the warning if the power is off. Devices like portable, battery-powered megaphones can be used for this purpose. All alarm systems should be tested at least once a month. Tests of the system, however, should be announced in advance so that employees will not panic and think that a real emergency has occurred.

Warning systems used in connection with nuclear attacks should be used only for that purpose. The 3- to 5-minute wavering sound on a siren means that an enemy attack against the United States has commenced. That signal has its counterpart on whistles or horns, with a series of short blasts. In the case of an impending disaster, a 3- to 5-minute steady blast is used to get attention. This usually means that the local government wants to broadcast important information on radio or television.

Personal Protective and Rescue Equipment

Both personal protective and rescue equipment should be clearly marked on maps, as discussed earlier in this chapter. Exactly what type of equipment is needed and where it should be located will be a matter of considerable discussion, since the nature of the emergency can only be guessed at, and each emergency may require somewhat different equipment.

Emergency equipment must be both readily available and well protected. This is a tough assignment, and the solution may require considerable forethought. Not only should emergency equipment be stored where it can be easily reached, but it should be reliable and capable of being stored for considerable periods of time without deterioration. (It should, of course, be checked, inspected, and cleaned on a regular basis, even when it has not been used, to guard against such deterioration or malfunction due to age or disuse.) Select equipment which will meet the requirements of the worst possible situation.

In addition to being stored in locations around the facility, some equipment may be kept at the emergency operations center. Such things as sandbags, tools, generators, pumping equipment, and battens should be located at a convenient central location or at the alternative control center.

Coordination

The entire plan will require continual coordination among all the parties involved. It is no coincidence that the plan director is most often referred to as the "emergency/disaster plan coordinator." Refer back to the section on "Authority" for a further discussion of this vital plan element.

Vital Records

The destruction of records essential to the business can be as disastrous as the damage to personnel or facilities. So vital are these records that an emergency/disaster plan must provide fully for their protection. If the corporation has a risk manager, he or she has probably already developed plans and taken action to ensure vital document survival. A risk manager will probably divide the records loosely into three broad categories:

1. Records pertaining to the corporation's existence.
2. Other financial and legal documents.
3. Operating records.

The categories used do not matter, so long as the importance of the records is recognized and the records are secured. Figure 17-5 provides a list

Accounts Payable	Patent and Copyright Authorizations
Accounts Receivable	
Audits	Payroll and Personnel Data
Bank Deposit Data	Pension Data
Capital Assets List	Plans: Floor, Building, Etc.
Charters and Franchises	Policy Manuals
Constitution and Bylaws	Purchase Orders
Contract	Receipts of Payments
Customer Data	Sales Data
General Ledgers	Service Records and Manuals for Machinery
Incorporation Certificates	
Insurance Policies	Social Security Receipts
Inventory Lists	Special Correspondence
Leases	Statistical and Operational Data
Legal Documents	
Licenses	Stock Certificates
Manufacturing Process Data	Stockholders' List
Minutes of Directors' Meetings	Stock Transfer Books
Notes Receivable	Tax Records

Figure 17-5. Examples of vital records.

of some of the vital records that should be included in developing emergency and disaster plans.

Plans for protecting vital records might include storing duplicate records at another location. This helps to avoid what is often referred to as "corporate amnesia," that is, the loss of records considered vital to tracing past operations and continuing present ones. It does not matter what form the records are in (microfilm, tape drums, or photocopies) or what medium is used for the duplicates. The important thing is that a record is retained somewhere, and the company's operations are not adversely affected by the loss of vital information.

While not a vital record, another essential item for carrying out business is money. Emergency funds should be stored at an accessible place, perhaps the alternative headquarters in a large corporation. There should be sufficient funds to ensure the resumption of operations. Prior arrangements should be made for credit lines and advance funds for emergency actions.

A final word on the protection and continuance of vital documents concerns proprietary or confidential data. Such things as formulas, patented processes, and proprietary specifications should be given special consideration in terms of preserving them physically and protecting their integrity. This also applies to the company's confidential material and other security documents. If the company is involved in operations that fall under government security regulations, it will already be aware of the requirements and guidelines established to protect documents, drawings, and so on from breach of security during an emergency or disaster.

PLAN FUNCTIONARIES

In discussing the plan elements, we have spoken repeatedly of the need to designate the persons responsible for certain functions and actions. One of the objectives of a disaster plan is to allow responsible individuals to concentrate on solving the major problems created by the emergency, rather than trying to bring order out of chaos due to simple lack of planning or preparedness. If plans provide for predictable and routine operations, those responsible can then concentrate on the unusual or unpredictable.

Some managers have difficulty grasping the requirements and rigors of emergency/disaster planning. They do not fully realize that the enterprise could face a disaster of massive proportions at any time, so they give little or no thought to developing a thorough plan. The damage from this shortsighted management attitude could be irreparable.

It is essential that top management recognize the magnitude of the problem and concentrate on developing a workable, effective plan. They should also recognize that even the best plan will not function on its own; the proper people and agencies are needed to carry it out. The following section deals with these main functionaries and their roles in the emergency plan. They are, ultimately, the people on whom corporate survival rests.

Executive Board

An emergency/disaster situation calls for a tremendous effort from the entire organization. Corporate survival is at stake. Emergency planning should therefore have early and complete backing from the executive board. in fact, the plan may well originate with the board. They should play a key role in setting the policy and purpose of the plan, and may well be the prime movers in some areas, such as determining the line of succession or emergency finances.

Chief Operating Officer

The chief operating officer has overall responsibility for carrying out the directions of the board. This is particularly true in plan development. This person will also, if available, activate the plan in the event of an emergency. In some corporations (particularly smaller ones), he or she may direct the execution of the plan as well. In larger corporations, however, the chief operating officer may simply direct the coordinator/director to activate and execute the plan. He or she will normally deliver the succession list directly to those involved and occupy the company control center or alternative headquarters. Once the emergency is over, this person will put into operation plans for repair and restoration in order to get back into production or operation.

Director/Coordinator

Ordinarily, the director/coordinator (either title may be used) has a key role in both the development and execution of the plan. Generally, he or she is responsible for the overall direction of the emergency plan. Upon the direction of the chief operating officer or by some prior arrangement, this person authorizes the giving of the alarm. On a work day, he or she then proceeds to the scene. During nonduty hours, he or she still puts the emergency plan in operation, either by phone or by delegation. Unless the disaster itself interferes, this person then proceeds to the site of the emergency. In a smaller corporation, these duties may be carried out by the head of the organization, as pointed out earlier.

Functional Committee

The functional committee is made up of representatives of various key departments. Their duty is to assist and advise the director/coordinator in developing and organizing the plan. Forming a committee of existing department heads is essential. Existing organizations, departments, and functions, after all, can serve as the best base for handling an emergency.

Under the guidance of the director/coordinator, the functional committee works to:

Make continuing studies of the company's vulnerability and preparedness.
Recommend actions based on the results of these studies.
Establish close liaison with local governments on the plan.
Prepare and publish the emergency/disaster plan.
Assign emergency duties as provided in the plan.
Develop emergency financial procedures.
Designate postemergency assembly points.
Test and revise the emergency plan.
Organize employees into special groups for self-help.
Arrange for training or briefings by appropriate local agencies.
Review and update the plan (with particular attention to internal company changes and outside influences).

These are only a few of the committee's duties. Whenever a consensus or specialized knowledge is needed, the committee can be a vital asset to the director/coordinator. In addition to these development-phase functions, the committee may serve special functions during an emergency or disaster. For example, they may:

Send advance parties to activate shelters, alternative headquarters, and evacuation points.
Instruct wardens and monitors to start the orderly movement of employees to shelters or evacuation points.
Commence direction of emergency operations from the control center or alternative headquarters.

These duties are frequently mentioned as committee tasks, but considerable thought should be given to them. If key management representatives were diverted from regular duties to handle committee affairs, great confusion could result. These emergency functions might be better performed by the director/coordinator or specifically designated assistants, leaving committee members free to attend to other equally pressing duties.

Unions

In a disaster or emergency situation, management and labor have the same primary objective: to protect the employees' welfare. For this reason, the union should be represented on the functional committee to ensure its involvement and cooperation. Union representatives and shop stewards can serve as leaders in an emergency. The union or unions involved should be fully informed about emergency plans, particularly as they affect employee safety and welfare.

Neighbors

Neighboring industries and businesses may well interact with the company's problems or emergencies. A release of dangerous vapors or a fire at a nearby establishment could well create a secondary emergency at the company's facilities. Likewise, the company's emergency could create an emergency situation for them. In the section on "Mutual Aid," we will further explore the supporting roles of neighbors. At this point, we will simply indicate that they are both an additional responsibility and a valuable resource in times of disaster or emergency, and cannot be ignored in developing an emergency/disaster plan.

Fire Agencies

Each year, some 60,000 fire losses are reported by industry. The concern about fire emergencies is therefore well founded. Of the businesses that suffer a major fire loss, *only 23 percent make a full recovery.* Of the remainder, 43 percent never resume operations and 28 percent fail within 3 years of the fire. These figures should speak for themselves and alert the manager to the probability that a fire could have grave consequences for the firm. Fire planning is obviously a matter of corporate survival.

Every possible effective use of in-house fire capabilities must be considered, and the use of public fire departments and other facilities promptly secured and coordinated. This is normally the task of the fire chief or the person who has the responsibility when a fire occurs. However, special instructions and plans may be needed when the fire emergency is accompanied by any other emergency situation, including widespread disaster.

Security

An emergency or disaster always throws an extra burden on the existing security system or department. They may have additional duties, such as directing ambulances, guiding medical assistance teams to the proper locations, guarding access and exit gates against stampede or intruders, patrolling facilities to protect them against vandals or looting, and taking care of additional traffic or parking problems. They may also be needed to provide extra guard or security protection in plant areas where confidential or secret papers, formulas, or items are kept to prevent a breach of security or compromise of integrity. This means that during any serious emergency, the normal security staff will need to be augmented. Prior arrangements should be made for this, either by contracting with an outside agency or by providing extra backup from the ranks of existing employees. Some employees can be readily assigned to assist in the less sensitive or hazardous aspects of security work, at least until outside reinforcements can be obtained. Whatever arrangements are made, they should be fully documented in the emergency plan.

Medical Services

In any emergency, we are naturally concerned about emergency medical care. This includes first aid, handling, and transport of an injured person from the time he or she is found until definite treatment is begun. This goes beyond the temporary care of the injured to include the prevention of further injuries, secondary trauma, or death.

Planners almost always tend to overestimate the patient load which can be handled by a hospital. A 200-bed hospital, for example, probably cannot handle more than 20 seriously injured persons at one time unless unusual foresight, planning, and training have taken place beforehand. This places a greater burden on plant medical facilities, particularly in cases of serious or widespread emergencies or disasters.

First aid and medical services should be headed by the company doctor. He or she should advise on the plan, the emergency equipment, and the facilities needed to handle an emergency. If the company does not have a full-time doctor, the contract physician can be used in this advisory capacity. In addition, all personnel should be encouraged to take first aid training, which could prove crucial in an emergency. At the very least, the director/coordinator should find out which employees already have basic first aid training or experience and incorporate them into the emergency plan.

The medical department or unit could also be responsible for coordinating the activities of rescue units. For planning purposes, a rescue unit of 5 or 6 people might serve a plant of 500 employees. This multitalented crew would not ordinarily include medical personnel, though all rescue unit members should have some basic first aid training, as they may be the first to find or contact injured employees. Their main job is to reach and free personnel who are trapped rather than render extensive medical aid.

Their tools should include such things as shovels, block and tackle, ropes, cutting torches, portable telephones and radios, safety goggles, hard hats, linesman's gloves, and so on. None of this is medical equipment, but all of it is necessary to the emergency medical rescue effort.

The plan should also consider those situations that demand fast action. These include:

1. *Breathing stoppages.* This demands the fastest possible action, since the victim can die within minutes.
2. *Hemorrhage or severe bleeding.* This, too, requires immediate action, controlling the bleeding until medical help can be secured.
3. *Poisoning.* Internal poisoning calls for speedy action, since many poisons work rapidly and can cause permanent damage even when they

are not fatal. Poisoning is frequently accompanied by shock, which should also be checked.
4. *Burns.* The seriousness of burns can quickly be determined. For severe burns, medical help should be sought as soon as possible. In all cases, shock is an additional danger.
5. *Snakebite.* This requires both immediate action and professional assistance.
6. *Insect bites.* Stings from bees, wasps, and hornets can be dangerous or even fatal if the victim is sensitive to the poison.

City, County, and State Governments

City, county, and state governments have a variety of emergency and disaster plans, often backed by federal funding and direction. The emergency assistance available from these agencies varies from nonexistent to plentiful and expert. The functional committee and the coordinator/director should evaluate available assistance well in advance. In some cases, the local government can furnish guidance about every aspect of corporate emergency plan development.

A corporation with high local visibility may likewise be obligated to furnish local governments with leadership support and assistance in order to ensure community survival in cases of widespread disaster or emergency. This places an extra burden on these company leaders and lessens their availability to the company itself. The community's expectations and the company leaders' probable role in a widespread disaster should be clearly understood beforehand and plans made accordingly.

Federal Government

The Federal Emergency Management Agency's (FEMA) program is designed to aid in times of disaster, whether nuclear or nonnuclear, natural or artificial. Following recent fires, floods, hurricanes, and earthquakes, this agency has functioned successfully, helping the area to recover and return to a normal life. Although it is a national program, it is carried out through state organizations. Each state functions somewhat differently, though under overall federal guidance. It is important for top management to know the details of its own state's program.

Red Cross

The American Red Cross is one of the best sources of emergency care training, providing both standard and advanced courses. On the whole, this may be the best local source of training and a valuable resource for the company in overall emergency preparedness.

The Red Cross Trauma Program has had a great impact on emergency care training, encouraging companies to secure appropriate knowledge and equipment. The emergency medical technician (EMT) course exceeds normal first aid training, consisting of some 80 hours of study and practice. This course has been widely subscribed to, largely by sponsoring companies that wish to enlarge their in-house capability. The Red Cross Ambulance Program is particularly applicable for companies that provide their own ambulance service.

Both the EMT and Ambulance Program courses are given through trauma centers, usually in local hospitals. The local trauma center itself may be an excellent source of guidance when developing corporate plans, particularly for items such as medical kits and emergency shelter supplies.

Mutual Aid

A mutual aid association is a cooperative organization of firms in a given industrial and business community. These firms have a voluntary agreement to assist each other by providing materials, equipment, and personnel during a disaster or emergency. Any company has a better chance of surviving an emergency if its knowledge and capabilities are combined and coordinated with those of neighboring firms. It is nearly impossible to have adequate emergency supplies, equipment, and personnel available for every possible disaster, so the best defense is the combined resources of several concerns, each providing some of the personnel and equipment.

Another arrangement is to contract for emergency services. This standby aid is usually prearranged through an annual retainer. It may include a wrecking company, a truck rental firm, or an electrical contractor.

ANNEX AND APPENDICES

In order to keep the emergency plan "lean," only essential information should be included. Listings of equipment, addresses of government agencies, or charts and maps might be placed in appendices or annexes in order to keep the main plan simple and easy to read.

Annexes to a basic emergency/disaster plan can also be used to outline actions for any particular type of emergency. For example, if earthquakes have a low probability of occurrence in your location, the details of earthquake amelioration might be placed in an appendix. Perhaps no radioactive materials are handled at your facility, but you are within the danger area of a nuclear power facility. The problems of evacuation or shelter during an emergency at the power plant might be covered in an annex.

In deciding what to include in an annex or appendix to the plan, keep in mind this simple rule: In an emergency, it is better to have information and not need it than to need it and not have it. Because emergency preparedness is such a complex and demanding matter, a mere verbal description is not adequate. To make the manager's task easier, and to provide a checklist for him, an actual example is helpful. It not only serves as a useful reference but may suggest to the manager some areas where the company's existing emergency/disaster plan can be improved. Several of the sources cited in the Bibliography for this chapter have useful checklists for this purpose.

CONCLUSION

Even this rather lengthy examination of emergency/disaster planning is far from complete. It is merely intended to give the senior manager an overview of what is involved when planning for the unexpected. Numerous experts have written excellent books and articles on the subject, and much aid can be secured from governmental agencies for further study. The manager is urged to seek out such sources of expertise.

Keep in mind that nearly all of the things to be covered in an emergency disaster planning effort are a part, or extension, of normal activities. The plan need only draw together available information in a manner suitable to the company's particular needs. Very little has to be researched or developed from scratch.

Just as good planning is essential to any business operation, it is vital to effective emergency and disaster handling. The company's response will be no more adequate than the plan that directs it. The importance of sound planning cannot be too strongly stressed, but it can perhaps be summarized in one simple statement: It is literally a matter of corporate life or death.

BIBLIOGRAPHY

Aaron, James E., et al., *First Aid and Emergency Care,* New York, Macmillan, 1979.
Accident Prevention Manual for Industrial Operations, 8th edition, Chicago, National Safety Council, 1980.
Accident Prevention Manual for Industrial Operations, 9th edition, Chicago, National Safety Council, 1988.
Bird, Frank E., Jr., and Robert G. Loftus, *Loss Control Management,* Loganville, GA, Institute Press, 1976.
Fine, William, "Mathematical Evaluations for Controlling Hazards," *Selected Readings in Safety,* ed. J. Widner, Macon, GA, Academy Press, 1973.
"Fire Protection," *Occupational Hazards Magazine,* June 1977, p. 31.
Fundamentals of Industrial Hygiene, Chicago, National Safety Council, 1971.
Grimaldi, John V., and Rollin H. Simonds, *Safety Management,* Homewood, IL, Irwin, 1984.
Hammer, Willie, *Product Safety Management and Engineering,* Englewood Cliffs, NJ, Prentice-Hall, 1980.
Handley, William, *Industrial Safety Handbook,* 2nd edition, London, McGraw-Hill, 1977.
Heinrich, H. W., Dan Petersen, and Nestor Roos, *Industrial Accident Prevention,* 5th edition, New York, McGraw-Hill, 1980.

"Industrial Emergency Plan Outline," *Protection of Assets Manual,* Santa Monica, CA, Merritt Company, ongoing.
Johnson, William G., *MORT Safety Assurance Systems,* New York, Marcel Dekker, 1980.
Nertney, R. J., *Occupational Readiness Manual,* Springfield, VA, National Technical Information Service (NTIS), 1976.
Petersen, Dan, *Analyzing Safety Performance,* New York, Garland STPM Press.
Planer, Robert G., *Fire Loss Control.* New York, Marcel Dekker, 1975.
"Step 1 in Disaster Planning, Put It in Writing," *Occupational Hazards,* June 1975, p. 39.
Thygerson, Alton L., *Accidents and Diseases,* Englewood Cliffs, NJ, Prentice-Hall, 1977.
Walsh, Timothy J., and Richard J. Healy, *Protection of Assets Manual,* Santa Monica, CA, Merritt Corporation, current.
Worick, Wayne W., *Safety Education,* Englewood Cliffs, NJ, Prentice-Hall, 1975.
Whitman, Lawrence E., *Fire Prevention,* Chicago, Nelson-Hall, 1979.

Chapter 17 / Part 3

Safeguarding Computer Systems

INTRODUCTION

In the "Information Age," computers play an increasingly important role in nearly all aspects of business, both strategically and operationally. Consider the value of computers in your organization. The investment in electronic data processing (EDP) equipment may range from a few thousand dollars for a small personal computer to several million for large mainframe systems.

Whatever the cost of the hardware, it is reasonable to assume that the value of the software and data stored and accessed by computers is considerably greater. Personnel records, budgets, accounts receivable, records, proprietary software, and other important data are seldom viewed monetarily, yet their loss could be devastating. For example, production and inventory control, process control, and scheduling functions are often entrusted to a computer; loss of these functions could cripple operations for days or even months. Computers and the information associated with them are often among an organization's most vital assets.

Generally, data processing systems may be subject to loss from six types of threat:

1. Fire damage.
2. Water damage.
3. Unfavorable environmental conditions.
4. Electrical power systems.
5. Inadequate physical security.
6. Inadequate data/software security.

This chapter will address the first five problems, leaving the complex topic of data/software security for discussion in EDP technical journals and texts. A brief discussion of contingency planning completes the chapter, followed by a Computer Protection Checklist.

FIRE

Microcomputers are relatively inexpensive, are easily replaced, and, because of their widespread use, are not amenable to specialized fire protection. Portable fire extinguishers offer cost-effective protection.

Larger systems, however, may consist of central processors, consoles, tape and disk memory units, high-speed printers, and other sophisticated equipment. Such systems are usually placed in a separate computer room on a raised floor, under which power, communications, and interconnecting cables are run. Rooms surrounding the main computer room are typically used for recorded media storage; mechanical and electrical equipment; stationery supplies; offices for programmers, operators, and keypunch operators; and other computer support functions.

Computer equipment is not inherently combustible. Major computer fires usually involve combustible records materials, i.e., paper, cards, plastic of various types, microfilm, and storage boxes. For that reason, the design of a fire protection system should include adjacent computer support areas, as well as the main computer room itself.

Room Construction

Fire protection begins with site selection and room construction. Adjacent buildings, floors, and rooms should contain low-hazard occupancies. Computer rooms with exterior windows are susceptible to arson and to other outside sources of fire. The building should be constructed of fire-resistive material, and all walls, ceilings, and floors should be noncombustible or fire resistant in accordance with National Fire Protection Association (NFPA) standards.

Detection

Automatic fire detection in computer rooms and support areas is very important. Fire detection equipment should be installed throughout computer operation and support areas, including those around air conditioning units and electrical equipment. It should also be used in conjunction with a fire suppression system in paper and tape storage areas, maintenance areas, data entry and office areas, and overhead and underfloor areas.

Control

If a fire begins, containment and suppression provisions must be implemented to minimize losses. Electrical power to the processing equipment should be shut off and the heating-ventilation-air conditioning (HVAC) shut down. This action helps minimize equipment damage from heat and prevents fans from dispersing smoke and spreading fire. Emergency power-down controls should be located at both a readily accessible location and a remote location to facilitate rapid de-energization.

Many computer centers use a Halon gas fire suppression system for fire control. Halon is noncorrosive, does not affect sensitive electronic equipment, and can generally be used in Class A, B, and C fires. Most Halon applications require a concentration in the range of 4 to 6 percent. The gas is relatively safe for human exposure at a concentration of less than 7 percent. However, in a total flooding system, human exposure should be limited to 1 minute if the concentration is as high as 7 to 10 percent. Light-headedness occurs at about 10 percent, and above 15 percent the symptoms can be severe. Underfloor and ceiling areas should also be protected with Halon, carbon dioxide or water sprinklers, or a combination thereof.

If the rooms housing data processing equipment are sprinklered, it is critical that electrical power be automatically shut off before the sprinklers operate. Computer professionals may be apprehensive about having sprinkler piping overhead and under the floor, but in the event of a fire, an open sprinkler head discharges significantly less water, and over a smaller area, than a fire fighter's hose. Support areas housing combustible materials should also be sprinklered. Salvage covers should be kept readily available for covering computer hardware as added protection against fire and water damage.

Computer rooms should be equipped with Class A and Class C portable fire extinguishers which are light enough to be handled by all personnel. Halon 1211, Halon 1301, and carbon dioxide are the agents of choice for electrical fires, and water is recommended for use on paper, plastics, and other Class A combustibles. Use of dry chemical extinguishers creates tremendous clean-up problems, and the chemical agents may damage or corrode sensitive electronics; they are not recommended for this application.

Employee training is an important part of fire safety in a computer center. Employees must be familiar with emergency shutdown procedures. They should have a basic understanding of the automatic and manual fire detection/protection systems and know how to actuate extinguishing systems. Employees should receive hands-on fire extinguisher training and be familiar with the location of extinguishers and exits. Emergency shutdown of equipment should be practiced (simulated) as an integral part of emergency evacuation drills.

Finally, responsible management should insist that storage of combustibles be minimized in hardware areas. Housekeeping should receive close attention, and smoking should be strictly prohibited in both computer rooms and support areas.

The Center for Fire Research at the Institute for Computer Sciences and Technology has made these suggestions to protect against catastrophic damage:

Hire a computer security expert to identify vital computer records and draw up a protection plan.
Install sprinklers and other fire-fighting equipment in critical areas.
Do not locate all telecommunications facilities on the same site or next to each other.

Identify the top ten tasks done on your computer system and pick an alternative site that can handle these functions in an emergency.
Make sure that the site has the equipment to do the required telecommunications functions.
Duplicate the computer software programs for those top ten functions and store them off-site.
Duplicate daily—and store off-site—important computer system data.
Make a list of employees and their phone numbers to call in an emergency and keep it off-site.
Keep master copies of essential documents off-site.
Have a qualified fire protection engineer inspect your offices once a year.
Make your employees aware of what you have done.

WATER

Water can seriously damage or destroy the sensitive electrical systems contained in data processing equipment. Water may enter a computer facility from a number of sources, including natural floods, pipeline leaks or ruptures, fire extinguishers, and beverages. Site design and other precautions can effectively cope with the threat of water damage from each of these sources.

Note: If water does enter your electronic equipment, be sure that all circuits have thoroughly dried prior to turning on the power.

Natural Flooding

Natural floods resulting from heavy storms can easily destroy data processing hardware and can seriously damage supplies of tapes, cards, and other recording media. Damage may be inflicted not only by the water but also by suspended contaminants. Protection against natural flood damage is best achieved by locating computers and other electrical equipment on a floor above ground level. Even though computer room floors are often raised several inches to provide underfloor space for cables, water and accompanying contaminants can cause substantial damage if allowed to enter even the underfloor space.

Fire Extinguishment

Even though the temperatures required to activate a closed sprinkler head can cause serious equipment damage before water is ever discharged, water from an open head can run into nearby equipment and damage it. Use of a secondary water sprinkler system to back up a primary Halon extinguishing system reduces the chances that the water system will ever be activated. If a fire does occur which may require water extinguishment, all equipment should be de-energized and, if possible, covered before water is discharged.

Water and water flow sensing systems which activate automatic shut-off controls can help reduce water damage when manual shutdown is not possible. Adequate drainage to prevent flooding is essential.

Beverage Spills

An often overlooked threat to electronic equipment is the presence of beverages. Water, coffee, soda, and other beverages can damage powered circuits and should not be allowed near electronic equipment. Although the losses resulting from a beverage spill may be small compared to the damage sustained in a flood, they are unnecessary and preventable. Enforcement of the rule prohibiting beverages in hardware areas is required.

ENVIRONMENT

Electronic data processing equipment is very sensitive to environmental conditions, and careful installation is essential for proper equipment performance. Environmental considerations include HVAC, vibration, and radiofrequency and magnetic interference. The installation or site preparation manual supplied by the computer manufacturer should be consulted for guidelines on acceptable and ideal environmental conditions.

Heating, Ventilation, Air Conditioning

The HVAC system for a computer facility is so critical to the computer's performance as to be considered an integral part of the data processing system. Failure of HVAC systems often results in computer downtime as well. The HVAC system must control temperature, humidity, and air quality.

Inadequate air conditioning can result in uncomfortable working conditions, frequent breakdowns of overtaxed cooling systems, increased heat stress on equipment, and loss of productivity. To avoid these problems, redundant air conditioning systems are recommended. The computer's air conditioning system should be completely separate from the building's system. Humidity and air quality problems associated with central air-handling systems frequently cost more in the long run than installation of separate systems. Dedicated air-handling systems also facilitate rapid shutdown if emergency power-down of computers becomes necessary. An automatic restart feature on such HVAC systems can ensure air system reactivation when the electronic equipment is reenergized.

One solution to air conditioning problems is the use of process cooling systems. These are free-standing units which control humidity, wide temperature swings, and air distribution, and are specifically designed for use with computer hardware.

Humidity control in a computer operating environment is also important. Too much humidity can cause deterioration of electronic circuitry and erratic performance. A relative humidity which is too low can result in a buildup of static electricity. Static electricity results from the widespread use of plastics and paper in computer operations, and static discharges can do serious damage to delicate circuits and storage media. Relative humidity in the 40 to 50 percent range should be maintained for optimum performance and component life. Humidity control is achieved through the use of humidifiers and dehumidifiers and by sealing a room against moisture entry.

In addition to temperature and humidity considerations, air in computer rooms and around industrial and process-control computers must meet certain quality requirements. Dirt and particulates can cause failure of circuits and contacts in electronic equipment. Corrosive gases such as sulfur dioxide, nitrogen dioxide, ozone, and acidic gaseous chlorine can cause electronic component failures.

Meeting optimum air quality standards in a typical office environment may only require dust filtering. In industrial locations, however, particle and chemical filtration systems may be necessary to control contaminants. Maintaining a slightly positive pressure in computer and control rooms is also effective in keeping contaminants out.

Prior to installation of data processing equipment, an HVAC professional should be consulted to ensure that temperature, humidity, and air quality controls will be adequate for the system being installed. Failure to provide a favorable operating climate can result in reduced productivity, equipment failure, shortened equipment life, erratic performance, and even bad data.

Vibration

Although vibration is an infrequent problem for computer hardware, it may be a concern in some industrial locations. Minor vibration may cause equipment to move unless it is anchored or placed on pads. Small personal computers and industrial control equipment may "walk" off tables and fall or be otherwise damaged. If vibration is a serious problem and the computer equipment cannot be moved to another area, it should be placed on some type of shock-absorbing material. The electronic equipment manufacturer should be able to provide some guidance for installation.

Radio Waves/Magnetism

The presence of radio waves or magnetic fields may create special problems which need to be addressed. Nearby radio transmitting antennae can cause interference. To protect vital data from outside tampering or interference, a radiofrequency (RF) shield is recommended. RF shields are made of conductive metal sheets that are built into computer room walls, ceilings, and floors. They effectively block all radio waves and communications frequencies.

Magnetic fields may also be present in computer rooms. Magnets can quickly destroy data on magnetic tapes or disks and should not be allowed in sensitive areas. Most common offenders are personal property such as magnetic key chains, small magnets used to post notices, and so on. Phones are also a source, but are generally far enough from the equipment not to be a problem. Signs warning personnel against bringing magnets into these areas can help reduce data losses from magnetic intrusion, along with adequate employee training.

ELECTRICAL POWER PROTECTION

EDP equipment and the data contained therein are also threatened by electrical power supply problems. Power line problems, including power failure, can cause frequent and expensive part replacements, damaged operating systems, random modification of data, read-write errors, and data loss. Losses stem from two factors: sensitive microcircuitry and "dirty" electricity (unreliable power supply, including power interruption). Unfortunately, many computer manufacturers will not honor warranties if equipment malfunction is traced to power line irregularities. Because incoming power is seldom "clean" enough to preclude processing problems, the computer owner usually must install power protection devices to prevent losses.

Power Supply

The frequency of power line disturbances depends on the time of year, the condition of the electrical environment, geographical location, and other uncontrollable factors. Power protection needs can be met by a wide range of electrical hardware, from inexpensive passive filters and surge suppressors to uninterruptible power supplies and motor-generator sets costing thousands of dollars. A qualified electrical engineer should be consulted to determine which power disturbances are likely to cause problems for your application and to recommend the appropriate power conditioning equipment. Potential costs associated with data loss, equipment damage, loss of process control, and reduced productivity should all be considered when making auxiliary equipment decisions.

Other Considerations

If computers are connected to common grounding systems with other electrical equipment, computer malfunctions and disturbances may result. To provide noise immunity, an independent, single point ground is required which is isolated from building steel, water pipes, and other equipment.

Receptacles and connectors should be UL listed and consistent with the equipment manufacturer's specifications. Installation should be in accordance with the National Electrical Code. To facilitate rapid and complete shutdown, incoming power should pass through a single power panel, with each circuit breaker and power disconnect clearly labeled. Computer room floors should be raised high enough to accommodate air flow around cables.

Physical Security

Numerous terrorist attacks in Europe have targeted computer centers, the heart of the larger national and multinational corporations. In the United States, damage is more likely to be inflicted by disgruntled employees. If your computer center is vital to your organization, physical security measures should be implemented which control access to computer rooms and sensitive support areas.

Computer rooms should be located so that they are not visible from outside the building and are away from areas frequented by visitors. In addition, access routes and entrances should be clearly visible to data processing personnel. More positive protection is afforded by the use of codes, keys, badges, pass cards, equipment locks, and surveillance.

Code locks require an employee to punch in an alphanumeric code on a keyboard located outside the room to be entered. If the code entered is correct, the door is electrically unlocked, providing access. Codes may be changed by simple reprogramming. Code locks do have drawbacks, however. Unauthorized personnel may learn the code by watching other employees enter. Also, during an emergency, excited personnel may forget memorized codes and be unable to gain access.

Door keys are a simple security tool. However, they are readily copied, easily lost, and difficult to control in most business settings. If tight security is to be maintained, locks must be changed frequently and new keys issued each time a key is lost or stolen. The trouble and expense of frequent lock changing discourages the use of keys for large computer room applications.

Pass cards are growing in popularity because of their simplicity, cost effectiveness, and ease of use. Pass cards contain a coded magnetic strip and look like a credit card. When an employee wishes to enter a secure area, the card is simply inserted into a slot beside the door where it is "read." A small processing unit then either permits or denies entry. Card codes can be easily canceled or changed, an important advantage in dealing with personnel changes or lost cards.

Around large installations with security guards, personal identification badges may be used. Although many organizations issue badges and have security guards for controlling access to their property, the use of dedicated security personnel for computer rooms is not practical for most businesses. Where maximum security requires surveillance, closed-circuit television or presence-sensing systems may be used in lieu of additional guards.

Finally, computer room security can be significantly enhanced by educating employees on the need for security. Employees should be taught to recognize everyone working in sensitive areas. Prior to being taken into secure areas, new employees should be introduced to all other data processing personnel. The presence of strangers should not be tolerated, and employees should be instructed to escort strangers out of secure areas and take them to appropriate security or management personnel for positive identification and authorization. Maintenance and custodial workers should also pass security checks. Employees who are discharged, or who may otherwise become disgruntled, should be required to turn in their pass cards and/or keys immediately.

CONTINGENCY PLANNING

When computer functions are critical for ongoing operations of an organization, contingency plans should be developed for use when protective measures fail. Such plans should include provisions for both hardware and software recovery. The average time needed to reestablish full-scalke computer operations after a disaster is 3 months, according to a 1981 survey. Effective contingency planning can allow restoration of critical functions within hours, and full operations within a few weeks or even days.

Hardware Loss

As with any physical asset, an organization may choose to manage the risk of computer hardware loss by simply accepting it (a gamble), by insuring against the risk, or by providing for a backup system (often in addition to insurance). The last approach is recommended.

Alternative computer facilities are typically established in one of three ways. First, computer rooms may be leased in a separate building in which equipment identical to that used in the main computer center is installed, ready for backup use. A second option utilizes the facilities of a neighboring firm under a mutual aid arrangement. For this to be feasible, the data processing equipment of both firms must be compatible and emergency scheduling should be agreed upon in advance. The third alternative depends on commercial time-sharing firms. These companies lease computer equipment and computer time to clients for backup purposes. Of course, the data processing equipment must be compatible with that used by the client. Whichever method is selected, alternative facilities should be periodically checked and equipment operated to ensure readiness and compatibility.

Contingency plans should include provisions for recovery, including the names, addresses, and phone numbers of responsible personnel. Sources of spare parts and replacement equipment should be located and periodically reviewed to protect against obsolescence and ensure availability. In addition, time estimates for obtaining, installing, and bringing up replacement equipment should be reviewed to ensure minimal downtime. Service and repair information should always be accessible. Finally, the entire contingency plan for hardware loss should be compiled in a well-organized reference manual. Copies of the manual should be kept on- and off-site.

Software Loss

Software and data are vulnerable to loss from a variety of threats, and contingency plans should include provisions for maintaining backup data. Three sets of all critical software materials should be kept. One set if used for normal operations. The second set should be kept in a vault on the premises, isolated from the first set. This set can be quickly retrieved if the first one is damaged or destroyed by an operator error. The third set should be stored remotely for use at a backup computer center.

Backup copies of computer data, software, and operating systems should be kept current and should be run periodically to ensure their integrity. As in the primary computer center, backup operating media, such as tapes and diskettes, should be stored in a secure room and protected against fire, water, heat, mechanical, and magnetic damage.

COMPUTER VIRUS

The "computer virus," seemingly a programmer's glitch, emerged as a crime on a night in November 1988 when two of the nation's premier science and research centers were struck. As other agencies and organizations such as NASA, the University of Pittsburgh, and Johns Hopkins University used the data, they too were hit by the virus. It coursed through networks, high-speed communication lines linking key university and government computers from coast to coast. Once inside the computers, it devoured space used to store information and brought operations to a halt. The only immediate solution was to shut down the computers, bringing everything from space operations to telephone calls to a halt.

It wasn't the first time that a computer "hacker" had created mischief, but it was soon apparent to those trying to clear up the problem locally that this operation was much bigger. Scientists raced frantically to their computers in the middle of the night, responding to calls from associates. In 24 hours the virus seemed to be under control, although untold person-hours were used to purge the virus. Some 52,000 outside researchers using NASA data had to spend 142 person-years of time to be certain of purging their systems. This bit of mischief by a hacker didn't destroy much data in the 6000 computers it attacked, since it only used up space. However, the virus could have destroyed everything it reached.

This particular attack of computer virus stunned and frightened the computer world. If computers can be sabotaged so easily and swiftly, the systems are obviously vulnerable to high-technology terrorists, whether foreign nationals or disgruntled employees. This particular virus traveled on an unclassified research and defense network used by institutions to share data. Computer security experts have increased their efforts to protect data through the use of safeguarding passwords and new security measures.

Because some of their methods are new, confidential, and not described in this chapter, such experts, whose business has surged, should be used.

One Company's Program

The information presented so far in this chapter suggests that the computer security covers two areas. One is the physical protection of equipment and data from catastrophes such as fire, power outages, sabotage, and other disasters. The other is protection against unauthorized use or misuse of the data stored in the computer. Here is one company's approach.

Physical Protection

The insurance company has a computer facility on the fourth floor of the building. This prevents unauthorized employees and visitors from going near the equipment. The computer facility and library are in two separate rooms.

The computer facility also has its own fire, water, and heat detection system. There is a separate plant that supplies water to the computers, diesel generators, and a battery backup system. The computer facility can function independently of the utilities in the rest of the building.

Protection of stored data on magnetic tapes is important. As part of the company's disaster recovery program, daily tapes of financial and record-keeping functions are shipped to a rented underground storage facility. This allows for the recovery of vital records in case of a complete disaster at the facility. Each department is also responsible for ensuring that vital data are properly backed up.

Every 6 months, the firm holds a disaster drill to test the data recovery procedures. They pick a date and then assume that everything is lost as of that date. They then try to see how quickly they can be up and running again.

Information Security. Computer access is controlled by a software program that prevents unauthorized persons and equipment from tapping into the computer. Everyone has to remember only one password and sign-on procedure. The password changes automatically every 90 days. The system keeps people from guessing at passwords. If the users do not follow the correct procedures in a few tries, the terminal shuts down.

All data are controlled, which means that more people can use the system, but only for their own authorized purpose. Users sign off from the computer if they must be away from it. They don't share their password. If someone uses the password and signs on under someone's names, that person is held responsible. Writing or saying the password where others can find or detect it is a bad practice. The password is treated the same way as a credit card.

Success Stories

Several large firms, including banks and insurance companies, have had catastrophes in or near their computer center and have survived nicely. These companies focus on computers centers in daily operation. Tens of thousands of other firms and individuals depend on them for information. As a result of planning, they are in a position to continue operating under emergency conditions at the same location or to activate a plan placing them in an emergency center in near-normal operation. One firm, following its emergency plan, moved to a branch location and continued business at near-normal speed. If the computer facilities of any of these firms had been out of operation for 2 weeks, it is estimated that all of them, giants though they were, would have been forced into bankruptcy within a few months, certainly in less than a year.

COMPUTER CRIME

Earlier in this chapter, we briefly discussed computer security and the computer virus, which is very close to the subject of computer crime. "Computer crime" generally refers to acts by employees against their employers, but persons outside the company may also be the perpetrators. Strategies to control computer crime include many of those covered under security, but also much more. For a detailed discussion of strategies for controlling computer crime, see Ralston (1990).

CONCLUSION

The value of computers to industry is rapidly increasing, despite declining purchase prices. As our dependence on EDP systems grows, our vulnerability to substantial losses resulting from system failures also increases. Threats to computer centers include fire, water, hostile environmental conditions, unreliable power sources, and security breaches. Measures can and should be taken to thwart these hazards and prevent hardware losses. In addition, a sound contingency plan should always be ready in the event of a disaster. Careful analysis of your organization's exposure to loss, followed by periodic audits, can avert a disaster and ensure continued data processing operations.

COMPUTER PROTECTION CHECKLIST

FIRE

_____ Are computers and associated hardware located in fire-resistant or noncombustible areas?

_____ Do adjacent buildings, floors, and rooms contain low-hazard occupancies?

_____ Is an automatic fire detection system installed in all computer operation and support areas, including air conditioning and electrical equipment, transformers, switchgear, and underfloor spaces?

_____ Is the fire detection system tested regularly?
_____ Are emergency power-down controls located at both an easily accessible location and a remote location?
_____ Are all computer operators familiar with emergency power-down procedures?
_____ Are the HVAC and data processing equipment tied together electrically to facilitate emergency shutdown?
_____ Are the automatic fire suppression systems inspected and tested regularly?
_____ Are salvage covers (placed over equipment to protect it from debris, water, etc.) on hand and readily accessible?
_____ Are all portable fire extinguishers of a type that will not damage the equipment, and are they accessible?
_____ Have all employees received hands-on training with each type of portable fire extinguisher used?
_____ Have all employees participated in an emergency drill, including shutdown, salvage covering, and evacuation?
_____ Is a no-smoking policy strictly enforced?
_____ Do waste containers have fire-suppressant lids?
_____ Is housekeeping consistently good in all computer and support areas?
_____ Are supplies of paper, cards, and other combustibles minimized in hardware areas?

WATER

_____ Is the floor over the electronic equipment waterproofed?
_____ Is the equipment located on a noncombustible, raised floor?
_____ Are the age, condition, and contents of overhead pipelines known?
_____ Are computers located above ground level? If not, are watertight corridor dams accessible?
_____ Has an automatic water detection system been installed? When was it last tested? Did the alarms work?
_____ Are there multiple drains, and do they all have backflow prevention?
_____ Are rules prohibiting beverages around electronic equipment observed and enforced?

ENVIRONMENT

_____ Are temperature, humidity, and air quality within the equipment manufacturer's recommended limits?
_____ Is the HVAC system for the computer room independent of the central air-handling system for the building?
_____ Does the computer room's HVAC system have an automatic restart feature?
_____ Is the computer room sealed to help maintain optimum environmental conditions?
_____ Are air filters serviced regularly?
_____ Does the computer/control room have a slightly positive air pressure?
_____ Is vibration a problem?
_____ Are signs present which warn against the use of magnets?
_____ Are radio-transmitting antennae or high-voltage electrical power lines located close enough to cause interference?

ELECTRICAL POWER

_____ What power protection devices have been installed to protect electronic circuitry?
_____ Are power protection devices inspected regularly?
_____ Are computers and other sensitive equipment grounded independently of the rest of the building's equipment?

SAFEGUARDING COMPUTER SYSTEMS

_____ Is the equipment installation in accordance with the National Electrical Code?
_____ Are electrical receptacles and connectors listed by a recognized testing laboratory?
_____ Is ventilation adequate in underfloor spaces and around uninterruptible power supplies?
_____ Does incoming power pass through a single common power panel, and is that panel labeled?

SECURITY

_____ Are computer rooms located away from routes frequented by visitors, and are they invisible from outside the building?
_____ Depending on the type of security system used, are codes, keys, and pass cards tightly controlled and changed as recommended in the text?
_____ Are equipment cabinets kept locked?
_____ Have employees been trained in the need to maintain security and security procedures?
_____ Is the presence of strangers not tolerated?

CONTINGENCY PLAN

_____ Are written lease and/or mutual aid agreements for backup hardware current?
_____ Is backup hardware checked periodically to ensure compatibility?
_____ Does a contingency plan exist which includes the names and responsibilities of key personnel?
_____ Is the contingency plan updated when necessary?
_____ Are several copies of the contingency plan kept off-site with key personnel?
_____ Have software and supporting data been copied, and is one set kept off-site at a properly protected location?
_____ How often are backup copies made of critical data?
_____ Is the integrity of off-site data checked periodically?

BIBLIOGRAPHY

Bernhard, Robert, "Breaching System Security," *IEEE Spectrum,* June 1982, pp. 24-31.
Bhagat, D. O., H. E. Brandmaier, and J. E. Ford, "More Power to You," *Datamation, Vol. 29, No. 10,* October 1983, pp. 121-130.
Coates, Joseph F., "The Future of Computer Data Security," *Vital Speeches of the Day,* February 15, 1982, pp. 280-284.
Courtney, Robert H., Jr., "Computer Security: The Menace Is Human Error," *The Office,* March 1984, pp. 119-120.
Dietel, Harvey M., *An Introduction to Operating Systems,* Reading, MA, Addison-Wesley, 1984, Chapter 17.
Fernandez, Edwardo O., Rita C. Summers, and Christopher Wood, *Database Security and Integrity,* Reading, MA, Addison-Wesley, 1981.
Haack, James L., "Upgrading a DP Center," *Datamation,* October 1983, pp. 108-116.
IBM Series/I Customer Site Preparation Manual, International Business Machines Corporation, 1983.
Koth, George C., "Electronic Computer/Data Processing Equipment," *Fire Protection Handbook,* 15th edition, Quincy, MA, National Fire Protection Association, 1981, Section 11, Chapter 1.
NFPA Codes, Standards and Recommended Practices, "NFPA 75, Standard for the Protection of Electronic Computer/Data Processing Equipment," Quincy, MA, National Fire Protection Association, 1981.
Physical Security, Field Manual No. 19-30, Washington, DC, U.S. Department of the Army, March 1979.

"Power Sources," *Product Review, EDN,* July 11, 1985, pp. 189-220.

"Protect Your Computer, Protect Your Business," *The Sentinel,* Industrial Risk Insurers, First Quarter, 1983.

Pryor, Douglas, and Amy Smith, "Power Line Protection," *Business Computer Systems,* December 1983, pp. 131-133.

Ralston, August, "Controlling Computer Crime," *Risk Management Manual,* Volume 1, Santa Monica, CA, Merritt Company, 1990.

Richards, Thomas C., "Improving Computer Security Through Environmental Controls," *SIGSAC Review,* Summer 1982, pp. 18-24.

Riley, Robert, "Computer Centers," *Industrial Fire Hazards Handbook,* Quincy, MA, National Fire Protection Association, 1979, Chapter 17.

Sippl, Charles J., and Richard J. Sippl, *Computer Dictionary and Handbook.* 3rd edition, Indianapolis, Howard W. Sams, 1982.

Spindle, Les, "Computer Corner: Computer Security," *Radio-Electronics,* February 1984, pp. 111, 114.

"Spreading a Virus," *Wall Street Journal,* November 7, 1988, pp. A1, A6.

Steinbrecher, David, "Automation Outlook," *Office Administration and Automation,* April 1984, p. 8.

Tucker, Elizabeth, "After the Fire: Picking Up the Pieces," *The Washington Post,* December 28, 1987.

"Water, Water Everywhere—but Shipyard Stays Afloat," *Computer World (staff report),* October 29, 1984, p. 25.

Chapter 18 / Part 1

Hazardous Waste Management: A Problem of Growing Concern

INTRODUCTION

What is meant by the term "hazardous wastes"? This is a complex and confusing issue, at best, since some wastes are dubbed "hazardous" simply because the EPA has placed them on its hazardous substances list. In addition, some states and local governments have added their own agents to the EPA list. To make the situation even more confusing, the general public, through the impact of generally exaggerated press and television coverage, has concluded that every chemical represents a toxic hazard. Actual data suggest that about 1 percent of hazardous waste generators account for about 90 percent of the waste created in the United States.

Hazardous wastes range from synthetic organic chemicals to heavy or toxic metals, to inorganic sludges, to solvents, to dilute aqueous waste streams. They may be solid, liquid, or gaseous; they may be pure materials, complex mixtures, residues and effluents from operations, discarded products, or contaminated containers or soil. Most hazardous wastes are managed on the site where they are generated, more so in some states and industries than others, and more likely by larger plants and generators than by small ones.

Scope of the Problem in the United States

The Office of Technology Assessment (OTA) estimates the total amount of hazardous waste generated annually to range from 255 to 275 million metric tons. Recent EPA data put the figure at 150 million metric tons, but it could rise. In the United States, federal and state databases for this problem are being developed.

The scope of the hazardous waste problem is enormous. It is estimated by the EPA that as many as 150 million metric tons of hazardous wastes are generated in the United States annually, primarily by the chemical industry and other heavy manufacturing sectors. About 95 percent of these wastes are managed on-site by the generator. The rest, about 2 billion gallons annually, are managed off-site.

Waste Treatment and Storage Facilities

Some 8500 hazardous waste treatment storage and disposal facilities are currently in operation. At the close of fiscal year 1981, the EPA and states with interim authorization from the EPA to run their own hazardous waste management programs had "called in" for review applications for final Resource Conservation and Recovery Act permits for more than 1700 storage and treatment facilities, 150 incinerators, and 100 land disposal sites. Currently, approximately 200 final and draft storage and treatment permits have been issued by the EPA.

Early Seeds of a Problem to Be

The hazardous waste problem in the United States began to develop as early as the 1950s at Love Canal in New York State. At that time, however, people had no idea of just how uncontrolled the situation could or would become. As a result of public outrage over such sites as Love Canal and Woburn, Massachusetts, Congress developed and enacted the Resource Conservation and Recovery Act (RCRA) of 1976 and the Comprehensive Environmental Response, Compensation, and Liability Act (CERCLA) of 1980. Until these laws were passed, if the problem was handled at all, it was handled in a very disorganized fashion.

The identification of priority uncontrolled waste disposal sites has been a slow and tedious process, and remedial action at such sites is finally taking place. The complexity of the situation has significantly contributed to the disorganization involved in carrying out remedial action. Liberally every site is unique. This fact has made progress difficult, because it has not been possible to come up with an easy "cookbook" approach to corrective action at any given site.

Flexible Action Plans Are Needed

In light of the overall uncertainty and ambiguity in this situation, it is imperative that a formal plan of action be developed. The plan must be flexible enough to apply to the unique characteristics of a given site, yet structured enough to apply certain practices to every situation. What is needed is a framework which is universal, but with space to adapt to subtle differences in its application. The answer, in this instance, is a systems approach to remedial action. By its nature, this approach will provide the structural framework which applies to every site, while allowing the flexibility needed to perform the remedial action in a variety of situations, using a variety of available technologies.

LEGAL ASPECTS OF COMPLIANCE

In addressing the legal issues associated with remedial action, much of the legislative and regulatory concern occurred first at the federal government level. Accordingly, the major area of concern is compliance with federal regulations. Increasing attention at state levels, however, has prompted 31 states to pass Superfund-type laws and regulations. In addition to federal and state laws, an organization must be cognizant of the many local ordinances and referendums now being passed which affect its locations.

For the most part, federal and state efforts follow along similar lines. There are some problems, though, which will be addressed at the end of this section. The local level, however, is a separate consideration. Many municipalities and other subdivisions have taken this problem to heart and devised rules deemed to be in the best interest of the community. Unfortunately, a common result of such efforts is confusion and, often, conflict.

A complete discussion of all the federal and state regulations pertaining to hazardous wastes would require a lengthy textbook, as well as an experienced attorney to assist in their interpretation. A summary of the federal legislation follows. (Also see Figure 18-1).

AGENCY	ACT	Main Provisions, Applications, Effects of Act
EPA Environmental Protection Agency	**RCRA** Resource Conservation and Recovery Act	Applies to: 　Generators (producers) of Hazardous Waste 　Transporters 　Treatment Facilities 　Storage Facilities 　Disposal Facilities
	CERCLA The Comprehensive Environmental Response, Compensation and Liability Act (Superfund)	Establishes Liability for Costs related to Cleanup of Hazardous Substances Releases, and also uncontrolled or abandoned Hazardous Waste Sites. Maintains a National Priorities List (NPL) which identifies sites as being of high priority for cleanup action. Establishes a Hazardous Substances Response Trust Fund 　Taxing structure and authorization for Superfund which acquires money from these sources: 　a.　Tax per barrel of crude oil 　b.　Tax per ton of specifically listed chemicals used or sold 　c.　Money from general revenue funds as appropriated by Congress 　d.　Post closure tax per dry ton of material.
OSHA Occupational Safety and Health Agency, U.S. Dept of Labor	Occupational Safety and Health Act	General industry standards CFR 1910 apply to workers employed by hazardous waste treatment, transporters, disposal site operators.

Figure 18-1. Federal agencies and legislative acts aimed at hazardous waste management.

Remedial Action Under Federal Legislation

Remedial action for hazardous waste facilities falls under three major federal laws:

1. The Resource Conservation and Recovery Act (RCRA) of 1976.
2. The Comprehensive Environmental Response, Compensation, and Liability Act (CERCLA) of 1980 which was combined with other legislation under the Superfund Act of 1985 (See Chapter 18-3.).
3. The Occupational Safety and Health Act (OSHA) of 1970.

While RCRA and CERCLA specifically address hazardous waste management, OSHA provides for protection of the workers carrying out remedial actions at hazardous waste sites. Each of these laws is complex. In general, the laws are independent of each other. They are somewhat interrelated by reference, but for the most part, each addresses separate parts of the waste management situation. Collectively, these laws are the most comprehensive regulatory effort thus far in attempting to control the hazardous waste problem in the United States.

RCRA. The primary objective of RCRA is to provide for a comprehensive means of regulating hazardous waste. The concept of "cradle to grave" management was derived from RCRA because of the intended scope of the law and its respective regulations. It was the intent of Congress to protect the environment through every stage of the material's life. To this end, RCRA provided for development of regulations which would cover hazardous waste generators, transporters, and treatment, storage and disposal facility operators.

1. Definition of Hazardous Waste
   a. RCRA defines hazardous waste as "a solid waste, or combination of solid wastes, which because of its quantity, concentration, or physical, chemical, or infectious characteristics may
      (1) cause, or significantly contribute to an increase in serious irreversible, or incapacitating reversible illness; or
      (2) pose a substantial present or potential hazard for human health or the environment when improperly treated, stored, transported, or disposed of, or otherwise managed."
   b. RCRA defines solid waste as "any garbage refuse, sludge from a wastewater treatment plant, water supply treatment plant, or air pollution control facility, and other discarded material, including solid, liquid, semi-solid or contained gaseous materials resulting from industrial, commercial, mining, and agricultural operations, and from community activities, but does not include solid or dissolved material in domestic sewage, or solid or dissolved materials

in irrigation return flows or industrial discharges which are point sources subject to permits under Section 402 of the Federal Water Pollution Control Act, as amended (86 Stat. 880), or source, special nuclear, or by-product material as defined by the Atomic Energy Act of 1954, as amended (68 Stat. 923)."

2. Identification of Criteria for and Listing of Hazardous Waste (III)

Section 3001 of RCRA required the development of specific criteria for identification and the creation of a list of materials classified as hazardous waste. The criteria developed to determining if a solid waste is hazardous include the following:

a. Ignitability
b. Corrosivity
c. Reactivity
d. Toxicity

In addition to these criteria, which are defined in detail in the regulations, a material may be listed as a hazardous material as a result of its source of origin. Examples include waste water treatment sludges, plating materials, specific industrial processes and others.

Much of the burden of RCRA and subsequent regulations falls upon the hazardous waste generators. Determining of whether a waste is hazardous, providing manifest documentation, and ensuring that a waste shipment is sent to an EPA-permitted disposal facility are among the specific obligations of the generators.

CERCLA. CERCLA, better known as the "Superfund Act," deals with a number of problems related to the uncontrolled release of hazardous substances into the environment. A subset of this is uncontrolled hazardous waste sites, which may include abandoned dumps, inactive sites, or still operating facilities. CERCLA was enacted to deal with hazardous substance releases. In addition, this law designates who is liable for the costs of cleanup activities. Superfund came about, for the most part, because of widespread recognition of the many uncontrolled hazardous waste sites throughout the nation and the need to provide a mechanism to accelerate their control and cleanup. Funding for the CERCLA was set to expire in 1985. Congress was then to examine and consider extending and perhaps changing the law. This they did, and major changes were made. These changes were so great, and the impact was so widespread, that a separate chapter of this book (Chapter 18, Part 3) covers the Superfund Act of 1986. The main differences in the revised law is the right-to-know provision and the emergency response to accidental chemical releases. However, this act is based largely on the 1980 legislation, which is covered here as a basis for the later chapter.

Diverse Supervision by EPA. Most aspects of this law are implemented by the EPA, although a number of other federal agencies also have defined responsibilities. Moreover, the states, if they choose, can play a major role in implementing these programs under the guidance and supervision of EPA.

The progress of the RCRA and Superfund programs has been surrounded by controversy, inspired in part by broader events within EPA. The hazardous waste programs have been criticized by virtually all interested and affected groups—ranging from citizens, industry, and Congress to environmental groups and states—for moving too slowly, for being ineffective in protecting public health and the environment, and for being too complex, costly, and sporadic.

Liability Established by EPA. Implementation of CERCLA has proved to be a complex and multifaceted process. Essentially, Congress provided government with the authority to address hazardous waste emergencies, established a trust fund to provide for response and cleanup activities, and created a taxing mechanism to generate revenue for the trust fund. CERCLA established broad liability for all those who generate, transport, store, treat, and dispose of hazardous wastes. The program is still new, and our experience to date with its full range of authorities is limited.

Priorities for Cleanup and Removal. Under the removal program, over 1,000 actions have been approved so far. The EPA expects to spend $8.5 billion for Superfund cleanup from 1986 to 1981, about $1.25 billion a year, a significant increase over the $25 million spent in 1983.

On the remedial side of the Superfund program, similar growth has been taking place and will continue in the years to come. More than 32,000 sites already have been identified. Of these, 1,218 are on EPA's National Priorities List (NPL); the agency proposes to add another 900 sites. There will be more.

1. Hazardous Substance Response Trust Fund (Section 211)
   This section establishes the taxing structure that generates the Superfund. In general, the $1.6 billion fund will acquire money from three sources:
   a. A tax of 79 cents per barrel of crude oil received at a U.S. refinery and for all petroleum products entering the United States for consumption, use, or warehousing.
   b. A tax levied per ton of forty two listed chemicals if sold or used by a manufacturer, producer, or importer thereof. There are some exemptions to this rule.
   c. General tax revenues in the amount of $44 million per year between 1981 and 1985. At the time of enactment, this tax structure was to remain in effect until 1985 under the stipulation that EPA submit an assessment of future needs, which might require extension.

A postclosure tax and trust fund (Sections 231 and 232) was to be established. This section of CERCLA provides for a tax on hazardous waste and for development of a trust fund for expenses related to postclosure activities at a permitted disposal facility. The tax in this case amounts to $2.13 per dry ton of waste received at a qualified hazardous waste disposal facility. The owner or operator of the facility is responsible for payment of the tax.

RCRA vs. CERCLA. Having provided a detailed breakdown of these laws, a comparison is necessary to explain the relationship between them. One distinct difference between RCRA and CERCLA is the noticeable absence of regulatory activity associated with CERCLA. For the most part, only the NCP and certain requirements for liability required additions/revisions to the Code of Federal Regulations. RCRA, however, required extensive efforts on the part of EPA for the development of regulations.

Proactive vs. Reactive Functions. The second major difference lies in the concepts of proactive and reactive functions. RCRA functions in a proactive manner, addressing present and future activities in hazardous waste management. CERCLA remedies problems now and in the future for improper disposal methods in the past. Finally, both laws establish trust funds for postclosure activities for disposal sites. The one item that both specifically preclude, though, is the compensation of persons whose health is impaired by exposure to released hazardous substances.

OSHA. OSHA was enacted to provide a safe, healthy workplace for all working Americans. To this end, specific standards were promulgated in 29 CFR 1910 and under specific state programs as applicable.

Protecting employees who engage in response activities requires knowledge of the types of hazards to which the worker is exposed, the protective measures applicable to specific hazards, and applicable OSHA standards. Apart from specific OSHA standards which provide at least minimal protection, each hazardous waste response team should have on staff, or within reach through consultation, a professional health and safety specialist who is familiar with good industrial hygiene and safety practices. In addition, it is advisable for this person to be experienced in the unique problems of hazardous waste management.

Safety Programs Need Constant Updating. In this chapter, site hazards and protection measures are addressed generically. This approach allows us to make general statements that are applicable to response activities at any hazardous waste site. There is no doubt, however, that within given time constraints, the safety program should be reviewed and updated for each remedial action site. This is recommended because, although the types of hazards may be very similar from site to site, the specific materials present and the physical characteristics of each site will vary.

Associated Problems. Apart from the sheer complexity of these laws and regulations, there are several other problems associated with the legal issues surrounding hazardous waste. The primary concern is to address the needs of people who suffer health problems as a result of exposure to released materials near hazardous waste sites. Neither RCRA nor CERCLA provides a means of compensating these victims for their illness and suffering associated with exposure. At present, this issue is handled in court under tort laws. A variety of laws have attempted to address this problem, and certain state laws do provide compensation for victims of toxic waste.

Dichotomy of Laws. Some of the other legal issues involved in hazardous waste include general liability, bankruptcy as a means of avoiding liability, state vs. federal programs, and local referendums with the attitude of "not in my backyard." Environment law and the concept of toxic torts are still relatively new to the legal profession. It will be many years before the evolutionary process of law provides answers to the complex legal issues associated with hazardous waste.

SOCIAL AND LEGAL INFLUENCES

As this brief overview of the major federal laws shows, management of hazardous waste can be a major undertaking, requiring significant resources and personnel. Such a management scheme is beyond the scope of this review.

The hazardous waste problem itself is very complex. When considered in combination with other social and legal influences, this problem is compounded. Given this complexity, and the unique factors involved in addressing Superfund sites, any attempts to deal with hazardous waste must take an integrated systems approach. It is imperative that all of the factors influencing remedial activities be identified and accounted for. In addition, it must be recognized from the beginning that any solution is the result of compromises and trade-offs. This approach is used in the hope of satisfying the needs of all the parties or system parameters which must be addressed.

Remedial Actions Quantified

A systems approach to remedial action brings together all of the influencing factors, and accounts for the system parameters and any outside constraints placed upon the system. To understand this approach, several concepts need to be explained. Malasky (1982) defines a system as "a composite, at any level of complexity of personnel, materials, tools, equipment, facilities, environment, and software. The elements of this composite entity are used together in the intended operational or support environment to perform a given task, to achieve a specific production, or to support a mission requirement."

Site System Defined

Based upon this definition, a hazardous waste site may, as a whole, be considered a system. In addition, any of its components (soil groundwater, liner, etc.) may be considered a system. For the purposes of this project, the entire Superfund site, including any remedial actions, will be considered one system. Expanding upon the concept of a system, Malasky defines system safety as: "an optimum degree of safety, established within the constraints of operational effectiveness, time, and cost, and other applicable interfaces to safety, that is achievable throughout all phases of the system lifecycle."

This definition provides the starting point for the systems approach to remedial action. Because of the stipulation of optimal safety within certain constraints, this approach falls in line with the realistic understanding that there must be trade-offs and compromises in achieving a workable solution. Furthermore, the systems approach presupposes an understanding of, and accounting for, all of the interfacing parties.

IDENTIFICATION OF STAKEHOLDERS AND INTERFACING CONSIDERATIONS

The first step in the systems approach (see Figure 18-2) is to identify the interfacing concerns, formally called "stakeholders." In the case of Superfund actions, the major stakeholders include:

1. Site owner/operator.
2. Hazardous waste transporters using the site.
3. Hazardous waste generators using the site.
4. Surrounding area populations.
5. Future generations to be located at or near the site.
6. Shareholders in all companies concerned.
7. Regulatory groups: DOT, EPA, OSHA, state and local agencies.
8. Employees of the companies involved and their families.
9. Employees of the regulatory agencies (inspectors).
10. Consumers of products manufactured by generators.
11. Lawmakers and other professionals.

With the exception of shareholders and consumers, the interests and concerns of the stakeholders have been addressed in previous sections. For the most part, shareholders' and consumers concerns are financial, unless they also happen to fall into one of the other groups. To a certain extent, the financial interests are covered under the consideration of cost in evaluating the feasibility of a given remedial measure.

1	Hazardous Waste Site Owner/Operator
2	Hazardous Waste Transporters Using Site
3	Hazardous Waste Generators Using Site
4	Population in Areas Nearby or Surrounding Site
5	Future Populations in Areas
6	Shareholders of Companies Concerned
7	Federal, State and Local Agencies: DOT, EPA, OSHA
8	Employees and their Families of Companies Concerned
9	Employees of Regulatory Agencies (Inspectors, Compliance Officers)
10	Consumers of Products Manufactured by Generators
11	Legislators
12	Professionals Involved with Hazardous Waste Management Cleanup and Removal

Figure 18-2. Stakeholders in Superfund hazardous waste cleanup actions.

The System Safety Program

The final concern, from a systems standpoint, is the process of bringing everything together in a workable framework which covers the situation as thoroughly as possible. The mechanism for accomplishing this task is a system safety program which provides for a formal method of dealing systematically with the problems. This program is specifically designed for a given site, tailored to the unique needs of the site, and structured in such a way as to include the entire life cycle of the system. The system safety approach for hazardous waste management will be discussed in more detail in Chapter 18, Part 2.

CONCLUSION

The hazardous waste problem is perhaps one of the most complex problems facing society today. Many conflicts and interfaces occur simultaneously at a given disposal site. When one considers the thousands of disposal sites located throughout the United States and the unique characteristics of each, the enormity of the task of dealing with the hazardous waste becomes apparent. In an attempt to gain some control over the situations, or at least begin to gain control, society must deal with all of the factors influencing this problem. Satisfaction of all the needs of all stakeholders may never be possible. However, by accounting for those needs and recognizing the requirement to prioritize them, society begins to attain some semblance of balance and control. The systems approach provides the mechanism necessary to identify and account for all of the stakeholders and their needs. It also provides the flexibility needed to adapt to the unique characteristics of each site, as well as to the changes occurring outside of the site, while maintaining a structural framework which allows the control of hazardous waste site.

BIBLIOGRAPHY

Code of Federal Regulations. 40 CFR 260, p. 356, Parts 190-199, revised July 1, 1983. ibid, p. 357; ibid, 40 CFR 261.21-261.24, 261.31-261.33, 262-267, and 270; ibid, 40 CFR 300.

Comprehensive Emergency Response, Compensation and Liability Act of 1980, PL 96-510, December 11, 1980.

EPA Interim Standard Operating Guides, Revised September 1982. Washington, DC, Office of Emergency and Remedial Response, Hazardous Response Support Division.

Handbook for Remedial Action at Waste Disposal Sites, USEPA, 625/6-82-006. Washington, DC, June 1982.

Handbook for Evaluating Remedial Action Technology Plans, USEPA, 600/2-83-076. Washington, DC, August 1983.

Malasky, S. W. *System Safety, Technology and Application,* New York, Garland STPM Press, 1982.

Chapter 18 / Part 2

Hazardous Waste: Responding to Chronic and Acute Episodes— A Systems Safety Approach

INTRODUCTION

In Chapter 18, Part 1, attention was drawn to the complex issues associated with the problems of hazardous waste: the extent and magnitude of the problem; and the various complexities of overlapping and sometimes duplicative federal and state laws, whose purpose is to regulate and control hazardous waste and their subsequent safe disposal and handling.

It is necessary to bring everything together in a workable framework which will cover a situation as thoroughly as possible. Attention must be given to all of the diverse and conflicting variables so as to reach an appropriate decision based upon sound logic and judgment. Such a mechanism is a system safety approach providing a format for dealing systematically with the problem. Such a program must be designed for each given site, tailored to its unique needs, and structured so as to include the entire life cycle of the system.

Public Anxieties

The hazardous waste problem is one of the most complex issues facing society today. This is so especially in view of the tremendous public concern created by the extensive press coverage, and the resultant fears and anxieties of an uninformed or improperly educated population.

The purpose of this chapter is to introduce the concept of a system safety approach to hazardous waste and to discuss its potential as a management tool that provides the flexibility needed to adapt to the unique characteristics of each site. The changes occurring outside the site must also be considered while maintaining the structural framework to accomplish the overall objective: to bring a hazardous waste site under control.

The term used throughout this chapter is "system safety." This should not be confused with the science of systems or fault tree analysis, as used in the academic sense. Rather, it should be viewed as a systematically organized, administrative approach to hazardous waste management within a structural framework.

GENERAL CONSIDERATIONS FOR A SYSTEM SAFETY PROGRAM

Given the consequences of potential public or occupational health exposures at uncontrolled hazardous waste disposal sites, a comprehensive program of control needs to be established. It takes the form of a system safety program because of the broad nature of any effort to address the full scope of the problem. The specific elements of such a program are outlined below.

The system safety program should be developed before any remedial actions are taken at the site. It must address the entire life cycle of the project in order to provide for the complete integration of all functions. This integration should be accomplished in a cost-effective manner, within the *constraints of technological feasibility*. It is also necessary to maintain a level of safety which protects the general public and the environment.

Commonality of Requirements

The requirements specified in the program must apply to all personnel and activities at a given site. The owner/operator must maintain full control of and responsibility for the management and implementation of the program in cooperation with a management committee representing all of the major functions. If the owner/operator of the site is not known, the remedial action contractor, in cooperation with the local EPA emergency response coordinator, shall assume control of the project and the system safety program. A basic assumption of this program must be complete voluntary compliance with all applicable acts, codes, standards, and regulations. Performance reviews and contract renewal evaluations include an assessment of activities related to safety and health.

The federal laws involved in remedial action for hazardous waste facilities include RCRA, CERCLA, and OSHA. While RCRA and CERCLA specifically address hazardous waste management, OSHA provides for protection of the workers carrying out remedial actions at hazardous waste sites. Each of these laws, as previously stated, is complex in its own right. In general, the laws are independent of each other but are interrelated by reference. An overview of these regulations was provided in Chapter 18, Part 1.

Other Codes and Regulations

In addition to these laws, other codes and regulations must be considered in designing a systems safety approach to the hazardous waste problem. These include the following:

1. Title 29, Code of Federal Regulations, Part 1910—General Industry Standards and all consensus standards incorporated therein; or specific State regulations, as applicable.
2. Title 40, Code of Federal Regulations, Parts 260 to 270—Hazardous

Waste Regulations (RCRA); and/or specific state and local regulations, as applicable.
3. Title 49, Code of Federal Regulations, Hazardous Materials Transportation; and/or specific state and local regulations, as applicable.

An additional source of information that serves as a guideline in the development of such a program is found in MIL-STD 882A, System Safety Programs for Systems and Associated Subsystems and Equipment (MIL Regulations, June 28, 1977).

RESPONSIBILITIES AND AUTHORITY

The system safety organization should be positioned in the overall management structure so that direct access to top management is ensured. Representatives of the system safety staff should participate in top-level decision-making groups so that decisions are based on solid information. This will help to minimize oversights and/or omissions in the program.

Each project should have a director of health and safety as well as a director of environmental engineering. The following professionals and groups need to be included:

1. A system safety group, the size of which depends on the size of the site and any variables known prior to planning for remedial action. This may vary as activity at the site varies.
2. Occupational health and safety staff, to be determined by site parameters.
3. Environmental engineering staff, to be determined by the same parameters.

Each of these groups should be composed of professionals whose qualifications meet or exceed those prescribed by the certifying body in the given field.

Specific Responsibilities

The primary responsibilities of the system safety group are (1) to ensure proper development of the plan in order to meet the basic assumptions of the program and (2) to implement it after it is endorsed by top management. Specific provisions for authority and accountability need to be developed which support the responsibilities of the system safety organization.

The responsibilities of the system safety group are as follows:

1. Development of all training requirements, including but not limited to occupational health and safety, hazard awareness, toxicology, performance of emergency procedures, accident investigation, record keeping, and supervisory training.
2. Review of all applicable engineering plans, with input into analysis and upgrading of the system.

3. Hazard analysis and reporting.
4. Development of occupational, public health, and safety programs.
5. Input to management decision making.
6. Providing adequate health and safety staffs to meet the needs of the site.
7. Interface with other departments to ensure cooperation between contractors in achieving system safety objectives.
8. Development and implementation of system safety milestones, and evaluation of the program at each milestone.
9. Development and implementation of emergency action plans; accident investigation procedures; environmental and personnel biological exposure monitoring procedures; public relations programs; site security procedures; record-keeping requirements; database provisions; and provisions for site closure and postclosure procedures.
10. Acquisition of all applicable operating licenses in compliance with their requirements.
11. Review of the program, as necessary, to coincide with site changes made in reference to remedial activities and needed program revisions.
12. System safety audits of all integrated functions to ensure compliance with, and upgrading of, applicable programs and activities.
13. Assurance that all company, contractor, and visiting personnel are made aware of, and comply with, all applicable regulations and policies concerning activities at the site.
14. Power to require all contractors to submit documentation of system safety efforts related to the actual and potential hazards created by their contribution to the remedial project; and provision for interfacing personnel by the owner/operator or managing contractor.
15. Development of system safety programs, in addition to those generally stated, which address specific needs at a given milestone.

Program Criteria

There are a number of criteria that need to be defined in the program. These include hazard level categories, system safety precedents, and analytical techniques. For definition purposes, these are described as follows:

1. Hazard Level Categories
   For remedial actions at uncontrolled waste disposal sites, the hazard categories are:
   a. Class IV: *Safe*—does not result in injury and/or damage to the equipment.
   b. Class III: *Marginal*—results in release of hazardous waste to the soil, which is detected and corrected before public exposure may occur.
   c. Class II: *Catastrophic*—results in release and subsequent contamination of soil and ground or surface water. There may or may not be

subsequent public health exposures resulting in illness, injury, or death. Release would require specific remedial measures capable of correcting the problem.
   d. Class I: *Critical*—results in release with subsequent contamination and public health effects. Remedial actions may not be capable of mitigating future exposures. There is significant present and future risk of illness, death, and/or genetic damage.
2. System Safety Precedents
   Management should ensure both the authority and accountability of all departments so that a Class I or II hazard is prevented, as fully as possible within the limits of the technology for hazardous waste management. If these hazards cannot be eliminated, control methods will be implemented to minimize the severity and probability of occurrences. Methods will be employed to counteract the occurrence of Class II, III, and IV hazards. If such methods are or become unavailable, the order of priority for control is as follows:
   a. Engineering to meet counteractive needs.
   b. Engineering controls to prevent release of hazardous waste to the point of public health exposure.
   c. Work practices and administrative controls or training.
   d. Personal protective equipment to provide protection from released materials.
3. Analytical Techniques
   The owner/operator will ensure completion of an environmental impact evaluation in choosing a specific remedial technology for the site, and will take all technologically feasible steps to mitigate further degradation of the environment. In addition, an environmental impact statement will be prepared upon completion of the evaluation. It is also the responsibility of the owner/operator to prepare preliminary hazard and fault tree analyses that determine the various means by which a release and occupational or public health exposure to hazardous waste may occur. The analyses should also define the means of control which may be available. Finally, to the extent possible, the owner/operator will perform a quantitative analysis and sensitivity study of the fault tree for potential exposures to hazardous waste at the site.

Four Phases of the Task

In conjunction with the system safety program as specified above, and assuming that the criteria previously described have been successfully met, the following system safety tasks can now be accomplished and divided into four distinct phases: concept formulation, development and production, operations, and, disposal.

Formulation. In the concept formulation phase, there are usually three major functions. First is a review of all environmental studies, with particular attention to the hydrological, geological, and meteorological factors. During the investigation for environmental impact statement evaluations, the appropriate remedial technologies to be used should also be determined. Second, top management must be informed about the system safety program and its requirements. Coordination and interfacing activities of management functions now begin. Third, management must be made aware of the identity of all the input groups, and each group must be assigned its own responsibilities within the overall project. This is essentially a coordination activity.

Development and Production. In the development and production phase, a number of tasks must be completed. These include:

1. Review of all purchasing documents and approval of all materials and equipment to be purchased.
2. Assurance of all quality control testing and documentation procedures, as well as all necessary actions to be taken in connection with them.
3. Development and implementation of site control and decontamination procedures, such as those described by the EPA.
4. Development of operations manuals and procedures for the chosen remedial action.
5. The conduct of inspections required by specific regulations and appropriate recommendations.

Operations. In the operations phase, four tasks need to be completed:

1. Checks for the completion of preplacement of periodic physical examinations, job training, safety training, emergency procedures; determination of the needs, and scheduling for, refresher training in these areas.
2. Confirmation of installation, inspection, and maintenance procedures, as well as all warning and monitoring systems to be used on site.
3. Review and approval of all engineering changes proposed and implemented at the site.
4. Development and use of accredited methods for sampling, analysis, and documentation of materials in conjunction with remedial action taken.

Disposal. The disposal phase is the final part of the system safety program in relation to the hazardous waste disposal site. During this phase, a closure and postclosure plan is developed and implemented, which includes monitoring and maintenance at the site. It also allows for the establishment of a postclosure fund or insurance coverage for future unanticipated accidents or release of hazardous waste materials from the site after closure.

Documentation. The final step in the design of the system safety approach is the documentation phase. All procedures developed during the planning, development, and action periods of the project need to be fully documented to comply with the regulations of OSHA and EPA.

Remedial Activities

It is imperative that all of the factors influencing remedial activities be identified and accounted for. In addition, it must be recognized from the beginning that any solution is the result of compromises and trade-offs. This approach is used in the hope of satisfying the needs of each of the parties or system parameters which must be addressed. A systems approach to remedial action brings together all of the influencing factors, and accounts for both the system parameters and any outside constraints placed upon the system.

The system safety or structured management approach is applicable to any given hazardous waste site which is amenable to a comprehensive, methodical approach where time is not of the essence. But about an "environmental incident" that requires a quick response?

The knowledgeable manager must be aware of the potential hazards of such an incident and the methods of protecting employees who respond to it. The next section will cover the basic concepts that apply to both a systematic, planned approach to a hazardous waste problem and the emergency response to an incident.

Incident Response

An environmental incident involves a release or a threat of release of hazardous substances; alternatively, it involves hazardous substances that pose an imminent and substantial danger to public health and to the welfare of the environment. Each incident presents special problems. Response personnel must evaluate them and determine an effective course of action.

Any incident represents a potentially hostile environment. Chemicals that are combustible, explosive, corrosive, radioactive, toxic, or biologically active can affect the general public or the environment, as well as workers. Workers may fall, trip, be struck by objects, or be subject to danger from water, electricity, and heavy equipment. Injury and illness may occur due to physical stress and weather. Conditions created by an incident vary widely, causing a broad range of risks to public health and welfare, the environment, or the response personnel. While each incident is unique, there are many commonalities. One of them is that all incident responses require protection of the health and safety of the workers from any hazards present.

Workers' Protection

Toxic or chemically active substances present a special hazard because they can be inhaled, ingested, and absorbed or be destructive to the skin. They may exist in the air or, due to site activities, become airborne or splash on the skin. When ingested or inhaled, the substances may cause no apparent illness or they may kill. On the skin they may cause no demonstrable effects; alternatively, they can damage the skin, be absorbed, and lead to systemic poisoning.

Two types of potential exposure exist:

1. *Acute:* Concentrations of toxic air contaminants are high relative to the type of substance and the criteria for protection against it. Substances may contact the skin directly through splashes, immersion, air, and so on, with damaging results. Exposures occur for relatively short periods of time.
2. *Chronic:* Concentrations of toxic air contaminants are relatively low. Direct skin contact is with substances that have low dermal activity, and exposures occur over longer periods of time.

In general, acute exposures to chemicals in air are more typical in transportation accidents, fires, or releases at chemical manufacturing or storage facilities. Such exposures do not persist for long periods of time. Acute skin exposures occur when workers must be close to the substances in order to control the release or contain and treat spilled material. Once the immediate problems have been handled, exposures tend to become chronic as cleanup progresses.

Chronic exposures usually are associated with longer-term remedial operations. Contaminated soil and debris from emergency operations may be involved, soil and ground water may be polluted, or containment systems may hold diluted chemicals. Abandoned waste sites represent chronic problems. As activities start at these sites, however, personnel engaged in sampling, handling containers, bulking compatible liquids, and so on face an increased risk of acute exposures to splashes, mists, gases, or particulates.

Acute and chronic exposure to toxic substances is one type of hazard. Others are materials that burn, explode, react, emit radiation, or cause disease. All can create life-threatening situations. In any specific incident, the hazardous properties of the materials may be only a potential threat. For example, if a tank car of liquefied natural gas involved in an accident remains intact, the risk from fire and explosion is low. In other incidents, hazards are real and risks high, as when toxic or flammable vapors are released. The health and safety of response personnel requires that the real or potential hazards at an episode be characterized and appropriate preventive measures instituted.

CREATING AN EFFECTIVE PROGRAM

To reduce the risks to workers who respond to hazardous substance incidents, an effective health and safety program must be implemented. At a minimum, this would include the following:

1. Safe work practices.
2. Engineered safeguards;
3. Medical surveillance.
4. Environmental and personnel monitoring.
5. Personal protective equipment.
6. Education and training.
7. Standard operating safety procedures.

Standard Operating Procedures

As part of an integrated program, standard operating safety procedures provide instructions on how to accomplish specific tasks in a safe manner. In concept and principle, standard operating safety procedures are independent of the type of incident. Their applicability at a particular incident must be determined and necessary modifications made to match prevailing conditions. For example, personal protective equipment, in principle, is an initial consideration for all incidents; however, its need and/or the type of equipment required are based on a case-by-case evaluation. Likewise, someone must make the first entry to a site. The approach to be used can only be determined after the conditions prevailing at the specific incident have been assessed.

The purpose of this discussion is to provide standard operating safety guides related to site control and entry. It is not meant to be a comprehensive treatment of the subjects.

Management Guidelines

This introduction can be used as a guideline to familiarize management with the basic procedures used by the safety professional. It is complemented by his or her professional training, experience, and knowledge.

There are many guides or procedures for performing the many tasks associated with responding to environmental episodes involving hazardous substances. These instructions may refer to administrative, technical, or safety matters and may include such items as filing an expense voucher, hiring a contractor, renting a car, using an instrument, following safety procedures, and collecting samples. All these procedures are intended to provide uniform instructions for accomplishing a specific task. In addition to other types of procedures, safety-oriented operating procedures are needed for incident responses. Another purpose of this section is to provide selected standard guides which can be used to develop more specific procedures.

Response Personnel

In responding to accidental releases of hazardous substances or incidents involving abandoned hazardous waste sites, a major consideration is the health and safety of response personnel. Not only must a variety of technical tasks be performed efficiently to mitigate the incident, but they must be accomplished safely. Appropriate equipment and trained personnel, combined with standard operating procedures, help reduce the possibility of harm to response workers.

Some procedures involved in response activities are primarily concerned with general health and safety. In concept and principle, these are independent of the type of incident, but they are adapted to meet site-specific requirements. Other response activities are unique. Each incident must be evaluated to determine its hazards and risks. Various types of environmental samples or measurements may be needed initially to determine the hazards or to provide additional information for continuing assessment. Personnel have to go on site to accomplish specific tasks. Efforts must be made to prevent or reduce harmful substances from leaving the site due to natural or human activities. Containment cleanup and disposal activities may be required. In each of these tasks, response personnel must be protected from existing or potential hazards.

Standard Operating Safety Guides

Standard operating safety guides, which are available through state and federal agencies, cover primarily site control and entry. They illustrate the technical factors to be considered in developing standard instructions. For a given incident, the procedures recommended should be adapted to the conditions of the situation while other safety procedures related to incident response operations are being developed.

Personnel who respond to environmental episodes involving chemical substances encounter conditions that are unsafe or potentially unsafe. In addition to the danger due to the physical, chemical, and toxicological properties of the material(s) present, other types of hazards—such as electricity, water, heavy equipment, falling objects, loss of balance, or tripping—can have an adverse effect on worker health and safety.

This part discusses only the safety measures and precautions associated with the hazardous nature of chemical compounds. Measures are established for reducing the risks associated with hazardous substance response operations. Safety measures to prevent accidents from other conditions must also be considered.

Basic Safety Practices

A number of basic safety practices must be considered in any response to a hazardous waste site or to an episode. These include:

1. Eating, drinking, chewing gum or tobacco, smoking, or any practice that increases the probability or hand-to-mouth transfer and ingestion of material is prohibited an any area designated as contaminated.
2. Hands and face must be thoroughly washed upon leaving the work area and before eating, drinking, or performing any other activities.
3. Whenever decontamination procedures for outer garments are in effect, the entire body should be thoroughly washed as soon as possible after the protective garment is removed.
4. No excessive facial hair, which interferes with a satisfactory fit of the mask-to-face seal, is allowed on personnel required to wear respiratory protective equipment.
5. Contact with contaminated or potentially contaminated surfaces should be avoided. Whenever possible, don't walk through puddles, mud, and other discolored surfaces; don't kneel on the ground; don't lean, sit, or place equipment on drums, containers, vehicles, or the ground.
6. Medicine and alcohol can potentiate the effects from exposure to toxic chemicals. Even medically prescribed drugs should not be taken by personnel on response operations where the potential for absorption, inhalation, or ingestion of toxic substances exists unless such use is specifically approved by a qualified physician. Alcoholic beverage intake should be minimized or avoided during response operations.

As in the systems approach, one should begin to develop a protocol similar to the one referred to in the earlier part of this chapter. However, detail and a long time for planning will not be available due to the necessity for prompt action.

Developing a Protocol

For operational instructions, the following tasks are minimal:

1. The entrance and exit must be planned and emergency escape routes delineated. Warning signals for site evacuation must be established.
2. Personnel should practice unfamiliar operations prior to doing the actual procedure.
3. Personnel on-site must use the "buddy" system when wearing respiratory protective equipment. At a minimum, a third person, suitably equipped as a safety backup, is required during initial entries.
4. During continual operations, on-site workers act as safety backup for each other. Off-site personnel provide emergency assistance.

5. Communications, using radios or other means, must be maintained between initial entry workers at all times. Emergency communications should be prearranged in case of radio failure, necessity for evacuation of the site, or other reasons.
6. Visual contact must be maintained between the pairs on-site and safety personnel. Entry team members should remain close together to assist each other during emergencies.
7. Wind indicators visible to all personnel should be strategically located throughout the site.
8. Personnel and equipment in the contaminated area should be minimized, consistent with effective site operations.
9. Work areas for various operational activities must be established.
10. Procedures for leaving a contaminated area must be planned and implemented prior to going on-site. Work areas and decontamination procedures must be established based on prevailing site conditions.

Medical Program

To safeguard the health of response personnel, a medical program must be developed and maintained. This program should have the two essential components of routine health care and emergency treatment.

Routine health care and maintenance should consist, at least, of a preemployment medical exam to establish the individual's state of health, baseline physiological data, and the ability to wear personal protective equipment. Arrangements should also be made to provide special medical examinations, as well as care and counseling, in case of known or suspected exposures to toxic agents.

Emergency Preparations

Emergency medical care and treatment or response personnel should include the address and telephone number of the nearest medical facility and directions on how to get there. The facility's ability to provide care and treatment for personnel exposed, or suspected or being exposed, to toxic substances should be determined. Information on other emergency facilities such as ambulance, fire and police should be readily available.

All personnel who respond to environmental episodes must be trained to carry out their functions. Training must be provided in the use of all equipment, including respiratory protective apparatus and protective clothing; safety practices and procedures; general safety requirements; advanced first aid; and hazard recognition and evaluation.

Safety training must be a continuing part of the total response program. Periodic retraining and practice sessions not only create a high degree of safety awareness, but also help to maintain proficiency in the use of equipment and knowledge of safety requirements.

Selecting Personnel—A Major Consideration

Personnel who respond to chemical incidents must make many complex decisions regarding safety. To make these decisions correctly requires more than elementary knowledge. For example, selecting the most effective personal protective equipment requires not only technical expertise about respirators, protective clothing, air monitoring, and physical stress, but also experience and professional judgment. Only a competent, qualified specialist has the technical know-how to evaluate a particular incident and determine the appropriate safety requirements. This individual will have acquired expertise through a combination of professional education, on-the-job experience, specialized training, and continual study, enabling him or her to make sound decisions.

Health and safety personnel are major considerations in all response operations. Planning must be done to prepare for a response. Safety practices should include an analysis of the hazards involved and procedures for preventing or minimizing the risk to personnel. Given the toxic nature of the chemical compounds present, all precautions should be taken to prevent unnecessary exposure.

One of the most important practices to be employed when entering any hazardous waste or incident site is the use of proper personal protective equipment.

Protective Equipment

Personnel must wear personal protective equipment when response activities involve known or suspected atmospheric contamination; when vapors, gases, or particulates may be generated; or when direct contact with skin may occur. Respirators can protect lungs, prevent gastrointestinal absorption, and shield eyes against airborne contaminants. Chemical-resistant clothing can protect the skin from contact with potentially dangerous substances. Good personal hygiene habits will limit or prevent ingestion hazardous of materials.

It should be pointed out that the misuse or improper selection of personal protective equipment can be a significant danger to the employee. Therefore, a knowledgeable and trained individual should make this decision.

Four Categories of Protection

Equipment to protect the body against contact with known or anticipated chemical hazards is usually divided into four categories according to the degree of protection afforded. The level of protection should be based upon the type and concentrations of the chemical substances thought to be present; their respective toxicities; and the potential exposure to substances in air, splashes of liquids, or other direct contact with materials due to the work being performed.

In situations where the types of chemicals, their concentrations, and the possibilities of contact are not known, the appropriate level of protection must be selected, based on professional experience and judgment, until the hazards can be characterized.

It is always better to overprotect the individual than to risk undue exposure. While personal protective equipment reduces the potential for contact with hazardous substances, ensuring the health and safety of response personnel also requires safe work practices, decontamination, and site entry protocols. Together, these will establish a combined approach for reducing potential harm to workers.

Protective Clothing—Evaluation and Selection

The fully encapsulating suit provides the highest protection to the skin, eyes, and respiratory system is its material is resistant to the chemical(s) of concern during the time it is worn and/or at the measured or anticipated concentration. While this type of suit provides maximum protection, its material may be rapidly permeated and penetrated by certain chemicals as a result of extremely high air concentrations, splashes, or immersion of boots or gloves in concentrated liquids or sludges. These limitations should be recognized when specifying the type of chemical-resistant garment needed. Whenever possible, the suit material should be matched with the substances to be encountered.

Protective Clothing: Effects on the Wearer. The use of this type of protection and other chemical-resistant clothing requires evaluation of the problems of physical stress, particularly the heat stress associated with wearing impermeable protective clothing. Response personnel must be carefully monitored for physical tolerance and recovery.

Protective equipment, being heavy and cumbersome, decreases dexterity, agility, visual acuity, and so on, increasing the probability of accidents. This probability decreases as less protective equipment is required. Thus, the increased probability of accidents should be considered when selecting the level of protection.

Many toxic substances are difficult to detect or measure in the field. When such substances (especially those readily absorbed by or destructive to the skin) are known or suspected to be present, and personnel contact is unavoidable, protection should be worn until more accurate information can be obtained.

Guidelines for Correct Use of Equipment

The proper use of personal protective equipment is not a simple matter. There are many pitfalls, and management would be wise to rely on the professional judgment and experience of safety and health experts. NIOSH, for example has recently published a series of guidelines for personal protective equipment used in cleaning up hazardous materials. This publication presents step-by-step guides for the selection of respirators, chemical protective clothing, and ancillary equipment, as well as model training, administrative, and emergency programs.

CONCLUSION

In today's complex society, in which we depend upon hundreds of thousands of chemicals for our comfortable existence, the issue of hazardous wastes is discussed almost daily in the public media. Management needs to be aware of the extent and magnitude of the problem and of how to provide a structural management program in order to deal with the cleanup of existing hazardous waste sites and to respond effectively to environmental incidents. This chapter has attempted to outline an approach to these situations. Of primary concern in any hazardous waste response program is the protection of employees and the surrounding population to prevent undue exposure and to minimize risk.

BIBLIOGRAPHY

Birns, Carl, "Grappling with the Problem of Uncontrolled Hazardous Waste Sites," *Hazardous Materials and Waste Management,* November-December 1983, pp. 18-25.

Bixler, B., B. Hanson, and G. Lagner, "Planning Superfund Remedial Actions." Presented at the American Industrial Hygiene Conference, Philadelphia, 1983.

Dumpsite Cleanups: A Citizen's Guide to the Superfund Program, Washington, DC, Environmental Defense Fund.

Ehrenfeld, J., and J. Bass, *Handbook for Evaluating Remedial Action Technology Plans,* USEPA 625/6-82-066. Washington, DC, June 1982.

Federal Register 7/16/82, 40 CFR 300 Part V EPA, National Oil and Hazardous Substances Contingency Plan—Final Rule.

Handbook: Remedial Action at Waste Disposal Sites, USEPA 625/6-82-066, Washington, DC, June 1982.

Malasky, S. W., *System Safety: Technology and Application,* New York, Garland STPM Press, 1982.

Mallow, Alex, *Hazardous Waste Regulations—An Interpretive Guide,* New York, Van Nostrand Reinhold, 1981.

Petak, W., and A. Atkisson, *Natural Hazard Risk Assessment and Public Policy,* New York, Springer-Verlag, 1982.

Sawyer, C. J., and K. A. Stormer, "Environmental Health, Safety, and Legal Considerations for Successful Evaluation of a Dioxin Contaminated Waste Site," Presented at the American Industrial Hygiene Conference, Philadelphia, 1983.

"The Hazardous Waste Training Manual for Supervisors," *Hazardous Waste Bulletin,* 1982, pp

Title Code 29 of Federal Regulations—Occupational Safety and Health, 29 CFR 1910—General Industry Safety Standards, March 1983.

Title 40 Code of Federal Regulations—Environmental Protection, 40 CFR 190-399, October 1983.

Title 49 Code of Federal Regulations—Transportation 49 CFR 100-177, October 1982.

Title 49 Code of Federal Regulations—Transportation, 49 CFR 178-399, October 1981.

HAZARDOUS WASTE: RESPONDING TO CHRONIC AND ACUTE EPISODES   549

Titles 22 and 23 California Administrative Code, Health and Safety Sections.

Turpin, R. D., "Safety Considerations for Hazardous Site and Emergency Spill Operations," Presented at the American Industrial Hygiene Conference, Philadelphia, 1983.

U.S. Congress, Comprehensive Environmental Response, Compensation, and Liability Act, P.L. 96-510, December 11, 1980.

U.S. Congress, Resource Conservation and Recovery Act, P.L. 94-580, October 21, 1976.

U.S. Environmental Protection Agency, *Interim Standard Operating Safety Guides,* revised. Washington, DC, Government Printing Office, September 1982.

U.S. Environmental Protection Agency, *Hazardous Waste Site National Priorities List,* 0-414-955, Washington, DC, Government Printing Office, August 1983.

Warner, Bill, "Emergency Response Company Employs Arsenal of Protective Clothing and Equipment to Safeguard Its Workers," *Hazardous Materials and Waste Management,* November-December 1983, pp. 18-25.

Warner, Bill, "Workers Suited for the Job," *Hazardous Materials and Waste Management,* May-June 1983, pp. 31-32.

Weinberg, D. B., G. S. Goldman, and S. M. Briggum, *Hazardous Waste Regulation Handbook—A Practical Guide to RCRA and Superfund,* New York, Executive Enterprise Publications, 1981.

West's Annotated California Codes, St. Paul, MN, West Publishing, 1979.

Chapter 18 / Part 3

SARA—Superfund Amendments and Reauthorization Act

TITLE III: THE EMERGENCY PLANNING AND COMMUNITY RIGHT-TO-KNOW LAW

Title III of the Superfund Amendments and Reauthorization Act (SARA) of 1986—specifically, Section 313 of the act—establishes the federal, state, and local government requirements and guidelines for businesses and industry for the gathering and submission of information on the release of toxic chemicals into the environment. The intent of this law is to inform the general public and communities surrounding covered facilities about releases of toxic chemicals, aid in research, develop regulations and standards, and coordinate emergency response planning.

This act builds on other regulatory acts and attempts to set uniform requirements that dovetail with those of the Hazard Communication Standard of the OSHA Act.

Whereas the original CERCLA or Superfund legislation of 1980 attempted to deal with emergency responses and cleanup of hazardous waste problems of the past, this new amendment focuses on present and anticipated problems in dealing with toxic materials and, hopefully, will minimize future cleanup costs.

SARA is quite complex, presenting detailed information and reporting requirements, as well as the responsibilities of public agencies at all levels. The purpose of this discussion is to give the senior manager an overview of the act sufficient to understand its implications for his or her organization and the steps to be taken to meet the requirements. No detailed discussion of the chemicals on the EPA list of toxic substances, material safety data sheets, resources, and specific training requirements will be presented. The manager is not expected to be a hazardous materials expert, but rather to have sufficient understanding of internal and external resources and to be able to discuss the matter with other managers and department heads who may be affected.

SARA—SUPERFUND AMENDMENTS AND REAUTHORIZATION ACT

SPECIAL TERMS ASSOCIATED WITH SARA
Some of the terms used in this chapter are as follows:

CERCLA	Comprehensive Environmental Response, Compensation, and Liability Act of 1980 Superfund, administered by EPA.
CGL	Comprehensive general liability (insurance) policy.
EHS	Extremely hazardous substances. Some 406 chemicals are listed under SARA Title III as EHSs.
FDA	Food and Drug Administration.
FIFRA	Federal Insecticide, Fungicide, and Rodenticide Act.
HAZ CHEM	Hazardous chemical as defined under the OSHA Hazard Communications Standard. Includes any chemical for which a material safety data sheet is required
HS	Hazardous substance. One of 717 chemicals specified by CERCLA.
LEPC	Local Emergency Planning Committee. Appointed by the State Emergency Response Commission (SERC).
MSDS	Material safety data sheet.
NRC	National Response Center.
NRT	National Response Team.
RQ	Reportable quantity. Applies to HSs and EHSs.
RRT	Regional Response Team.
SARA	Superfund Amendments and Reauthorization Act of 1986. Revises and extends the authority of CERCLA.
SERC	State Emergency Response Commission.
SIC	Standard Industrial Classification. Created by the Bureau of Labor Statistics. All SICs are listed in the *Standard Industrial Classification Manual* (available from the Government Printing Office).
TPQ	Threshold planning quantity. Applies only to EHSs.

TITLE III—COMMUNITY RIGHT-TO-KNOW (CRTK) ACT: SECTIONS 301-304 AND 311-313 OF SARA
Background of CERCLA and SARA

CERCLA provides authority for federal cleanup of hazardous waste sites. It also requires companies to report to the National Response Center (NRC) the release of a hazardous substance into the environment above certain quantity. The act lists 717 hazardous substances that could, upon release, pollute the environment and cause public harm.

CERCLA, which was set to expire in 1985, did not address the emergency response to accidental chemical releases. The incidents in Bhopal, India, and Institute, West Virginia, showed that highly toxic chemicals are a special emergency response problem. Often, by the time emergency personnel arrive, the spill or released cloud of toxic gas has already damaged public health or the environment and may have dissipated. For such chemicals, early warning and emergency response planning are needed for effective public and environmental protection.

To meet this need, SARA was enacted in 1986. SARA reauthorizes the 1980 law, originally funded at $1.6 billion for 5 years. The new amendments mandate an $8.5 billion Superfund for the 5-year period 1986-1991. SARA authorizes the EPA to order responsible parties to clean up hazardous substances at sites or vessels where there is a release or threatened release into the environment. CERCLA funds are used when the parties responsible for hazardous releases do not clean up the substances voluntarily or refuse to participate in cleanup procedures, and the government proceeds with the task. The government then attempts to collect the funds it has spent.

SARA does not change the basic structure of CERCLA, but greatly expands and defines it. It also establishes new settlement procedures with responsible parties.

Part of SARA is Title III, the Emergency Planning and Community Right-to-Know Act of 1986. Title III sets requirements for industry and federal, state, and local governments for emergency planning and community right-to-know regarding the presence and releases of certain hazardous and toxic chemicals in use at and by facilities subject to the act. It focuses upon chemical releases which would immediately affect public health. SARA establishes a new category of 406 extremely hazardous substances (EHSs) that can cause acute effects upon release.

This legislation builds upon the EPA's Chemical Emergency Preparedness Program (CEPP) and similar state and local government programs.

Title III's aims are separate from the original hazardous waste cleanup provisions. Incorporated in the new law are requirements for coordinated emergency response planning and a chemical release inventory reporting and notification system. This system gives the public information about the chemical substances in use at and released from a facility.

Failure to Notify Invites Penalties

It is important to understand that for reporting purposes, Title III does not supersede CERCLA. If notification is required under CERCLA but not under Title III, appropriate notification procedures are to be followed. Civil and criminal penalties built into the law can be imposed for compliance failure.

Title III has three subtitles that interest us:

1. Framework for emergency planning—Sections 301-303—and emergency notification—Section 304. These sections focus on developing state and local government emergency response and preparedness capabilities and emphasize better coordination and planning, especially within the local community.
2. Community right-to-know reporting requirements—Sections 311-312—and toxic chemical release reporting (emissions inventory)—Section 313.
3. Trade secret protection, citizen suits, and the availability of information to the public—Sections 321-330.

SARA's Acronyms

SARA regulations are filled with a confusing collection of acronyms. This overview provides an informal guide through the regulations and their background.

The right-to-know provisions are a separate title to the Superfund legislation, not merely an amendment. The new law does not preempt state or local community right-to-know laws. Four major components of the community right-to-know provisions are as follows:

1. Emergency planning and response (Sections 301-303).
2. Emergency release and notification (Section 304).
3. Community right-to-know: hazardous chemicals inventory reporting (Sections 311-312).
4. Toxic chemical emissions and reporting inventory (Section 313).

Material Safety Data Sheet (MSDS). SARA refers often to MSDSs. They have been used in many states but only in recent years have become a federal requirement. At first, only manufacturers were required to provide MSDS information. Now, however, most businesses that handle hazardous/toxic materials must have them.

The idea behind an MSDS is that if a company knows the hazards associated with a chemical, it will safeguard those who use it. Also, if a worker knows what the hazards are, he or she will more likely use the safeguards provided by the employer. We can reduce the problem of meeting reporting requirements and avoid some work by knowing which materials do *not* require an MSDS:

1. Pesticides that are already labeled as required by the EPA under the Federal Insecticide, Fungicide, and Rodenticide Act.
2. Food, drugs, and cosmetics labeled as required by the FDA.

3. Any distilled spirits not intended for nonindustrial uses when labeled as required by the Federal Alcohol Administration Act.
4. Consumer products or hazardous substances regulated by the Consumer Product Safety Commission.
5. Hazardous wastes covered by the Solid Waste Disposal Act.
6. Tobacco or tobacco products.
7. Wood or wood products.
8. Manufactured items configured in specific shapes so that their use depends on those shapes and, when used, do not release or cause exposure to a hazardous chemical.
9. Food, drugs, or cosmetics intended for personal consumption by employees while in the workplace.

Emergency Planning. The purpose of the Emergency Planning section of SARA is to develop community chemical emergency response plans and capabilities. The plans must list the available resources in preparing for and responding to a chemical emergency.

Companies that produce, use, or store extremely hazardous substances (EHSs) over a specified quantity (the threshold planning quantity, TPQ) are subject to the requirements of this section. The company must advise the State Emergency Response Commission (SERC) and the Local Emergency Planning Committee (LEPC) of its facilities that are subject to the requirement. The planning activity focuses on facilities which must report. In April 1987, there were 256 EHSs which were not so classified under CERCLA as originally enacted in 1980. This section of the act is sometimes called the "Bhopal Effect."

For emergency planning (Sections 301-303), Title III requires designated state commissions to develop a method of receiving and processing public requests for information collected under the other sections of Title III. States must also coordinate the activities of LEPCs and review local chemical emergency plans. The following are key dates and requirements which should have been met or for which plans are to be in place:

1. October 1987—Designate a facility representative to work with local planning groups on this subject.
2. March 1, 1988—Submitt (if required) emergency inventory forms.
3. July 1, 1988 (and by July 1 of each year thereafter)—Facilities covered by the law submit initial toxic chemical reporting forms to EPA and to state officials.
4. October 17, 1988—LEPCs complete emergency plans.

Emergency Release Reporting and Verification. Covered facilities must immediately notify the LEPC and the SERC if there is a release of certain

EHSs (Section 304). Notification is also required for any EHSs or any substance subject to emergency notification requirements under CERCLA (Section 103[a]) if the release is over a specified TPQ.

The requirements of the Emergency Release Reporting section apply to facilities where hazardous substances (HSs) and/or EHSs are produced, used, or stored. If either an HS or an EHS is released above a TPQ, a report must be made immediately to the LEPC, the SERC, and the National Response Center (NRC). The initial report can be by telephone, radio, or in person. A follow-up written report is must be made to the SERC and the LEPC. *Note:* Section 304 (emergency notification) is not subject to trade secret protection.

Community Right-to-Know: Hazardous Chemical Inventory. It is important to understand what the EPA considers a hazardous chemical. For the purpose of these regulations, a hazardous chemical is any chemical which has a material safety data sheet (MSDS) requirement. MSDSs are a central source of information and data needed to comply with the reporting and training requirements of the EPA, as well as state and local government agencies.

The sections of SARA most dependent upon MSDSs are Sections 311 and 312—the community right-to-know reporting requirements.

Section 311 calls for preparing or maintaining MSDSs. In an effort to ensure uniform information gathering and, consequently a useful and uniform database, the act provides that all state and local laws enacted after August 1, 1985, which require MSDSs shall have or require MSDSs similar to those required by OSHA's Toxic and Hazardous Substances standards and its Hazard Communication Standard. In practice, a correctly prepared MSDS can form the basis for and provide the information needed for much of the reporting, labeling, and right-to-know communications and training programs required by EPA and OSHA.

When the complexities and far-reaching effects of EPA regulations are realized, many companies seek quick ways to meet the requirements. As a result, there has been an increased demand for off-the-shelf preprinted chemical labels, MSDSs, and training programs. The easiest and least expensive way to prepare an MSDS is to use or modify a ready-made MSDS or to extract information from other data sheets. Several commercial producers of ready-made MSDSs offer a wide range of data sheets covering the more well-known chemicals and chemical substances. Some trade associations (e.g., the Printing Industry Association) have compiled and published a book of MSDSs on chemicals and substances associated with a specific industry. Some of the collections contain a massive amount of data and could be a one-stop information source.

Do not confuse these commercial data sheets with the empty templates and blank fill-in forms offered by commercial stationers and some mail order safety product programs.

Companies must submit a list of hazardous chemicals, by location, to the SERC, LEPC, and local fire department. Section 312 requires annual submission of an emergency and hazardous chemical inventory, using a two-tier approach.

Tier I contains information about the average quantities and locations of the chemicals.

Tier II includes, for each chemical, the name as it appears on an MSDS, amount present, storage and use processes, and trade secret claims.

Both Tier I and Tier II information must be available to the public, on request from local and state governments during normal working hours, subject to trade secret protection.

Toxic Chemical Release. OSHA has expanded the Hazard Communication Standard beyond the Standard Industrial Classification (SIC) manufacturing sector (SICs 20-39) to all workplaces where hazardous chemicals are manufactured, used, or stored. More than 4.5 million nonmanufacturing establishments are subject to the new regulations. These establishments include:

Agricultural crop services, landscape and horticultural services, and forestry service. (There are many exemptions in the farming area.)
Oil and gas field services.
Building, highway, and trade contractors.
Transportation facilities and services.
Wholesale and retail trade establishments.
Health, educational, and social services.

Many manufacturing facilities now must submit emission inventories to inform EPA, state officials, and the public about releases to the environment of certain toxic chemicals (Section 313). These reporting requirements apply to facilities with ten or more full-time employees that manufacture, process, or otherwise use or process, and which use more than 10,000 pounds of a listed chemical during a calendar year. Under Section 313, anyone can petition the EPA to add or delete chemicals from the mandated toxic chemicals inventory list.

The chemicals involved generally fall into three categories:

1. Those known to cause cancer or serious reproductive disorders, genetic mutations, or other chronic health effects.
2. Those that can cause significant health effects outside the facility.
3. Those that can cause adverse effects on the environment. When there is an in-transit release and no notification is made, both the owner and the operator of the means of transport can be held liable. The owner

may be held liable even if he or she didn't know of the incident. The EPA holds the owner responsible throughout the entire process of transportation, handling, and disposal, and believes that the owner should have known.

SARA Makes Industry Accountable

Section 313 of the law, the heart of community right-to-know legislation, affects more than 33,000 businesses, many of which have little knowledge of the EPA requirements and their effective dates. Business owners and managers should be aware of the following:

The EPA set July 1, 1988, as the date to submit or have submitted data on a broad range of chemical usage. Its list of 329 substances includes such commonly used chemicals as acetone, toluene, methanol, and chromium compounds. The purpose is to assemble data to provide needed information on total annual environmental releases of toxic chemicals from manufacturing processes. Public access to this information is a key requirement of the legislation.

Manufacturers, processors, and importers who are not users have less stringent requirements. The law recognizes difficulties which may be encountered in initially fulfilling the reporting requirements and working with the specified reporting forms, and has mandated a 3-year phase-in period that provides for diminishing quantities of chemical use that trigger the reporting requirement.

Reporting is necessary for firms that manufacture or process more than the following amounts in a calendar year:

YEAR	QUANTITY (LB)	REPORT DUE
1987	75,000	July 1, 1988
1988	50,000	July 1, 1989
1989	25,000	July 1, 1990

Facilities which use the chemicals, however, must report at the 10,000-pound level.

Since a 55-gallon drum of liquid weighs about 500 pounds, the requirement is triggered if a firm uses about 20 drums a year and is subject to the 10,000-pound TPQ.

Bottom-line consideration—A full range of information is required. A covered firm should provide information on each chemical, answering these questions:

How is the chemical used?
What quantity of the chemical is present at the facility at any time during a calendar year?

For each waste stream, what waste treatment or disposal methods are employed?

What quantity of the toxic chemical is entering the environment by release to the air, ground, and water?

The expanded standard is essentially the same as the original standard: Training, labeling, a written program, and MSDS requirements are still in effect. There are, however, these significant new provisions:

Employers whose employees handle only sealed containers of chemicals must maintain copies of any MSDSs received. However, such employers do not have to obtain MSDSs if they are not received from the manufacturer or supplier unless there is an employee request.

Storage tanks, but not tanks on vehicles, must be labeled if they contain fuels.

Labels must be affixed only once to shipments of solid metal to the same employer when the metal is not an article. Labels, however, must be provided with each shipment of a solid for any hazardous chemical or chemicals present as a coating on the metal (e.g., cutting fluids, lubricants, or compounds).

Retail distributors (such as hardware stores, auto supply stores, and agricultural chemical dealers) must post a sign or otherwise inform commercial customers that MSDSs are available upon request.

Employers in a multiemployer workplace must exchange or make available MSDSs for chemicals to which the other employer's employees may be exposed; inform each other of the precautionary measures that employees must take to protect themselves under normal operating conditions and in foreseeable emergencies; and exchange information about the labeling systems that each employer is using.

Remedial Investigations and Feasibility Studies. The EPA must begin at least 275 new remedial investigations and feasibility studies within the next 3 years or, alternatively, begin 650 studies in 5 years. Actual remedial action must begin on 175 new facilities in 3 years and another 200 in the next 2 years. If all goes as planned, these studies should now be well underway.

Agency for Toxic Substances and Disease Registry (ATSDR). The amendments expand the ATSDR's authority by requiring the agency to prepare 275 toxicological profiles for hazardous substances. ATSDR must do health assessments at all sites proposed for inclusion on the National Priorities List in 1988 and at all new sites.

Depending on the results of the health assessment, ATSDR can decide to conduct health surveillance programs, conduct pilot or epidemiological studies, relocate, or provide alternative water supplies. The costs of all these procedures are recoverable from the responsible parties.

HAZCOM: Where Do We Go from Here?

In 1987, when OSHA expanded the scope of the Hazard Communication Standard (HAZCOM) to cover all employees exposed to hazardous chemicals, the intent was to ensure that such workers would have the right-to-know identities and hazards of chemicals in their workplace. OSHA's expanded approach should greatly increase worker protection, since it is expected that (1) knowledgeable workers will modify their behavior to handle chemicals more carefully and (2) employers will use the added data to design and use better protective programs. Once the first stage of the standard is in place, safety and health professionals must evaluate the programs they designed, search for ways to use the information generated, and improve its quality and communication methods.

After the crucial first efforts to comply, a hazard communication program becomes an ongoing activity wherever workers are exposed to hazardous chemicals. The long-term effort should go beyond standard compliance in terms of changing the workplace.

When the standard is fully implemented and all hazardous chemicals data are available to workers, various government agencies, and the public, the job of the industrial hygienist is expected to be easier. The reduced time needed to find information and the knowledge already acquired will improve the quality of the data and the decisions based on it.

The sheer volume of data indicates the need to find ways to standardize or collate it. Some plants have as many as 30,000 data sheets. Perhaps 70 million different MSDSs may eventually be in use. In larger operations, computerizing the data is the only way to deal with the vast amount of information that has to be assembled, maintained, and accessed.

The rush for large-scale MSDS development highlights the lack of information on many chemicals and casts doubt on the adequacy and accuracy of some data. One benefit of this massive new effort is the push to find and disseminate information on the hazards of chemicals rather than rely on earlier facts.

Hazard Communication Standard as a Cost-Cutting Aid

Some companies use the hazcom standard to find out exactly what chemicals are on hand, and in what containers and quantities. This survey approach helps identify chemicals no longer needed which can be disposed of. One benefit of this approach has been a significant decrease in workplace hazards in some cases. It also provides a clearer picture of the chemical inventory size and purchasing practices. Careful review of existing inventories has led to the removal of as many chemicals as possible from use, a reduction in permissible inventory, and improved chemical inventory control.

While many employers have had some type of hazcom program for many years, the success of such programs has not been widely studied or known. A fair evaluation will be hard to complete. One of the difficulties of such an evaluation is that occupational illnesses have been largely unreported. It is also difficult to separate occupational exposures from off-the-job, home, and recreational exposures.

Trade Secrets. The Superfund Act allows firms to withhold certain proprietary data on products which they claim as trade secrets or vital to their stake in the market for their product. While trade secret protection will save industry money and may sometimes be justified, it has greatly upset certain groups, such as emergency response personnel. They insist that there must be a way to identify such products for emergency workers without compromising the legitimate proprietary data and interests of the company.

The new law has a built-in method to ensure that reporting requirements are met. Any citizen can begin a civil lawsuit if the owner/operator of a covered facility does not follow the law's data reporting requirements. The same holds true for local and state governments which believe that a certain industry is not supplying the required MSDSs. The law specifies the federal district courts as the place for such actions to originate.

New Liability Provisions. Companies have reason to worry about the potential for litigation related to public access to this information. Community and environmental groups are anxiously awaiting it. There are also concerns about marketing products in the face of anticipated adverse consumer reactions. It has been suggested that some chemical products now widely in use may be perceived as laden with toxic substances and shunned.

Of primary concern to business and industry is the need for senior managers to certify information in the reporting system. Few environmental laws clearly specify top management responsibility. In addition to management responsibility, senior executives will have to face up to the cost of compliance. The EPA estimates that compliance will average $12,000 for each firm. Compliance cost is just one of the financial considerations. Another, less clearly recognized cost factor is the added public relation expenditures in dealing with the community and the labor force.

The public ranks violations of environmental law just below murder on the list of serious crimes, according to a Justice Department spokesman. That may be a slight exaggeration, but the department has pushed for criminal actions on environmental cases. We can take this action as a clear reflection of the public mood.

Response Action Contractors. Because liability insurance coverage was unavailable for engineering and environmental consultants working on Superfund cleanups, the amendments limit their potential liabilities and allow the EPA to indemnify them. Response action contractors are not strictly, jointly, or severally liable for response costs, personal injury, property damage, or other liability resulting from releases or threatened releases of hazardous substances under federal law (but state laws can apply). The contractor is liable for negligence or intentional misconduct. Indemnification agreements can be used if the contractor has made "diligent efforts" to obtain liability insurance but cannot find it at a fair price.

Public Participation. The EPA must provide for public notice and comment, as well as a public meeting near the Superfund site, on the remedy to be used. The EPA must document its choice of a remedy and make it available to the public. SARA provides for citizen suits to be brought against the federal government, an agency, or an individual who is alleged to violate provisions of the Superfund law and its amendments.

Underground Storage Tanks. SARA requires the EPA to establish financial responsibility standards for owners and operators of underground storage tanks. For large-volume tanks, at least $1 million per event is required to ensure settlement for cleanup and third-party claims. The EPA can take corrective action and sue the tank's owner/operator to recover cleanup costs, or it can require the owner to do the cleanup. SARA also creates a $500 million Leaking Undergound Storage Tank Trust Fund, financed by taxes on motor fuels.

Site Cleanup. The EPA has designed 850 to 900 sites for the National Priorities List (NPL) for cleanup. The number varies as sites are added or removed from the NPL.

CERCLA Natural Resource Claims procedures became effective in 1986. The claims cover the cost of restoring, rehabilitating, replacing, or acquiring the equivalent amount of natural resources injured as a result of the release of a hazardous substance, as well as the costs for assessing injury to such natural resources. This creates a new defense for "innocent landowners" who buy property without knowing that it was used for the disposal of hazardous substances. The new owner, however, must have adequately evaluated the site. Civil penalties for violations of up to $25,000 per day may be imposed and can increase to $75,000 per day for later violations. Criminal penalties also increase under the amendments.

Insurers Face Liability. More and more commercial and industrial property owners now own property contaminated with hazardous waste. It is often found that the contamination was caused by prior occupants or by operations on adjoining property. The dilemma faced by property owners is that in many states the current legal owner must clean up the contamination. Failure to do so can lead to severe monetary and criminal penalties.

Property owners invariably seek help from their comprehensive general liability (CGL) insurance carrier, but often find it unwilling to provide the coverage. The carriers frequently rely on a "pollution exclusion" or "on-site property exclusion" clause in denying coverage. However, the courts seem to be coming to the aid of property owners who seek coverage. If the CGL policy does cover an owner, it is important to notify the carrier of claims against the owner. Failure to give timely notification can cause the coverage to be canceled. When notified as required, the carrier will investigate the claim, usually reserving the right to disclaim coverage later. The decision of whether or not to give coverage is so complex, and involves so many legal challenges and court decisions, that the subject cannot be covered in depth here. While accident clauses are specifically tied to sudden, unforeseen, and unexpected occurrences to support a claim, the courts have recently settled several claims by eliminating "sudden," making the event only unforeseen and unexpected.

As noted, insurance companies attempt to use a pollution exclusion clause to reduce their exposure to pollution claims. However, when vandals released 14,000 gallons of oil from an insured's tank into a nearby river, the court required the insurer to pay the cost, even though the insurer claimed that it was not "sudden and accidental." The court made it plain that it was both sudden and unexpected.

In any event, a company faced with a hazardous waste problem should immediately notify its CGL insurance carrier of the contamination problem and file a claim for full coverage under the applicable insurance policy. If the insurer disclaims coverage under a policy exclusion clause, the property owner should carefully consider the options available in light of recent court decisions.

The OMB Takes Aim at Excess Paperwork. It should be noted that the new right-to-know law is a separate title to the Superfund legislation. It does not preempt state or local right-to-know laws. It does, however, require that MSDS requirements be identical to those of OSHA in order to reduce the paperwork burden.

OMB's concerns center on duplication of paperwork collection provisions dealing with MSDSs at some sites, as well as coverage of consumer products and drugs.

Hazcom Program Quality Control. Senior managers, particularly safety managers, must keep in mind that when a project is set up to carry out a certain responsibility, such as a new compliance program, it is often considered finished once the initial tasks are completed. Activities are then directed to a new project. Within a year or two, the lack of a programmatic approach is felt. Often within 3 to 5 years, few traces of the original effort remain, and the project is a shambles or has been forgotten.

Quality control is one way to ensure that projects become programs and that needed programs are ongoing. "Quality control" here means reviewing the contents of a program for accuracy, upgrading whenever possible, and having methods to measure its effectiveness. Building a quality control plan into a hazcom program will trigger timely and consistent actions and provide evidence of the program's success. It is best to start from the ground up—that is, to make sure that the basics are in place and then to set up a plan whereby the program continues to be upgraded as needed.

Figure 18-3 shows the major information flow requirements, and Table 18-1 summarizes the chemicals involved. Both list the corresponding section numbers of SARA, Title III.

Table 18-2 lists the basic elements of a hazcom program as required by OSHA standards under 29 CFR 1910.1200. Included are suggestions for ways to build upon the basics and set up a series of checks and balances that will continue to make the program more effective.

More suggestions for quality control of a hazcom program can be found in Ness (1988).

RIGHT-TO-KNOW OVERVIEW

Because HAZCOM deals with employees' rights in regard to hazardous chemicals, it quickly became known as the "Right-to-Know" standard. HAZCOM was broadened by more laws and regulations, most notably SARA Title III.

HAZCOM focuses on employees' right to know about chemicals in the workplace. Title III of SARA expands this right in several new directions:

1. SARA says that not only employees have a right to know about chemicals in the workplace. The community at large and the general public have the same right.
2. SARA recognizes that not all the effects of hazardous chemicals occur in the workplace. Chemical handling and disposal practices (and accidents) often affect the surrounding community. As a result, Title III is also known as the "Community Right-to-Know" standard.

Overall, community right-to-know legislation is under the EPA. It is managed (enforced) at all government levels, usually by a state, county, or city agency.

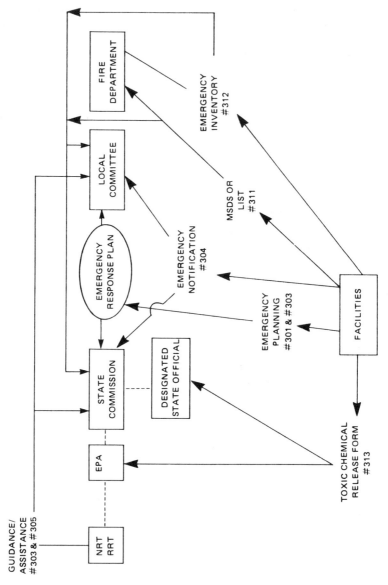

Figure 18-3. Title III — Major information flow requirements.

Table 18-1. Summary of chemical lists.

LIST	SECTION	PURPOSE
List of Extremely Hazardous Substances (402 substances)	#302: Emergency Planning #304: Emergency Notification	— Certain facilities must notify state commission — Initial focus to prepare local emergency plans — Some release of substances trigger Section 304 notification to state & local agencies
Substances requiring notification under Section 103(a) of CERCLA (717 substances)	#304: Emergency Notification	— Some substance releases also trigger notification to National Response Center
Hazardous Chemicals considered physical or health hazards under OSHA's Hazard Communication Standard. (This is a performance standard, there is no specific list of chemicals.)	#304: Emergency Notification #311: Material Safety Data Sheets #312: Emergency Inventory	— Lists those subject to emergency notification — Lists chemicals subject to state/local notification — Lists who must provide site-specific data
Toxic Chemicals identified as of concern NY, NJ, and Maryland.	#313: Toxic Chemical Release Reporting	— Chemicals reported on emissions inventory to inform government & public on releases of toxic chemicals in environment

Community right-to-know legislation requires those who use hazardous chemicals to provide information (either a list of the chemicals or copies of the MSDSs for those chemicals) to designated organizations. The public or community then has free access to that information through these organizations. The organizations are government agencies at appropriate levels.

To summarize:

1. HAZCOM is under OSHA through either a federal or a state plan.
2. Title III of SARA, hatched and overseen by the EPA, is run through state or local agencies.
3. HAZCOM and SARA are partners in right-to-know processes. Both are the responsibility of the employer, even though this discussion addresses OSHA regulations and practices.

Table 18-2. Hazcom program.

INITIAL TASKS	UPDATING AND UPGRADING	INITIAL TASKS	UPDATING AND UPGRADING
Written Program Develop a written program to address who, what, where, and when	**Written Program** Review program annually Disseminate tasks and delegate	**Training** Train employees in standard control to read & understand MSDSs, & in hazards of chemicals they work with	**Training** Train transfers & new employees Upgrade trng. for chemicals used Drill employees in hazards & safeguards need for emergencies
Chemical Lists Develop chemical lists	**Chemical Lists** Update chemical lists Put lists in computer Review use of each chemical for adequate controls Substitute for high-hazard chemicals Minimize inventories Test suspect products for unwanted types, amounts of contaminants Decrease # of chemicals in use		Test employees oral & written Review effectiveness of trainers Upgrade education of trainers Develop & upgrade training tools Develop and/or revise outlines
		Recordkeeping Develop documentation	**Recordkeeping** Upgrade trng. lists & schedules Create master file, include letters to manufacturers, monthly insp. reports & trng. outlines
Nonroutine Uses Identify nonroutine uses	**Nonroutine Uses** Expand chemical lists Review operations for contaminating by-products Review lists annually for accuracy Integrate w/emergency drills		
MSDSs Acquire MSDSs Make available to employees	**MSDSs** Update MSDS files Review for completeness & accuracy Notify maker of inaccuracies	**Contractors** Develop method to notify of chemicals in their work area	**Contractors** Require them to provide lists of all chemicals to be used on job Require MSDSs on their chemicals Review contractor's operations Participate in their briefings

SARA—SUPERFUND AMENDMENTS AND REAUTHORIZATION ACT

Table 18-2. (continued)

INITIAL TASKS	UPDATING AND UPGRADING	INITIAL TASKS	UPDATING AND UPGRADING
MSDSs	**MSDSs** Acquire MSDSs before purchase	**Contractors**	**Contractors** Restrict contractor chemical storage to a specific location, do post-job inspect. on unused materials & empty containers
	Acquire MSDSs for all possible hazards		
	Replace faded & torn MSDSs		
	Put MSDSs in computer		**Monitor Exposures**
Labeling Label in-house containers	**Labeling** Update in-house labels		**Review Policies** Purchase control, disposal & reuse
	Replace worn labels		**Audit** Develop monthly walkthrough & yearly review
	Review labels for accuracy		
	Notify manufacturers with inadequate labels		
	Review usefulness of plan		
	Label for byproducts hazards		

4. Unless specifically exempted, nearly every business and industry must take part in right-to-know activities. Employers under OSHA must prepare MSDSs must also submit them to state and local agencies as required.
5. Right-to-know legislation is here to stay.
6. The safety practitioner cannot deal with the OSHA aspects of right-to-know laws without being involved with SARA as well.

SUMMARY AND OBSERVATION

The cleanup regulations of SARA have only begun to address the problem of cleaning up toxic wastes. Unless the wastes now produced are better managed, new threats will simply replace the old ones. New SARA cleanup standards favor chemical waste treatment and reduction and discourage untreated disposal on land. SARA is one more step in an ongoing effort to deal with the difficulties and complexities of hazardous materials and hazardous waste management.

The experience of several European countries offers valuable lessons in improving industrial waste management, though few have adequately tackled the problem of controlling the total volume of waste generated. Without concerted efforts to reduce, recycle, and reuse more industrial waste, even the best treatment, disposal, and information systems will be overwhelmed.

BIBLIOGRAPHY

The information presented in this chapter is only a general overview. It is not to be considered an exact or detailed guide. A listing of right-to-know resources follows.

Arbuckle, Gordon J., et al., *Environmental Law Handbook,* 9th edition, Rockville, MD, Government Institutes, 1987.

Clark, Ann N., "Expanded HazCom Standard Covers 18 Million Additional Employees," *Occupational Health and Safety News Digest,* February 1988.

Marsick, Daniel J., "Resources for Right-to-Know Compliance," *Applied Industrial Hygiene,* January 1988, pp. F34-F40.

Miller, Lynn M., Environmental Strategies Corp., 8521 Leesburg Pike, Suite 650, Vienna, VA 22180.

Musselman, Victoria Cooper, "Community Right-to-Know Law Makes Industry Accountable," *Environmental Management News,* January-February 1988.

Naughton, Michael J., "Insurers Face Liability for Pollution Cleanup," *Environmental Health News,* January-February 1988, pp. 1-3.

Ness, Shirley A., "Quality Control in an Employer's Hazard Communication Program," *Applied Industrial Hygiene,* January 1988.

OSHA Reporter, Bureau of National Affairs.

Port, Eugene A., "Material Safety Data Sheets," *OSHA NEWS,* Santa Monica, CA: Merritt Company, February 1988, pp. 4-5.

Silk, Jennifer C., "Hazard Communication: Where Do We Go from Here?" *Applied Industrial Hygiene,* January 1988, pp. F27-F28.

Chapter 19

Women in the Workplace

INTRODUCTION

The number of women working outside the home has increased dramatically within the last few years. In 1954, 35 percent of all women had paid employment. In June 1984, this figure had risen to 53.9 percent. The reasons for this increase are many, including the following:

1. Delay of or decision against marriage.
2. Increased divorce rate.
3. Increased number of female heads of household.
4. Federal antidiscrimination legislation.
5. Inflation.
6. Increased educational level of women.
7. Realization of the potential hazard of relying solely on a husband's salary.

As a matter of interest, the stereotypical average American family with an employed father, mother at home, and two children applies to only about 7 percent of all families in the United States.

These social changes have made the working mother and the pregnant employee a normal part of everyday life in all types of industries. The shifts in family structure, changes in the composition of the labor force, and needs of all workers must be taken into account in any consideration of the employment and social policies of the nation.

The move into the workplace by women, even though mostly into female-dominated occupations, has also involved a substantial number of jobs traditionally held by men. As these traditional barriers fall, more women than ever work in all types of jobs, including crane operation, coal mining, and various machine operations.

There are specific considerations concerning the safety and health of women in the workplace, not the least of which is the dearth of evidence on potential safety and health hazards associated with places where women generally work. However, job health is not defined solely as scientific or medical in nature. There has been a great deal of emphasis on the pregnant employee and the health status of the fetus, but these are not the only topics of concern to the employer of the working woman, no matter how important they are. There are also economic and political considerations that are intimately connected to job health. For that reason, these issues are explored in depth in this chapter.

Employers realize that the health of the work force—physical, mental, psychosocial, and economic—influences efficiency and productivity and makes a significant contribution to the bottom line. The following sections explore many of the working woman's considerations and concerns. To improve the work environment, these become the employer's considerations and concerns as well.

HISTORY OF WOMEN IN THE WORK FORCE

The issue of safe and healthy working conditions for women is not new. From the mid-1800s to the present, attention has focused on the special considerations of women in the workplace. Dr. Alice Hamilton, one of the most important occupational health physicians in the United States, headed the Women's Bureau in the early 1900s. At the turn of the century, many occupational health reformers were women. The organizations they founded, such as the Women's Trade Union League and the Women's League for Equal Opportunity, were active in the fight for improved occupational health conditions for women.

During World War II, when many men were in the armed services, women were called upon to fill the employment vacancies that occurred. After the war, the women were removed from mining, construction, transportation, and all types of basic industry. Not until the 1970s and 1980s did they begin returning to these traditionally male areas of employment.

In the mid-1960s, federal laws were passed which granted equal opportunity to and treatment of women workers in the United States.

The Equal Pay Act, an amendment to the federal minimum wage law, was passed in 1963. This act stipulated that men and women who work in the same establishment must receive the same pay if their jobs require equal skill, effort, and responsibility.

Growing out of the civil rights struggles, Title VII of the Civil Rights Act was passed in 1964. This act outlawed all discriminatory employment practices, including those connected with hiring and firing; wages, fringe benefits; classifying, referring, assigning and/or promoting; training, retraining, or apprenticeship; or any other terms, conditions, or privileges of employment.

The Equal Employment Opportunity Commission (EEOC) is the federal agency that enforces these two laws.

In 1965, President Lyndon Johnson issued Executive Order 11246 to counteract employment discrimination based on race, color, religion, and national origin. It was amended in 1967 to address sex discrimination as well.

In 1978, Title VII was amended by Congress to ban discrimination based on pregnancy. It addressed women affected by pregnancy, childbirth, or related medical conditions. It stipulated the same treatment for all employment-related purposes, including benefits, given to other persons with the same ability or inability to work.

By 1978 at least sixty-five federal laws and orders had been passed on equal employment opportunity. All state protective legislation has been struck down as discriminatory, and women have been placed on an equal footing with men.

Men's work has traditionally been viewed as the norm. More attention has been paid to their working conditions than to those occupations dominated by women and/or those workplaces where most women work. Obvious hazards such as back injuries from lifting among healthy workers and possible carcinogenic exposures of textile and laundry workers have not been equally addressed. When women have been considered, the primary topic has been possible dangers to the fetus should pregnancy occur. This consideration must be broadened in fairness to all workers and employers. Neither sex should have a monopoly on effecting changes designed to produce safer and healthier working conditions.

DEMOGRAPHICS — WHO IS THE WORKING WOMAN?

Age

In 1983, 53 percent of all women were 16 years of age or older. Of these, 48.5 million were working or looking for work.

Of those women 18 to 64 years of age in 1983, approximately 46 million, or 63 percent were in the civilian labor force. Those in the 20 to 29 age group had the highest representation, with 70 percent working.

Since 1970, nearly half of the increase of women in the labor force has been among the 25 to 34 age group. At the time of this writing approximately one in four women workers is within this age group.

Education

The more education a woman has, the greater the likelihood that she will seek employment. In March 1984, 78 percent of women 25 to 64 years old with 4 or more years of college were working. The average female worker is nearly as well educated as the average male worker. The median number of years of school completed for women as of March 1984 was 12.7 years; the comparable figure for men was 12.8 years. To enter new fields or to get out of dead-end jobs, more and more women have identified the need to return to school for additional education and training.

Wages, however, are not generally on a par with education. In 1983, female workers with 4 or more years of college had an average income slightly above that of men who had 1 to 3 years of high school; these average figures were $14,679 and $12,117, respectively. In the same year, the average income of female high school graduates was lower than that of men who had completed fewer than 8 years of school; these figures were $13,787 and $14,093, respectively.

Marital Status

The largest group of female workers are married and living with their husbands. From 1972 to 1983, the labor force participation of married women increased from 40 to 52 percent. In March 1982, 26 million wives, or 51 percent of married women, were working, whereas 20 years earlier, only one-third were employed. By early 1984, more than 27 million wives were in the labor force.

As of March 1984, nearly two-thirds of the women in the civilian labor force were single (61 percent), divorced (11 percent), widowed (5 percent), separated (4 percent), or were married to men with 1983 incomes less than $15,000 (19 percent).

Divorced women are more likely to be in the labor force than women of any other marital status. The growth in the number of female-headed families has far outpaced that of other type of family. Between 1972 and 1982, the number of female-headed households increased by 57 percent; at the same time, there was an increase of 10 percent for other types of families. In March 1984, about one out of six of all families, or 17 percent, were maintained by a woman.

In 1982, the proportion of poor families maintained by women was 43 percent; by 1983, this figure was 47 percent. In 1982, women as heads of household had, on average, completed fewer years of school than wives and were concentrated in lower-skilled and lower-paying jobs.

In general, many of the women who are divorced, separated, or widowed, often at midlife, are at an economic disadvantage because they have low or no job skills.

In 1984, there were record levels of women with children in the labor force. Of those women with children less than 18 years old, approximately 61 percent, or 19.5 million, were working in March 1984. About 52 percent of mothers with preschool children, or 8 million women, were working. Nearly three-fifths (56 percent) of all children less than 18 years old (32.7 million) had working mothers; 48 percent (9.3 million) of children less than 6 years of age were in this category. From these figures, one can readily see that adequate day care is a primary concern to the working woman who is also a mother. Many companies have responded by providing quality day care at their facilities, to the mutual advantage of employer and employee.

Union Membership

Women do not constitute a significant segment of organized labor. Generally, women are concentrated in female-dominated occupations which tend to be at the low end of the pay scale, with fewer benefits than professional or union workers.

Despite the increase in the number of women in the labor force, there has been no concurrent growth in the number of organized women. Currently, fewer than 13 percent of women workers are union members. Women who are union members earn nearly 30 percent more than unorganized women, although the weekly earnings of organized men exceed those of organized women by about 60 percent.

Many trade unions have established women's sections or divisions which address special interest areas such as child care, pregnancy benefits, maternity leave, nonsexist language, equal pay, nondiscrimination, entrance into apprenticeship programs, and the like.

When OSHA held hearings on the proposed lead standard in March 1977 concerning exclusionary policies toward fertile women, women's groups joined with labor unions in calling for standards strict enough to ensure the safety and health of all workers. One outcome of this joint venture was the establishment of the Women's Occupational Health Resource Center (WOHRC), a project of the American Health Foundation. This organization maintains a library; conducts workshops and forums; and publishes educational literature and a bimonthly newsletter on women's occupational health issues. One of its major goals is the dissemination of information and coordinated action, through a network of organizations and individuals, to make women's occupational health issues a research- and policy-making priority.

Future Employment

The Bureau of Labor Statistics projects that large numbers of women, especially in the 20 to 54 age group, will be entering the work force in the 1980s. It is estimated that women will constitute 70 percent of the labor force growth during this decade. As this is written in 1990, the projection, although not fully verified, appears to have taken place.

The reasons for this influx include more variety in jobs and opportunities, a more relaxed societal attitude toward working mothers, more openings in nontraditional jobs, and inflation.

The total number of payroll jobs has risen by 5 million since November 1982. Whereas the number of jobs has increased in the construction and manufacturing sectors, most of the job growth (3.1 million) has occurred in the service areas. Projections to 1995 include a continued increase in employment in the latter category. This trend may mean that movement into skilled and semiskilled blue-collar jobs by women will slow considerably. It also may mean that more employment opportunities will be available in those same low-paying job categories that currently employ the largest number of women: secretarial, administrative support, technical work, and sales.

EMPLOYMENT—WHERE DO WOMEN WORK? HOW MUCH DO THEY EARN?

Types of Occupations

During the past few years, with increasing safety and health and antidiscriminatory legislation, employment opportunities for women have increased in breadth and scope. Jobs have been obtained by women in almost every field of employment.

The majority of women, however, continue to work in traditional professional, clerical, and service jobs. Most of the occupations in which 90 percent or more of the workers are female are in the clerical category, which includes five of the top ten occupations that employ women. These clerical occupations are represented by secretaries, cashiers, typists, bookkeepers, and nonapparel sales workers. Four other examples in these female-intensive occupations include nurses' aides, waitresses, and the relatively low-paying, professionally educated elementary school teachers and registered nurses.

Between 1972 and 1982, women accounted for 68 percent of the 14 million increase in employment in white-collar occupations as the shift to service industries developed. Women are especially concentrated in this sector, constituting one-half of the employees in 1982. By contrast, women made up a little more than one-fourth of the employees in the goods-producing industries in that same year.

About one-half of the 12 million workers employed in the nation's plants and factories in 1983 were women. They make up the majority in the electronics (890,000), apparel (995,000), and textile production industries (890,000). One-third of furniture workers are female. Approximately 100,000 women work in the automotive industry, and another 1.4 million are employed in the plastics, chemical, rubber, and machinery industries.

The concept of "men's jobs" and "women's jobs" is still relatively prevalent in our society. This attitude affects the behavior of employers, employment agencies, legislators, the armed services, and the entire educational and job training system, from elementary and secondary schools to vocational schools, training programs, and graduate schools. Unfortunately, the expectations of women are often likewise affected.

Even though there has been an increase of women working in nontraditional jobs, the employment patterns of women have changed little since 1950. Of the more than 450 occupations listed in 1950, women were concentrated primarily in 20. By 1978, 9.9 percent of women were employed in traditionally male jobs, 21.6 percent were in non-sex-stereotyped jobs, and 68.5 percent were in traditionally female jobs. A predominantly male job is defined as having 25 percent or fewer women; a predominantly female job as having 55 percent or more women; and an integrated or non-sex-stereotyped job as employing 25 to 55 percent women.

Less traditional female jobs generally offer higher pay. They also tend to differ from female-intensive occupations in providing greater responsibility and more opportunities for advancement. However, studies have demonstrated that although more women now hold supervisory and managerial positions, their occupational status and earnings are generally less than those of their equally educated male coworkers.

According to the Department of Labor, women workers in 1983 were divided into the following occupational categories:

Category	Percent
Managerial/professional	21.9%
Technical	3.3%
Sales	12.8%
Administrative support/clerical	29.7%
Private household	2.1%
Service	16.8%
Craft	2.3%
Operatives (machinery)	9.7%
Farm	1.3%

Many of the women in the managerial/professional category tend to be registered nurses, school administrators, or teachers.

Currently, job growth patterns reflect a greater increase in employment opportunities in traditional female areas than in male-dominated occupations.

Earnings of Women

It is well known that women generally earn less money than men. In 1983, the median income of all women was $14,192 compared to $22,410 for men. The average woman earns only $.63 for every $1.00 earned by the average man when both work full-time all year.

Fewer women work full-time than men, and they are concentrated in relatively few occupations. Women continue to work in female-intensive occupations rather than in other types of jobs. In fact, those occupations with high earnings for women do not have a large percentage of women workers. The most highly paid occupations, for women as for men, are in the professional and managerial categories. However, the income of women in these groups is far less than that of men. In 1981, the highest median weekly salary for men was $619.99; for women it was $422.00.

In 1960, women's salaries were 60.8 percent of men's; in 1982, the figure was 61.7 percent.

Women who maintain families face particularly serious financial problems, since their earnings continue to average only about 60 percent of that of men. The average annual income of a family headed by a woman is less than half that of a married couple. In 1982, the median income of single-parent families headed by a woman was $12,070. In two-parent families with children, the figure was $30,030 when both parents worked and $23,900 when only the father was employed.

Families headed by women are five times as likely to be living in poverty as are those headed by married couples. In 1981, approximately 3.4 million families maintained by women, or one out of three, were living in poverty, whereas only one out of sixteen two-parent families were in this category. In 1983, 61 percent of those with incomes below the poverty level were women 16 years of age or older.

The issue of comparable worth, or equal pay for work of comparable value, is an important policy issue. In the perception of equal value, the Equal Pay Act addresses only those situations where men and women perform substantially the same work in the same establishment. The issue of comparable worth becomes a matter of concern when jobs traditionally performed by women result in lower earnings than jobs requiring less skill but traditionally performed by men. This issue has been addressed through studies, hearings, conferences, and proposed regulations. Lawsuits and complaints have been filed by nurses, librarians, clerical workers, and female jail keepers. Several national unions have adopted the concept of equal pay for work of comparable value; the AFL-CIO passed a resolution supporting this principle.

For more than 30 years, the concept of comparable worth has been accepted internationally. Eighty countries supported the concept at the International Labor Conference in Rome in 1951. Since that time, a number of countries have passed legislation related to this issue. The United States has begun to recognize the fairness of equal pay for jobs of equal value, but strategies to implement it are not widespread.

OCCUPATIONAL SAFETY AND HEALTH CONSIDERATIONS OF THE WORKING WOMAN

Women as High-Risk Employees

Traditionally, women have been viewed by our society as physically weak and frail, as well as more susceptible to emotional difficulties and to chemical and physical stressors. There is no scientific evidence to support any of these generalizations. Except for pregnancy and childbirth, men and women are more similar than dissimilar.

Most hypersusceptibility in women, as in men, is due to hypersensitivity, allergy, genetic deficiencies, personal habits and lifestyle, nutritional status, age, medical regimen, or underlying disease.

Occupational Hazards

According to a study by Kaplan and Knutson (1980), the difference in male and female injury and illness patterns appears related more to different job placements and types of job tasks than to biological differences. In fact, the injury and illness rates of women in nontraditional fields of employment have been demonstrated to approximate those of men.

There are a limited number of epidemiological studies concerning workers in traditional female-dominated occupations, partly because there is a widespread assumption that women do not work in hazardous jobs. In reality, there are many potentially hazardous exposures inherent in these types of work. Examples of these occupations and some of the associated exposures include the following:

Nurses, laboratory technicians, dental hygienists	Infections, puncture wounds, lacerations, radiation, anesthetic gases, back injuries, various chemicals
Waitresses	Varicosities, back injuries
Clerical workers	Chemicals from indoor pollution due to poor ventilation and efficient insulation, pesticides due to building extermination work, ozone from reproduction machines
Textile workers	Cotton dust
Sewers and stitchers	Chemicals from synthetic materials
Garment workers	Formaldehyde residues from permanent press fumes
Electrical machinery assemblers	Trichloroethylene
Hairdressers and cosmetologists	Aerosols, dyes, detergents
Social workers, teachers, child care attendants	Infections
Laundry workers	Heat stress, infection, burns, falls on wet floors

Nonetheless, most occupational health studies are concerned with male-dominated occupations. For example, epidemiological studies of longevity, morbidity, and mortality among radiologists have been going on for more than 20 years. Yet in the United States, there have been no studies of x-ray technologists, nurses, or nuclear medical technologists dealing with nuclear therapy. Almost all occupational health studies of women are directed to the female reproductive system and pregnancy.

Proper Worker-Job Fit

Job Analysis. Job titles and stereotyped job descriptions are often misleading. Therefore, job analyses should be performed, and periodically updated, to ensure a proper job-worker fit, whether the potential worker is male or female.

Written job analyses should include the following:

1. Monitoring of job tasks and the work environment. In this manner, materials and/or activities that might be harmful will be identified. Changes to eliminate or control the hazard(s) can then be recommended.
2. Analysis of the circumstances under which the job is performed. Direct observation, with attention to infrequently performed tasks and potential emergencies, will provide a definitive picture of the necessary capabilities of a potential worker.

The federal government has the responsibility to ensure a safe, healthy workplace for all workers, both male and female, while at the same time guaranteeing all workers equal employment opportunities.

Equality of opportunity should come from the equality of effort and skill applicable to the task; this does not necessarily mean the same effort and skill. For example, men have more muscle tissue, relative to weight, than women. Men also are stronger than women per unit of body weight. By performing job analyses, the amount of strength necessary to perform specific job tasks can be quantified; this will help ensure a proper worker-job fit.

Preplacement Assessment. As women enter nontraditional fields of work, questions arise concerning the relationship between job placement and physical strength. Therefore, in addition to job analyses, there must be an assessment of the worker's capabilities and characteristics. Observation and appropriate tests of strength, physical fitness, and aptitude will help to determine a person's fitness for a given job. Such an assessment is crucial to protect the potential worker from undertaking work which may be harmful and to protect the employer from hiring a worker who would be physically incapable of performing required job tasks.

A preplacement assessment of a worker also provides baseline data which can be utilized later to determine any physical, mental, or biological changes which may occur.

Job Training. All workers, both men and women, must be able to recognize and cope with the hazards of their work environment. They also need to develop the skills to perform their jobs efficiently and safely with the proper tools and equipment. Sound safety and health education and job training programs, which usually go hand in hand, apply equally to men and women.

Ergonomics. Ergonomics is a relatively new science that explores the relationship between the worker and the environment and promotes workplace design that is conducive to the physical and emotional well-being of the worker. A correctly designed workplace can increase worker productivity, help prevent occupational disease, and help alleviate stress.

The employment of women in nontraditional roles means that a new group is performing tasks and using tools and equipment that were originally designed for men.

Equipment and tools should be selected or designed and arranged, as far as is practical, to be safe and free of health risks. They should coincide with the anatomical and physiological structures of workers. Workers should be classified in terms of their lifting capabilities and strengths through an effective preplacement procedure, and not on the basis of sex. Even though it is known that, on the average, men are taller and one-third stronger than women, there are many overlaps between the strength scores of men and women. Men also, on the average, are thinner than women of similar stature, and have broader shoulders, narrower hips, greater leg strength, longer sarms, less adipose tissue, and more skeletal muscle than women. What these statements mean, when comparing men to women, has to be explored in greater depth to understand relative work capabilities more fully. In the meantime, much more data are needed to clarify the capabilities and limitations of women, since most current data concern men.

When discussing ergonomic factors, acknowledgment of sex differences works to everyone's interests in terms of worker selection, equipment design, and a good fit between the worker and the job. In actuality, most jobs make little or no physical demands on the worker, and if jobs fall within the capabilities of most men and women, sex differences are of no consequence.

Many traditionally female jobs also contain potential ergonomic problems. Hundreds of thousands of women either sit or stand for the majority of their workday. To date, there are inadequate data on the long-term effects of excessive sitting or standing. However, there does appear to be a relationship between these activities and disorders of the circulatory system such as varicose veins. Chairs should be designed to provide adequate back support and to avoid impeding circulation. It has been estimated that as much as 40 minutes of work time can be lost daily by poor seating arrangements.

Tenosynovitis and carpal tunnel syndrome are relatively common among workers who make rapid finger and hand movements. Among hospital workers, back injuries from lifting and turning patients are fairly common.

An additional injury risk factor may be introduced if the female employee, after her workday, performs housework with muscles already tired from occupational activities.

Stress. Currently, a great deal is being written about stress in the workplace. This stress may emanate from two primary sources: stressors inherent in a job, such as deadlines, and non-work-related stressors such as financial concerns or family problems. The manner in which the worker copes with these stressors is, in part, a function of that person's personality.

Many, if not most, working women have dual roles: at work and at home. As role demands increase, the woman may experience role overload, the occurrence of too many demands related to too many roles. This is not uncommon for working women.

Financial concerns, quality child care, the decision to have children or not, responsibility for housework and child care, lack of leisure time, no time to continue education or train for a better job—these are all sources of stress for many women in the work force. In 1977, the International Labor Office estimated that the typical employed woman works an average of 75 to 80 hours per week in all industrialized countries. In contrast, men were estimated to work approximately 50 hours per week. It was also estimated that the employed married woman has 17 percent less free time than the employed married man. A Quality-of-Environment Survey in 1977 revealed that nearly half of the women surveyed spent an additional 3.5 hours on housework on the days they worked on the job.

For a job to be satisfying and fulfilling, a number of factors must be considered. The physical and psychological conditions of work, the nature of the work itself, the amount of responsibility and autonomy provided, and the sense of accomplishment after completing a task are all inherent in job satisfaction.

Many women express dissatisfaction with their jobs or occupations because of several types of work-related stress. Some of these include underutilization of skills; boredom; repetitive, paced work; lack of recognition for accomplishments; low pay; unfulfilled job expectations; powerlessness or lack of control over the job situation; authoritarian, verbally abusive bosses; sexual harassment; lack of job mobility; jobs ranked low in the social structure; and lack of promotional opportunities.

Reproductive Hazards. The issue of reproductive hazards in the workplace is extremely complex. Most current standards for occupational exposures to chemical and physical agents do not consider reproductive effects. The teratogenicity (ability to cause birth defects while the embryo/fetus is developing) of most industrial chemicals has not yet been investigated. It is generally believed that the teratogenic ability of a substance depends upon the degree of exposure, not just the potential of the substance.

There are millions of different chemical compounds currently in use, and many more are introduced weekly. Various chemicals are probably present in every type of industry. In 1966, the Food and Drug Administration issued guidelines for the testing of new drugs for teratogenic potential. No rapid screening processes are available; teratogenic testing is a time-consuming, complex process in which many exacting criteria must be met.

An increasing number of women are being exposed to toxic substances during the childbearing years. Teratogenic effects of a substance can be experienced during the first few weeks after conception, when many women do not realize that they are pregnant.

It is estimated that 50 percent of first-time mothers are employed at some time during their pregnancies and that 85 percent of women in the work force will be pregnant at some point during their working lives. In 1983, 30 percent of all babies in the United States (more than 1 million) were born to mothers who were employed while pregnant.

Even though most studies have focused on birth defects caused by the pregnant woman's exposures, a more comprehensive view is needed to identify exposures of both men and women which are suspected of leading to reproductive failure. Generally, a substance which endangers the fetus or a woman's reproductive capacity also poses a danger to the male reproductive system. There is evidence that a number of substances can cause impotence, lowered sperm counts, infertility, abnormal sperm, and/or adverse fetal outcome after exposure of the male. Chemicals with adverse effects on the male reproductive system include vinyl chloride, dibromochloropropane (DBCP), hydrocarbons, chloroprene, polychlorinated biphenyls (PCBs), antineoplastic drugs, anesthetic gases, toluene diamine, organic mercury, and hexachlorobenzene. Another source of danger is radiation. Unfortunately, the exposure history of the father is rarely assessed in relation to the birth of a child.

It has been estimated that as many as 20 million jobs may expose men and women to reproductive hazards.

There may be an artificial distinction between protecting the fetus and protecting the reproductive organs of men and women. Adults are just as valuable as infants.

In some industries, women of childbearing age have been excluded or restricted from certain types of employment because of concern about liability associated with damage to a woman's reproductive capacity or to a fetus. Depending upon the sources consulted, various solutions are offered as possible answers to this problem. In general, such exclusions of women should be permitted only if the reproductive health of women is shown to be in danger of significant harm, as documented by reputable scientific evidence. In such cases, research on the effects on male reproductive systems should also be conducted.

The practice of isolating women who are or might become pregnant is inadequate to protect the human population. In addition, such women are often excluded from equal job opportunities are are economically penalized, while men and nonfertile women remain overexposed, in contravention to the purpose of occupational safety and health regulations.

To help reduce the reproductive hazards possibly attributable to the workplace, working conditions must be improved for all employees by reducing or eliminating exposure to all toxins. Occupational and environmental exposures must be more clearly defined and measured. All industrial hygiene practices must be improved and maintained so that exposure is kept to a minimum.

Women are concerned about their reproductive health and the health and normality of their unborn children. They should be informed of all the known and suspected risks of their work environment to enable them to make informed decisions about accepting certain jobs; they must be able to react knowledgeably according to new research findings.

The occupational safety and health problems of women are no different from those of men except when women are pregnant. Any temporary restrictions that may be necessitated by pregnancy are analogous to those of a middle-aged man with cardiovascular disease who needs temporary or permanent partial disability in order to continue working.

There are normal physiological changes which occur during pregnancy. There is an increase in cardiac volume and a decrease in the residual capacity of the lungs; this may result in increased harmful effects from the inhalation of toxins. As pregnancy progresses, there will be weight gain and a change in posture, and the woman may experience more fatigue. The pregnant worker, like all workers, should be evaluated and treated on an individual basis, depending upon the specific situation.

Personal Protective Equipment. When women first entered nontraditional jobs, there was virtually no appropriate personal protective equipment designed for their needs. Therefore, women wore what was available for men and tried to modify it accordingly. Men's hard hats and safety shoes were stuffed with tissue or socks; hearing protectors, too large for many women's ear canals, caused irritation and infection; respirators which were too large to provide an airtight seal allowed inhalation exposures; sleeves that were too long were in danger of being caught in moving machinery; and gloves that extended beyond the fingertips reduced dexterity and were a safety hazard.

Women are not scaled-down versions of men. Therefore, the personal protective equipment of smaller men will not fit women. Generally, women are not proportionally smaller than men in three body dimensions: chest depth, hip circumference, and back curvature at the hip. Also, with every height-weight combination, men have significantly larger shoulders. Coveralls designed for a man, for example, will generally be too snug in the hips and too large in the shoulders for a woman.

Personal protective equipment must conform to a worker's body dimensions to be effective. Properly fitting protective equipment is a critical safety issue. Otherwise, the equipment is ineffective and may be hazardous. For example, ANSI has revised its safety-toe footwear requirements to include women's footwear because men's feet tend to be larger than women's in length, breadth, and circumference.

A number of manufacturers now offer personal protective equipment for women which provides adequate design, comfort, and fit. However, this market is relatively new and not all sizing problems have been solved, so the buyer must carefully check the claims of the manufacturer prior to purchase.

SUMMARY

The history, demography, employment experiences, and occupational safety and health considerations of women have been explored in this chapter. The subject of working women is complex. Even though there are more similarities between men and women than differences, there are some special considerations when the working woman is being discussed.

To accommodate the specific needs of working women in relation to the dual roles of work and family, reentry into the job market, child care, or other considerations, a number of employers have instituted changes in various company policies.

For example, most part-time workers in this country are women. Part-time work generally allows more flexibility to accommodate both family and work responsibilities. Alternatives to part-time work that may be offered by companies include compressed work weeks, job sharing, and flextime. A number of companies are experimenting with alternative work schedules. A survey by the American Management Association in the late 1970s revealed that 13 percent of all private employers with more than fifty employees offered flexible work schedules. Since that time, flexible scheduling has become increasingly available through a broader range of employers.

Offering supplemental child care at work is another option. There is a positive correlation between worker productivity and quality child care. There is no concern about the adequacy of the child's care to interfere with the woman's job responsibilities.

Other changes companies have adopted or are exploring include counseling for working women, training programs, and modification of job-related benefits.

In the work environment, there are many unanswered questions about the effect of chemicals on the fetus; there is also little knowledge about the effects of chemicals on the reproductive systems of men and women. Ergonomic considerations of the female worker, to ensure a proper worker-environment fit, are equally applicable to the male worker. Sex differences in strength and endurance are not significant in most jobs. Stress is relatively important in the lives of most working women; its causes are primarily social, political, and economic. Nonetheless, all such factors play a contributory role in the concept and the reality of occupational safety and health.

Refer also to "Job Safety Analysis" in Chapter 8.

BIBLIOGRAPHY

Bingham, Eula, ed., *Proceedings of the Conference on Women and the Workplace, June 17-19, 1976,* Washington, DC, Society for Occupational and Environmental Health, 1977.

Celentano, E. J., J. W. Nottrodt, and P. L. Saunders, "The Relationship Between Size, Strength, and Task Demands." *Ergonomics,* 1984, pp. 481-488.

Chavkin, Wendy, "Walking a Tightrope: Pregnancy, Parenting, and Work." In *Double Exposure,* ed. Wendy Chavkin. New York, Monthly Review Press, 1984, pp. 196-213.

Cox, S., T. Cox, and Jill Steventon, "Women at Work: Summary and Overview." *Ergonomics,* 1984, pp. 597-605.

Cuthbert, Jean W., "The Health of Women in the Pharmaceutical and Chemical Industries," *Journal of Occupational Medicine,* 1978, pp. 601-604.

Fatt, Naomi, "Women's Occupational Health and the Women's Health Movement," *Preventive Medicine,* 1978, pp. 366-371.

Hatch, Maureen, "Mother, Father, Worker: Men and Women and the Reproductive Risks of Work." In *Double Exposure,* ed. by Wendy Chavkin, New York, Monthly Review Press, 1984, pp. 161-179.

Heifetz, Ruth, "Women in the Workplace: How Are They Different? Who Should We Protect? What Mode of Protection?" Presented at the American Society of Safety Engineers Professional Development Conference, San Diego, California, June 12, 1985.

Hricko, Andrea, "Social Policy Considerations of Occupational Health Standards: The Example of Lead and Reproductive Effects," *Preventive Medicine,* 1978, pp. 394-406.

Hughes, Mary, "Good Health Is Good Business," *Occupational Health Nursing,* September 1984, pp. 463-464.

Hunt, Vilma R., "Occupational Radiation Exposure of Women Workers," *Preventive Medicine,* 1978, pp. 294-310.

Johnson, Dave, "Women's Safety—An Update on Meeting the Challenge," *Industrial Safety and Hygiene News,* Radnor, PA, Chilton Co., May 1985, p. 59.

Mackay, Colin J., and Carole M. Bishop, "Occupational Health of Women at Work: Some Human Factors Considerations," *Ergonomics,* 1984, pp. 489-498.

Malloy, Gail, "Stress, Pregnancy, and the Workplace," *Occupational Health Nursing,* September 1984, pp. 474-479.

Manson, Jeanne M., "Human and Laboratory Animal Test Systems Available for Detection of Reproductive Failure," *Preventive Medicine,* 1978, pp. 322-331.

McCann, Michael, "The Impact of Hazards in Art on Female Workers," *Preventive Medicine,* 1978, pp. 338-348.

National Commission on Working Women, *National Survey of Working Women: Perceptions, Problems, and Prospects,* Washington, DC, National Manpower Institute, 1979.

"Never Done—Hours of Work Around the World," *WOHRC Fact Sheet,* Vol. 6, No. 6, New York, School of Public Health, Columbia University.

Punski, Charlotte R., "Do Blue Collars Protect Working Women?" *National Safety News,* October 1983, pp. 34-38.

Randolph, Susan A., "Stress, Working Women, and an Occupational Stress Model," *Occupational Health Nursing,* December 1984, pp. 622-625.

Redgrove, June S., "Women Are Not from Lilliput or Bedlam," *Ergonomics,* 1984, pp. 469-473.

Rytina, Nancy F., "Earnings of Men and Women; A Look at Specific Occupations," *Monthly Labor Review,* April 1982, pp. 25-31.

Scilken, Mary P., "Pregnancy in the Workplace," *Occupational Health Nursing,* September 1984, pp. 465-468.

Scott, Judith A., "Keeping Women in Their Place; Exclusionary Policies and Reproduction," in *Double Exposure,* ed. Wendy Chavkin, New York, Monthly Review Press, 1984, pp. 180-195.

Stellman, Jeanne M., "Occupational Health Hazards of Women: An Overview," *Preventive Medicine,* 1976, pp. 281-293.
Trebicock, Anne M., "OSHA and Equal Opportunity Laws for Women," *Preventive Medicine,* 1978, pp. 372-384.
U.S. Department of Labor, *Working Women and Public Policy,* Report 710, Washington, DC, Government Printing Office, 1984.
U.S. Department of Labor, *Women at Work: A Chartbook,* Bulletin 2168, Washington, DC, Government Printing Office, 1983.
U.S. Department of Labor, Women's Bureau, *Employment Goals of the World Plan of Action: Developments and Issues in the United States, Report for the World Conference on the United Nations Decade for Women 1976-1985,* Washington, DC, Government Printing Office, 1980.
Waldron, Ingrid, Joan Herold, Dennis Dunn, and Roger Staum, "Reciprocal Effects of Health and Labor Force Participation Among Women: Evidence from Two Longitudinal Studies," *Journal of Occupational Medicine,* February 1982, pp. 126-132.
Ward, Joan S., "Sex Discrimination Is Essential in Industry," *Journal of Occupational Medicine,* 1978, pp. 594-596.
Ward, Joan S., "Women at Work—Ergonomic Consideration," *Ergonomics,* 1984, pp. 475-480.
Warshaw, Leon J., "Employee Health Services for Women Workers," *Preventive Medicine,* 1978, pp. 385-393.
"Women in the Workforce—Collective Action—Their Most Potent Weapon," Washington, DC, International Labor Organization, May 1985.
Women, Work and Health Hazards: A Fact Sheet, Washington, DC, National Commission on Working Women, National Manpower Institute, 1981.
Women's Bureau, Office of the Secretary, U.S. Department on Labor, *Twenty Facts of Women Workers,* Washington, DC, Government Printing Office, 1984.
Zenz, Carl, "Reproductive Risks in the Workplace," *National Safety News,* September 1984, pp. 33-46.

Chapter 20

Legal Aspects

SAFETY AND RISK

Before the executive or manager can truly understand his or her role in safety and liability prevention, it is necessary to understand the relationship between safety and risk. Safety is judged; risk is measured. Risk is the probability and potential severity of harm to people and property. A product or plant is reasonably safe if society judges the risks involved to be acceptable. A determination of safety, then, involves the acceptability of the risk. Liability exposure is the penalty placed on a company for failure to achieve this required level of safety. This penalty is usually economic, but it can also involve criminal sanctions.

The senior manager's role in safety is twofold. First, he or she is responsible for achieving, predicting, and maintaining a desired safety level. Second, the manager must evaluate the acceptability of the risk and the desired safety levels. Once these standards are set, the safety and engineering staff can apply formal design or process reviews to achieve them. Safety and engineering play an important role even before the manager sets the safety levels, however. Traditional safety engineering techniques, such as fault tree analysis, hazard analysis, and failure modes and effects analysis, are useful in computing and predicting the probability and severity of potential harm in any process. The manager may rely heavily on this input in making his or her initial decisions. Industrial safety and quality control programs enter the process later. They are used to maintain the desired safety level both in the plant and in the products that leave the plant.

SETTING ACCEPTABLE SAFETY LEVELS

In setting acceptable safety levels, the manager must consider several factors. First, he or she must evaluate the company's ability to achieve a desired level of safety. This involves a knowledge of both technological and economic feasibility. If the manager's level of safety meets society's desire for safety (i.e., the risks presented by the company's products or activities are acceptable), there is no excessive liability exposure. Society's desires are based on the benefits society wishes to obtain, the price society is willing to pay for these benefits, and the risk society is willing to accept in order to obtain them.

When the company's level of risk exceeds what society is willing to accept, the difference can be defined as the liability exposure the company will have. Stated as a formula, this would be:

Company's risk level — society's acceptance of risk = liability exposure

In response to this exposure, the company must either decrease the riskiness of the product, remove the product from the market, cease production of the product, or (in the extreme case) go out of business.

This decision is more complex and unpredictable than it may seem. For one thing, society can disagree with the manufacturer's conclusion on how much risk is acceptable, rendering legal judgments against the company in a liability suit. For another, society is continually altering its view of what constitutes an acceptable risk. The definition of acceptable risk may vary from year to year, industry to industry, court to court, even juror to juror.

Even with all the complexities and uncertainties involved, the manager still has to decide on an acceptable level of risk for his or her plant and products. Fortunately, the manager doesn't have to make this decision in a vacuum. He or she can draw on federal and state statutes, safety rules and regulations, standards, and prior court decisions as criteria. Some of these, such as standards, establish only the *lower* limits for risk acceptability. If the product or plant falls below the standards' level of safety, it can be considered unreasonably dangerous. Mere compliance with such standards, however, is not accepted as an absolute defense in a liability action. Studying previous appellate court decisions may offer the company greater (though by no means absolute) certainty. Unless conditions have changed drastically or new law have been enacted, the court will follow its previous decisions. In other words, the manager should review important legal precedents and use them as a basis for establishing corporate guidelines. For example, recent court decisions have held that where industrial workers or consumers are exposed to hazardous or toxic substances, there must be adequate warnings to alert them to the hazards. Further, if the substances are to be used in the southwestern United States, the warnings must be written in both Spanish and English, because a large portion of that population is Hispanic. Many manufacturers use international warning symbols to deal with this problem, especially when products are exported.

Managers can best carry out their responsibilities if they understand the major sources and types of liability to which their companies can be exposed. Generally speaking, liability may arise from four major sources:

1. Regulatory penalties.
2. Workers' compensation.
3. Product liability.
4. Contractual agreements.

FEDERAL OCCUPATIONAL SAFETY AND HEALTH ACT (OSHA)

Under OSHA, a company is responsible for the health and safety of its employees in the workplace. The act's stated purpose is "to assure, so far as possible, every working man and woman in the nation safe and healthful working conditions." This applies directly to employers and employees and indirectly to the manufacturers of products that are used in the industrial workplace. The act authorizes OSHA to promulgate, modify, or revoke occupational safety and health standards according to a prescribed rulemaking procedure.

Under OSHA, individual states are given the choice of either creating their own equivalent occupational safety and health programs or being preempted by the federal programs. State plans must be at least as effective as the federal program and must be approved by the U.S. Department of Labor. State plans, once approved and enacted, supersede the federal act. The federal government, however, continues to monitor the state plans to ensure their effectiveness.

Record Keeping

Most employers covered by OSHA are required to maintain records on occupational injuries and illnesses. These records must be retained for 5 years following the end of the year to which they relate. Exposure and data analysis records, on the other hand, must be kept for 30 years. Medical records must be kept for the duration of employment plus 30 years. A more detailed discussion of the records to be kept, and sound management approaches to this function, will be found in Chapters 14 and 21.

Inspections

Department of Labor compliance officers, under the supervision of OSHA area directors, are responsible for conducting inspections of companies to determine if they are in compliance with the OSHA Act. Compliance officers advise the employer, his or her representatives, and employee representatives of all conditions and practices that do not meet the standards. The compliance officer may also recommend that citations be issued for alleged violations and may propose penalties for such violations. The authority for actually issuing the citations and proposed penalties, however, rests with the area director or his or her representative.

Violations

There are four types of violations: (1) imminent danger, (2) serious, (3) nonserious, and (4) de minimis (very minor).

Imminent Danger. An imminent danger is one that has a high probability of causing death or serious physical harm immediately or before it can be corrected (e.g., a gasoline can next to a furnace). If a compliance officer finds what he deems to be an imminently dangerous situation, he will immediately attempt to have the employer correct it and remove employees from the danger zone. If the employer refuses, the compliance officer can recommend a court order to shut down the operation. (Note that the compliance officer himself has no authority to order a shutdown or to direct employees to leave the danger zone.)

Serious Violation

A serious violation is one that involves a substantial probability of death or serious physical harm. If the compliance officer finds that the employer knew or should have known of a hazard, he may cite it as a serious violation. The compliance officer will also consider how severe the most likely injury due to the hazard would be.

Nonserious Violation

A nonserious violation is one whose likely consequences are less than death or serious physical harm. Also, if the employer did not know of the hazard, or could not reasonably be expected to know of it, the violation would be considered nonserious.

De Minimis Violation

De minimis violations are those that have no immediate or direct relationship to safety and health. Technically, they do not comply with the letter of the law, however, and may still be brought to the employer's attention. As an example, OSHA specifies letters of a certain height and legibility for all exit signs; exit signs with slightly smaller lettering would be de minimis violations.

There are two other types of noteworthy violations: (1) a willful violation and (2) a repeated violation. Both of these can bring significant increases in penalties.

A willful violation exists where the evidence shows that either:

1. The employer committed an intentional violation of the OSHA Act, knowing that it was a violation; or
2. The employer was not intentionally violating the Act, but was aware that a hazardous condition existed and made no attempt to eliminate it.

A repeated violation occurs when a second citation is issued for a violation of the OSHA Act. This is not the same as failure to correct a previous violation. A repeated violation exists when the employer has once corrected the condition or violation but later violates the same standard.

Penalties

A compliance officer is not required to propose a penalty for nonserious violations but must do so for a serious violation. In either case, the maximum penalty that may be proposed is $1000. In the case of willful or repeated violations, a civil penalty of up to $10,000 may be proposed. Criminal penalties may also be imposed on an employer whose willful violation of a standard results in the subsequent death or injury of an employee.

Contested Violations

An employer has the right to contest an OSHA action if he or she feels that it is not justified. The first step in this process is an informal conference between OSHA and the employer. If the dispute is not resolved satisfactorily, an OSHA Review Commission judge will be assigned to the case. He, or she, in turn, conducts a hearing to determine the case's merits. The judge then submits the record and his or her report to the Review Commission. If no commissioner orders a review of the judge's recommendation, that recommendation becomes the Review Commission's decision. However, if any commissioner orders a review of the case, the Commission itself must render a decision. If an employer still disagrees with the Commission's decision, he or she may take steps to obtain a review in the U.S. Court of Appeals.

WORKERS' COMPENSATION

In addition to regulatory penalties, the company may be exposed to workers' compensation liability for employee injuries. Workers' compensation is a no-fault system offering speedy recovery for employee injuries. Under it, the employee accepts a fixed schedule of benefits and relinquishes the right to sue the employer in a civil lawsuit. Under this system, an employee must rely on workers' compensation as the exclusive remedy. He or she cannot file a civil lawsuit against the employer to collect for the same injury. There are, however, several exceptions to this general rule. One of them is the situation where the employee is injured while using a machine or substance manufactured by the employer which is also sold and distributed to the general public. This is known as the "dual capacity doctrine." The employee, in that case, is functioning in the dual capacity of employee and consumer. Other exceptions exist where the employer or a fellow employee owes a separate duty to the injured employee. For example, the employer may face additional liability if a company-employed doctor is found guilty of medical malpractice.

Another exception to the exclusive remedy doctrine exists where the employer has *intentionally* inflicted harm on the employee. These cases fall into three categories: (1) physical assaults, (2) intentional infliction emotional distress, and (3) fraud or deceit. An employee who has been assaulted by the employer or a fellow employee can pursue a claim for civil damages, as for any other physical assault. Employees have also been allowed to recover against their employers and fellow employees for the intentional infliction of emotional distress. In one such instance, the employee alleged that he had suffered discrimination, as well as rude and degrading treatment.

One case under consideration involves the question of whether the exclusive remedy rule bars employee lawsuits against employers for the intentional torts of fraud, deceit, or concealment. In this case, the employee alleges that the employer intentionally concealed from him the dangers of working with asbestos, and subsequently concealed the nature of the disease the employee had contracted. According to his allegations, company doctors diagnosed him as having asbestosis but failed to inform him of the fact.

Whether or not the employee has an independent civil claim against the employer, he or she still retains the right to sue the manufacturer of the equipment that caused the injury. This is based on traditional product liability concepts, as discussed below. When the employee is not bound by the exclusive remedy provision of workers' compensation, he or she may also pursue these products liability suits against the employer. The interaction of workers' compensation and products liability is illustrated in Figure 20-1.

PRODUCT LIABILITY

Product liability has been very much in the news lately. No manager needs to be told that the company can be held liable if a product it manufactures or sells causes personal injury or property damage to buyers or third parties. This liability is not lessened if the firm produces finished products whose components were manufactured by someone else. If a component causes injury, the firm that assembled and sold the article can still be held liable. In fact, the manager may find, to his or her dismay, that the sales agreement for the components contains provisions that hold the original manufacturer of the components harmless and limit or prevent recovery of damages. Obviously, this is a complex field, and it is growing more complex every day. The following summary does not present all the details. Precedents will have changed even before this book is printed. However, a brief summary may still be useful to the executive.

Parties to a Product Liability Lawsuit

Product liability was and continues to be a result of complex social policies and pressures. That is, it is based on current ideas of what is best for society and the individuals that make it up. Product liability law has always changed

592 SAFETY AND HEALTH MANAGEMENT

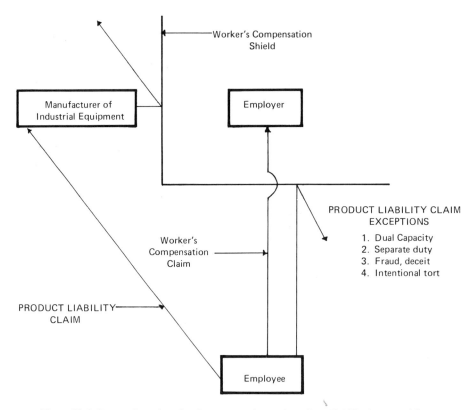

Figure 20-1. Interaction of workers' compensation and products liability in an accident.

along with changes in society and its value system. The problem is merely more noticeable today, when change is so rapid, radical, and unpredictable.

Much of the growing concern about product liability is due to the fact that today more people can recover for injuries or property damage from defective products. Courts have expanded the list to include buyers, guests, users, bystanders, neighbors, trespassers—in fact, anyone who can foreseeably be injured by the product. In other words, nowadays almost anyone can sue.

Product liability is no longer the sole headache of the manufacturer or seller. The number of potential defendants has also been expanded. It now includes all persons in the chain of distribution (supplier, manufacturer, distributor, and retailer), as well as certifiers of safety (such as independent testing laboratories, insurance company loss control engineers, government safety inspectors, patent licensors, sellers of used products, design and safety engineers, repairers, installers, and government entities).

Standards of Responsibility

The basis on which a defendant is judged legally liable is called the "standard of responsibility." There are three such legal standards. These are based on the theories of:

1. Negligence.
2. Strict liability.
3. Warranty.

Negligence. Under the theory of negligence, a manufacturer must take due care in supplying products that do not present an unreasonable risk of injury or harm to the public. Negligence focuses on the conduct of the manufacturer and asks whether he or she acted as a reasonable or prudent person would. In the event of a claim, the manufacturer will be evaluated on the basis of quality control procedures, safety and design analyses or reviews, and other efforts to ensure the safety of the product. The manufacturer who can prove that he or she has taken due and reasonable care may present a successful defense against liability charges.

Strict Liability and Implied Warranty. Under the theory of strict liability and implied warranty, the question is whether the product is defective or not. That is, strict liability focuses on the product itself and asks whether it is defective, regardless of how or why it became defective (that is, whether due to negligence or not).

The term "implied warranty" is easy to understand. In the act of selling an item, a company implies that it is capable of safely performing the job or operation for which it was intended. The implied warranty on a radio, for example, is that it will play. If it explodes when turned on, it has not performed as the warranty implied.

Express Warranty and Misrepresentation. An express warranty is one that is specifically stated (generally in writing) rather than merely implied. Under the theory of express warranty, the question is whether the manufacturer's warranty accurately represented the capability of a product. If the product does not live up to the representations, there can be liability even when the product itself is not defective and when due care was taken by the manufacturer. For example, a cleaning fluid manufacturer may take all due care in producing a high-quality product and may print "Guaranteed safe for all fabrics" on the label (an express warranty). If the product later damages new synthetic fabrics, the manufacturer can be held liable under the theory of express warranty.

Other Legal Concepts

There are certain other legal terms and concepts that help to explain the basis of product liability. These include

1. Product defects.
2. Levels of technology.
3. Foreseeability.
4. Conduct of the user.

Product Defects. There are three types of product defects:

1. Manufacturing or construction defects.
2. Design defects.
3. Defects due to a failure to warn.

A product contains a manufacturing defect if it does not conform to the design. A production error, for instance, may cause the product to fall short of the manufacturer's own design. It is more difficult to establish that a product is defective in design. There is no absolute standard by which to measure the adequacy or inferiority of a design. Similarly, the manufacturer may provide what he or she genuinely considers to be adequate warning but may fail to alert users to the potential risk of damage or injury. These may involve highly subjective judgments.

Level of Technology. There are two essential concepts to consider in discussing a product's level of technology: (1) custom and practice within the industry and (2) state-of-the-art technology.

These two terms have very different meanings, although their distinction can sometimes be difficult. "Custom and practice" refers to the level of scientific and technological knowledge that is actually applied to products that are sold to customers—in other words, what manufacturers in a particular industry are doing today. To take an example, research may have revealed a newer material that is safer for use in an item, but manufacturers have not yet developed an economical way to utilize it. Custom and practice would be to continue manufacturing the item using the traditional, less safe material.

"State of the art," on the other hand, is whatever level of scientific and technological knowledge exists, regardless of whether it has been transformed into marketable hardware. To the best of the manufacturer's knowledge, there may be no safer way to make a particular item; the state of the art has not gone so far.

Evidence that a product complied with the state of the art has always been sufficient to relieve the manufacturer of liability. What the manufacturer proves in this instance is that there is no scientific knowledge or technology available to make the product economically safer. Evidence of custom or practice, on the other hand, is generally not sufficient, since the manufacturer and the entire industry may simply have chosen to ignore economically feasible and available safety devices and options by mutual agreement.

Generally, the level of technology is related to the date of manufacture. That is, a product can reflect only the state of the art and the custom and practice at the time of its manufacture. However, the trend is not to impose a postmanufacturing duty on the manufacturer to monitor and upgrade the safety of his products.

Foreseeability. Many of the requirements imposed by the courts on manufacturers are based on the concept of "foreseeability." The manufacturer is required to foresee (1) the environment in which the product will be used, (2) the applications of the product, (3) the age and skill of the product's users, and (4) the expected types of modifications and/or misuses to which the product may be subject. The manufacturer must design a product that meets not only its *intended uses* but also its *foreseeable misuse*. One example of this is the type of guard installed on a chain saw intended for home use. The manufacturer might foresee that if the guard is detachable, a future user might easily remove it and later be injured. The manufacturer's failure to make the guard an integral part of the saw's design might therefore make him liable under the concept of foreseeability.

Conduct of the User. Traditionally, if a claimant knew that an item was dangerous or defective, and used it despite the knowledge that he or she might be injured, the claimant was considered to assume the risk. Under the doctrine of "assumption of risk," a claimant may be barred from recovery for damage or injury suffered. Similarly, if willful or deliberate misuse of an item cause the user harm, he or she was considered to have contributed to the injury. Under this concept, known as "contributory negligence," the user could also be barred from recovery (except in some cases of strict liability).

A growing number of states have adopted a new concept, "comparative negligence," to replace the older idea of contributory negligence. Comparative negligence involves the relative responsibility that the plaintiff and defendant had in the case of an injury. The plaintiff's recovery is reduced by the percentage of responsibility attributed to him or her. To take our earlier example of the manufacturer who installs a detachable guard on a chain saw, here is how it would work. Suppose that a customer buys the chain saw, removes the guard, and is subsequently injured. The court might find that the manufacturer was negligent under the concept of foreseeability. However, it might also find that the customer's injury was at least 50 percent due to his own willful action in removing the guard, knowing that to do so might be dangerous. So, instead of awarding him the $100,000 he asked for, the court would award him only 50 percent of that amount, or $50,000.

Damages

A major concern of both manufacturers and insurers is the amount of damages that injuries have assessed in recent years. This amount has been steadily increasing. There have been some proposals to regulate or limit certain types and amounts of damages, but to date the situation remains unchanged.

There are four types of damages for which a manufacturer may be liable: specific damages, general damages, punitive damages, and consequential losses. "Specific damages" are those to which an exact amount or cost can be assigned: exactly how much the plaintiff had to pay for medical care after an injury, exactly how many dollars' worth of her property was destroyed, exactly how much he lost in wages due to an injury. "General damages," on the other hand, are those to which an exact dollar amount cannot be fixed because of their subjective nature: pain and suffering, mental anguish, and so on. In general, liability insurance will cover both specific and general damages. In both cases, the aim is to compensate the injured party for the loss, injury, or damage.

Unlike specific and general damages, "punitive damages" are generally not insurable. For this reason, although they are not often assessed, they present a serious problem for the manufacturer. Punitive damages are damages assessed over and above specific or general damages as punishment for a manufacturer's willful or wanton disregard for the safety of the user. Because such damages are regarded as a punishment for outrageous acts against the public, the insurance industry believes that it would not be in the public interest to insure against them.

In addition to damages as such, a plaintiff may sue to recover "consequential losses," such as loss of profits or other resultant business losses. For example, if a product used in a small business fails, that failure may result in a loss or profits, custom, and goodwill during the period of time in which the business cannot function. Manufacturers should be aware of the provisions of the Uniform Commerical Code for such business losses. Properly drafted disclaimers of liability for such business losses can protect the manufacturer.

Hold Harmless Agreements and Disclaimers

To a certain extent, business entities have the freedom to shift liability from themselves to another by contract. In this sense, product liability has a certain economic value. As an example, a commercial buyer may be willing to accept a certain amount of liability in return for a reduced purchase price. A user may believe that he is adequately covered by his own property and business insurance. A distributor may require indemnification or hold harmless agreements as an incentive to market and distribute a particular manufacturer's products. Manufacturers, on the other hand, may require indemnification or hold harmless agreements from the suppliers of components to protect themselves if those components prove to be defective. There are certain limitations on what liability can be shifted via these agreements, however. The law does not allow disclaimers of liability for personal injury. Only disclaimers relating to property damage or consequential losses are allowable as a legal defense.

Under the Magnuson-Moss Federal Consumer Warranty Act, a manufacturer does not have to provide warranties. If he or she does, however, such warranties must comply with the technical requirements of the act. A manufacturer should be awre of these requirements to make sure that his or her representations or warranties do comply with the act.

Statute of Limitations

Except for contractual warranties, the statute of limitations for product liability actions does not begin until an injury has occurred. Thus, a manufacturer's or insurer's potential liability is often of unlimited or uncertain duration. In Utah, for example, it is 10 years from the date of the first sale. Other states shift the burden to the plaintiff to prove defectiveness after a certain period of time. In Colorado, for instance, this is a 10-year period.

Other schemes use the concept of "useful safe life" to determine how long a manufacturer should be subject to liability. A product is presumed not to be defective if the accident occurs after the expiration of its useful safe life. This is the scheme proposed by the Department of Commerce in its Model Product Liability Act. The useful safe life concept, however, will not apply if the manufacturer has warranted the product for a longer period.

HOW TO MINIMIZE LIABILITY

Minimizing Design Liability

The first significant step in minimizing liability is to minimize design liability. Product safety must be evaluated, reviewed, and adequately documented for each product, new design, and significant design revision. All products should be conceived and designed to meet or exceed all applicable industry and government safety standards, codes, and regulations. Product safety can be attained by:

1. Product designs which contain no safety hazards.
2. Designs which shield personnel or property from the hazards.

DESIGN REVIEW CHECKLIST

1. Describe the scope of product uses and applications.

2. Identify the environments within which the product will be used.

3. Describe the types of potential buyers and users.

4. Postulate all possible hazards, including estimates of probability of occurrence and seriousness of resulting harm.

5. Describe alternative design features or production techniques, including warnings and instructions, that can be expected to reduce or eliminate the hazards.

6. Evaluate the expected performance of the alternatives, including the following:

    a. Characteristics and comparisons of different products that might serve the same purpose.

    b. Other hazards that may be introduced by the alternatives.

    c. Their effect on the subsequent usefulness of the product.

    d. Their effect on the ultimate cost of the product.

Figure 20-2. Design review checklist for products liability.

3. Operational procedures which recognize hazards and instruct personnel on how to perform the tasks so as to avoid them.
4. Warnings which identify hazards and provide specific instructions on their avoidance.

Figure 20-2 is a checklist for design review.

Maintenance

Proper maintenance features, procedures, and instructions can also help minimize liability risks. First, the product or process must be designed to ensure the safety of personnel during maintenance operations. The product should also provide wear, fatigue, or malfunction indicators as required. In addition, there should be a means of verifying that required maintenance has been performed (such as tags, logs, and so on). These measures will reduce safety problems stemming from lack of maintenance, as well as decreasing the hazards involved in the maintenance operation itself.

Word Control

Poorly selected words and illustrations in printed matter create an undesirable and unintentional product risk for manufacturers. There are three areas in which proper word control is essential:

1. *Sales and Marketing Claims.* Sales and marketing claims must be consistent with the actual capability of the product. Product advertising, sales promotion, and publicity materials are considered part of the company's product warranty. Words, photographs, and pictures must promote safe and proper use of the product. These representations must be technically accurate and must not misrepresent or otherwise overstate the product's capabilities.
2. *Warnings.* Warnings are not a substitute for safe design or adequate quality control. However, they may be required to make the installer, repairer, or user aware of certain limitations or dangers associated with the equipment. Although design, failure to warn, or inadequate warning allegations are sometimes separately alleged in a lawsuit, they are not really independent. The final design review should include a review of the need for and content of warnings, instructions, and manuals. Warnings should be considered part of the design phase. A design is incomplete unless it specifies proper and adequate warnings or cautions. In order to be adequate, warnings must be prominent, intense, and permanent. The user must be fully aware of the magnitude and possible consequences of the danger. A product that can be used for many years must have a warning that will be legible for that entire period of time.
3. *Contractual Arrangements.* Purchase and sales agreements should be carefully reviewed so as not to expose the company to unnecessary product liability. Many product liability loss prevention programs are compromised because the company has accepted unnecessary product liability from its suppliers and distributors. The company should make conscious choices on whether to seek or give hold harmless agreements, rather than routinely accepting those contained in boilerplate contracts.

Field Support and Monitoring

Another essential area in which liability can be minimized is field support and operation. There must be an effective monitoring system to ensure that products perform adequately and safely in service. When a complaint or other user feedback is received, an evaluation should be made of the cause of the problem and the steps to be taken to prevent similar occurrences in the future. Where warranted, a recall campaign or retrofit program may be required.

Accident and Failure Investigation

In the event of an accident or mishap, adequate accident investigation and failure analysis may be essential in defending product liability lawsuits. Manufacturers should develop stringent policies and procedures for reporting and processing accident and incident information. Moreover, there should be firm instructions regarding disclosure of such information, and all company employees should be familiar with these guidelines. It is essential that one individual be assigned sole responsibility for communicating with the public regarding accidents, incidents, and hazards. Such communications should be made only after consultation with corporate management and/or corporate counsel. *All other employees must understand that they are not to provide such information to the public.* Employees should not give interviews or written statements or sign any statements in connection with an accident, incident, or hazard. No admissions of liability should be made. Written reports pertaining to an accident should be confined to the facts and based on the personal knowledge of the company and the employees. Such reports should not venture opinions of conclusions unless specifically requested by legal counsel.

The reason for these cautions should be obvious: if there is a product liability lawsuit, the plaintiff has broad discovery rights. He or she may give written questions (known as "interrogatories") to the manufacturer and may orally depose, or ask questions, of designers and other employees. Records and documentation may become part of court evidence, so the manufacturer should be aware that compromising or misleading statements in written reports may eventually do great harm.

Postmarketing Duties

Under the proposed Department of Commerce Product Liability Model Act and a number of court decisions, a manufacturer has a duty to warn users of dangers that come to light after manufacture. Under these circumstances, the manufacturer should make reasonable efforts to inform users of the risk and of the action needed to avoid it. In order to do this, the manufacturer should maintain a thorough feedback system in order to become aware of newly discovered dangers associated with the use of the product. The manufacturer must also have adequate sales records, which will allow him to contact the ultimate user of the product.

Record Keeping and Documentation

All critical information relating to design decisions, manufacturing, and quality control should be adequately documented. Complete and proper records are the company's first and best line of defense in a liability suit. Even strict liability is not absolute liability. Many cases are won—or liability is greatly reduced—when a company can document the reasons for its decisions.

CONCLUSION

This is not a complete or exhaustive discussion of the subject. That would require a book, or possibly a library. However, this brief overview should give managers some insight into the legal implications of the decisions they make every day. With these insights, product safety and liability prevention programs can be greatly strengthened and given their proper place in the company's overall safety and health programs.

BIBLIOGRAPHY

Weinstein, Alvin S., et al., *Product Liability and the Reasonably Safe Product,* New York, Wiley, 1978.

Chapter 21

Regulatory Considerations

INTRODUCTION

Today the term "regulation" is apt to conjure up images of every manager's headache: a network of confusing and constraining rules and standards; costly modification of existing installations to meet new legal demands; inspections, fines, or time-consuming legal hearings; and, above all, an increasingly burdensome task of record keeping and paperwork.

This, however, is a shortsighted and often defeatist view. While no one can deny that regulatory considerations play an increasing role in business today, senior managers have far more control than they realize. Their attitudes, and the policies they set for the organization, can either make regulatory needs a costly and cumbersome appendage or can integrate them into an overall safety and health program that ultimately means more effective and profitable management. Indeed, many profit-minded managers are finding that self-regulation, setting even higher standards than the minimum required by law, is in their company's best interests — part of a long-term effort to reduce costly accidents, soaring workers' compensation costs, and the irreplaceable loss of the company's valuable human resources. However, in order to achieve this ideal working system, the manager first needs a solid grasp of the overall regulatory picture. This chapter attempts to provide that foundation.

ADMINISTRATIVE AND CIVIL LAW

Understanding the function of regulatory agencies requires a clear grasp of the differences between civil and administrative law. These differences are summarized in Figure 21-1. Stated briefly, a court will only decide real and present issues. An issue that has already been resolved by the parties may be considered "moot," and the courts will not state an opinion on it; an issue that has not yet developed into a controversy is not "ripe" for review and therefore will not be decided by a court. When deciding a case, the court is careful to *resolve only the present controversy.* Courts do not give advisory opinions; they do not try to decide what would happen in a hypothetical situation. They leave such decisions to future courts that may actually be faced with the issue.

In some of the things it does, an administrative agency acts like a court of

COURT (CIVIL LAW)	AGENCY (ADMINISTRATIVE LAW)
Will only decide real, present issues	Develops rules and procedures to decide future issues or events
Evaluates facts regarding past events only	Does not have to wait for an event to occur before acting
May decide cases on basis of existing laws only	Has power to investigate situations and make new rules as a result
Hold trials	Holds hearing and makes findings
Has no power to make laws (this belongs to government)	Has legislative powers to make rules and regulations
Has no power to enforce laws (this belongs to civil authorities)	Has the power to regulate and control entities that fall under its jurisdiction

Figure 21-1. Civil vs. administrative.

law. OSHA, for example, may rule for or against an individual or a business. Unlike courts, however, agencies do not have to wait until a case (in this instance, a worker's injury) has occurred before they act. Agencies are also legislative in nature; they have the power to investigate situations which come to their attention, and to formulate rules and standards. As an example, the Consumer Product Safety Commission may investigate a series of accidents involving a particular consumer article. On the basis of its findings, it may decide that a standard is needed for that product and formulate one. The main function of an agency is to develop and implement rules to carry out its legislative mandate, while the main function of courts is to decide on past events.

Administrative Agency Activities

Administrative agencies have three primary activities:

1. Holding hearings.
2. Making rules.
3. Monitoring activities.

In carrying out these functions, the agency is usually given broad discretionary powers. However, if an agency abuses this discretion, the affected party has the right to seek a court's review of the agency's actions. If the agency's findings and actions are confirmed by substantial evidence, if it is supported by a reasonable and rational basis, and if its procedures were followed fairly, the court will generally defer to the agency's expertise and discretion. In keeping with procedural fairness, agencies seldom make rules without first giving adequate notice that such rules are being considered. Ordinarily, a proposed rule of a federal agency is published in the *Federal Register*. Thus, those who would be affected by the rule have an opportunity to study the proposed regulation and to present their viewpoints, usually at hearings announced well in advance. Publication in the *Federal Register* is assumed to be adequate notice to those who would be affected.

Regulatory Decision Making

In arriving at their decisions, most regulatory agencies use some form of risk/benefit analysis. "Risk/benefit analysis" is defined as an evaluation of all the benefits and costs of a proposed action. It is a much broader concept than the traditional cost/benefit analysis, which involves only economic factors. The risk/benefit analysis technique is not really new. Most people use it all the time, making decisions by weighing the advantages (benefits) and disadvantages (risks) of an action. The major risk considered is usually personal injury or harm. Benefits usually involve either economic or health considerations. The average person tends to think of risk avoidance in terms of dollars ("How much is it worth to me to avoid injury while still getting the benefits I want?").

Company managers must also make these choices. There are significant differences, however, between the risk/benefit analyses performed by consumers, company managers, and government agencies. Consumers determine what benefits they want, what price they are willing to pay to obtain them, and what risks they are willing to accept. Company managers, on the other hand, must step outside their personal risk/benefit judgments and attempt to predict what benefits society (the consumer) wants and what risks it will consider acceptable, as well as what is a reasonable price to ask. Managers are also subject to legal and business constraints that must be taken into account. Similarly, government safety regulators must go beyond their personal risk/benefit assessments. Regulators, do not commit their own resources for their own benefits. They, too, are subject to legal and political constraints. They must decide how best to use the public resources available to provide the greatest public benefits.

Some governmental functions are used to regulate private activities in the public interest, rather than to bring benefits directly to the public. Consumer and market buying patterns are used as indicators of society's desires. This form of regulatory decision making must consider whether the government should interfere with risks that the public wishes to accept in order to achieve some other public benefit.

REGULATORY CONSIDERATIONS

The government regulator must assess the risks and benefits from the standpoint of three different parties:

1. The provider of the product or service.
2. The individuals who would enjoy the benefits of the product or service.
3. Society as a whole.

Actually, "society as a whole" is not as uniform as that term implies. There are numerous groups that compose society, including:

1. Individuals who knowingly and voluntarily choose to accept certain risks in order to certain benefits (e.g., a worker who deliberately chooses a hazardous occupation because it offers a higher rate of pay).
2. Individuals who desire certain benefits but do not understand the true scope or nature of the associated risks. They would probably choose to forego the benefits if they knew of the risks (e.g., workers in good-paying jobs that involve habitually handling carcinogenic materials, which they are unaware can cause cancer).
3. Individuals who enjoy or would enjoy the benefits of the product or service and might require protection against their own wants (e.g., eager weight watchers who will purchase and use a diet pill even if it could seriously harm their health).
4. Innocent bystanders who may be injured by activities related to the product or service, but who do not want or receive any of the benefits (e.g., a family living near a manufacturing plant that discharges toxic fumes into the atmosphere).
5. Members of the public who have no direct interest in the matter, either from a risk or a benefit standpoint (e.g., individuals or groups who do not want or use a particular product or service, such as nonsmokers, people who do not own home power tools, and so on).
6. Individuals who know of the scope and nature of the risk but do not believe that they have the ability to make a voluntary choice or judgment on whether to accept it or not (e.g., factory workers who handle carcinogenic agents and cannot afford to give up their jobs, and thus feel that they cannot refuse to handle these agents).

Understanding the criteria that an agency will use in decision making is vital. Such criteria are based on the agency's statutory mandate and its primary concern. OSHA's primary concern, for instance, is worker safety and health. OSHA's criteria, therefore, concentrate on increasing worker safety, even if this means reducing plant productivity.

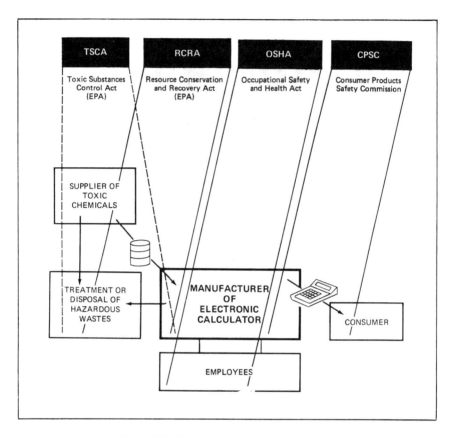

Figure 21-2. Interaction of regulatory agencies.

Interaction of Regulatory Agencies

The manager who is concerned with safety and health also has the difficult task of dealing with several different agencies and personnel. Various regulatory agencies can interact, overlap, or even contradict each other. Consider a simple example: the manufacture of an electronic calculator.

1. The vendor who supplies toxic chemicals to the calculator manufacturer is regulated by the Toxic Substances Control Act.
2. The safety and health of an electronics worker at the calculator manufacturer's plant is regulated by OSHA.
3. The treatment, shipment, storage, or disposal of hazardous wastes incident to the manufacture of the calculator are regulated by the Resource Conservation and Recovery Act.

REGULATORY CONSIDERATIONS 607

AGENCY	AREA OF RESPONSIBILITY
Consumer Product Safety Commission (CPSC)	All consumer products except foods, drugs, cosmetics, medical devices, motor vehicles, boats, airplanes, radiation devices, tobacco, firearms, explosives, alcohol, and pesticides.
Department of Transportation Materials Transportation Bureau	Hazardous materials, pipelines safety.
Environmental Protection Agency (EPA)	Pesticides; air, water, noise pollutants; solid wastes; toxic substances which may affect the environment.
Food and Drug Administration (FDA)	Foods, drugs, biological products, cosmetics, therapeutic devices.
Nuclear Regulatory Agency (NRA)	All nuclear facilities, equipment, materials, and their operations.
Occupational Safety and Health Administration (OSHA)	Industrial and construction equipment and workplace, except for mines, nuclear facilities, and railroads, and their operations.

Figure 21-3. Major regulatory agencies concerned with safety.

4. The calculator in the hands of the consumer is regulated by the Consumer Product Safety Commission.

Figure 21-2 shows this in graphic terms.

Figure 21-3 shows the major government agencies that are concerned with safety. As you can see from the descriptions of their areas of responsibility, it is possible for some of them to overlap.

RESOURCE CONSERVATION AND RECOVERY ACT (RCRA)
Requirements and Provisions

Hazardous wastes are one of the most serious environmental issues we face today. To deal with this problem, Congress passed the Resource Conservation and Recovery Act (RCRA) in 1976. This law authorizes the EPA to develop a nationwide hazardous waste management system covering the period from the time a hazardous waste is generated until its final disposal. On May 19, 1980, the *Federal Register* published its key regulations, which lay the foundation for the entire Federal Hazardous Waste Regulation Program. These regulations became effective November 19, 1980. The regulatory program governs all aspects of the hazardous waste cycle: generation, transportation, treatment, storage, and disposal.

The first step in understanding the EPA's program is to determine who will be regulated. Any firm that generates, transports, treats, stores, or disposes of hazardous wastes is subject to the regulations. There are basically three major groups of regulations:

1. Regulations for generators of hazardous wastes.
2. Regulations for transporters of hazardous wastes.
3. Regulations for treatment, storage or disposal facilities.

In addition, upon EPA approval of its hazardous waste management program, each state will take over the operation and enforcement of the RCRA program. The states are free to prescribe additional rules, provided that they are at least as stringent as the federal regulations. Each company generating hazardous wastes will be assigned an identification number and will be required to keep records allowing the progress of the wastes to be carefully controlled from generation to disposal. Companies which transport, store, treat, or dispose of hazardous wastes will need to obtain permits for these activities, and will be subject to strict safety, health, and environmental controls and record-keeping rules.

The primary question is whether the waste is, in fact, hazardous under the EPA definitions. First, the company must determine if the waste meets the EPA criterion for being a solid waste. (A solid waste is not necessarily solid; it may take the form of a liquid or sludge, according to the EPA definition.) The RCRA also names several specific exemptions from the definition, including:

1. Domestic sewage (untreated sanitary wastes that pass through a sewer system).
2. Domestic sewage mixed with other waste which passes through a sewer to a publicly owned treatment works for treatment.
3. Industrial waste discharges regulated under the Clean Water Act as point source discharges.
4. Nuclear waste regulated under the Atomic Energy Act.

If a company decides that the material in question is not a solid waste according to this definition, it is unregulated. If, however, the material is a solid waste as defined, the company must then decide if it is also a *hazardous waste.*

One method of determining whether or not a material is a hazardous waste is to see if it is listed in the EPA regulations. Over 400 wastes and waste streams have been listed as hazardous by the EPA. If the material is not on the EPA list, it must be tested to see if it exhibits any one of the four characteristics of hazardous wastes. These characteristics are:

1. Ignitability.
2. Corrosivity.
3. Reactivity.
4. Toxicity.

Under the RCRA, the burden is on the generator to determine whether or not the waste is hazardous.

Keep in mind one obvious fact, however: the RCRA notification *applies only to wastes*. Even if a substance appears on the EPA list, notification is *not* required if it has not entered the manufacturing process, if it is being actively used in the manufacturing process, or if it is a *product* (rather than waste).

Even if the material is a hazardous waste, the company does not necessarily have to notify the EPA. If less than 1000 kg of all hazardous wastes is generated in a calendar month and accumulated at any time, the company is not required to notify the EPA of hazardous waste generation. This is also true if the waste is recycled. It should be noted, however, that for certain listed wastes, the small-quantity exemption is less than 1000 kg; in some cases, notification must be made if as little as 1 kg of waste is generated. Quantities of waste are measured at the time the material becomes a waste. Subjecting a waste to a process which decreases its volume would constitute "treatment" under the RCRA and would not meet the small-quantity limits.

If a company is a generator of hazardous waste, it should have notified the EPA of its status as of August 19, 1980, and subsequently received an EPA identification number to be used in all hazardous waste transactions.

One of the responsibilities of the generator is the creation of a manifest system to track the waste from the point of generation to the point of disposal. The generator must prepare this form with enough copies to each subsequent handler. The form itself must accompany the shipment at all times.

Waste handlers are also required to comply with Department of Transportation regulations governing the use of appropriate containers and proper labeling of those containers for hazardous wastes.

A hazardous waste generator may accumulate waste on-site without a permit for 90 days, provided that the waste is packaged, labeled, located, and monitored as specified in the regulations. Generators who accumulate hazardous wastes for more than 90 days are considered to be operating a treatment, storage, or disposal facility. Regulations for these facilities are considerably more complex and stringent than those that govern generators.

If the company is a transporter, it is required to notify the EPA of its existence and to receive an EPA identification number. In addition, it is required to:

- Keep records concerning the origin and destination of the waste transported.
- Accept hazardous waste for transport only if it is properly described, labeled, and packaged.
- Properly load and stow wastes before shipment.
- Comply with the manifest system.
- Deliver all hazardous wastes to a permitted disposal facility.
- Comply with regulations on proper procedures for dealing with accidental spills.
- Comply with regulations governing the marketing of transport vehicles formerly used for transporting hazardous wastes.

Transport of hazardous wastes is the movement of wastes by air, rail, highway or water, and essentially refers to off-site movement of wastes. A generator may move waste onto contiguous property without a permit if it involves merely crossing a public right of way (i.e., driving it directly across the street to the company's facilities). However, a generator who moves the waste along a public road must have a transporter's permit and is subject to the regulations.

If the company is a treater or a disposer, it is also required to apply for permits for its operations. The definition of "disposal" in the regulations is extremely broad, covering not only conventional disposal methods but spills and leaks as well. The regulations' equally broad definition of "treatment" includes processes to make wastes safe, safer, or smaller; more convenient for storage, transportation, and disposal; or more amenable to resource recovery. A generator who also disposes of or treats hazardous wastes is required to have a permit for these operations.

State hazardous waste programs may contain stricter regulations than those of the EPA. For example, California does not have a small-quantity exemption from hazardous waste regulations. Careful scrutiny of applicable state regulations is absolutely necessary.

Penalties for Noncompliance

The regulations required that notification of hazardous waste activity be given by August 18, 1980. Unless a company notifies the EPA of hazardous waste activity, it may not legally transport, treat, store, or dispose of waste. The EPA is authorized to use an enforcement order or court injunction to halt such illegal activities.

The regulations also provide for penalties in cases of noncompliance. Criminal penalties may apply to any person or company that intentionally:

1. Transports hazardous wastes to a facility without a permit.
2. Disposes of hazardous wastes without a permit.

3. Makes a false representation in an application, label, manifest, report, notification, or other RCRA document.

The penalty for the first offense is a maximum of $25,000 per day of violation and/or maximum imprisonment of 1 year. The penalty for a subsequent offense is a maximum daily fine of $50,000 and/or maximum imprisonment of 2 years. Finally, RCRA violations may also be in violation of state law, so the company may be subject to additional penalties imposed by the state.

Plan of Action

Waste generators should first determine whether their waste is subject to the RCRA notification requirements. If so, the generator should promptly complete the notification form and return it to the EPA. Generators should also determine whether their operations involve other hazardous waste activities which require a permit. If so, interim permit status should be sought. Generators should also ensure the availability of a permitted facility to receive their waste, and should prepare for manifest, record-keeping, packaging, and other requirements.

This is merely a brief synopsis of the RCRA regulations. The actual regulations are far more complex, and the consequences of noncompliance are severe. For that reason, the manager should not hesitate to secure technical advice and legal counsel whenever it may be needed to clarify the regulations or to determine which regulations apply. In addition, manufacturers should monitor any changes in the regulations, whether federal or state, to evaluate how these revisions will affect their operations.

TOXIC SUBSTANCES CONTROL ACT (TSCA)

Requirements and Provisions

In 1971, the President's Council on Environmental Quality developed a legislative proposal for coping with the increasing problems of toxic substances. Over the next 5 years, Congress held many hearings, debated, and finally passed the Toxic Substances Control Act (TSCA) of 1976. This act was signed by President Gerald Ford on October 11, 1976, and became effective on January 1, 1977. The law grants the EPA significant new authority to anticipate and address chemical risks and to regulate the chemical industry.

The purpose of the TSCA is to regulate chemical substances that present a health or environmental hazard. The act authorized the EPA to obtain information from industry on the protection, use, health effects, and other matters concerning chemical substances and mixtures. After considering the costs, risks, and benefits of a substance, the EPA may regulate its manufacture, processing, distribution, commerce, use, and disposal. Pesticides, tobacco, nuclear material, firearms and ammunition, food, food additives, drugs, and cosmetics are exempted from this act, as they are currently regulated by other laws or agencies.

The primary aim of the TSCA is to regulate chemicals in commerce, without regard to their specific use or area of application. The EPA is thus able to control a wide range of chemical hazards, particularly at their source, before they present a danger to the environment.

The major provisions of the TSCA are as follows:

1. *Testing.* Manufacturers and processors of potentially harmful chemical substances and mixtures are required to test them so that their effects on health and environment may be evaluated.
2. *Notification.* Manufacturers of new or existing chemical substances must notify the Administrator of the TSCA 90 days in advance of a new commercial production or application.
3. *Risks.* The manufacture of new chemical substances can be restricted or delayed when there is inadequate information to evaluate the health or environmental effects involved, if the substance may involve an unreasonable risk.
4. *Rules.* The Administrator may adopt rules governing the manufacture, processing, distribution, labeling, or disposal of a harmful chemical substance or mixture.
5. *Imminent Hazard.* The Administrator can obtain an injunction from a U.S. District Court to protect the public and the environment from a chemical substance or mixture that presents an imminent hazard.
6. *Record Keeping.* Manufacturers and processors may be required to submit reports and maintain records on their commercially produced chemical substances and mixtures, and to provide available safety, health, and environmental data on them.
7. *Information.* Manufacturers and processors must immediately notify the Administrator of information indicating that one of their substances or mixtures may cause or contribute to health or environmental risks.
8. *Inspections.* Inspections may be conducted to enforce the TSCA, and to authorize court actions for seizures of chemical substances and mixtures which have been manufactured or processed in violation of the act.

Prohibited Acts

Under the TSCA, the following are prohibited acts:

1. Failure or refusal to comply with any rule or order issued under the act.
2. Commercial use of any chemical substances or mixtures which the user knows were manufactured, processed, or distributed in violation of the act.

3. Failure or refusal to:
   a. Establish or maintain records.
   b. Submit reports, notices, or other information.
   c. Permit access to or copying of records as required by the law.
   d. Permit entry or inspection as authorized by the regulations.

Penalties

The TSCA includes civil and criminal penalties, which apply both to the corporation and to its employees. Specific civil and criminal penalties are:

1. *Civil.* Any person who commits one of the prohibited acts faces a fine of up to $25,000 for each violation. Each day such a violation continues constitutes a separate violation.
2. *Criminal.* Any person who knowingly or willfully commits one of the prohibited acts faces a fine of up to $25,000 for each day of violation and/or imprisonment for up to 1 year, in addition to any civil penalties.

The TSCA is an extremely complex law, as well as one that involves severe penalties for noncompliance. Any company that determines that it manufactures or processes chemical substances or mixtures as defined by the TSCA should seek both legal and technical expertise to interpret the act and its most recent regulations.

FEDERAL OCCUPATIONAL SAFETY AND HEALTH ACT (OSHA)

General Provisions

The most widely known regulatory act and agency is probably OSHA. This is the overall act that holds a company responsible for the safety and health of its employees. It applies directly to employers and employees and indirectly to manufacturers of products that are used in the industrial workplace. OSHA is authorized to promulgate, modify, or revoke occupational safety and health standards according to a prescribed set of rule-making procedures. Details on record-keeping requirements, inspections, and violations under OSHA are presented in Chapter 20.

CONSUMER PRODUCT SAFETY ACT (CPSA)

Public awareness of consumer safety, consumer rights, and consumer protection has never been higher than it is today. In response to this growing social concern, Congress passed the Consumer Product Safety Act (CPSA) in 1973. It had concluded that there were too many consumer products which presented unreasonable risks of injury and that existing state, local, and federal controls were inadequate. The act authorizes the Consumer

Product Safety Commission (CPSC), a five-member commission, to regulate more than 10,000 consumer products.

The purpose of the CPSA is fourfold:

1. To protect the public against unreasonable risks of injury.
2. To assist consumers in making comparative safety evaluations.
3. To develop uniform standards and minimize conflicting state and local regulations.
4. To promote research and investigation into the cause and prevention of product-related death, illness, or injury.

What Is a Consumer Product?

The term "consumer product" refers to any article (or its components) produced for use, consumption, or enjoyment in the home or school or for recreation. The definition does not include:

Any article which is not customarily produced or distributed for sale to a consumer.
Tobacco or tobacco products.
Motor vehicles or motor vehicle equipment.
Pesticides.
Aircraft.
Boats.
Drugs, devices, or cosmetics covered by the Federal Food, Drug and Cosmetics Act.
Food.

The definition, you will note, is surprisingly broad. Courts have held, in recent cases, that such things as aluminum branch current wiring, children's bicycles, and amusement park rides are consumer products. On the other hand, there have been cases in which courts have held that alcoholic beverages, guns, and ammunition do not fall within the meaning of the CPSA. Obviously, some manufacturers may have to seek legal advice to determine whether they are producing consumer products or not.

The Commission attempts to assist manufacturers by providing information on injuries so that hazards can be avoided or controlled. The National Electronic Injury Surveillance System (NEISS), for example, was set up to provide these accident data. NEISS's source of information is 119 hospital emergency rooms located throughout the country. The electronically gathered data makes it possible to compile records on the type and nature of injuries produced by a product or product group. The CPSC's Bureau of Epidemiology also makes in-depth investigations, describing the circumstances of selected accidents. This includes the number and severity of injuries associated with specified products. However, these data do not indicate whether the mishap was caused by the product itself or by a particular feature of the product.

Banned Hazardous Products

Whenever the CPSC finds that a consumer product presents an unreasonable risk of injury, and no standard under the act adequately protects the public, the Commission may declare such a product a banned hazardous product. In addition, any interested persons, including a consumer organization, may petition the Commission to issue, amend, or revoke a consumer product safety rule. In this case, the Commission would first determine the need or validity of such a rule and then take action as indicated.

Imminent Hazards

The Commission may also act in cases of imminent hazard. An imminent hazard is defined as one that presents an imminent or unreasonable risk of death, serious illness, or injury. In these cases, an action may be filed in a U.S. District Court for the seizure of the product and/or an action against the manufacturer. The CPSC has employed this authority to seize firecrackers which contained charges in excess of the regulation limits and carpeting which did not comply with flammability standards.

Substantial Product Hazards

One of the most important sections of the CPSA deals with "substantial product hazards." This section enables the Commission to regulate products that are already in the hands of consumers, including products manufactured before the enactment of the CPSA.

The term "substantial product hazard" means a hazard that creates a substantial risk of injury to the public. This includes a product defect which, because of the number of defective products distributed, creates a risk that a substantial number of people will be injured. A substantial product hazard is considered less grave than an imminent hazard.

Any manufacturer, distributor, or retailer who becomes aware that a product may fail to comply with CPSA safety rules or involves a substantial product hazard must immediately notify the Commission. If the Commission concludes that a substantial product hazard does exist, it may order the manufacturer, distributor, or retailer to:

1. Give public notice of the defect or failure to comply.
2. Mail a notice to each manufacturer, distributor, or retailer of the product.
3. Mail a notice to every known person to whom the product was sold or delivered.

4. Bring the product into conformity with the applicable standard by repairing or replacing it. If it must be withdrawn from use entirely, the purchase price or value must be refunded.

Record Keeping

Like the other acts governing safety and health, the CPSA requires manufacturers, distributors, and retailers to establish and keep certain records and make certain reports to the Commission.

Prohibited Acts

Under the provisions of the CPSA, a variety of activities are specifically prohibited. These prohibited acts are:

> Manufacturing, selling, distributing, or importing any product that does not conform to CPSA standards.
> Manufacturing, selling, distributing, or importing any product that has been declared a banned hazardous product.
> Failing or refusing to permit access to or copying of records.
> Failing or refusing to establish and maintain records, to make reports, to provide information, or to permit entry and inspection as required under the act.
> Failing to furnish information that indicates that a product contains a substantial product hazard.
> Failing to comply with an order relating to notification, repair, replacement, or refund involving a substantial product hazard.
> Failing to furnish any certificate required by the act.
> Issuing a false certificate or one that contains false information.
> Failing to comply with any rule related to labeling.
> Failing to comply with any rule related to stockpiling.
> Failing to comply with any rule related to giving prior notice and a description of new consumer products.
> Failing to submit required performance and technical data to the Commission.

Civil Penalties

Any person who knowingly performs one of these prohibited acts may be subject to a civil penalty of up to $2000 for each separate violation and up to $500,000 for any related series of violations.

Criminal Penalties

In addition to civil penalties, the CPSA provides for criminal penalties. These can apply when a person willfully or knowingly performs a prohibited act after receiving notice of noncompliance from the Commission. He or she can be penalized with a fine of up to $50,000 and/or 1 year in prison. The same penalties apply both to corporations and to their directors, officers, and agents. These criminal penalties are in addition to any civil penalties under common, federal, or state law. Moreover, compliance with CPSA rules and standards alone does not protect a manufacturer from liability suits or actions involving product safety.

CONCLUSION

From this brief overview alone, the reader can readily see that regulatory considerations are far too complex and detailed for the average manager to master. He or she needs expert help in interpreting the regulations, determining which standard apply to the company's operations or products, and understanding the full implications of his or her responsibilities and rights. Dealing with such far-reaching and complex codes on a "do-it-yourself" basis involves grave risks, and the manager would do well to seek the services of an expert.

Author Index

Aaron, James E., 506
Aljian, George W., 328
Ammer, Dean D., 328
Anton, Thomas J., 48, 129, 350, 368
Arbuckle, Gordon J., 568
Ashford, Nicholas A., 368
Atkisson, A., 548
Aubeie, James P., 450

Bass, J., 548
Bass, Lewis, 307, 369,
Benner, Ludwig E. Jr., 267
Bernhard, Robert, 521
Bhagat, D. O., 521
Biancardi, M. F., 57
Bingham, Eula, 584
Bird, Frank E. Jr., 208, 306, 328, 506
Birns, Carl, 548
Bishop, Carole M., 584
Bixler, B., 548
Blum, John D., 157
Boyle, William G., 306
Brandmaier, H. E., 521
Briggum, S. M., 549
Briscoe, G. J., 86, 208
Bullock, M. G., 174, 208, 306
Bunk, Kenneth H., 350
Buys, J. R., 189

Cammack, Emerson, 282
Caplain, W., 412
Celetano, E. J., 584
Chaffin, Don, 254, 306
Chavkin, Wendy, 584
Cheny, Carol, 12
Clark, Ann N., 568
Clarke, Richard, 306
Coates, Joseph H., 521
Cohen, Alexander, 157
Corn, Morton, 157

Courtney, Robert H. Jr., 521
Cox, S., 584
Cuthbert, Jean W., 584

Dardis, R. A., 57
Davidson, Cynthia, 398
DeReamer, Russell, 450
Dietel, Harvey M., 521
Donaldson, James H., 282
Donnelly, James H., 48, 74
Donvito, P. A., 57
Dowie, M., 57
Drucker, Peter F., 28
Dunn, Dennis, 585

Eddy, C., 350
Ehrenfeld, J., 548
Eicher, R. W., 12, 28

Farrell, Paul V., 328
Fatt, Naomi, 584
Fernandez, Edwardo O., 521
Ferry, Ted S., 74, 129, 267-268, 306
Fielding, J. R., 26
Findlay, James V., 12, 28, 74, 86, 129, 306
Fine, William T., 28, 57, 506
Ford, J. E., 521
Freidman, Milton, 74, 412
Freidman, Rose, 74

Gagne, F. Jr., 57
Garner, Charlotte A., 12
Gibson, James L., 48
Gilbert, Thomas F., 157
Glieck, James, 12
Gokaydin, Tug, 110
Goldman, G. S., 549
Goldsmith, Phillip E., 12
Good, Carter V., 129

619

AUTHOR INDEX

Grandjean, E., 306
Gray, Irwin, 306, 368
Greenberg, Leo, 110, 254
Grimaldi, John V., 57, 129, 208, 306, 350, 368, 459, 506

Haack, James L., 521
Hammer, Willie, 129, 307, 506
Handley, William, 350, 506
Hanson, B., 548
Harvey, C. M., 57
Hatch, Maureen, 584
Healy, Richard J., 507
Heath, Earl D., 74, 129
Heifitz, Ruth, 584
Heinrich, H. W., 12, 350, 506
Heinritz, Stuart F., 328
Hendrick, Kingsley, 208
Herold, Joan, 585
Horn, Ronald C., 282
Hricko, Andrea, 584
Hughes, Mary, 584
Hulet, Michael W., 307
Hunt, Vilma, 584

Invancevich, James M., 48

Jacobson, E., 412
Jefferies, J. W., 268
Johnson, Dave, 584
Johnson, W. G., 189, 254, 306, 434, 507
Johnstone, R. T., 398

Kelly, A. B., 57
Kelman, S., 57
King, Ralph W., 12, 129, 189, 208, 307
Klarman, H. E., 57
Know, N. W., 12, 28
Koth, George C., 521
Kroemer, K., 350
Kuhlman, Raymond L., 86, 328, 434

Lagner, G., 548
Leary, Margaret M., 368
Lenz, Matthew, Jr., 368
Letecky, C. R., 57
Lofthouse, J. H., 86, 208, 328, 506
Lowrance, W. H., 57

McCall, Brenda, 450
McCann, Micheal, 584
McCormick, Ernest J., 307

Mackay, Colin J., 584
Mackie, Jack B., 328
Malasky, S. W., 533, 548
Malecki, Donald S., 282
Mallow, Alex, 548
Malloy, Gail, 584
Mann, Jim, 307
Manson, Jeanne M., 584
Margolis, B., 412
Marsick, Daniel J., 568
Mehr, Ronald I., 282
Miller, Lynn M., 568
Miller, S. E., 398
Moll, K. D., 57
Musselman, Victoria Cooper, 568

Naquin, Arthur J., 225
Naughton, Michael J., 568
Nertney, R. J., 86, 174, 189, 208, 268, 307, 507
Ness, Shirley A., 568
Nottrodt, T. T., 584

Odom, S. S. G., 350
Olson, R. E., 48
O'Neill, B., 57
Opatz, Joseph P., 398
Ossler, C. C., 57

Patty, Frank A., 398, 450
Pavlov, Paul V., 208
Petak, W., 548
Peters, George A., 307
Peters, Tom, 12
Petersen, Dan, 12, 28, 307, 350, 368, 506, 507
Planer, Robert G., 507
Polakoff, Phillip D., 398
Pope, William C., 12, 28, 74, 86, 208, 268, 369
Port, Eugene A., 568
Poulton, E. C., 412
Pryor, Douglas, 522
Punski, Charlotte R., 584

Rahe, Richard H., 412
Raheja, Dev, 189
Ralston, August, 522
Randolph, Susan A., 584
Recht, J. L., 57
Redgrove, June S., 584
Reid, J. L., 484
ReVelle, Jack B., 129
Richards, Thomas C., 522
Riggs, J. L., 57

Riley, Robert, 522
Roos, Nestor, 12, 350, 506
Rosenfield, Harry N., 369
Rosenman, R. H., 412

Saunders, P. L., 584
Sawyer, C. J., 548
Schulman, Brian, 12
Scilken, Mary P., 584
Scott, Judith A., 584
Selders, Thomas, 450
Selye, Hans, 412
Sheridan, Peter I., 282
Silk, Jennifer C., 568
Simonds, R. H., 57, 305, 350, 368, 450, 506
Sippl, Charles J., 522
Sippl, Richard J., 522
Smith, Amy, 522
Spence, Lowell C., 110
Staum, Roger, 585
Steinbrecher, David, 522
Stellman, Jeanne M., 585
Steventon, Jill, 584
Stormer, K. A., 548
Strachta, Bruce J., 398
Summers, Rita C., 521

Tepper, Lloyd B., 157
Thygerson, Alton L., 507
Tihansky, D. P., 57
Trebicock, Anne M., 585
Tucker, Elizabeth, 522
Tuinstra, Bernard J., 350
Tytina, Nancy F., 584

Waldron, Ingrid, 585
Walsh, Timothy J., 507
Ward, Joan S., 585
Warner, Bill, 549
Warshaw, Leon J., 585
Weaver, D. A., 12, 306
Webber, Ross A., 12
Weinberg, D. B., 549
Weinstein, Alvin S., 601
Whipple, C., 57
Whiting, Basil, 211
Whitman, Lawrence E., 369, 507
Wood, Christopher, 521
Worick, Wayne W., 507
Wynhold, Hans, 307, 369

Zent, C., 57
Zenz, Carl, 585

Subject Index

Accident, 471-472
  cause, 237-239
Accident classification system, 227
Accident Facts, 316, 465
Accident information, vehicle, 366
Accident Investigation, 45-46, 255-268, 600
  committee, 263
  ensuring good, 267
  management stake, 255-268
  OSHA report, 353
  priorities, 265-267
  stakeholders, 259-264
    line manager, 261-262
    safety professional, 262
    senior management, 260
    staff managers, 263
    supervisor/foreman, 260-261
Accident Prevention Manual for Industrial Operations, 312
Accident record, 316
Accidents
  cost, 256-257
  direct costs, 256-257
  impact, 255-256
  indirect cost, 256-257
  maintenance-related, 185
  management inefficiencies, 257
  morale, 257
  public concern, 258-259
  preventive actions, 185
Accidents, materials handling. *See* Materials handling accidents
Accountability, 160
ACGIH, 10-11
Adequate, less than. *See* LTA
AFL-CIO, 576
Age, 410
Agency for Toxic Substances and Disease Registry, 558
AIDS, 10
Airborne particulate levels, 214

ALARA guidelines, 456
Alcohol, 404-405
Alcoholics Anonymous, 382
Alcoholism, 380-382
American Conference of Governmental Hygienists. *See* ACGIH
American Health Foundation, 573
American Heart Association, 376
American Industrial Hygiene Association (AIHA), 435, 447, 474
American National Standards Institute. *See* ANSI
American Society of Safety Engineers. *See* ASSE
ANSI, 4, 251, 447, 448
ASSE, 25, 99, 124, 127, 128, 251, 447
Audit
  basic steps, 81-82
  functions, 80
  management, 80
  outside, 79
  process, 79-80
  types, 81
Audit, regulatory, 39
Avoidance of loss, 15-16

Back injuries, 330-331
  costs, 331
Basic readiness, 159
Best's Safety Directory, 128, 312
Beyer, David Stewart, 445
Bingham, Eula, 448
Bismark, 2, 271, 445
BLS, 8, 132, 153, 352, 573
Books, safety and health, 156
Budget, 56
  administration, 44
  contingency, 33
  developing, 32-34
  estimated, 36
  fixed costs, 150

623

Budget *(Continued)*
  industrial hygiene, 35, 39
  methods of determining, 30-32
  optimum, 30
  reduction, 33
  task, 34, 44
  training, 149-150
  typical effort, 35
Budgeting, 29-48
  practical reasons for, 30
  time factors, 29-30
  total effort, 36
Building maintenance, 186-188
  lighting systems, 187
  lubrications, 188
  piping systems, 187
Bureau of Labor Statistics. *See* BLS

Canadian Centre for Occupational Safety and Health, 93
Cardiopulmonary resuscitation. *See* CPR
Cargo Security Advisory Standards, 366
Center for Fire Research, 510
CERCLA, 524-530, 550-552, 555, 561
  liability, 528
  priorities, 528-529
  stakeholders, 531-532
  supervision, 527
CFR, 529, 535-536, 367
Chamber of Commerce, U.S., 6
Change(s), 482
  body, 400
  physical, 400
  psychological, 400
  role of, 169-171
Checklist
  computer protection, 519-521
  design review, 598
  manager's, 173-174
  required OSHA reports, 359-362
Chemical control group, 323-324
Chemical limits, 399
Chemical lists, 565
Chemical spill, 472
Chemical substances, 10
Chemical Transportation Center, 472
CHEMINFO, 93
Chemtox, 93-94
CHEMTREC, 93
Civil preparedness, 477-479
Civil strife, 470-471

Civil War, 2
Coal Mines Act, 444
Code locks, 514
Code of Federal Regulations (CFR) 112, 138
College and University Safety Courses, 123
Committees, 82-86
  advantages, 82-83
  chairperson, 84-85
  disadvantages, 83-84
  membership, 84
  operating levels, 88
  organizing, 84
  practical considerations, 85
Common law doctrine, 446
Communicating hazards, 324-325
  input, 76
Community Right-to-Know Act. *See* SARA
Compliance, 24-27
  hazardous waste, 524-526
Comprehensive Environmental Response, Compensation and Recovery Act. *See* CERCLA
Computer
  laptop, 105
  pocket, 105
Computer-assisted learning, 88
Computer crime, 519
Computer protection checklist, 519-521
Computer resources, 99
Computer systems, safeguarding 508-522
  beverage spills, 512
  contingency planning, 516-517
    hardware loss, 516
    software loss, 517
  electrical power protection, 514
    power supply, 514
  fire, 508-509
    control, 509-510
    detection, 509
    extinguishment, 511-512
    room construction, 509
  natural flooding, 511
  physical security, 515
  water, 511-512
Computer virus, 517-519
  information security, 518
  physical protection, 518
Construction safety, 40-41
Construction safety orders, 40-41
Consultant, training, 124

SUBJECT INDEX 625

Consumer product safety, 362-364
Consumer Product Safety Act, 362, 613-617
   consumer product, definition, 614
   hazardous products, 615
   imminent hazards, 615
   penalties, civil, 616
   penalties, criminal, 616-617
   prohibited acts, 616
   record keeping, 616
   substantial product hazards, 615-616
Consumer Product Safety Commission. *See* CPSC
Control center, 487-488
Control charts, 235
Control Data Corporation, 395
Control, purchasing, *See* Purchasing controls
Conveyors, 341-342
   gravity, 342
   power, 342
Corn, Morton, 137
Corrosive chemicals, 215
Cost/Benefit Analysis, 49-57, 224-229
   cost centers and benefits, 53
   factors, 50
   misuses, 55
   origin, 50
   problems/controversy, 53
   steps, 52
   tracking, 56
   uses/adaptions, 55
Cost-effective protection, 209, 228
Cost/effectiveness analysis, 51, 228
   cost centers, 52
   units of effectiveness, 52
Costs, 269-282
   controlling, 281-282
   direct, 226
   identification, 184
   indirect, 226
   safe practices, 252-253
Countermeasures, 130-132
CPR, 40, 376
CPSC, 362-363, 364, 614
   liability, 363-364
   required reports, 363-364
Crane Manufacturers Association of America, 336
Crane safety, 41
Cranes, 336-338
Criticality safety, 166
Cryogenic liquids, 348-349

Database management, 92-93
Davidson, Cynthia, 392
De Morbis Arfactum Diatriba, 442
Department of Defense policy, 61-62
Department of Energy. *See* DOE
Department of Labor. *See* DOL
Department of Health and Human Services. *See* HSS
Department of Health Education and Welfare, 51
Department of Transportation (DOT), 366
DeReamer, Russell, 445
Design, 283-307
   committees, 288
   idealized procedure, 288-289
   maintenance personnel, 287-288
   management role, 285-286
   safety personnel, 286-287
   supervisors, 286
Design checklists, 300-301
Design costs, 299
   scaling device, 300
Design defects, 284
Design phases, 283, 289-299
   concept, 290-291
   design phase, 291-292
      experience and feedback, 296-297
      hazard identification, 292-294
      human considerations, 294-296
   disposal, 299-300
   operations, 298-299
   production, 297
Design process, 284
   management obligation, 302
Design review, 291
   savings, 252-254
Design team, 302
Designers, 284-285
Dictionary of Safety Related Computer Resources, 99
Disaster. *See also* Emergency
Disease detection, 378
Disposal, 321-322
   program, 322
   salvage equipment, 321-322
   toxic waste, 321
Doctrine of common employment, 270
DOE, 175, 244
DOL, 575, 588
Donvito, R. A., 51
Dose limits, 454

626 SUBJECT INDEX

Dosimeter, 456
Dosimetry, 456
Douglas v. Gallo Winery, 277
Drug abuse, 382-383
Drugs, 405
Dual capacity doctrine, 590
duPont company, 155, 413

Earthquake, 469-470
Educational Resource Centers, 123
Electricity, static, 240
Electronic data processing (EDP), 508
Emergency and disaster planning, 485-507, 464-465
  alarms, 496-497
  annex, 505-506
  appendix, 505-506
  authority, 487
  communications, 490-491
  considerations, 464-465
  control center, 487-488
  coordination, 497
  dealing with the media, 483-484
  elements, 485-499
  emergency description, 488
  emergency lighting, 496
  employee welfare, 496
  fire equipment chart, 488
  functionaries, 485, 486, 499-505
    chief operating officer, 500
    director/coordinator, 500
    executive board, 500
    federal government, 504
    fire agencies, 502
    functional committees, 500-501
    local governments, 504
    medical services, 503-504
    mutual aid, 505
    neighbors, 502
    Red Cross, 504-505
    security, 502-503
    unions, 501
  maps, 488-490
    emergency equipment map, 488-489
    evacuation map, 489-490
    plant map, 488
    shelter map, 489
  personal protective equipment, 497
  personnel assignment, 495-496
  policy, 485-486
  purpose, 486-487
  security guards, 493
  shutdown, 491-493
  records, vital, 498-499
  rescue equipment, 497
  restoration, repair and salvage, 493-494
  training, 495
  transportation, 494
  visitor control, 493
Emergencies and disasters
  preventing, 463-464
  steps to meet emergencies, 466
  types, 465-477
    accident, 471-472
    chemical spill, 472
    civil strife, 470-471
    earthquake, 469-470
    energy, 476
    explosion, 468
    extortion, 474
    fire, 467-468
    flood, 468
    kidnaping, 476
    nuclear events, 475
    hurricane, 469
    radiation, 474
    sabotage and terrorism, 471
    strikes, 471
    tornado, 469
    vapor release, 472
    weather, extreme, 474
Emergency and exposure limits (EELS), 474
Emergency evacuation, 46
Emergency preparedness, 171-173
Employee safeguards, 212
Energy, 746
  accident cause, 237-239
  energy consumption, 235
  hazard, 239-240
  sources, 238
  transfer, 238
Energy, barriers, 244-244
  strategies, 244
Engineered time standards, 183
Engineering controls, 209-236
  alternatives, 222-223
    isolation, 224
    substitution, 223
    ventilation, 224
  application, 231-234
    energy control and cost savings, 233-234

SUBJECT INDEX 627

hardware design and redesign, 231-232
  process design and redesign, 232-233
basic decision, 210-211
basic question, 210-211
cost factors, 213
human factors, 242
machine safeguarding, 217-218
  devices, 220
  feeding and ejection methods, 221
machine tools, 215
mechanical hazards, 216
personal protective equipment, 219-222
practical applications, 237-254
vs. protective equipment, 210-224
Environment, 169, 512-514
  heating, ventilation, air conditioning, 512-513
  off-the-job, 414
  on-the-job, 413-414, 415
  radio waves, 513-514
  vibration, 513
Environment, work, 401-404
  exercise, 406
  job demands, 401
  lighting, 402
  morale, 403
  noise, 402
  physical demands, 401
  rest, 406
  stress, 402
  temperature, 402
Environmental impact statement, 459
Environmental protection, 167-168
Environmental Protection Agency. *See* EPA
Environmental safety, 42-43
EPA, 224, 229-230, 365, 523-525, 527-529, 535, 556, 557, 558, 561, 607-611
Equal Employment Opportunity Commission, 570
Equal Pay Act, 570, 576
Equipment design, 297
Equipment, materials handling, 335-344
  conveyors, 341-342
  cranes, 336-338
  forklifts, 338-341
  hoisting apparatus, 341
  industrial trucks, 338-341
  rules, 337
Ergonomic optimization, 331-332
Essentials of Engineering Economics, 50

Essentials of a Planning-Programming-Budgeting System, 51
European Community (EC), 1
Evaluation, 56, 195
EWP. *See* Wellness program
Exclusive remedy, 590-591
Explosion, 468
Extortion, 474
Eye protection, 38-39

Factories Act, 178, 248
Factory Act, 444
Failure modes and effects analysis, 293
Family Activity Survey, 415-417
Family Safety Survey, 418-419
Fatigue, 406
Fault tree analysis, 293
Federal Aviation Administration, 378
Federal Coal Mine Safety and Health Act, 447, 448
Federal Emergency Management Administration (FEMA) 478, 504
Federal Register, 158, 300, 607
Federal Vocational Rehabilitation Act, 274
Feedback, 77
Fire, 467-468
Fire agencies, 502
Fire protection, 163
Fire reports, 367
First Cooperative Safety Congress, 4, 447
First-line supervisor, 22
Flammable Fabrics Act, 362
Flood, 468
Food and Drug Administration (FDA), 580
Ford, Gerald, President, 611
Forklifts, 338-341
Function, safety. *See* Safety function
Funds, allocation, 56

Gas tank evaluation, 50
Germany, 271

Halon, 509, 510
Hamilton, Alice, 570
Hardware, 197, 199
Hardware requirements, 195-201
  concept and design, 196-198
  decommission, 199
  disposal, 199
  fabrication, 198
  installation, 198

628 SUBJECT INDEX

Hardware requirements *(Continued)*
  life cycle study, 198
  occupancy-use, 198
  operation, 199
  modification, 199
Hazard
  classification, 240-241
  unwanted energy, 239-240
Hazard communication, 40
Hazard communication standard, 138, 365-366, 550, 555
  cost cutting-aid, 559-560
  direction, 559
  quality control, 562-563
HAZARDLINE, 104
Hazardous materials, 42
Hazardous Materials Emergency Action Guide, 473
Hazardous materials handling, 347-349
Hazardous materials, purchasing. *See* Purchasing hazardous materials
Hazardous mechanical motions, 216
Hazardous Substances Act, 362
Hazardous Substance Response Trust Fund, 528
Hazardous waste, 42, 523-533, 608-609
  agencies involved, 525
  criteria for listing, 527
  definition, 526-527
  legal aspects, 524-526, 530
    remedial action, 526
  noncompliance, 610
  penalties, 610
  problem, scope of, 523-524
  remedial actions, 530
  responding, 534-549
  site system, 531
  social influences, 530
  stakeholders, 531-532
  storage, 524
  system safety program, 532-533
  treatment, 524
Hazardous waste program, 542-548
  elements, 542
  emergency preparations, 545
  management guidelines, 542
  medical program, 545
  personnel selection, 546
  protective categories, 546-548
  protective equipment, 546, 547-548
  protocol, 544-545
  response personnel, 543
  safety practices, 544
  standard operating procedures, 542, 543
Hazardous waste, system safety. *See also* System safety in hazardous waste
  exposures, 541
    acute, 541
    chronic, 541
  requirements, commonality, 535
  worker protection, 541
HAZ-MAT, 97-98
Health and Moral Apprentices Act, 443
Health habits, personal, 404
Health physicist, 38
Hearing conservation, 39
Heating-ventilation-air conditioning, (HVAC), 509-511
Heinrich, H. W., 4
HES-Line
Heredity, 400
Hispanic, 587
Hoisting apparatus, 341
HSS, 123
Human behavior modifications, 112-113
Human factors, 164, 197, 241-243, 296
  research, 295
Hurricane, 469

Industrial Accident Prevention, 4, 445
Industrial hygiene, 164, 435-450
  and stress, 439
  employee responsibility, 438
  environmental factors, 440
  historical notes, 441-446
    child labor, 443
    common law, 446
    labor legislation, 442-443
    middle ages, 442
    mine safety inspection, 444
    powered machinery, 443
    women, 443-444
  management interaction, 435-436, 438
  safety professional, and the, 436-437
  supervisor's responsibility, 438
  work processes, 440
Industrial hygienists, 371
  evaluator, as, 440
  functions, 440-441
  protector, as, 440
Industrial injuries, treatment, 375
Industrial revolution, 269
Industrial trucks, 338-341
Industrialization, 3

SUBJECT INDEX 629

Information and instruction, as countermeasures, 130-132
Institute for Computer Sciences and Technology, 510
Insurance carrier, 562
International Harvester Company, 445
International Labor Organization, 1, 93, 155, 580
International Loss Control Institute, 122-123, 413
International Safety Academy, 155, 317
International Safety Rating System, 317
Investigation, 22-24
Ionizing radiation, 451, 454, 459
   education and training, 458-459

Job ambiguity, 409
Job hazard analysis, 140
Job process control, 190-210
   and safety, 190-210
   elements, 191
   interfaces, 206
      hardware to procedures, 207
      people to hardware, 207
      procedures to people, 207
Job safety analysis, 200-202
   steps, 201
Johns-Manville Products Corp. v. Superior Court, 277, 517
Journal of the American Industrial Hygiene Association, 137, 435

Kidnaping, 476

Labor unions. *See* Unions
Law, administrative, 602-604
   hearings, 603-604
   monitoring, 603-604
   rulemaking, 603-604
   vs. civil law, 603
Law, civil, 602-604
   vs. administrative law, 603
Leaking Underground Storage Tank Trust Fund, 561
Learning activities, 142-144
Legal aspects, 586-601
   damages, 596
   disclaimers, 596-597
   hold harmless, 596-597
   statute of limitations, 597
Legal concepts, 594-595
   conduct of the user, 598-599

forseeability, 595
level of technology, 594
product defects, 594
Less than adequate. *See* LTA
Lesson plans, 146-147
Level, of safety, 586-588
Liability exposure, 587
Liability, minimizing, 597-598
   accident investigation, 600
   design review, 597-598
   documentation, 600
   field support, 599-600
   maintenance, 598-599
   monitoring, 599-600
   postmarking, 600
   record keeping, 601
   word, control, 599
Liability, sources, 587
Life cycle, 195, 198
Lighting, emergency, 496
Liquefied petroleum gas (LPG), 472
Litigation, 44
Local Emergency Planning Committee (LEPC), 554, 556
Loss control, 18, 26
Loss exposure, 414-415
   employee interview, 415-416
   family activity survey, 418-419
   family safety survey, 416-417
   major categories
      electrical safety, 421
      home safety, 420
      public safety, 422
      transportation safety, 421
Love Canal, 524
LTA, 20

Machine guarding, 247-250
   basic requirements, 250
   color coding hazards, 251
   common hazards, 247
   machine design, 247-248
   objectives, 249
   selection, 249
   types of guards, 250-251
Machine safeguarding, 217-218
Machine tools, 215
Maintenance, 185
   and accidents, 185
   building, 186-188
   mandated preventive, 179
   on-call, 179

Maintenance Manager's Guide, 175
Maintenance program, 175-189
   and safety, 175-189
   back up, 180-181
   backlog control, 182
   control vs. actual performance, 183
   cost identification, 184
   coverage, 175
   elements, 176
   equipment records, 178
   job planning, 180
   mandated preventive maintenance, 178
   performance improvement, 182
   performance measurement, 182
   personnel training, 185
   policies, 175
   prioritization, 183
   repair vs. replacement criteria, 184
      scrapping, 185
      standards, 184
   scheduling, 181
Major Hazard Incident Data Service (MHIDAS), 93
Management commitment, 243-244
Management failure, 13, 22
Management function, 1-12, 14
Management responsibilities, 6
Management roles, 17
Management vs. supervision, 17
Managerial control system, 203
   accountability, 204
   authority, 203
   feedback, 205-206
   policy, 202
   responsibility, 203
   requirements, 201-204
Manhattan Project, 451
Manualization, 71-73
Marshall, F. Ray, 448
Massachusetts, Act, 444
Material Safety Data Sheet. *See* MSDS
Materials, 168
Materials control, 177
Materials flow, 345-347
Materials handling, 329-350
   efficiency, 329
   functions, 333-334
   principles, 334-335
   safety, 329
Materials handling accidents, 330-331
   preventing, 346-347
Materials storage, 177, 345
Measurement standards, 78-79

Medical case management, 391-392
Medical department, 375
   company-community interaction, 389-391
Medical examinations, 373
   certifying, 378
   disease detection, 378
   evaluations, 378
   key personnel, 377-378
   symptoms, 383-384
   termination, 379
Medical facilities, 388-389
   first aid stations, 388
   hospital, 388-389
Medical functions, special, 380-387
   research programs, 387
   testing, 383, 384-385
      benefits, 386
      legal considerations, 386
      testing policy, 385-386
Medical records, 364, 376
Medical services, 370-397, 503-504
   basic functions, 374-376
      diagnosis, 374-375
      inoculations, 377
      key personnel, 377-378
      monitoring, 375-376
      recordkeeping, 376
      treatment, 374-375
   emotional problems, 380
   management's role, 397
Medical staff, 370-372
   administrative, 372
   industrial hygienists, 371
   insurance coordinator, 327
   nurses, 371-372
   secretarial, 372
Medical standards, 373
MEDLINE, 104
Metallic and Nonmetallic Safety Act, 448
Metropolitan Insurance Company, 465
Microcomputers, 87-110
   applications, 87-88
   buying, 104-105, 108-109
   central processing unit, 103
   compact disks, 109-110
   floppy disk, 100
   hard disk, 100
   integrated packages, 95
   operating system, 102
   peripherals, 101
   safety and health concerns, 89-90
   substituting for specialists, 88
   terminology, 99-100

SUBJECT INDEX   631

Microsafe, 95-96
Milestones, 29
MIL-STD 882A, 536
Mines Act, 443
Modern Safety and Health Technology, 445
Motivators, 49-51
Motor vehicle injuries, 54
MSDS, 42, 138, 153, 323, 533, 555, 558-559, 565
  not required, 553-554
Multiple causation, 13-14

NASA, 517
National Academy of Sciences, 131, 137
National Association of Insurance Commissioners, 278
National Commission on Workers' Compensation, 280
National Conference of Industrial Hygienists, 447
National Council of Industrial Safety, 447
National Council on Compensation Insurance, 278
National Directory of Safety Consultants, 124
National Fire Protection Association (NFPA), 367, 448
National Institute for Occupational Safety and Health. *See* NIOSH
National Labor Relations Act, 448
National Library of Medicine, 474
National Priorities List (NPL), 528
National Safety Congress, 12
National Safety Council (NSC), 4, 114, 122, 128, 155, 256, 312, 316, 349, 447
National Science Research Council, 9
Negligence
  comparative, 595-596
  contributory, 595
New installations, 158
  accountability, 160
  managerial control, 159, 160
  personnel, 160
  responsibility, 160, 162
    criticality safety, 166
    environmental protection, 167-168
    fire protection, 163
    human factors, 164
    industrial hygiene, 164
    industrial safety, 162
    radiation protection, 169-170
    vehicular safety, 165-166
  structures, 159, 160

New York Telephone Company, 397
NIOSH, 6, 93, 123, 128, 138, 448, 449
Noise levels, 213-214
Noncompliance, 610-611
NPL, 561
NSC, 413
Nuclear events, 475
Nuclear Regulatory Commission (NRC), 367, 452, 453, 460
Nurses, 371-372
Nutrition, 404

Occupant-user, 160-161
Occupational Hazards, 316
Occupational Safety and Health, 129
Occupational Safety and Health Act. *See* OSHA
Occupational Safety and Health Review Commission (OSAHRC), 136
Office of Management and Budget (OMB), 562
Office of Technology Assessment (OTA), 523
Off-the-job safety, 11, 413-434
  justifying, 413
Oil spill control, 43
On-the-job safety, 11
Organizing and operating functions, 62-73
  locating, 68-69
  responsibility and authority, 62
  staffing, 69-70
OSHA, 5, 6, 7, 25, 75, 88, 112, 132-137, 142, 178, 211, 224, 229-230, 239, 248, 300, 304, 331, 352, 353, 359, 366, 377, 448, 449, 525, 529, 535, 550, 556, 573, 588-590, 605, 606, 613
  contested violation, 590
  de minimus violation, 589-590
  imminent danger, 589
  inspections, 588
  penalties, 590
  record keeping, 588
  serious violation, 589
  standards, 353
  willful violation, 589
OSHA Reference Manual, 179, 347, 359, 392
Oversights and omissions, 20-21

People, 168
People performance sciences, 232
Periodicals, safety and health, 156-157
Perkins, Frances, 4
Personal protective equipment, 219-222, 497, 582

Personnel, 192-193
  placement, 193
  requirements, 191, 192
  selection, 193
Planning, 58-59
  emergency and disaster, 463-484
    assess hazards, 480
    assess probability, 479
    coordinator, 480, 482-483
    plan development and approval, 481
    revision, 482
    steps, 479
    testing, 481
    train personnel, 481-482
Poison Prevention Packing Act, 363
Policy, 59-62, 243, 485-486
  communication tool, 61
  corporate statement, 61
  good policy, 60
  problems, 60
  relationships, 64
  training, 148-149
Positive performance, 232
President's Council on Environmental Quality, 611
PRESTEL, 94
Preventive maintenance, 178
Procedural requirements, 204
  participation, 204-205
  testing, 205
Product liability, 301-304, 591-597
  documentation, 301-304
  interaction with workers' compensation, 592
  liability lawsuit, 591-592
  liability, strict, 593
  misrepresentation, 593
  negligence, 593
  reasonable care, 301
  records, 303
  records retention, 303-304
  responsibility, standards of, 593
  warranty, express, 593
  warranty, implied, 593
Product safety, 43-44
  litigation, 44
  marketing, 44
  recall, 44
  system safety, 43
Professional development, 126
Professional Safety, 25
Program, disposal. *See* Disposal program

Program, purchasing. *See* Purchasing program
Program, off-the-job safety, 413-434
  agenda, 426
  campaign, 425-426
  developing, 429
  elements, 429-430
  employee, 432-434
  guidelines, 423-424
  planning, 422-423
  project teams, 424
  team approach, 424
  training, 432-433
  steering committee, 424-425
Program, radiation safety, 451
Program, recordkeeping, 457-458
Program, wellness, 392-397
Protective systems, 169
Purchasing, 251-252, 304-306
  agent, 304-305
  responsibilities, 304-305
  specifications, 305
  standards, 305
Purchasing concepts, 308-309
  change order, 310
  instruments, 309-310
  order notice, 310
  policies, 309-310
  receiving report, 310
  requisition, 310
Purchasing controls, 308-328
  managing, 325
Purchasing general materials, 325-326
  specifications, 325-326
  standards, 325-326
Purchasing hazardous materials, 322-325
  chemical control group, 323-324
  chemicals inventory, 324
  communicating hazards, 324-325
  identifying hazards, 322
  limiting exposure, 324
  material safety data sheets, 323
Purchasing program, 308
Purchasing supplies, 311-321
  loss exposure, 312
  on-site surveys, 317
  prospective suppliers, 314-315
  responsibilities, 311
  selection, 318-319
  self-sourcing, 319-320
  sourcing, 311
  specifications, 312
  supplier criteria, 313

SUBJECT INDEX 633

supplier's performance, 317
user needs, 311-312
vendor information, 315-316
Purchasing tools and equipment, 326-327
plan, 327
program, 326-327

Qualitative severity index, 293
Quality control, 298
Quality-of-Environment Survey, 580
Quick reference, OSHA reports, 359-362

Radiation, 474
Radiation protection, 169-170
Radiation safety, 38-39, 451-462
administration, 452-453
glossary, 460-461
pregnancy, and, 457
program, 451
elements, 451-452
environmental discharges, 450, 456
external exposure control, 454-455
internal exposure control, 455-456
monitoring, 456
recordkeeping, 458-459
Radiation safety committee, 452
Radiation Safety Officer. See RSO
Radioactive contamination, 455
Radioactive material, 451
Rail cars, 343-344
Ramazzini, Bernado, 442
Random access memory (RAM), 99-100, 103
RCRA, 229-230, 524-526, 529-530, 607-611
generator(s), 609, 611
provisions, 607-608
requirements, 607-608
transporter, 609-610
Read only memory (ROM), 100-101
Record-keeping, 351, 458-459, 601
definitions, 358-359
function, 351-369
statistics, 359
under OSHA, 352
Records and forms, 352
access, 362
location, 362
retention, 362
types, 352
OSHA form 101, 352-358
OSHA form 200, 352-358
Records, equipment, 178
Records, vital, 498-499

Red Cross, 376, 504-505
Refrigerator Safety Act, 363
Registry of Toxic Materials (RETECS), 93
Regulatory agencies, 607
interaction, 606
Regulatory considerations, 602-617
Regulatory decision making, 604-607
Rehabilitation, 274
Rescue units, 503
Resource Conservation and Recovery Act.
See RCRA
Resources, training, 153-156
NIOSH, 154
nongovernment sources, 154-157
OSHA, 154
other government, 154
Respiratory protection, 41
Response, body, 399
Responsibilities, by position, 65-66
Responsibility, 160, 410
RCRA (Resource Conservation and Recovery Act), 229-230
Riggs, James I., 50
Right-to-know, 10, 137, 563-567
Risk, 245-246
and safety, 586
employees at, 153
Risk Management Information Systems (RMIS), 107
Robots, 233
Role conflict, 409
Role, manager, 173-174
Roles
communicating, 75-76
responsibilities, 75-76
ROM (rough-order of magnitude), 226-228
Roosevelt, Theodore, 3
Ropes, chain and slings, 342-343
Royal Society for the Prevention of Accidents, 88
RSO, 452, 453

Sabin, 214
Sabotage and terrorism, 471
Safety and Health, 129
Safety and health
emphasis, 5
ideas, 5
issues, 6
uncertain times, 5-6
Safety and Health Executive, 93
Safety and health manual, 71-73

634 SUBJECT INDEX

Safety and health movement, 1-12
  beginnings, 2
  early twentieth century, 3-4
  Victorian period, 2-3
Safety and health program, elements, 70-71
Safety and health roles, 73
Safety, and risk, 586
Safety circles, 427-428
  common denominators, 427
  facilitator, 427
  organizing, 428-429
Safety expertise, 14
Safety function, 13-28
  evaluation, 15
Safety level, acceptable, 586
Safety Management Newsletter, 437
Safety objectives, 17
  multiple, 17-19
Safety organizations, 447
Safety problems, 13-17, 19-20
Safety professional, 26, 436-437
  industrial hygiene responsibilities, 437-438
Safety program
  equipment acquisition, 45
  major categories, 31
  materials, 45
Safety program controlling, 75-86
Safety specialist, 186
Salvage equipment, 321
SARA, 231, 550-568
  accountability, 557-558
  acronyms, 553
  community right-to-know, 555
  emergency planning, 554
  emergency reporting, 554-555
  HAZCOM provisions, 563-565, 566-567
  inventories, annual, 556
  liability provisions, 560, 561
  public participation, 561
  response contractors, 561
  right-to-know provisions, 563-565
  site cleanup, 561
  terms, 551
  toxic chemicals inventory list, 556
  trade secrets, 560
  underground storage tanks, 561
Secretary of Transportation, 366
Senior manager, 11
SIC, 151-152, 353, 556
Small Business Regulatory Flexibility Act, 367
Smoking, 405
Software, 90-92, 101-102
  bundled, 106-107

  office management, 91
  word processing, 91-92
Sourcing. *See* Purchasing supplies
Specific control factors, 21
Specifications, 325-327
Spreadsheets, 94
Staffing, medical, 370-372
Standard Industrial Classification Code. *See* SIC
Standards, 315, 325-326, 353
  health, 449
    legal challenges, 449
  medical, 373
State Emergency Response Commission (SERC), 554
Statistics, 359
Stevens, Thaddeus, 449
Stress, 389-412, 439, 579-589
  defined, 399
  domestic, 406
  management, 409-412
    causative factors, 409-410
  occupational, 10
  predisposing factors, 400
  preventive measures, 407-408
    assignment, 407
    counseling services, 411
    detection, 408
    exercise facilities, 411
    health evaluations, 411
    management factors, 407
    matching to job, 407
    motivation, 406
    placement, 407
    preemployment examinations, 407
    preventive medical programs, 408-409
  recognizing, 410-411
  social, 406
  toleration, 400
Stressor, 399, 402
Strikes, 471
Structures, 158
Superfund. *See* CERCLA; SARA
Superfund Amendments and Reauthorization Act. *See* SARA
Supervisor, 63, 64
Supreme Court, 211
Surveys, 317-318
System
  accident classification, 227
  analysis, 228
  business organization, 13
  communication, 75

SUBJECT INDEX 635

database management, 92-93
feedback, 77
information, 75, 90-91
hardware control, 196
job process, 196
management, 21, 159
piping, 187
protective, 169
work order, 176
System safety, 43, 46-47, 532-533
　in hazardous waste, 534-549
　　analytical techniques, 538
　　authority, 536
　　defined, 534
　　development and production, 539
　　disposal, 539
　　documentation, 540
　　formulation, 539
　　general considerations, 535
　　hazard level categories, 537-538
　　military guideline, 536
　　operations, 539
　　precedents, 538
　　program criteria, 537
　　remedial activities, 540
　　responsibility, 536
　　task phases, 538-540
Systems, computer, 508-522

Tension, chronic, 400
Tests, 193-194
　aptitudes, 194
　occupational interests, 194
　motor ability, 194
　standard, 193
Testing and qualification, 195
Testing, medical, 383-385
Threshold limit value. *See* TLV
Title III-Community Right-to-Know Act. *See* SARA
TLV, 10-11
Tools and equipment, 326-327
Tornado, 469
Toxic chemical release, 556-557
Toxic chemicals inventory list, 556
Toxic Substances Control Act (TSCA), 229-230, 606, 611-613
　penalties, 613
　prohibited acts, 612-613
　provisions, 611-612
　requirements, 611-612
Toxic Substance List, 93

Toxic waste, 321
TOXLINE, 104, 474
Toy and Child Protection Act, 362
Trade names, 93
Traffic fatalities, 54
Training, 39-40, 194-195, 495, 578
　and education, 111-129, 130-157
　basic requirements, 113
　budget, 149-150
　business schools, 127
　conducting, 144-145
　difference, 111-112
　documentation, 136
　effectiveness, 145
　employees at risk, 153
　engineering schools, 127
　external, 121-125
　functions
　　collateral duty, 118
　　employee, 117-118
　　executive, 116
　　line management, 117
　　senior management, 116
　　staff, 116
　　supervisor, 117
　government requirements, 112
　improving, 145
　internal, 118-120
　　advantages, 119
　　disadvantages, 119
　　resources, 119
　matching to employees, 151-153
　need, 111
　new employee, 121
　OSHA requirements, 132-135
　penalties for non-compliance, 135-136
　personnel, 185
　resources, 153-156
　responsibilities, 116
　state requirements, 134
　strategy, 120
　voluntary guidelines, 138
　when needed, 139
Training needs
　goals and objectives, 142-143
　identifying, 140
　learning activities, 143
Training, off-the-job, 432-433
Training policy, 148-149
Training program
　control, 114-115
　development, 114
　staff, 115-116

Training requirements, safety professional, 125
Transportation of Goods, 320-321
Trends, 76-77
Triangle Shirtwaist Factory fire, 3
Trucks, 343-344
TSCA. *See* Toxic Substances Control Act

Union(s), 3, 27, 67, 501
United States, 1
Unplanned events, 16
Unwanted events, 16
U.S. Army, 341
U.S. government, 123
U.S. Postal Service, 386

Vapor release, 472
Vehicular safety, 165-166
Ventilation, 224
Veterans Administration, 382
Violations. *See* OSHA
Virus, computer. *See* Computer virus

Waste disposal, 177
Waste, hazardous, 42-43. *Also see* Hazardous waste
  radioactive, 457
  solid, 608
Weather, extreme, 474
Wellness program, 392-397
  benefits, 395
  case history, 395
  costs, 393
  effectiveness, 394
  starting, 394-396
Woburn, Mass., 524
Women in the workplace, 569-585
  considerations, safety and health, 576-583
    ergonomics, 578
    high-risk, 576-577
    occupational hazards, 577
    reproductive hazards, 580-582
    stress, 579-589
    worker-job fit, 578-580
  job analysis, 578
  preplacement assessment, 578
  training, 578
  demographics, 571-574
    age, 571
    education, 571-572
    marital status, 572
    union membership, 573
  earnings, 575-576
  future employment, 573-574
  history, 570-571
  occupations, 574-575
  personal protective equipment, 582
Women's Bureau, 570
Women's League for Equal Opportunity, 570
Women's Occupational Health Resource Center, 573
Women's Trade Union League, 570
Work overload, 409
Work sampling, 181
Worker selection matrix, 152
Workers' compensation, 9, 269-282, 374, 446, 590-591
  administration, 278-280
  benefits, 272-273
  compensable injuries, 275-276
  cost of administration, 278-279
  cost of safety, 269-282
  development, 269-272
    liability and compensation, 269-271
  disability, types, 273-274
  early developments, 271
  employer's liability, 276
  exclusive remedy, 276
  objectives, 272
  rehabilitation, 274
  second-injury funds, 274-275
  sole remedy, 276
  trends, 280-282
  United States, 271-272
Workshop Regulation Act, 444